*Model Selection and Inference*
A Practical Information-Theoretic Approach

# Springer
*New York*
*Berlin*
*Heidelberg*
*Barcelona*
*Hong Kong*
*London*
*Milan*
*Paris*
*Singapore*
*Tokyo*

Kenneth P. Burnham    David R. Anderson

# Model Selection and Inference

## A Practical Information-Theoretic Approach

With 21 Illustrations

Springer

Kenneth P. Burnham  
Colorado Cooperative Fish  
and Wildlife Research Unit  
Colorado State University  
Fort Collins, CO 80523-1484, USA

David R. Anderson  
Colorado Cooperative Fish  
and Wildlife Research Unit  
Colorado State University  
Fort Collins, CO 80523-1484, USA

Cover photograph by Edgar Mueller, Mayville, WI.

Library of Congress Cataloging-in-Publication Data  
Burnham, Kenneth P.  
   Model selection and inference : a practical information-theoretic  
approach / Kenneth P. Burnham, David R. Anderson.  
     p.    cm.  
   Includes bibliographical references and index.  
   ISBN 0-387-98504-2 (hardcover : alk. paper)  
   1. Biology—Mathematical models.   2. Mathematical statistics.  
I. Anderson, David Raymond, 1942–     II. Title.  
QH323.5.B87   1998  
570´.1´51—dc21                                                  98-13046

Printed on acid-free paper.

© 1998 Springer-Verlag New York Inc.  
All rights reserved. This work may not be translated or copied in whole or in part without the written permission of the publisher (Springer-Verlag New York, Inc., 175 Fifth Avenue, New York, NY 10010, USA), except for brief excerpts in connection with reviews or scholarly analysis. Use in connection with any form of information storage and retrieval, electronic adaptation, computer software, or by similar or dissimilar methodology now known or hereafter developed is forbidden.  
The use of general descriptive names, trade names, trademarks, etc., in this publication, even if the former are not especially identified, is not to be taken as a sign that such names, as understood by the Trade Marks and Merchandise Marks Act, may accordingly be used freely by anyone.

Production managed by Terry Kornak; manufacturing supervised by Jeffrey Taub.  
Typeset by Bartlett Press, Marietta, GA.  
Printed and bound by Maple-Vail Book Manufacturing Group, York, PA.  
Printed in the United States of America.

9 8 7 6 5 4 3 2 (Corrected second printing, 2000)

ISBN 0-387-98504-2 Springer-Verlag New York Berlin Heidelberg   SPIN 10752617

To my mother and father, Lucille R. (deceased) and J. Calvin Burnham, and my son and daughter, Shawn P. and Sally A. Burnham.

To my parents, Charles R. (deceased) and Leta M. Anderson; my wife, Dalene F. Anderson; and my children, Tamara E. and Adrienne M. Anderson.

# Preface

We wrote this book to introduce graduate students and research workers in various scientific disciplines to the use of information-theoretic approaches in the analysis of empirical data. In its fully developed form, the information-theoretic approach allows inference based on more than one model (including estimates of unconditional precision); in its initial form, it is useful in selecting a "best" model and ranking the remaining models. We believe that often the critical issue in data analysis is the selection of a good approximating model that best represents the inference supported by the data (an estimated "best approximating model"). Information theory includes the well-known Kullback–Leibler "distance" between two models (actually, probability distributions), and this represents a fundamental quantity in science. In 1973, Hirotugu Akaike derived an estimator of the (relative) Kullback–Leibler distance based on Fisher's maximized log-likelihood. His measure, now called *Akaike's information criterion* (AIC), provided a new paradigm for model selection in the analysis of empirical data. His approach, with a fundamental link to information theory, is relatively simple and easy to use in practice, but little taught in statistics classes and far less understood in the applied sciences than should be the case.

We do not accept the notion that there is a simple, "true model" in the biological sciences. Instead, we view modeling as an exercise in the approximation of the explainable information in the empirical data, in the context of the data being a sample from some well-defined population or process. Selection of a best approximating model represents the inference from the data and tells us what "effects" (represented by parameters) can be supported by the data. We focus on Akaike's information criterion (and various extensions) for selection of a parsimonious model as a basis for statistical inference. Later chapters offer formal methods to promote inference from a suite of *a priori* models and give approaches to estimate uncertainty without conditioning on a specific model. Model selection based on information theory represents a quite different approach in the statistical sciences, and the resulting selected model may differ substantially from model selection based on some form of statistical hypothesis testing.

We recommend the information-theoretic approach for the analysis of data from "observational studies." In this broad class of studies, we find that all the various hypothesis testing approaches have no theoretical justification and may often perform poorly. For experiments (control-treatment, with randomization and replication) we generally support the traditional approaches (e.g., analysis of variance); there is a very large literature on this classic subject. However, even for experiments, we advocate an emphasis on fitting explanatory models, hence on estimation of the size and precision of the treatment effects and on parsimony, with far less emphasis on "tests" of null hypotheses, leading to the arbitrary classification "significant" versus "not significant."

We do not claim that the information-theoretic methods are always the very best for a particular situation. They do represent a unified and rigorous theory, an extension of likelihood theory, an important application of information theory, and are objective and practical to employ across a very wide class of empirical problems. Inference from multiple models, or the selection of a single "best" model, by methods based on the Kullback–Leibler distance are almost certainly better than other methods commonly in use now (hypothesis testing of various sorts or merely the use of just one available model). In particular, subjective data dredging leads to over-fitted models and the attendant problems in inference and is to be strongly discouraged.

Parameter estimation has been viewed as an optimization problem for at least eight decades (e.g., maximize the log-likelihood or minimize the residual sum of squared deviations). Under the information-theoretic paradigm, model selection and model weighting also becomes an optimization problem. The value of AIC is computed for each *a priori* model to be considered, and the model with the minimum AIC is used for statistical inference. Akaike viewed his AIC and model selection as "... a natural extension of the classical maximum likelihood principle." This extension brings model selection and parameter estimation under a common framework—optimization. However, the paradigm described in this book goes beyond merely the computation and interpretation of AIC to select a parsimonious model for inference from empirical data; it refocuses increased attention on a variety of considerations and modeling *prior* to the actual analysis of data. Model selection, under the information-theoretic approach presented here, attempts to identify the (likely) best model, orders the models from best to worst, and measures the plausibility ("calibration") that each model is really the best as an inference. Used properly, model selection methods allow inference from more than a single "best" model (model averaging), although such methods are sometimes (incorrectly) described as useful only for selecting a "best" model. Several methods are given that allow the uncertainty as to which model is "best" to be incorporated into both parameter estimates (including prediction) and estimates of precision.

Our intention is to present and illustrate a consistent methodology that treats model formulation, model selection, estimation of model parameters, and their uncertainty in a unified manner, under a compelling common framework. We review and explain other information criteria (e.g., $AIC_c$, QAIC, and TIC) and present several new approaches to estimating model selection uncertainty and incorporating selection uncertainty into estimates of precision. An array of examples is given to illustrate various technical issues. Chapter 5 emphasizes model averaging as a way to avoid inference based on only a single best model. In addition, we provide references to the technical literature for those wishing to read further on these topics.

This is an applied book written primarily for biologists and statisticians using models for making inferences from empirical data. Research biologists working either in the field or in the laboratory will find simple methods that are likely to be useful in their investigations. Research people in other life sciences and medicine might also find the material useful but will have to deal with examples that have been taken largely from ecological studies of free-ranging vertebrates. Applied statisticians might consider the information-theoretic methods presented here quite useful and a superior alternative to the null hypothesis testing approach that has become so tortuous. We hope material such as this will find its way into classrooms where applied data analysis and associated philosophy of science are taught. This book might be useful as a text for a 3-credit-hour course for students with substantial experience and education in statistics and applied data analysis. A second, primary audience includes honors or graduate students in the biological, medical, or statistical sciences. People interested in the empirical sciences will find this material useful, as it offers an effective alternative to (1) the widely taught, yet often both complex and uninformative, hypothesis testing approaches and (2) the far less taught (and far less simple) Bayesian approaches.

Readers should ideally have some maturity in the quantitative sciences and experience in data analysis. Several courses in contemporary statistical theory and methods as well as some philosophy of science would be particularly useful in understanding the material. Some exposure to likelihood theory is essential, but people with experience only in least squares regression modeling will gain some useful insights. Biologists working in a team situation with someone in the quantitative sciences might also find the material to be useful. The book is meant to be relatively easy to read and understand, but the conceptual issues may preclude beginners. Chapters 1–5 are recommended for all readers. Chapters 1–3 are somewhat tutorial; while Chapters 4 and 5 present new research results and a wide variety of approaches, and these are illustrated with examples. Few readers will be able to absorb the concepts presented here after just one reading of the material; some rereading and second consideration will often be necessary to understand the deeper points. Chapter 6 covers much more advanced mathematical material, while a comprehensive summary of the book is provided in Chapter 7.

We intend to remain active in this subject area after this book has been published, and we invite comments from colleagues as an ideal way to learn more and understand differing points of view. We hope that the text does not appear too dogmatic or idealized. We have tried to synthesize concepts that we believe are important and incorporate these as recommendations or advice in several of the chapters. We realize that there are other approaches, and that some people may still wish to test hypotheses to build models of empirical data, and that others may have a more lenient attitude towards data dredging than we advocate here. We do not want to deny other model selection methods, such as cross validation, nor deny the value of Bayesian methods. However, in the context of objective science, we are compelled by the *a priori* approach of building candidate models, the use of information-theoretic criteria for selecting a best approximating model or model averaging when truth is surely very complex, the use of likelihood theory for deriving parameter estimators, and incorporating model selection uncertainty into statistical inferences. In particular, we recommend the use of Akaike weights as a basis for model averaging, hence for making inference from several models.

Several people have helped us as we prepared this book. In particular, we acknowledge C. Chatfield, C. Hurvich, B. Morgan, D. Otis, J. Rotella, R. Shibata, and K. Wilson for comments of earlier drafts of chapters of the developing book. We are grateful to three anonymous reviewers for comments that allowed us to improve the presentation. Early discussions with S. Buckland, R. Davis, R. Shibata, and G. White were very useful. S. Beck, K. Bestgen, D. Beyers, L. Ellison, A. Franklin, W. Gasaway, B. Lubow, C. McCarty, M. Miller, and T. Shenk provided comments and insights as part of a graduate course on model selection methods that they took from the authors. C. Flather allowed us to use his data on species accumulation curves as our first example, and we thank C. Braun and the Colorado Division of Wildlife for the data on sage grouse; these data were analyzed by M. Zablan under the supervision of G. White. C. Southwell allowed us to use his kangaroo data from Wallaby Creek. J. Kullback allowed us to use a photo of his father, and H. Akaike, R. Leibler, R. Shibata, and K. Takeuchi kindly sent us photos and biographical material that appear in the book. The American Mathematical Society allowed our use of the photo of L. Boltzmann from the book *Wissenschaftliche Abhandlungen von Ludwig Boltzmann* and the International Biometric Society authorized our use of a photo of R. Fisher (from *Biometrics* 1964, taken in 1946 by A. Norton). C. Dion, R. Fulton, S. Kane, B. Klein, and T. Sundlov helped obtain library materials.

We are happy to acknowledge the long-term cooperators of the Colorado Cooperative Fish and Wildlife Research Unit: the Colorado Division of Wildlife, Colorado State University, the Biological Resources Division of the U.S. Geological Survey, and the Wildlife Management Institute. Graduate students and faculty within the Department of Fisheries and Wildlife Biology at Colorado State University provided a forum for our interests in the analysis of empirical data. In particular, we extend our appreciation to several federal agencies within the Department of Interior for their support of our long-term research interests.

Kenneth P. Burnham<br>
David R. Anderson<br>
January, 1998

# Contents

**Preface** vii

**Glossary** xvii

**About the Authors** xx

1 **Introduction** 1
  1.1 Objectives of the Book . . . . . . . . . . . . . . . . . . . . . . 1
  1.2 Background Material . . . . . . . . . . . . . . . . . . . . . . 4
      1.2.1 Inference, Given a Model . . . . . . . . . . . . . . . . 4
      1.2.2 The Critical Issue: "What Model To Use?" . . . . . . . 5
      1.2.3 Science Inputs—Formulation of the Set of Candidate Models . . . . . . . . . . . . . . . . . . . . . . . . . . 7
      1.2.4 Models Versus Full Reality . . . . . . . . . . . . . . . 11
      1.2.5 A "Best Approximating Model" . . . . . . . . . . . . . 13
  1.3 Overview of Models in the Biological Sciences . . . . . . . . . 13
      1.3.1 Types of Models Commonly Used in the Biological Sciences . . . . . . . . . . . . . . . . . . . . . . . . . 14
      1.3.2 Likelihood and Least Squares Theory . . . . . . . . . . 15
      1.3.3 Data Dredging . . . . . . . . . . . . . . . . . . . . . . 17
      1.3.4 Some Trends . . . . . . . . . . . . . . . . . . . . . . . 20
  1.4 Inference and the Principle of Parsimony . . . . . . . . . . . . 21
      1.4.1 Over-fitting to Achieve a Good Model Fit . . . . . . . 21
      1.4.2 The Principle of Parsimony . . . . . . . . . . . . . . . 23
      1.4.3 Model Selection Methods . . . . . . . . . . . . . . . . 27

|   |     |                                                                                    |     |
|---|-----|------------------------------------------------------------------------------------|-----|
|   | 1.5 | Model Selection Uncertainty                                                        | 29  |
|   | 1.6 | Summary                                                                            | 30  |
| **2** | **Information Theory and Log-Likelihood Models: A Basis for Model Selection and Inference** | | **32** |
|   | 2.1 | The Distance or Discrepancy Between Two Models                                     | 33  |
|   |     | 2.1.1   $f$, Truth, Full Reality, and "True Models"                                | 36  |
|   |     | 2.1.2   $g$, Approximating Models                                                  | 36  |
|   |     | 2.1.3   The Kullback–Liebler Distance (or Information)                             | 37  |
|   |     | 2.1.4   Truth, $f$, Drops Out as a Constant                                        | 41  |
|   | 2.2 | Akaike's Information Criterion                                                     | 43  |
|   | 2.3 | Akaike's Predictive Expected Log-Likelihood                                        | 49  |
|   | 2.4 | Important Refinements to AIC                                                       | 51  |
|   |     | 2.4.1   A Second-Order AIC                                                         | 51  |
|   |     | 2.4.2   Modification to AIC for Overdispersed Count Data                           | 52  |
|   | 2.5 | A Useful Analogy                                                                   | 54  |
|   | 2.6 | Some History                                                                       | 56  |
|   |     | 2.6.1   The G-Statistic and K-L Information                                        | 56  |
|   |     | 2.6.2   Further Insights                                                           | 57  |
|   |     | 2.6.3   Entropy                                                                    | 58  |
|   |     | 2.6.4   A Summary                                                                  | 59  |
|   | 2.7 | Further Comments                                                                   | 59  |
|   |     | 2.7.1   A Heuristic Interpretation                                                 | 60  |
|   |     | 2.7.2   Interpreting Differences Among AIC Values                                  | 60  |
|   |     | 2.7.3   Nonnested Models                                                           | 63  |
|   |     | 2.7.4   Model Selection Uncertainty                                                | 64  |
|   |     | 2.7.5   AIC When Different Data Sets Are to Be Compared                            | 64  |
|   |     | 2.7.6   Order Not Important in Computing AIC Values                                | 64  |
|   |     | 2.7.7   Hypothesis Testing Is Still Important                                      | 65  |
|   | 2.8 | Comparisons with Other Criteria                                                    | 66  |
|   |     | 2.8.1   Information Criteria That Are Estimates of K-L Information                 | 66  |
|   |     | 2.8.2   Criteria That Are Consistent for K                                         | 68  |
|   |     | 2.8.3   Contrasts                                                                  | 70  |
|   | 2.9 | Return to Flather's Models                                                         | 71  |
|   | 2.10 | Summary                                                                           | 72  |
| **3** | **Practical Use of the Information-Theoretic Approach** | | **75** |
|   | 3.1 | Computation and Interpretation of AIC Values                                       | 76  |
|   | 3.2 | Example 1—Cement Hardening Data                                                    | 78  |
|   |     | 3.2.1   Set of Candidate Models                                                    | 79  |
|   |     | 3.2.2   Some Results and Comparisons                                               | 79  |
|   |     | 3.2.3   A Summary                                                                  | 82  |
|   | 3.3 | Example 2—Time Distribution of an Insecticide Added to a Simulated Ecosystem       | 83  |

|   |   |   | Contents | xiii |
|---|---|---|---|---|
|   |   | 3.3.1 Set of Candidate Models | | 84 |
|   |   | 3.3.2 Some Results | | 86 |
|   | 3.4 | Example 3—Nestling Starlings | | 87 |
|   |   | 3.4.1 Experimental Scenario | | 87 |
|   |   | 3.4.2 Monte Carlo Data | | 88 |
|   |   | 3.4.3 Set of Candidate Models | | 88 |
|   |   | 3.4.4 Data Analysis Results | | 92 |
|   |   | 3.4.5 Further Insights into the First 14 Nested Models | | 94 |
|   |   | 3.4.6 Hypothesis Testing and Information-Theoretic Approaches Have Different Selection Frequencies | | 96 |
|   |   | 3.4.7 Further Insights Following Final Model Selection | | 99 |
|   |   | 3.4.8 Why Not Always Use the Global Model for Inference? | | 100 |
|   | 3.5 | Example 4—Sage Grouse Survival | | 101 |
|   |   | 3.5.1 Introduction | | 101 |
|   |   | 3.5.2 Set of Candidate Models | | 102 |
|   |   | 3.5.3 Model Selection | | 104 |
|   |   | 3.5.4 Hypothesis Tests for Year-Dependent Survival Probabilities | | 106 |
|   |   | 3.5.5 Hypothesis Testing Versus AIC in Model Selection | | 106 |
|   |   | 3.5.6 A Class of Intermediate Models | | 110 |
|   | 3.6 | Example 5—Resource Utilization of *Anolis* Lizards | | 110 |
|   |   | 3.6.1 Set of Candidate Models | | 111 |
|   |   | 3.6.2 Comments on Analytic Method | | 112 |
|   |   | 3.6.3 Some Tentative Results | | 112 |
|   | 3.7 | Summary | | 114 |
| 4 | **Model-Selection Uncertainty with Examples** | | | **118** |
|   | 4.1 | Introduction | | 118 |
|   | 4.2 | Methods for Assessing Model-Selection Uncertainty | | 120 |
|   |   | 4.2.1 AIC Differences and a Confidence Set on the K-L Best Model | | 122 |
|   |   | 4.2.2 Likelihood of a Model and Akaike Weights | | 123 |
|   |   | 4.2.3 More Options for a Confidence Set for the K-L Best Model | | 127 |
|   |   | 4.2.4 $\Delta_i$, Model-Selection Probabilities and the Bootstrap | | 129 |
|   |   | 4.2.5 Concepts of Parameter-Estimation and Model-Selection Uncertainty | | 130 |
|   |   | 4.2.6 Including Model-Selection Uncertainty in Estimator Sampling Variance | | 133 |
|   |   | 4.2.7 Unconditional Confidence Intervals | | 137 |
|   |   | 4.2.8 Uncertainty of Variable Selection | | 140 |
|   |   | 4.2.9 Model Redundancy | | 141 |
|   |   | 4.2.10 Recommendations | | 144 |
|   | 4.3 | Examples | | 145 |
|   |   | 4.3.1 Cement Data | | 145 |

|  |  |  |  |
|---|---|---|---|
|  | 4.3.2 | *Anolis* Lizards in Jamaica | 151 |
|  | 4.3.3 | Simulated Starling Experiment | 151 |
|  | 4.3.4 | Sage Grouse in North Park | 152 |
|  | 4.3.5 | Sakamoto et al.'s (1986) Simulated Data | 152 |
|  | 4.3.6 | Pine Wood Data | 153 |
| 4.4 | Summary | | 157 |

# 5 Monte Carlo and Example-Based Insights  159
## 5.1 Introduction  159
## 5.2 Survival Models  160
### 5.2.1 A Chain Binomial Survival Model  160
### 5.2.2 An Example  163
### 5.2.3 An Extended Survival Model  168
### 5.2.4 Model Selection If Sample Size Is Huge, or Truth Known  171
### 5.2.5 A Further Chain Binomial Model  173
## 5.3 Examples and Ideas Illustrated with Linear Regression  176
### 5.3.1 All-Subsets Selection: A GPA Example  177
### 5.3.2 A Monte Carlo Extension of the GPA Example  182
### 5.3.3 An Improved Set of GPA Prediction Models  186
### 5.3.4 More Monte Carlo Results  189
### 5.3.5 Linear Regression and Variable Selection  195
### 5.3.6 Discussion  199
## 5.4 Estimation of Density from Line Transect Sampling  205
### 5.4.1 Density Estimation Background  205
### 5.4.2 Line Transect Sampling of Kangaroos at Wallaby Creek  206
### 5.4.3 Analysis of Wallaby Creek Data  206
### 5.4.4 Bootstrap Analysis  208
### 5.4.5 Confidence interval on $D$  208
### 5.4.6 Bootstrap Reps: 1,000 vs. 10,000.  210
### 5.4.7 Bootstrap vs. Akaike Weights: A lesson on $QAIC_c$  211
## 5.5 An Extended Binomial Example  213
### 5.5.1 The Durban Storm Data  213
### 5.5.2 Models Considered  214
### 5.5.3 Consideration of Model Fit  216
### 5.5.4 Confidence Intervals on Predicted Storm Probability  217
### 5.5.5 Precision Comparisons of Estimators  219
## 5.6 Lessons from the Literature and Other Matters  221
### 5.6.1 Use $AIC_c$, Not AIC, with Small Sample Sizes  221
### 5.6.2 Use $AIC_c$, Not AIC, When $K$ Is Large  222
### 5.6.3 Inference from a Less Than Best Model  224
### 5.6.4 Are Parameters Real?  226
## 5.7 Summary  227

| 6 | **Statistical Theory** | | **230** |
|---|---|---|---|
| | 6.1 | Useful Preliminaries | 230 |
| | 6.2 | A General Derivation of AIC | 239 |
| | 6.3 | General K-L–Based Model Selection: TIC | 248 |
| | | 6.3.1 Analytical Computation of TIC | 248 |
| | | 6.3.2 Bootstrap Estimation of TIC | 249 |
| | 6.4 | $AIC_c$: A Second-Order Improvement | 251 |
| | | 6.4.1 Derivation of $AIC_c$ | 251 |
| | | 6.4.2 Lack of Uniqueness of $AIC_c$ | 255 |
| | 6.5 | Derivation of AIC for the Exponential Family of Distributions | 257 |
| | 6.6 | Evaluation of $\text{tr}(J(\theta_o)[I(\theta_o)]^{-1})$ and Its Estimator | 261 |
| | | 6.6.1 Comparison of AIC vs. TIC in a Very Simple Setting | 261 |
| | | 6.6.2 Evaluation Under Logistic Regression | 266 |
| | | 6.6.3 Evaluation Under Multinomially Distributed Count Data | 273 |
| | | 6.6.4 Evaluation Under Poisson-Distributed Data | 280 |
| | | 6.6.5 Evaluation for Fixed-Effects Normality-Based Linear Models | 281 |
| | 6.7 | Additional Results and Considerations | 288 |
| | | 6.7.1 Selection Simulation for Nested Models | 288 |
| | | 6.7.2 Simulation of the Distribution of $\Delta_p$ | 290 |
| | | 6.7.3 Does AIC Over-fit? | 292 |
| | | 6.7.4 Can Selection Be Improved Based on All the $\Delta_i$? | 294 |
| | | 6.7.5 Linear Regression, AIC, and Mean Square Error | 296 |
| | | 6.7.6 AIC and Random Coefficient Models | 299 |
| | | 6.7.7 $AIC_c$ and Models for Multivariate Data | 302 |
| | | 6.7.8 There Is No True $TIC_c$ | 303 |
| | | 6.7.9 Kullback–Leibler Information Relationship to the Fisher Information Matrix | 304 |
| | | 6.7.10 Entropy and Jaynes Maxent Principle | 304 |
| | | 6.7.11 Akaike Weights, $w_i$, Versus Selection Probabilities, $\pi_i$ | 305 |
| | | 6.7.12 Model Goodness-of-Fit After Selection | 306 |
| | 6.8 | Kullback–Leibler Information Is Always $\geq 0$ | 307 |
| | 6.9 | Summary | 312 |
| 7 | **Summary** | | **315** |
| | 7.1 | The Scientific Question and the Collection of Data | 317 |
| | 7.2 | Actual Thinking and A Priori Modeling | 317 |
| | 7.3 | The Basis for Objective Model Selection | 319 |
| | 7.4 | The Principle of Parsimony | 321 |
| | 7.5 | Information Criteria as Estimates of Relative Kullback–Leibler Information | 321 |
| | 7.6 | Ranking and Calibrating Alternative Models | 324 |
| | 7.7 | Model-Selection Uncertainty | 325 |
| | 7.8 | Inference Based on Model Averaging | 326 |

7.9 More on Inferences . . . . . . . . . . . . . . . . . . . . . . 327
7.10 Final Thoughts . . . . . . . . . . . . . . . . . . . . . . . . 328

**References** **329**

**Index** **351**

# Glossary

Notation and abbreviations generally used are given below. Special notation for specific examples can be found in those sections.

| | |
|---|---|
| AIC | Akaike's information criterion. |
| $AIC_k$ | In any set of models (given a data generating context) one model *will* be the expected Kullback–Leibler, hence AIC, best model. We usually let $k$ index that best model (it is the concept, not the specific index, that is important). The true best model is analogous to the true value of an unknown parameter, given a model. With real data, $k$ is not known, so $AIC_k$ is not identified, even though the set of $AIC_i$ values is known. It must be clear that this "best" model $k$ is the same model over all possible samples (of which we have only one sample). |
| $AIC_c$ | A second-order AIC, necessary for small samples. |
| Akaike weights | Estimates of the likelihood of the model, given the data. These are normalized to sum to 1 and are denoted by $w_i$. |
| Bias | (of an estimator) Bias $= E(\hat{\theta}) - \theta$. |
| BIC | Bayesian information criterion. |
| $c$ | A simple variance inflation factor used in quasi-likelihood methods where there is overdispersion (e.g., extra-binomial variation). |
| $\Delta_i$ | AIC differences relative to the smallest AIC value in the set, hence AIC values scaled by a simple, additive constant such that |

| | |
|---|---|
| | the model with the minimum AIC value (or $AIC_c$ or $QAIC_c$) has $\Delta_i \equiv 0$. ($\Delta_i = AIC_i - \min AIC$). These values are estimates of the relative K-L distance between the best (selected) model and the $i$th model. |
| $\Delta_p$ | A "pivotal" value analogous to $(\theta - \hat{\theta})/\hat{se}(\hat{\theta})$; $\Delta_p = AIC_k - \min AIC$ (model $k$ is the true best model). Useful in understanding model selection uncertainty via Monte Carlo methods and extends to a bootstrap application. |
| df | Degrees of freedom as associated with hypothesis testing. The df = the difference in the number of parameters between the null and alternative hypotheses in standard likelihood ratio tests. |
| $E(\hat{\theta})$ | The statistical expectation of the estimator $\hat{\theta}$. |
| Estimate | The computed value of an estimator, given a particular set of sample data (e.g., $\hat{\theta} = 9.8$). |
| Estimator | A function of the sample data that is used to estimate some parameter. An estimator is a random variable and is denoted by a "hat" (e.g., $\hat{\theta}$). |
| $f(x)$ | Used to denote "truth" or "full reality"; the process that produces multivariate data; $x$. This conceptual probability distribution is often considered to be a mapping from an infinite-dimensional space. |
| $g_i(x)$ | Used to denote the set of candidate models that are hypothesized to provide an adequate approximation for the distribution of empirical data. The expression $g_i(x \mid \theta)$ is used when it is necessary to be clear that the function involves parameters $\theta$. Often the parameters have been estimated; thus the estimated approximating model is denoted by $\hat{g}_i(x \mid \hat{\theta})$. Often, the set of $R$ candidate models is represented as simply $M_1, M_2, \ldots, M_R$. |
| Global model | An over-parameterized model containing all the variables and associated parameters thought to be important as judged from an *a priori* consideration of the problem at hand. Other models in the set are special cases of this global model. The global model is often the basis for goodness-of-fit evaluation. |
| K | The number of estimable parameters in an approximating model. |
| K-L | Kullback–Leibler distance (or discrepancy, information, number). |
| LRT | Likelihood ratio test. |
| LS | Least squares method of estimation. |
| $\mathcal{L}(\theta \mid x)$ | Likelihood function of the model parameters, given the data $x$. |
| $\log(\cdot)$ | The natural logarithm ($\log_e$). |
| $\text{logit}(\theta)$ | the logit transform: $\text{logit}(\theta) = \log(\theta/(1-\theta))$, where $0 < \theta < 1$. |

| | |
|---|---|
| $M_i$ | Shorthand notation for the candidate models considered. See $g_i(x)$. |
| ML | Maximum likelihood method of estimation. |
| MLE | Maximum likelihood estimate (or estimator). |
| $n$ | Sample size. |
| Parsimony | The concept that a model should be as simple as possible concerning the included variables, model structure, and number of parameters. Parsimony is a desired characteristic of a model used for inference, and it is usually defined by a suitable trade-off between squared bias and variance of parameter estimators. Parsimony lies between the evils of under- and over-fitting. |
| Precision | A property of an estimator related to the amount of variation among estimates from repeated samples. |
| $\propto$ | A symbol meaning "proportional to." |
| QAIC or QAIC$_c$ | Versions of AIC or AIC$_c$ for overdispersed count data where quasi-likelihood adjustments are required. |
| $\pi_i$ | Model selection probabilities (or relative frequencies), often from Monte Carlo studies or the bootstrap. |
| $R$ | The number of models in the candidate set; $i = 1, \ldots, R$. |
| $\tau_i$ | Prior probability that model $i$ is the expected K-L best model. |
| $\theta$ | Used to denote a generic parameter vector (such as a set of conditional survival probabilities, $S_i$). |
| $\hat{\theta}$ | An estimator of the generic parameter $\theta$. |
| TIC | Takeuchi's information criterion |
| $w_i$ | Akaike weights. Used with any of the information criteria that are estimates of Kullback–Leibler information (AIC, AIC$_c$, QAIC, TIC). The $w_i$ sum to 1 and may be interpreted as the probability that model $i$ is the actual K-L best model for the sampling situation considered. |
| $\chi^2$ | A test statistic distributed as chi-squared with specified degrees of freedom, df. Used here primarily in relation to a goodness-of-fit test of the global model. |

# About the Authors

Drs. Kenneth P. Burnham and David R. Anderson have worked closely together for the past 25 years and have jointly published 7 books and research monographs and 54 journal papers on a variety of scientific issues. Currently, they are both in the Colorado Cooperative Fish and Wildlife Research Unit at Colorado State University, where they conduct research, teach graduate courses, and mentor graduate students.

Ken Burnham has a B.S. in biology and M.S. and Ph.D. degrees in statistics. For 26 years post-Ph.D. he has worked as a statistician, applying and developing statistical theory in several areas of life sciences, especially ecology and wildlife, most often in collaboration with subject-area specialists. Ken has worked (and lived) in Oregon, Alaska, Maryland (Patuxent Wildlife Research Center), North Carolina (U.S. Department of Agriculture at North Carolina State University, Statistics Department), and Colorado. He is the recipient of numerous professional awards including Distinguished Achievement Medal from the American Statistical Association, and Distinguished Statistical Ecologist Award from INTECOL (International Congress of Ecology). Ken is a Fellow of the American Statistical Association.

David Anderson received B.S. and M.S. degrees in wildlife biology and a Ph.D. in theoretical ecology. He is currently a Senior Scientist with the Biological Resources Division within the U.S. Geological Survey and a professor in the Department of Fishery and Wildlife Biology. He spent 9 years at the Patuxent Wildlife Research Center in Maryland and 9 years as leader of the Utah Cooperative Wildlife Research Unit and professor in the Wildlife Science Department at Utah State University. He has been at Colorado State University since 1984. He is the recipient of numerous professional awards for scientific and academic contributions.

# 1
# Introduction

## 1.1 Objectives of the Book

This book is about making valid inferences from scientific data when a meaningful analysis depends on a model. Our general objective is to provide scientists with a readable text giving practical advice for the analysis of empirical data under an information-theoretic paradigm. We first assume that an exciting scientific question has been carefully posed and relevant data have been collected, following a sound experimental design or probabilistic sampling program. Often, little can be salvaged if data collection has been seriously flawed or if the question was poorly posed (Hand 1994). We realize, of course, that these issues are never as ideal as one would like. However, proper attention must be placed on the collection of data (Chatfield 1991, 1995a). We stress inferences concerning understanding the structure and function of the biological system, estimators of relevant parameters, and valid measures of precision; we say somewhat less about formal prediction.

Of necessity such an analysis will often need to be based on a model to represent the information in the data. There are many studies where we seek an understanding of relationships, especially causal ones. There are many studies to understand our world; models are important because of the parameters in them and relationships between and among variables. These parameters have relevant, useful interpretations, even when they relate to quantities that are not directly observable (e.g., survival probabilities, animal density on an area, gene frequencies, and interaction terms). Science would be very limited without such unobservables as constructs in models. We make statistical inferences from the data, to a real or conceptual population or process, based on models involving such parameters.

Observables and prediction are often critical, but science is broader than these issues.

The first objective of this book is to outline a consistent *strategy* for issues surrounding the analysis of empirical data. Induction is used to make statistical inference about a defined population or process, given an empirical sample or experimental data set. Then, "data analysis" leading to valid inference is the integrated process of careful *a priori* model formulation, model selection, parameter estimation, and measurement of precision (including a variance component due to model selection uncertainty). We do not believe that modeling and model selection should be treated as activities that precede "the analysis"; rather, model selection is usually a critical and integral aspect of scientific data analysis that leads to valid inference.

A philosophy of thoughtful, science-based, *a priori* modeling is advocated when possible. One first develops a robust, global model (or set of models) and then derives several other plausible candidate (sub)models postulated to represent good approximations to information in the data at hand. This forms the *set of candidate models*. Science and biology play a lead role in this *a priori* model building and careful consideration of the problem. The modeling and careful thinking about the problem are critical elements that have often received relatively little attention in statistics classes (especially for nonmajors), partly because such classes rarely consider an overall strategy or philosophy of data analysis. A proper *a priori* model building strategy tends to avoid "data dredging," which leads to over-fitted models, that is, to the "discovery" of spurious effects. Instead, there has often been a rush to "get to the data analysis" and begin to rummage through the data and compute various estimates of interest or conduct hypothesis tests. We realize that these other philosophies may have their place, especially in more exploratory phases of investigation.

The second objective is to explain and illustrate methods developed recently at the interface of information theory and mathematical statistics for selection of an estimated "best approximating model" from the *a priori* set of candidate models. In particular, we review and explain the use of Akaike's information criterion (AIC) in the selection of a model (or small set of good models) for statistical inference. AIC provides a simple, effective, and objective means for the selection of an estimated "best approximating model" for data analysis and inference. Model selection includes "variable selection" as frequently practiced in regression analysis. Model selection based on information theory is a relatively new paradigm in the biological and statistical sciences and is quite different from the usual methods based on hypothesis testing or various Bayesian methods. Model selection based on information theory approaches is not the only reasonable approach, but it is what we are focusing on here.

The practical use of information criteria, such as Akaike's, for model selection is relatively recent (the major exception being in time series analysis, where AIC has been used routinely for most of the past two decades). The marriage of information theory and mathematical statistics started with Kullback's (1959) book. Akaike considered AIC to be an extension of R. A. Fisher's likelihood theory. These are all complex issues, and the literature is often highly technical and scattered widely throughout books and research journals. Here we attempt to bring this relatively new material into a short, readable text for people (primarily) in the biological and statistical sciences. We provide a series of examples, many of which are biological, to illustrate various aspects of the theory and application.

In contrast, hypothesis testing as a means of selecting a model has had a much longer exposure in science. Many seem to feel more comfortable with the hypothesis testing paradigm in model selection, and some even consider the results of a test as *the* standard by which other approaches should be judged (we believe that they are wrong to do so). Bayesian methods in model selection and inference have been the focus of much recent research. However, the technical level of this material often makes these approaches unavailable to many people in the biological sciences. A variety of cross-validation-based methods have been proposed for model selection, and these, too, seem like very reasonable approaches. The computational demands of many of the Bayesian and cross-validation methods for model selection are often quite high (often 1–3 orders of magnitude higher than information-theoretic approaches), especially if there are more than a dozen or so high-dimensional candidate models.

The third objective focuses on "model selection uncertainty"—inference problems that arise in using the same data for both model selection and the associated parameter estimation and inference. If model selection uncertainty is ignored, precision is often overestimated, achieved confidence interval coverage is below the nominal level, and predictions are less accurate than expected. Another problem is the inclusion of spurious variables, or factors, with no assessment of the reliability of their selection. Some general methods for dealing with model and variable selection uncertainty are suggested and examples provided. Incorporating model selection uncertainty into estimators of precision is an active area of research, and we expect to see additional approaches developed in the coming years.

Modeling is an art as well as a science and is directed toward finding a good approximating model of the empirical data as the basis for statistical inference from those data. In particular, the number of parameters estimated from data should be substantially less than the sample size, or inference is likely to remain somewhat preliminary (e.g., Miller [1990:preface page $x$]) mentions a regression problem with 757 variables and a sample size of 42—it is absurd to think that valid inference is likely to come from the analysis of these data). In cases where there are relatively few data per estimated parameter, a small-sample version of AIC is available (termed $AIC_c$) and should be used routinely rather than AIC. There are cases where quasi-likelihood methods are appropriate when count data are overdispersed; this theory leads to modified criteria such as QAIC and $QAIC_c$, and these extensions are covered in the following material.

Simple models with only 1–2 parameters are not the central focus of this book; rather, we focus on models of more complex systems. Parameter estimation has been firmly considered to be an optimization problem for many decades, and AIC formulates the problem of model selection as an optimization problem across a set of candidate models. Minimizing AIC is a simple operation with results that are easy to interpret. Models can be clearly ranked and calibrated, allowing full consideration of other good models, in addition to the estimated "best approximating model." Competing models, those with AIC values close to the minimum, are also useful in the estimation of model selection uncertainty. Inference should often be based on more than a single model, unless the data clearly support only a single model fit to the data. Thus, some approaches are provided to allow inference from several of the best models, including model averaging.

This is primarily an applied book; Shibata (in prep.) reviews and extends much of the theory relevant to the issues covered here. A person with a good background in mathematics and theoretical statistics would benefit from studying Shibata's book. McQuarrie and Tsai (1998) present both theoretical and applied aspects of model selection in regression and time series analysis, including the results of large-scale Monte Carlo simulation studies.

## 1.2 Background Material

Data and stochastic models of data are used in the empirical sciences to make inferences concerning both processes and parameters of interest (see Box et al. 1981 and Lunneborg 1994 for a review of principles). Statistical scientists have worked with people in the biological sciences for many years to improve methods and understanding of biological processes. This book provides practical, omnibus methods to achieve valid inference from models that are good approximations to biological processes and data. A broad definition of data is employed here. A single, simple data set might be the subject of analysis, but more often, data collected from several field sites or laboratories are the subject of a more comprehensive analysis. The data might commonly be extensive and partitioned by age, sex, species, treatment group, within several habitat types or geographic areas. In linear and nonlinear regression models there may be many explanatory variables. There are often factors (variables) with large effects in these information-rich data sets. Parameters in the model represent the effects of these factors. We focus on modeling philosophy, model selection, estimation of model parameters, and valid measures of precision under the relatively new paradigm of information-theoretic methods. Valid inference rests upon these four issues, in addition to the critical considerations relating to problem formulation, study design, and protocol for data collection.

### 1.2.1 Inference, Given a Model

R. A. Fisher (1922) discussed three aspects of the general problem of valid inference: (1) model specification, (2) estimation of model parameters, and (3) estimation of precision. Here, we prefer to partition model specification into two components: formulation of a set of candidate models and selection of a model (or small number of models) to be used in making inferences. For much of this century methods have been available to objectively and efficiently estimate model parameters and their precision (i.e., the sampling covariance matrix). Fisher's *likelihood theory* has been the primary, omnibus approach to these issues, but it *assumes* that the model structure is known and that only the parameters in that structural model are to be estimated. Simple examples include a linear model such as $y = \alpha + \beta x + \epsilon$ where the residuals ($\epsilon$) are assumed to be normally distributed, or a log-linear model for the analysis of count data displayed in a contingency table. The parameters in these models can be estimated using *maximum likelihood* (ML) methods. That is, if one assumes or somehow chooses a particular model, methods exist that are objective and asymptotically optimal for estimating model parameters and the sampling covariance structure, conditional on the model. A more challenging example might be to assume that data are appropriately modeled by a 3-parameter gamma distribution; one can routinely use the method of maximum likelihood to estimate these model parameters and the model-based $3 \times 3$ sampling covariance matrix. Given an appropriate model, and if the sample size is "large," then maximum likelihood provides estimators of parameters that are consistent (i.e., asymptotically unbiased with variance tending to zero), fully efficient (i.e., minimum variance among consistent estimators), and normally distributed. With small samples, but still assuming an appropriate model, ML estimators are often biased, where bias $\equiv E(\hat{\theta}) - \theta$. Such bias is usually a trivial consideration, as it is often substantially less than the $se(\hat{\theta})$, and bias-adjusted estimators can often be found if this is deemed necessary. The sampling distributions of ML estimators are often skewed with small samples, but profile likelihood intervals or log-based intervals or bootstrap procedures can be used to achieve asymmetric confidence intervals with good coverage properties. **In general, the maximum likelihood method provides an objective, omnibus theory for estimation of model parameters and the sampling covariance matrix,** *given an appropriate model.*

## 1.2.2 The Critical Issue: "What Model To Use?"

While hundreds of books and countless journal papers deal with estimation of model parameters and their associated precision, relatively little has appeared concerning model specification (what set of candidate models to consider) and model selection (what model(s) to use for inference) (see Peirce 1955). In fact, Fisher believed at one time that model specification was outside the field of mathematical statistics, and this attitude prevailed within the statistical community for several decades until at least the early 1970s. "**What model to use?**" is *the* critical question in making valid inference from data in the biological sciences.

If a poor or inappropriate model is used, then inference based on the data and this model will often be poor. Thus, it is clearly important to select an appropriate model for the analysis of a specific data set; however, this is not the same as trying to find the "true model." Model selection methods with a deep level of theoretical support are required and, particularly, methods that are easy to use and widely applicable in practice. Part of "applicability" means that the methods have good operating characteristics for realistic sample sizes. As Potscher (1991) noted, asymptotic properties are of little value unless they hold for realized sample sizes.

Sir Ronald Aylmer Fisher was born in 1890 in East Finchley, London, and died in Adelaide, Australia, in 1962. This photo was taken when he was approximately 34 years of age. Fisher was one of the foremost scientists of his time, making incredible contributions in theoretical and applied statistics and genetics. Details of his life and many scientific accomplishments are found in Box (1978). He published 7 books (one of these had 14 editions and was printed in 7 languages) and nearly 300 journal papers. Most relevant to the subject of this book is Fisher's likelihood theory and parameter estimation using his method of maximum likelihood.

A simple example will motivate some of the concepts presented. Flather (1992 and 1996) studied patterns of avian species-accumulation rates among forested landscapes in the eastern United States using data from the Breeding Bird Survey (Bystrak 1981). He derived an *a priori* set of 9 candidate models from two sources: (1) the literature on species area curves (most often the power or exponential models were suggested) and (2) a broader search of the literature for functions that increased monotonically to an asymptote (Table 1.1). Which model should be used for the analysis of these ecological data? Clearly, none of these 9 models are likely to be the "truth" that generated the data (full reality) from the Breeding Bird Survey over the years of study. Instead, Flather wanted an approximating model that fit the data well and could be used in making inferences about bird communities on the scale of large landscapes. In this first example, the number of parameters in the candidate models range only from 3 to 5. Which approximating model is "best" for making inferences from these data is answered philosophically by the principle of parsimony (Section 1.4) and operationally by several information-theoretic criteria in Chapter 2. Methods for estimating model selection uncertainty

TABLE 1.1. Summary of *a priori* models of avian species-accumulation curves from Breeding Bird Survey data for Indiana and Ohio (from Flather 1992:51 and 1996). The response variable ($y$) is the number of accumulated species, and the explanatory variable ($x$) is the accumulated number of samples. Nine models and their number of parameters are shown to motivate the question, "Which model should be used for making inference from these data?"

| Model structure | Number of parameters ($K$)[a] |
|---|---|
| $E(y) = ax^b$ | 3 |
| $E(y) = a + b\log(x)$ | 3 |
| $E(y) = a(x/(b+x))$ | 3 |
| $E(y) = a(1 - e^{-bx})$ | 3 |
| $E(y) = a - bc^x$ | 4 |
| $E(y) = (a + bx)/(1 + cx)$ | 4 |
| $E(y) = a(1 - e^{-bx})^c$ | 4 |
| $E(y) = a\left(1 - [1 + (x/c)^d]^{-b}\right)$ | 5 |
| $E(y) = a[1 - e^{-(b(x-c))^d}]$ | 5 |

[a] There are $K-1$ structural parameters and one residual variance parameter, $\sigma^2$. Assumed: $y = E(y) + \epsilon$, $E(\epsilon) = 0$, $V(\epsilon) = \sigma^2$.

and incorporating this into inferences are given in Chapter 4 and illustrated in Chapters 4 and 5.

## 1.2.3 Science Inputs—Formulation of the Set of Candidate Models

Model specification or formulation, in its widest sense, is conceptually more difficult than estimating the model parameters and their precision. Model formulation is the point where the scientific and biological information formally enter the investigation. Building the set of candidate models is partially subjective—that is why scientists must be trained, educated, and experienced in their discipline. The published literature and experience in the biological sciences can be used to help formulate a set of *a priori* candidate models. The most original, innovative part of scientific work is the phase leading to the proper question. Good approximating models in conjunction with a good set of relevant data can provide insight into the underlying biological process and structure.

Lehmann (1990) asks "where do models come from" and cites some biological examples (also see Ludwig 1989, Walters 1996). Models arise from questions about biology and the manner in which biological systems function. Relevant theoretical and practical questions arise from a wide variety of sources (see O'Connor and Spotila 1992). Traditionally, these questions come from the scientific literature, results of manipulative experiments, personal experience, or contemporary debate within the scientific community. More practical questions stem from resource management controversies, biomonitoring programs, quasi-experiments, and even judicial hearings.

Chatfield (1995b) suggests that there is a need for more *careful thinking* (than is usually evident) and a *better balance* between the problem (biological question), analysis theory, and data. Too often, the emphasis is focused on the analysis theory and data analysis, with too little thought about the reason for the study in the first place (see Hayne 1978 for convincing examples).

The philosophy and theory presented here must rest on well-designed studies and careful planning and execution of field or laboratory protocol. Many good books exist giving information on these important issues (Burnham et al. 1987, Cook and Campbell 1979, Mead 1988, Hairston 1989, Desu and Roghavarao 1991, Eberhardt and Thomas 1991, Manly 1992, Skalski and Robson 1992, and Scheiner and Gurevitch 1993). Chatfield (1991) reviews statistical pitfalls and ways that these might be avoided. Research workers are urged to pay close attention to these critical issues. Methods given here should not be thought to salvage poorly designed work. In the following material we will assume that the data are "sound" and that inference to some larger population is reasonably justified by the manner in which the data were collected.

Development of the *a priori* set of candidate models often should include a global model: a model that has many parameters, includes all potentially relevant effects, and reflects causal mechanisms thought likely, based on *the science of the situation*. The global model should also reflect the study design and attributes of the system studied. Specification of the global model should not be based on a probing examination of the data to be analyzed. At some early point, one should investigate the fit of the global model to the data (e.g., examine residuals and measures of fit such as $R^2$, deviance, or formal $\chi^2$ goodness-of-fit tests) and proceed with analysis only if it is judged that the global model is an acceptable fit to the data. Models with fewer parameters can then be derived as special cases of the global model. This set of reduced models represents plausible alternatives based on what is known or hypothesized about the process under study. Generally, alternative models will involve differing numbers of parameters; the number of parameters will often differ by at least an order of magnitude across the set of candidate models. Chatfield (1995b) writes concerning the importance of subject-matter considerations such as accepted theory, expert background knowledge, and prior information in addition to known constraints on both the model parameters and the variables in the models. All these factors should be brought to bear on the makeup of the set of candidate models, prior to actual data analysis.

The more parameters used, the better the fit of the model to the data that can be achieved. Large and extensive data sets are likely to support more complexity, and this should be considered in the development of the set of candidate models. *If a particular model (parametrization) does not make biological sense, it should not be included in the set of candidate models.* In developing the set of candidate models, one must recognize a certain balance between keeping the set small and focused on plausible hypotheses, while making it big enough to guard against omitting a very good *a priori* model. While this balance should be considered, we advise the inclusion of all models that seem to have a reasonable justification, prior to data analysis. While one must worry about errors due to both under-fitting and over-fitting, it seems that modest over-fitting is less damaging than under-fitting (Shibata 1989). We recommend and encourage a considerable amount of careful, *a priori* thinking in arriving at a set of candidate models (see Peirce 1955, Burnham and Anderson 1992, Chatfield 1995b).

Freedman (1983) noted that when there are many, say 50, explanatory variables ($X_1, X_2, \ldots, X_{50}$) used to predict a response variable ($Y$), variable selection methods will provide regression equations with high $R^2$ values, "significant" $F$ values, and many "significant" regression coefficients, as shown by large $t$ values, *even if the explanatory variables are independent of $Y$*. This undesirable situation occurs most frequently when the number of variables is of the same order as the number of observations. This finding, known as Freedman's paradox, was illustrated by Freedman using hypothesis testing as a means to select a model of $Y$ as a function of the $X$s, but the same type of problematic result can be found in using other model selection methods. Miller (1990) notes that estimated regression coefficients are biased away from zero in such cases. The partial resolution of this paradox is in the *a priori* modeling considerations, keeping the number of candidate models small, and achieving a large sample size relative to the number of parameters to be estimated.

It is not uncommon to see biologists collect data on 50–130 "ecological" variables in the blind hope that some analysis method and computer system will "find the variables that are significant" and sort out the "interesting" results. This shotgun strategy will likely uncover mainly spurious correlations, and it is prevalent in the naive use of many of the traditional multivariate analysis methods (e.g., principal components, stepwise discriminant function analysis, canonical correlation methods, and factor analysis) found in the biological literature. We believe that mostly spurious results will be found using this unthinking approach (also see Flack and Chang 1987 and Miller 1990), and we encourage investigators to give very serious consideration to a well-founded set of candidate models as a means of minimizing the inclusion of spurious variables and relationships. Ecologists are not alone in collecting a small amount of data on a very large number of variables. A. J. Miller (personal communication) indicates that he has seen data sets in other fields with as many as 1,500 variables where the number of cases is less than 40 (a purely statistical search for meaningful relationships in such data is doomed to failure).

After a carefully defined set of candidate models has been developed, one is left with the evidence contained in the data; the task of the analyst is to interpret this evidence from analyzing the data. Questions such as "What effects are supported by the data?" can be answered objectively. This modeling approach allows a clear place for experience (i.e., prior knowledge and beliefs), the results of past studies, the biological literature, and current hypotheses to formally enter the modeling process. Then, one turns to the data to see "what is important" within a sense of parsimony. Using AIC and other similar methods one can only hope to select the best model from this set; if good models are not in the set of candidates, they cannot be discovered by model selection (i.e., data analysis) algorithms.

We lament the practice of generating models (i.e. "modeling") that is done in the total absence of real data and yet "inferences" are made about the status, structure, and functioning of the real world based on studying the models so generated. We do not object to the often challenging and stimulating intellectual exercise of model construction as a means to integrate and explore our myriad ideas about various subjects. For example, Berryman et al. (1995) provide a nice list of 26 candidate models for predator–prey relationships and are interested in their "credibility" and "parsimony" (however, as is often the case, there are no empirical data available on a variety of taxa to pursue these issues in a rigorous manner). Such exercises help us sort out ideas that in fact conflict when their logical consequences are explored. Modeling exercises can strengthen our logical and quantitative abilities. Modeling exercises can give us insights into how the world *might* function, and hence modeling efforts can lead to alternative hypotheses to then be explored with real data. Our objection is only to the confusing of inferences about such models with inferences about the real world. An inference from a model to some aspect of the real world is only justified after the model has been shown to adequately fit relevant empirical data (this will certainly be the case when the model in its totality has been fit to and tested against reliable data). Gause (1934) had similar beliefs when he stated, "Mathematical investigations independent of experiments are of but small importance . . . ."

The underlying philosophy of analysis is important here. We advocate a conservative approach to the overall issue of *strategy* in the analysis of data in the biological sciences with an emphasis on *a priori* considerations and models to be considered. Careful, *a priori* consideration of alternative models will often require a major change in emphasis among many people. This is often an unfamiliar concept to both biologists and statisticians, where there has been a tendency to use either a traditional model or a model with associated computer software, making its use easy (Lunneborg 1994). This *a priori* strategy is in contrast to strategies advocated by others who view modeling and data analysis as a highly iterative and interactive exercise. Such a strategy, to us, represents deliberate data dredging and should be reserved for early exploratory phases of initial investigation. Such an exploratory avenue is not the subject of this book.

Here, we advocate the deliberate exercise of carefully developing a set of, say, 4–20 alternative models as potential approximations to the population-level information in the data available and the scientific question being addressed. Some practical problems might have as many as 70–100 or more models that one might want to consider. The number of candidate models is often larger with large data sets. We find that people include many models that are far more general than the data could reasonably support (e.g., models with several interaction parameters). There needs to be some well-supported guidelines on this issue to help analysts better define the models to be considered. This set of models, developed without first deeply examining the data, constitutes the "*set of candidate models.*" The science of the issue enters the analysis through the *a priori* set of candidate models.

## 1.2.4 Models Versus Full Reality

Fundamental to our paradigm is that none of the models considered as the basis for data analysis are the "true model" that generates the biological data we observe (see, for example, Bancroft and Han 1977). We believe that "truth" (full reality) in the biological sciences has essentially infinite dimension, and hence full reality cannot be revealed with only finite samples of data and a "model" of those data. It is generally a mistake to believe that there is a simple, "true model" in the biological sciences and that during data analysis, this model can be uncovered and its parameters estimated. Instead, biological systems are complex, with many small effects, interactions, individual heterogeneity, and individual and environmental covariates (most being unknown to us); we can only hope to identify a model that provides a good *approximation* to the data available. The words "true model" represent an oxymoron, except in the case of Monte Carlo studies whereby a model is used to generate "data" using pseudo-random numbers (we will use the term "generating model" for such computer-based studies). The concept of a "true model" in biology seems of little utility and may even be a source of confusion about the nature of approximating models (e.g., see material on BIC and related criteria in Section 2.8).

A model is a simplification or approximation of reality and hence will not reflect all of reality. Taub (1993) suggests that unproductive debate concerning true models can be avoided by simply recognizing that a model is not truth by definition. Box (1976) noted that "all models are wrong, but some are useful." While a model can never be "truth," a model might be ranked from very useful, to useful, to somewhat useful to, finally, essentially useless. Model selection methods try to rank models in the candidate set relative to each other; whether any of the models is actually "good" depends primarily on the quality of the science and *a priori* thinking that went into the modeling. Full truth (reality) is elusive (see de Leeuw 1988). Proper modeling and data analysis tells what inferences the data support, not what full reality might be (White et al. 1982:14–15, Lindley 1986). Models, used cautiously, tell us "what effects are supported by the (finite) data available." Increased sample size (information) allows us to chase full reality, but never quite catch it.

In using some model selection methods, it is assumed that the set of candidate models contains the "true model" that generated the data. We will not make this assumption, unless we use a data set generated by Monte Carlo methods as a tutorial example (e.g., Section 3.4), and then we will make this artificial condition clear. In the analysis of real data, it seems unwarranted to pretend that the "true model" is included in the set of candidate models, or even that the true model exists at all. Even if a "true model" did exist and if it could be found using some method, it would not be a good model for general inference (i.e., understanding or prediction) about some biological system, because its numerous parameters would have to be estimated from the finite data, and the precision of these estimated parameters would be quite low.

Often the investigator wants to simplify some representation of reality in order to achieve an understanding of the dominant aspects of the system under study.

If we were given a nonlinear formula with 200 parameter values, we could make correct predictions, but it would be difficult to *understand* the main dynamics of the system without some further simplification or analysis. Thus, one should tolerate some inexactness (an inflated error term) to facilitate a more simple and useful understanding of the phenomenon.

In particular, we believe that there are tapering effect sizes in many biological systems; that is, there are often several large, important effects, followed by many smaller effects, and, finally, followed by a myriad of yet smaller effects. These effects may be sequentially unveiled as sample size increases. The main, dominant, effects might be relatively easy to identify and support, even using fairly poor analysis methods, while the second-order effects (e.g., a chronic treatment effect or an interaction term) might be more difficult to detect. The still smaller effects can be detected only with very large sample sizes (cf. Kareiva 1994 and related papers), while the smallest effects have little chance of being detected, even with very large samples. Rare events that have large effects may be very important but quite difficult to study. Approximating models must be related to the amount of data and information available; small data sets will appropriately support only simple models with few parameters, while more comprehensive data sets will support, if necessary, more complex models.

This tapering in "effect size" and high dimensionality in biological systems might be quite different from some physical systems where a small-dimensioned model with relatively few parameters might accurately represent full truth or reality. Biologists should not believe that a simple, "true model" exists that generates the data observed, although some biological questions might be of relatively low dimension and could be well approximated using a fairly simple model. The issue of a range of tapering effects has been realized in epidemiology, where Michael Thun notes, "... you can tell a little thing from a big thing. What's very hard to do is to tell a little thing from nothing at all" (Taubes 1995). *Full reality will always remain elusive.*

At a more advanced conceptual level, these is a concept that "information" about the population (or process or system) under study exists in the data and the goal is to express this information in a more compact, understandable form using a "model." Conceptually, this is a change in coding system, similar to using a different "alphabet." The data have only a finite, fixed amount of information. The *goal* of model selection is to achieve a perfect one-to-one translation so that no information is lost; in fact, we cannot achieve this ideal. The data can be ideally partitioned into *information* and *noise*. The noise part of the data is not information. Conceptually, the role of a good model is to filter the data so as to separate information from noise.

Our main emphasis in modeling empirical data is to understand the biological structure, process, or system. Sometimes prediction will be of interest; here, however, one would hopefully have an understanding of the structure of the system as a basis for making trustworthy predictions. We recommend developing a set of candidate models prior to intensive data analysis, selecting one that is "best," and estimating the parameters of that model and their precision (using maximum likelihood or least squares methods). This unified strategy is a basis for valid inferences, and there are several more advanced methods to allow additional inferences and insights (Chapters 4 and 5). Statistical science is not so much a branch of mathematics, but rather it is concerned with the development of a practical theory of information using what is known or postulated about the science of the matter. In our investigations into these issues, we were often surprised by how much uncertainty there is in selecting a good approximating model; the variability in terms of what model is selected or considered best from independent data sets, for example, is often large.

## *1.2.5 A "Best Approximating Model"*

We consider some properties of a good model for valid inference in the analysis of data. Ideally, the process by which a "best" model is selected would be objective and repeatable; these are fundamental tenets of science. Furthermore, precise, unbiased estimators of parameters would be ideal, as would accurate estimators of precision. The best model would ideally yield achieved confidence interval coverage close to the nominal level (often 0.95) and have confidence intervals of minimum width. Achieved confidence interval coverage is a convenient index to whether parameter estimators and measures of precision are adequate. Finally, one would like as good an approximation of the structure of the system as the information permits. If prediction was the goal, then having the above issues in place might warrant some tentative trust in model predictions. There are many cases where two or more models are essentially tied for "best," and this should be fully recognized in further analysis and inference. In other cases there might be 4–10 models that have at least some support, and these, too, might deserve scrutiny in reaching conclusions from the data, based on inferences from more than a single model.

## 1.3 Overview of Models in the Biological Sciences

Models are useful in the biological sciences for understanding the structure of systems, estimating parameters of interest and their associated variance–covariance matrix, predicting outcomes and responses, and testing hypotheses. Such models might be used as "relational" or "explanatory" or might be used for prediction. In the following material we will review the main types of models used in the biological sciences. Although the list is not meant to be exhaustive, it will allow the reader an impression of the wide class of models of empirical data that we will treat under an information-theoretic framework.

## 1.3.1 Types of Models Commonly Used in the Biological Sciences

Simple linear and multiple linear regression models (Seber 1977, Draper and Smith 1981, Brown 1993) have seen heavy use in the biological sciences over the past four decades. These models commonly employ one to perhaps 8–12 parameters, and the statistical theory is fully developed (either based on least squares or likelihood theory). Similarly, analysis of variance and covariance models have been widely used and the theory underlying these methods is closely related to regression models and fully developed (both are examples of general linear models). Theory and software for this wide class of methods are readily available.

Nonlinear regression models (Gallant 1987, Seber and Wild 1989, Carroll et al. 1995) have also seen abundant use in the biological sciences (logistic regression is a common example). Here, the underlying theory is often likelihood based, and some classes of nonlinear models require very specialized software. In general, nonlinear estimation is a more advanced problem and is somewhat less well understood by many practicing researchers.

Other types of models used in the biological sciences include generalized linear (McCullagh and Nelder 1989, Morgan 1992) and generalized additive (Hastie and Tibshirani 1990) models (these can be types of nonlinear regression models). These modeling techniques have seen increasing use in the past decade. Multivariate modeling approaches such as multivariate ANOVA and regression, canonical correlation, factor analysis, principal components analysis, and discriminate function analysis have had a checkered history in the biological and social sciences, but still see substantial usage (see review by James and McCulloch 1990). Log-linear and logistic models (Agresti 1990) have become widely used for count data. Time series models (Brockwell and Davis 1987, 1991) are used in many biological disciplines. Various models of an organism's growth (Brisbin et al. 1987, Gochfeld 1987) have been proposed and used in biology.

Compartmental models are a type of state-transition in continuous time and continuous response and are usually based on systems of differential or partial differential equations (Brown and Rothery 1993). There are discrete state transition models using the theory of Markov chains (Howard 1971); these have found use in a wide variety of fields including epidemiological models of disease transmission.

Models to predict population viability (Boyce 1992), often based on some type of Leslie matrix, are much used in conservation biology, but rarely are alternative model forms given serious evaluation. A common problem here is that these models are rarely based on empirical data; the form of the model and its parameter values are often merely only "very rough guesses" necessitated by the lack of empirical data.

Biologists in several disciplines employ differential equation models in their research (see Pascual and Kareiva 1996 for a reanalysis of Gause's competition data and Roughgarden 1979 for examples in population genetics and evolutionary ecology). Many important applications involve exploited fish populations (Myers et al. 1995). Computer software exists to allow model parameters to be estimated using least squares or maximum likelihood methods (e.g., SAS and Splus). These are powerful tools in the analysis of empirical data, but also beg the issue of "what model to use."

Open and closed capture–recapture (Lebreton et al. 1992) and band recovery (Brownie et al. 1985) models represent a class of models based on product multinomial distributions (see issues 5 & 6 of volume 22 of the *Journal of Applied Statistics*, 1995). Distance sampling theory (Buckland et al. 1993) relies on models of the detection function and often employ semiparametric models. Parameters in these models are nearly always estimated using maximum likelihood.

Spatial models (Cressie 1991 and Renshaw 1991) are now widely used in the biological sciences, allowing the biologist to take advantage of spatial data sets (e.g., geographic information systems). Stein and Corsten (1991) have shown how Kriging (perhaps the most widely used spatial technique) can be expressed as a least squares problem, and the development of Markov chain Monte Carlo methods such as the Gibbs sampler allow other forms of spatial models to be fitted by least squares or maximum likelihood (Augustin et al. 1996). Further unifying work for methods widely used on biological data has been carried out by Stone and Brooks (1990). Geographic information systems potentially provide large numbers of covariates for biological models, so that model selection issues are particularly important.

Spatiotemporal models are potentially invaluable to the biologist, though most researchers model changes over space or time, and not both simultaneously. The advent of Markov chain Monte Carlo methods may soon give rise to a general but practical framework for spatiotemporal modeling; model selection will be an important component of such a framework. A step towards this general framework was made by Buckland and Elston (1993), who modeled changes in the spatial distribution of wildlife.

There are many other examples where modeling of data plays a fundamental role in the biological sciences. Henceforth, we will exclude only modeling that cannot be put into a likelihood or quasi-likelihood (Wedderburn 1974) framework and models that do not explicitly relate to empirical data. All least squares formulations are merely special cases that have an equivalent likelihood formulation in usual practice. We exclude what are effectively nonparametric models such as splines and kernel methods, which either do not have explicit parameters or cannot or generally are not put into a likelihood or least squares framework.

## *1.3.2 Likelihood and Least Squares Theory*

Historically, biologists have typically been exposed to least squares (LS) theory in their classes in statistics. LS methods for linear models are often relatively simple to compute and therefore enjoyed an early history of application. Fisher's likelihood methods often require iterative, numerical methods and were thus not popular prior to the widespread availability of personal computers and the development of easy-to-use software. LS theory has many similarities with likelihood theory and yields identical estimators of the structural parameters (but not $\hat{\sigma}^2$) for linear and nonlinear models when the residuals are assumed to be independent and normally distributed. Biologists can easily allow alternative error structures (i.e., nonnormal residuals such as Poisson or log-normal) for regression and other similar problems, but these extensions are often easy only in a likelihood or quasi-likelihood framework.

Biologists familiar with LS but lacking insight into likelihood methods might benefit from an example. Consider a multiple linear regression model where a dependent variable $y$ is hypothesized to be a function of $p$ explanatory (independent) variables $x_j$ ($j = 1, 2, \ldots, p$). Here the residuals $\epsilon_i$ of the $n$ observations are assumed to be independent, normally distributed with a constant variance $\sigma^2$, and the model is expressed as

$$y_i = \beta_0 + \beta_1(x_1) + \beta_2(x_2) + \cdots + \beta_p(x_p) + \epsilon_i, \qquad i = 1, \ldots, n.$$

Hence,

$$E(y_i) = \beta_0 + \beta_1(x_1) + \beta_2(x_2) + \cdots + \beta_p(x_p), \qquad i = 1, \ldots, n,$$

where $E(y_i)$ is a linear function of $p + 1$ parameters. The conceptual residuals,

$$\epsilon_i = y_i - (\beta_0 + \beta_1(x_1) + \beta_2(x_2) + \cdots + \beta_p(x_p)),$$
$$= y_i - E(y_i),$$

have the joint probability distribution $g(\underline{\epsilon} \mid \underline{\theta})$, where $\underline{\theta}$ is a vector of $K = p + 2$ parameters ($\beta_0, \beta_1, \ldots, \beta_p$, and $\sigma$). Here, corresponding to observation $i$ one has the model

$$g(\epsilon_i \mid \underline{\theta}) = \frac{1}{\sqrt{2\pi}\sigma} e^{-\frac{1}{2}\left[\frac{\epsilon_i}{\sigma}\right]^2}.$$

The likelihood is simply the product of these over the $n$ observations and is a function of the unknown parameters, given the data (and the linear model),

$$\mathcal{L}(\underline{\theta} \mid \underline{x}_i) = \prod_{i=1}^{n} \frac{1}{\sqrt{2\pi}\sigma} e^{-\frac{1}{2}\left[\frac{\epsilon_i}{\sigma}\right]^2},$$

$$= \left(\frac{1}{\sqrt{2\pi}\sigma}\right)^n e^{-\frac{1}{2}\sum_{i=1}^{n}\left[\frac{\epsilon_i}{\sigma}\right]^2}.$$

When the $\epsilon_i$ are normally distributed with constant variance $\sigma^2$, the maximum likelihood estimator (MLE) of $\beta$ is identical to the usual LS regression estimators (however, the estimator of $\sigma^2$ differs slightly). This formalism shows, *given the model*, the link between the data, the model, and the parameters to be objectively estimated, using either LS or ML.

In all fitted linear models the residual sums of squares (*RSS*) is

$$RSS = \sum_{i=1}^{n} \hat{\epsilon}_i^2,$$

where

$$\hat{\epsilon}_i = y_i - (\hat{\beta}_0 + \hat{\beta}_1(x_1) + \hat{\beta}_2(x_2) + \cdots + \hat{\beta}_p(x_p)),$$
$$= y_i - \hat{E}(y_i).$$

The ML estimator $\hat{\sigma}^2 = RSS/n$, while the estimator universally used in the LS case is $\hat{\sigma}^2 = RSS/(n-(p+1))$. This shows that ML and LS estimators of $\sigma^2$ differ by a factor of $n/(n-(p+1))$; often a trivial difference unless the sample size is small. The maximized likelihood is

$$\mathcal{L}(\hat{\theta} \mid x_i) = \left[\frac{1}{\sqrt{2\pi}\hat{\sigma}}\right]^n e^{-\frac{1}{2}n},$$

or

$$\log(\mathcal{L}(\hat{\theta})) = -\frac{1}{2}n \log(\hat{\sigma}^2) - \frac{n}{2}\log(2\pi) - \frac{n}{2}.$$

The additive constants can be discarded from the log-likelihood. Thus for all standard linear models, we can take

$$\log(\mathcal{L}(\hat{\theta})) = -\frac{1}{2}n \log(\hat{\sigma}^2).$$

Note that the log-likelihood is, and need only be, defined up to an arbitrary, additive constant. All uses of the log-likelihood are relative to its maximum, or to other likelihoods at their maximum, or to curvature of the log-likelihood function at the maximum.

The number of parameters $K = p + 2$ in these linear models must include the intercept (say, $\beta_0$), the $p$ regression coefficients $(\beta_1, \ldots, \beta_p)$, and the residual variance ($\sigma^2$). Often, one considers only the number of parameters being estimated as the intercept and the slope parameters (ignoring $\sigma^2$); however, in the context of model selection, the number of parameters must include $\sigma^2$ and thus $K = p + 2$. If the method of LS is used to obtain parameter estimators, one must use the regression-based estimate of $\sigma^2$ times $(n-(p+1))/n = (n-K+1)/n$ to obtain the ML estimator of $\sigma^2$. In LS estimation, we minimize $RSS = n\hat{\sigma}^2$, which for all parameters other than $\sigma^2$ itself, is equivalent to maximizing $-\frac{1}{2} \cdot n \log(\hat{\sigma}^2)$. Thus, there is a close relationship between LS and ML methods for linear and nonlinear models where the $\epsilon_i$ are assumed to be normally distributed. Likelihood methods allow easy extensions to the many other classes of models, and with the exploding power of computing equipment, likelihood methods are finding increasing use by both statisticians and researchers in other scientific disciplines (see Garthwaite et al. 1995 for background).

## 1.3.3 Data Dredging

The process of analyzing data with few or no *a priori* questions, by subjectively and iteratively searching the data for patterns and "significance," is often called by the derogatory term of "data dredging." Other terms include "post-hoc data analysis" or "data mining." Often the problem arises when data on many variables have been taken with little or no *a priori* motive or without benefit of supporting science. No specific objectives or alternatives were in place prior to the analysis; thus the data are "submitted for analysis" in the hope that the computer and several hypothesis test results will provide information on "what is significant."

A model is fit, and variables not in that model are added to create a new model, letting the data and intermediate results suggest still further models and variables to be investigated. Patterns seen in the early part of the analysis are "chased" as new variables, cross products, or powers of variables are added to the model and alternative transformations tried. These new models are clearly based on the intermediate results from earlier waves of analyses. The final model is the result of effective dredging, and nearly everything remaining is "significant." Under this view, Hosmer and Lemeshow (1989:169) comment that "Model fitting is an iterative procedure. We rarely obtain the final model on the first pass through the data." However, we believe that such a final model is probably over-fitted and unstable (i.e., likely to vary considerably if other sample data were available on the same process) with actual predictive performance (i.e., on new data) often well below what might be expected from the statistics provided by the terminal analysis (e.g., Chatfield 1996). The inferential properties of *a priori* versus post-hoc data analysis are very different. For example, no valid estimates of precision can be made following data dredging.

If data dredging is done, the resulting model is very much tailored (i.e., over-fitted) to the data in a *post hoc* fashion, and the estimates of precision are likely to be overestimated. Such tailoring over-describes the data and seriously diminishes the validity of inferences made about the information in the data to the population of interest. Many naive applications of classical multivariate analyses are merely "fishing trips" hoping to find "significant" linear relationships among the many variables subjected to analysis (Rexstad et al. 1988, 1990).

Computer routines (e.g., SAS INSIGHT) and associated manuals make data dredging both easy and "effective." Some statistical literature deals with the so-called *iterative process of model building* (e.g., Henderson and Velleman 1981). One looks for patterns in the residuals, employs various tests for selecting variables in their decreasing order of "importance," and tries all possible models. Stepwise regression and discriminant functions, for example, are used to search for "significant" variables; such methods are especially problematic if many variables (Freedman's paradox) are available for analysis (sometimes data are available on over 100 variables, and the sample size may often be less than the number of variables). These problems of over-fitting can escalate when flexible generalized linear or generalized additive models are employed.

Examples of data dredging include the examination of crossplots of all the variables or the examination of a correlation matrix of the variables. These data-dependent activities can suggest apparent linear or nonlinear relationships and interactions *in the sample* and therefore lead the investigator to consider additional models. These activities should be avoided, as they probably lead to over-fitted models with spurious parameter estimates and nonimportant variables as regards the *population*. The sample may be well fit, but the goal is to make a valid inference from the sample to the population. This type of data-dependent, exploratory data analysis might have a place in the earliest stages of investigating a biological relationship and should probably remain unpublished. However, such cases are not the subject of this book, and we can only recommend that the results of such procedures be treated as possible hypotheses. New data should be collected to address these hypotheses effectively and then submitted for a comprehensive and largely *a priori* strategy of analysis such as we advocate here.

Two types of data dredging might be distinguished. The first is that described above; a highly interactive, data dependent, iterative *post hoc* approach. The second is also common and also leads to likely over-fitting. In this type, the investigator also has little *a priori* information; thus "all possible models" are considered as candidates. Note that the "all possible models" approach usually does not include interaction terms (e.g., $x_2 * x_5$) or various transformations such as $(x_1)^2$ or $1/x_3$ or $\log(y)$. In even moderate-sized problems, the number of candidate models in this approach can be very large. At least this second type is not explicitly data dependent, but it is implicitly data dependent and leads to the same "sins." Also, it is usually a one-pass strategy, rather than taking the results of one set of analyses and inputting some of these into the consideration of new models. Still, in some applications, computer software often can systematically search all such models nearly automatically, and thus the strategy of trying all possible models (or at least a very large number of models) continues, unfortunately, to be popular. We wonder whether many situations might be substantially improved if the researcher tried harder to focus on the science of the situation before proceeding with such an unthoughtful approach.

Standard inferential tests and estimates of precision (e.g., ML or LS estimators of the sampling covariance matrix, given a model) are invalid when a final model results from the first type of data dredging. Resulting "$P$-values" are misleading, and there is no valid basis to claim "significance." Even conceptually there is no way to estimate precision because of the subjectivity involved in iterative data dredging and the high probability of over-fitting. In the second type of data dredging one might consider Bonferroni adjustments of the $\alpha$-levels or $P$-values. However, if there were 1,000 models, then the $\alpha$-level would be 0.00005, instead of the usual 0.05! Problems with data dredging are often linked with the problems with hypothesis testing (see Section 1.3.4). This approach is hardly satisfactory; thus analysts have ignored the issue and merely pretended that data dredging is without peril and that the usual inferential methods somehow still apply. Journal editors and referees rarely seem to show concern for the validity of results and conclusions where substantial data dredging has occurred. Thus, the entire methodology based on data dredging has been allowed to be perpetuated in an unthinking manner.

We certainly encourage people to understand their data and attempt to answer the scientific questions of interest. We advocate some examination of the data prior to the formal analysis to detect obvious outliers and outright errors (e.g., determine a preliminary truncation point or the need for grouping in the analysis of distance sampling data). One might examine the residuals from a carefully chosen global model to determine likely error distributions in the candidate models (e.g., normal, lognormal, Poisson). However, if a particular pattern is noticed while examining the residuals and this leads to including another variable, then we might suggest caution concerning data dredging. Often, there can be a fine line between a largely *a priori* approach and some degree of data dredging.

20   1. Introduction

Thus, this book will address primarily cases where there is substantial *a priori* knowledge concerning the issue at hand and where a relatively small set of good candidate models can be specified in advance of actual data analysis. Of course, there is some latitude where some (few) additional models might be investigated as the analysis proceeds; however, results from these explorations should be kept clearly separate from the purely *a priori* science. We believe that objective science is best served using *a priori* considerations with very limited peeking at plots of the data, parameter estimates from particular models, correlation matrices, or test statistics as the analysis proceeds. We do not condone data dredging in confirmatory analyses, but allow substantial latitude in more preliminary explorations. If some limited data dredging is done after a careful analysis based on prior considerations, then we believe that these two types of results should be carefully explained in resulting publications. For this philosophy to succeed, there should be more careful *a priori* consideration of alternative candidate models than has been the case in the past. As we admit in the Preface, we do not expect everyone to share our conservative philosophy regarding the fundamental strategy of data analysis and inference.

## 1.3.4   Some Trends

At the present time, nearly every analysis is done using a computer; thus biologists and people in other disciplines are increasingly using likelihood methods for more generalized analyses. Standard computer software packages allow likelihood methods to be used where LS methods have been used in the past. LS methods will see decreasing use and likelihood methods will see increasing use as we enter the twenty-first century. Likelihood methods allow a much more general framework for addressing statistical issues (e.g., a choice of link functions and error distributions as in log linear and logistic regression models). Another advantage in a likelihood approach is that confidence intervals with good properties can be set using profile likelihood intervals. Edwards (1976), Berger and Wolpert (1984), Azzalini (1996), and Royall (1997) provide additional insights into likelihood methods, while Box (1978) provides the historical setting relating to Fisher's general methods.

During the past twenty years, modern statistical activity has been moving away from traditional formal methodologies based on statistical hypothesis testing (Clayton et al. 1986, Jones and Matloff 1986, Yoccoz 1991, Nester 1996, Bozdogan 1994, Johnson 1995, Stewart-Oaten 1995). The historic emphasis on hypothesis testing will continue to diminish in the years ahead (e.g., see Quinn and Dunham 1983, Bozdogan 1994), with increasing emphasis on estimation of effects or effect sizes and associated confidence intervals (Graybill and Iyer 1994:35). It seems curious that the field of molecular phylogenetics has only recently "found" hypothesis testing and seems oblivious to its many shortcomings (Huelsenbeck and Rannala 1997).

In particular, hypothesis testing for model selection is often poor (Akaike 1981a) and will surely diminish in the years ahead. There is no statistical theory that supports the notion that hypothesis testing with a fixed $\alpha$ level is a basis for model selection; there are not even general, formal rules (or even guidelines) t... ously define how the various P-values might be used to arrive at a final m... How does one interpret dozens of P-values, from tests with differing power, to arrive at a good model? Only ad hoc rules exist in this case and generally fail to result in a final parsimonious model with good inferential properties. The multiple testing issue is problematic as is the fact that likelihood ratio tests exist only for nested models. Tests of hypotheses within a data set are not independent, making inferences difficult. The order of testing is arbitrary, and differing test order will often lead to different final models. Model selection is dependent on the arbitrary choice of $\alpha$, but $\alpha$ should depend on both $n$ and $K$ to be useful in model selection; however, theory for this is lacking. Testing theory is problematic when nuisance parameters occur in the models being considered. Finally, there is the fact that the so-called null is probably false on simple *a priori* grounds (e.g., $H_0$: the treatment had *no* effect, so the parameter $\theta$ is constant across years, $\theta_1 = \theta_2 = \cdots = \theta_k$). Rejection of such null hypotheses does not mean that the effect or parameter should be included in the (approximating) model! All of these problems have been well known in the literature for many years; they have merely been ignored in the practical analysis of empirical data. Nester (1996) provides an interesting summary of quotations regarding hypothesis testing.

Computational restrictions prevented biologists from evaluating alternative models until the past two decades or so. Thus, people tended to use an available model, often without careful consideration of alternatives. Present computer hardware and software make it possible to consider a number of alternative models as an integral component of data analysis. Computing power has permitted more computer-intensive methods such as the various cross validation and bootstrapping approaches and other resampling schemes (Mooney and Duval 1993, Efron and Tibshirani 1993), and such techniques will see ever increasing use in the future.

The size or dimension ($K$) of some biological models can be quite high, and this has tended to increase over the past two decades. Open capture–recapture and band recovery models commonly have 20–40 estimable parameters for a single data set and might have well over 200 parameters for the joint analysis of several data sets (see Burnham et al. 1987, Preface, for a striking example of these trends). Analysis methods for structural equations commonly involve 10–30 parameters (Bollen and Long 1993). These are applications where objective model specification and selection is essential to answer the question, "*What inferences do the data support about the population?*"

## 1.4 Inference and the Principle of Parsimony

### 1.4.1 Over-fitting to Achieve a Good Model Fit

Consider two analysts studying a small set of biological data using a multiple linear regression approach. The first exclaims that a particular model provides an excellent fit to the data. The second notices that 22 parameters were used

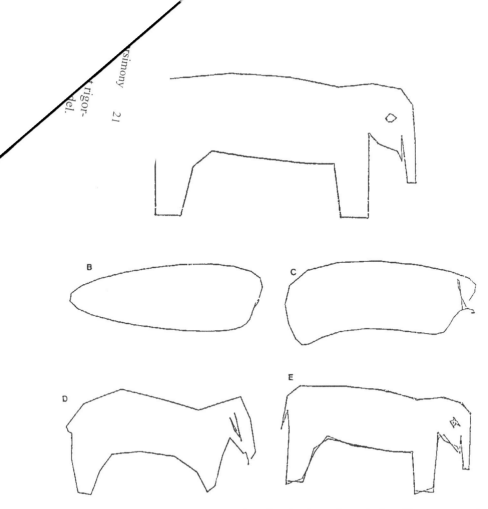

FIGURE 1.1. "How many parameters does does it take to fit an elephant?" was answered by Wel (1975). He started with an idealized drawing (A) defined by 36 points and used least squares Fourier sine series fits of the form $x(t) = \alpha_0 + \sum \alpha_i \sin(it\pi/36)$ and $y(t) = \beta_0 + \sum \beta_i \sin(it\pi/36)$ for $i = 1, \ldots, N$. He examined fits for $K = 5, 10, 20$, and 30 (shown in B–E) and stopped with the fit of a 30 term model. He concluded that the 30-term model "may not satisfy the third-grade art teacher, but would carry most chemical engineers into preliminary design."

in the regression and states, "Yes, but you have used enough parameters to fit an elephant!" This seeming conflict between increasing model fit and increasing numbers of parameters to be estimated from the data led Wel (1975) to answer the question, "How many parameters *does* it take to fit an elephant?" Wel finds that about 30 parameters would do reasonably well (Fig. 1.1); of course, had he fit 36 parameters to his data, he could have achieved a perfect fit.

Wel's finding is both insightful and humorous but deserves further interpretation for our purposes here. His "standard" is itself only a crude drawing—it even lacks ears, a prominent elephantine feature; hardly truth. A better target would have been a large, digitized, high-resolution photograph; however, this, too, would have been only a model (and not truth). Perhaps a real elephant should have been used as "truth," but this begs the question, "Which elephant should we use?" This simple example will encourage thinking about full reality, "true models," and approximating models and motivate the *principle of parsimony* in the following section. William of Occam suggested in the fourteenth century that one "shave away all that is unnecessary"—often referred to as *Occam's razor*. Occam's razor has had a long history in both science and technology and it is embodied in the principle of parsimony. Albert Einstein is supposed to have said, "Everything should be made as simple as possible, but no simpler."

Success in the analysis of real data and the resulting inference often depends importantly on the choice of a "best approximating model." Data analysis in the biological sciences should be based on a parsimonious model that provides an accurate approximation to the structural information in the data at hand; this should not be viewed as searching for the "true model." Model selection is essentially concerned with the "art of approximation" (Akaike 1974).

## *1.4.2 The Principle of Parsimony*

If the fit is improved by a model with more parameters, then where should one stop? Box and Jenkins (1970:17) suggested that the *principle of parsimony* should lead to a model with "... the smallest possible number of parameters for adequate representation of the data." Statisticians view the principle of parsimony as a bias versus variance tradeoff. In general, bias decreases and variance increases as the dimension of the model ($K$) increases (Fig. 1.2). Often, we may use the number of parameters in a model as a measure of the degree of structure inferred from the data. The fit of any model can be improved by increasing the number of parameters (e.g., the elephant-fitting problem); however, a tradeoff with the increasing variance must be considered in selecting a model for inference. Parsimonious models achieve a proper tradeoff between bias and variance. All model selection methods are based to some extent on the principle of parsimony (Breiman 1992, Zhang 1994).

In understanding the utility of an approximate model for a given data set, it is convenient to consider two undesirable possibilities: under-fitted and over-fitted models. Here, we must avoid judging a selected model in terms of some supposed "true model," as occurs when data are simulated from a known, often very simple, model using Monte Carlo methods. In this case, if the generating model had 10 parameters, it is often said that an approximating model with only 7 parameters is under-fitted (compared with the generating model with 10 parameters). This interpretation is often of little value, as it largely ignores the principle of parsimony and its implications and hinges on the misconception that such a simple generating model exists in biological problems. If we believe that truth is essentially infinite-dimensional, then over-fitting is not even defined in terms of the number of parameters in the fitted model. We will avoid this use of the terms under- and over-fitted because of the supposed existence of a low-dimensional "true model" as a "standard."

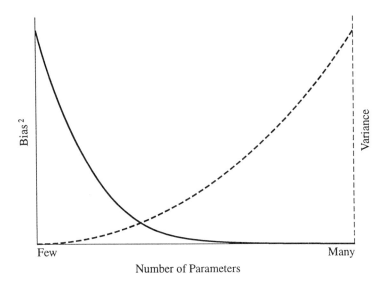

FIGURE 1.2. The *principle of parsimony*: the conceptual tradeoff between squared bias (solid line) and variance vs. the number of estimable parameters in the model ($K$). All model selection methods implicitly employ some notion of this tradeoff. The best approximating model need not occur exactly where the two curves intersect. Full truth or reality is not attainable with finite samples and usually lies well to the right of the region in which the best approximating model lies (the tradeoff region). Bias decreases and variance (uncertainty) increases as the number of parameters in a model increases.

Instead, we reserve the terms under- and over-fitted for use in relation to a "best approximating model" (Section 1.2.5). Here, an under-fitted model would ignore some important replicable (i.e., conceptually replicable in most other samples) structure in the data and thus fail to identify effects that were actually supported by the data. In this case, bias in the parameter estimators is often substantial, and the sampling variance is underestimated, both factors resulting in poor confidence interval coverage. Under-fitted models tend to miss important treatment effects in experimental settings. Over-fitted models, as judged against a best approximating model, are often free of bias in the parameter estimators, but have estimated sampling variances that are needlessly large (the precision of the estimators is poor, relative to what could have been accomplished with a more parsimonious model). Spurious treatment effects tend to be identified, and spurious variables are included with over-fitted models. Shibata (1989) argues convincingly that under-fitted models are a more serious issue in data analysis and inference than over-fitted models.

The concept of parsimony and a bias versus variance tradeoff is very important. Thus we will provide some additional insights. The goal of data collection and analysis is to make inferences from the sample that properly apply to the population. The inferences relate to the *information* about structure of the system under study as inferred from the models considered and the estimated parameters in each model. A paramount consideration is the repeatability, with good precision, of any inference reached. When we imagine many replicate samples, there will be some recognizable features common to almost all of the samples. Such features are the sort of inference about which we seek to make strong inferences (from our single sample). Other features might appear in, say, 60% of the samples yet still reflect something real about the population or process under study, and we would hope to make weaker inferences concerning these. Yet additional features appear in only a few samples, and these might be best included in the error term ($\sigma^2$) when modeling. If one were to make an inference about these features quite unique to just the single data set at hand, as if they applied to all (or most all) samples (hence to the population), then we say that the sample is over-fitted by the model (we have over-fitted the *data*). Conversely, failure to identify the features present that are strongly replicable over samples is under-fitting. The data are not being approximated; rather we approximate the structural information in the data that is replicable over such samples (see Chatfield 1996, Collopy et al. 1994). Quantifying that structure with a model form and parameter estimates is subject to some "sampling variation" that must also be estimated (inferred) from the data.

True replication is very advantageous, but this tends to be possible only in the case of strict experiments where replication and randomization are a foundation. Such experimental replication allows a valid estimate of residual variation ($\sigma^2$). An understanding of these issues makes one realize what is lost when observational studies seem possible and practical, and strict experiments seem infeasible.

A best approximating model is achieved by properly balancing the errors of under-fitting and over-fitting. Stone and Brooks (1990) comment on the "... straddling pitfalls of under-fitting and over-fitting." The proper balance is achieved when bias and variance are controlled to achieve confidence interval coverage at approximately the nominal level and where interval width is at a minimum. Proper model selection rejects a model that is far from reality and attempts to identify a model in which the error of approximation and the error due to random fluctuations are well balanced (Shibata 1983, 1989). Some model selection methods are "parsimonious" (e.g., BIC, Schwarz 1978) but tend to select models that are too simple (i.e., under-fitted); thus, bias is large, precision is overestimated, and achieved confidence interval coverage is well below the nominal level. Such instances are not satisfactory for inference—one has only a highly precise, quite biased result. We focus on confidence interval coverage of parameter estimators in the selected model, including a variance component for model selection uncertainty.

Sakamoto et al. (1986) simulated data to illustrate the concept of parsimony and the errors of under- and over-fitting models (Fig. 1.3). Ten data sets (each with $n = 21$) were generated from the simple model

$$y = e^{(x-0.3)^2} - 1 + \epsilon;$$

where $x$ varied from 0 to 1 in equally spaced steps of 0.05, and $E(\epsilon^2) = 0.01$, i.e., normal$(0, \sigma^2)$. Thus, in this case, they considered the generating model to have $K = 3$ parameters: $0.3, -1$, and $0.01$. They considered the set of candidate models (i.e., the approximating models) to be simple polynomials of degree 0 to 5,

| Order | Approximating Model |
|---|---|
| 0 | $E(y) = \beta_0$ |
| 1 | $E(y) = \beta_0 + \beta_1(x)$ |
| 2 | $E(y) = \beta_0 + \beta_1(x) + \beta_2(x^2)$ |
| 3 | $E(y) = \beta_0 + \beta_1(x) + \beta_2(x^2) + \beta_3(x^3)$ |
| 4 | $E(y) = \beta_0 + \beta_1(x) + \beta_2(x^2) + \beta_3(x^3) + \beta_4(x^4)$ |
| 5 | $E(y) = \beta_0 + \beta_1(x) + \beta_2(x^2) + \beta_3(x^3) + \beta_4(x^4) + \beta_5(x^5)$ |

Strong model bias occurs when an under-fitting (e.g., the mean model with $K = 2$ or the linear, 1st order, $K = 3$) model is employed (Fig. 1.3A). Here bias is obvious, the nonlinear structure of the generating model is poorly approximated, and confidence interval coverage and predictions from the model will be quite poor. Of course, there is *some* model bias for each of the 5 models because they are only simple polynomial approximations. Over-fitting is illustrated in Fig. 1.3B, where a 5th-degree polynomial ($K = 7$) is used as an approximating model. Here, there is little evidence of bias (an average quantity), precision is obviously poor, and it is difficult to identify the simple structure of the model. Prediction will be quite imprecise from this model, and it has features that do not occur in the generating model, particularly when extrapolating beyond the range of the data (always a risky practice). Both under- and over-fitting are undesirable in judging approximating models for data analysis.

If a second-degree polynomial ($K = 4$) is used as the approximating model, the fits seem quite reasonable (Fig. 1.3C), and one might expect valid inference from this model. Finally, *if* it were known *a priori* from the science of the situation that the function was nonnegative and had a minimum of zero at $x = 0.3$, then an improved quadratic approximating model could use this information very effectively (Fig. 1.3D). The form of this model is

$$E(y) = \beta_0(x + \beta_1)^2$$

with $K = 3$ (i.e., $\beta_0$, $\beta_1$, and $\sigma^2$), whereas the second-degree polynomial has 4 parameters. This example illustrates that valid statistical inference is only partially dependent on the analysis process; the science of the situation must play an important role through modeling. This particular example provides a visual image of under- and over-fitting in a simple case where the generating model and various approximating models can be easily graphed in 2 dimensions. Parsimony issues with real data in the biological sciences nearly always defy such a simple graphical approach because truth is not known; one rarely has 10 independent data sets on exactly the same process, and plots in high dimensions are problematic to produce and interpret. Therefore, objective and effective methods are needed that do not rely on simple graphics and can cope with the real-world complexities and high dimensionality.

### 1.4.3 Model Selection Methods

Model selection has most often been viewed, and hence taught, in a context of hypothesis testing. Sequential testing has most often been employed, either stepup (forward) or stepdown (backward) methods. Stepwise procedures allow for variables to be added or deleted at each step, and these remain popular in spite of their poor operating characteristics. Testing schemes are based on subjective $\alpha$ levels (commonly 0.05 or 0.01); however, Rawlings (1988) recommends 0.15 in the context of stepwise regression. The multiple testing problem is serious if many tests are to be made (see Westfall and Young 1993) and the tests are not independent. Tests between models that are not nested are problematic. A model is nested if it is a special case of another model; e.g., a 3rd-degree polynomial is nested within a 4th-degree polynomial. Generally, hypothesis testing is a poor basis for model selection (Akaike 1974 and Sclove 1994b).

Cross validation has been suggested and well studied as a basis for model selection (Mosteller and Tukey 1968, Stone 1974, 1977; Geisser 1975). Here, the data are divided into two partitions. The first partition is used for model fitting and the second is used for model validation (sometimes the second partition has only one observation). Then a new partition is selected, and this whole process is repeated hundreds or thousands of times. Also, then some criterion is chosen, such as minimum squared prediction error, as a basis for model selection. There are several variations on this theme, and it is a useful methodology (Craven and Wahba 1979, Burman 1989, Shao 1993, Zhang 1993a, and Hjorth 1994). These methods are quite computer intensive and tend to be impractical if more than about 15–20 models must be evaluated or if sample size is large. Still, cross validation offers an interesting alternative for model selection.

Some analysts would favor using a very general model in all cases (e.g., an over-fitted model). We believe that this is generally poor practice (Fig. 1.3B). Others have a "favorite" model that they believe is good, and they use it in nearly all situations. For example, some researchers always use the hazard rate model (Buckland et al. 1993) with 2 parameters ($K = 2$) as an approximating model to the detection function in line and point transect sampling. This might be somewhat reasonable for situations where a simple model suffices (e.g., $K = 2$–3), but will be poor practice in more challenging modeling contexts where $15 \leq K \leq 20$ or more is required. These ad hoc rules ignore the principle of parsimony and data-based model selection, in which the data help "select" the model to be used for inference.

If goodness-of-fit tests can be computed for all alternative models even if some are not nested within others, then one could use the model with the fewest parameters that "fits" (i.e., $P > 0.05$ or 0.10). However, increasingly better fits can often be achieved by using models with more and more parameters (e.g., the elephant-fitting problem), and this can make the arbitrary choice of $\alpha$ very critical. A large $\alpha$-level leads to over-fitted models and their resulting problems. In addition, other problems may be encountered such as over- or under-dispersion and low power if

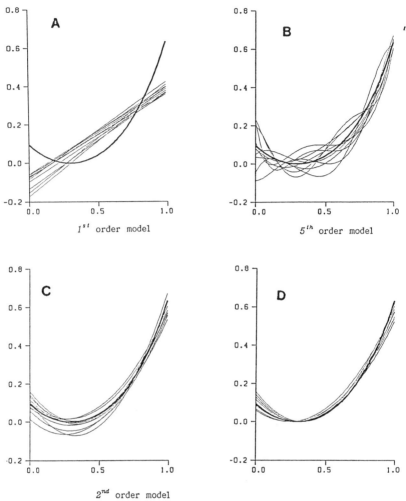

FIGURE 1.3. Ten Monte Carlo repetitions of data sets ($n = 21$) generated from the model $y = e^{(x-0.3)^2} - 1 + \epsilon$; $0 \leq x \leq 1$, $E(\epsilon^2) = 0.01$ (from Sakamoto et al. 1986:164–179). A 1st-degree polynomial (A) is clearly biased, misidentifies the basic nonlinear structure, and is under-fitted and unsatisfactory. A 5th-degree polynomial (B) has too many parameters, an unnecessarily large variance, and will have poor predictive qualities because it is unstable (over-fitted). Neither A nor B is properly parsimonious, nor do they represent a best approximating model. A 2nd-degree polynomial seems quite good as an approximating model (C). If it is known that the function is nonnegative and has its minimum at $x = 0.3$, then the approximating model that enforces these conditions is improved further (D). Average AIC values for the 10 reps are $-20.35$, $-27.86$, $-32.13$, and $-33.92$ for A–D, respectively. AIC is minimized for model D, which is obvious from "seeing" the figure. In more realistic situations, one lacks the benefit of simple plots and 10 independent data sets, such as those shown in A–D.

one must pool small expectations to assure that the test statistic is chi-square distributed. Perhaps, most importantly, there is no theory to suggest that this approach will lead to selected models with good inferential properties (i.e., an adequate bias vs. variance tradeoff or good achieved confidence interval coverage and width).

The adjusted coefficient of multiple determination has been used in model selection in an LS setting (the adjusted coefficient $= 1 - (1 - R^2)\left(\frac{n-1}{n-p}\right)$, where $R^2$ is the usual coefficient of multiple determination; Draper and Smith 1981:91–92). Under this method, one selects the model in which this adjusted statistic is largest. Mallows's $C_p$ statistic (Mallows 1973, 1995) is also used in LS regression with normal residuals and a constant variance and in this special case provides a ranking of the candidate models that is the same as the rankings under AIC (the numerical values, $C_p$ vs. AIC, will differ, see Atilgan 1996). The selection of models using the adjusted $R^2$ statistic and Mallows's $C_p$ are related for simple LS problems (see Seber 1977:362–369). Hurvich and Tsai (1989) provide some comparisons of $AIC_c$ vs. several competitors for simple regression problems.

Bayesian researchers have taken somewhat different approaches and assumptions, and have proposed several alternative methods for model selection. Methods such as CAIC, BIC, SIC, WIC, and HQ are mentioned in Section 2.8. Other Bayesian approaches to model selection and inference are at the current state of the art in statistics but seem to be very difficult to understand and implement and very computer intensive (e.g., Laud and Ibrahim 1995 and Carlin and Chib 1995). Draper (1995) provides a recent review of these advanced methods (also see Potscher 1991).

The general approach that we advocate here is one derived by Akaike (1973, 1974, 1977, 1978a and b, and 1981a and b), based on information theory and is discussed at length in Chapters 2 and 3. Akaike's information-theoretic approach has led to a number of alternative methods having desirable properties for the selection of best approximating models in practice (e.g., AIC, $AIC_c$, $QAIC_c$, and TIC—see Section 2.8 and Chapter 6). Our general advocacy concerning AIC and the associated criteria is somewhat stronger than that of Linhart and Zucchini (1986) but similar in that they also recommend objective procedures based on some well-defined criterion with a strong, fundamental basis.

## 1.5 Model Selection Uncertainty

The literature on model selection methods has increased substantially in the past 15–20 years; much of this has been the result of Akaike's influential papers in the mid-1970s. However, relatively little appears in the literature concerning the properties of the parameter estimators, *given* that a data-dependent model selection procedure has been used (see Rencher and Pun 1980, Hurvich and Tsai 1990, Goutis and Casella 1995). Here, data have been used to both select a proper model and estimate the model parameters and their precision (i.e., the conditional sampling covariance matrix, given the selected model).

If a best approximating model has been selected from a reasonable set of candidate models, bias in the model parameter estimators might be small. However, there is uncertainty in the best model to use (e.g., should $\beta_3$ even be in the model?), and this model uncertainty is a component of variance in the estimators. For example, we might denote the sampling variance of an estimator $\hat{\theta}$, *given a model*, by var($\hat{\theta}$ | model). However, generally, the sampling variance of $\hat{\theta}$ should have two components: (1) var($\hat{\theta}$ | model) and (2) a variance component due to not knowing the best approximating model to use (and, therefore, having to estimate this). Thus, if one uses a method such as AIC to select a parsimonious model, given the data, and estimates a conditional sampling variance, given the selected model, the estimated variance will likely be too small because one component of this variance is missing. Model selection uncertainty is the component of variance that reflects that model selection merely *estimates* which approximating model is best, based on the single data set; a different model (in the fixed set of models considered) may be selected as best for a different data set.

The problem with model selection uncertainty has a certain "parallel" with using the multiple coefficient of determination $R^2$ in multiple linear regression. If one has a response variable $Y$ and $p$ independent variables $X_1, X_2, \ldots, X_p$, then $R^2$ can be estimated using standard LS regression theory (or equivalently, ML theory), based on a sample of $n$ independent observations. However, if the estimated regression coefficients from this analysis were used on another independent sample of size $n$, the resulting $R^2$ value would almost surely be lower. In other words, the estimated value of $R^2$ for the first data set was optimistic because the regression coefficients tend to be tailored to the particular data set. Similarly, MLEs are best for the particular sample analyzed, and rarely perform as well with a new sample (the so-called shrinkage problem; Copas 1983, Lehmann 1983). Leamer (1978), Gilchrist (1984), Breiman (1992), Zhang (1992), and Chatfield (1995b, 1996) give insights into problems when the same data are used to both select the model and make inferences from that model.

Failure to allow for model selection uncertainty often results in estimated sampling variances and covariances being too low, and thus the achieved confidence interval coverage will be below the nominal value. Optimal methods for coping with model selection uncertainty are at the forefront of statistical research; better methods might be expected in the coming years, especially with the continued increases in computing power. Model selection uncertainty is problematic in making statistical inferences; if the goal is only data description, then perhaps selection uncertainty is a minor issue.

## 1.6 Summary

Truth in the biological sciences and medicine is extremely complicated, and we cannot hope to find the "true model" from the analysis of a finite amount of data. Thus, inference about truth must often be based on a good "approximating model."

Likelihood and least squares methods provide a rigorous inference theory if the model structure is "given." However, in practical science problems, the model is *not* "given." Thus, the critical issues is "what model to use." This is the model selection problem.

The emphasis then shifts to the careful, *a priori* definition of a set of candidate models. This is where the science of the problem enters the analysis. Critical thinking about the science question and modeling alternatives, prior to looking at the data, have been underemphasized in many statistics classes in the past. These are important issues, and one must be careful not to engage in data dredging, as this weakens inferences that might be made. Information-theoretic methods provide a simple way to select a "best" approximating model from the candidate set of models.

*Data analysis* is taken to mean the entire, integrated process of *a priori* model specification, model selection, and estimation of parameters and their precision. Scientific inference is based on this process. Information-theoretic methods free the analyst from the limiting concept that the proper approximating model is somehow "given."

Perhaps we cannot totally overcome problems in estimating precision, following a data-dependent selection method such as AIC (e.g., see Dijkstra 1988). This limitation certainly warrants exploration, as model selection uncertainty is a quite difficult area of inference. However, we must also consider the "cost" of *not* selecting a good parsimonious model for the analysis of a particular data set. That is, a model is just somehow "picked" independent of the data and used to approximate the data as a basis for inference. Then, one does not know (or care?) whether it fits, over-fits, under-fits, etc. This naive strategy certainly will incur substantial costs in terms of reliable inferences. Alternatively, one might be tempted into an iterative, highly interactive strategy of data analysis (unadulterated data dredging)—again there are substantial costs in terms of reliable inference using this approach. In particular, it seems impossible to objectively estimate the precision of the estimators following data dredging. So, which strategy incurs the lowest cost? The cost of not selecting a parsimonious model for data analysis is rarely considered.

Zhang (1994) notes that for the analyst who is less concerned with theoretical optimality, it is more important to have available methods that are simple, but flexible enough to be used in a variety of practical situations. The information-theoretic methods fall in this broad class and promote reliable inference.

# 2
# Information Theory and Log-Likelihood Models: A Basis for Model Selection and Inference

Full reality cannot be included in a model; thus we seek a good model to approximate the effects or factors supported by the empirical data. The selection of an appropriate approximating model is critical to statistical inference from many types of empirical data. This chapter introduces concepts from information theory (see Guiasu 1977), which has been a discipline only since the mid-1940s and covers a variety of theories and methods that are fundamental to many of the sciences (see Cover and Thomas 1991 for an exciting overview; Fig. 2.1 is produced from their book and shows their view of the relationship of information theory to several other fields). In particular, the Kullback–Liebler "distance," or "information," between two models (Kullback and Leibler 1951) is introduced, discussed, and linked to Boltzmann's entropy in this chapter. Akaike (1973) found a simple relationship between the Kullback–Liebler distance and Fisher's maximized log-likelihood function (see deLeeuw 1992 for a brief review). This relationship leads to a simple, effective, and very general methodology for selecting a parsimonious model for the analysis of empirical data.

Akaike introduced his "*entropy maximization principle*" in a series of papers in the mid-1970s (Akaike 1973, 1974, 1977) as a theoretical basis for model selection. He followed this pivotal discovery with several related contributions beginning in the early 1980s (Akaike 1981a and b, 1985, 1992, and 1994). This chapter provides several insights into a derivation of AIC (several rigorous derivations have appeared in the literature, and particularly Chapter 6); it intentionally lacks mathematical rigor and is written for biologists wanting a heuristic glimpse into the theory underlying AIC. We urge readers to understand the logic of this heuristic derivation, as without it, the simple and compelling idea underlying Kullback–Leibler information and AIC cannot be appreciated.

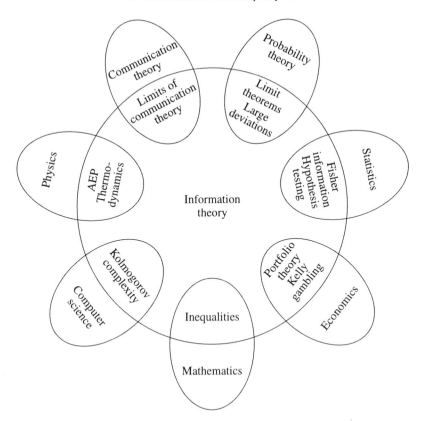

FIGURE 2.1. Information theory and its relationships to other disciplines (from Cover and Thomas 1989). Information theory began in the mid-1940s, at the close of WWII. In the context of this book, the most relevant components of information theory include Fisher information, entropy (from thermodynamics and communication theory), and Kullback–Liebler information.

Before going further we need to review some notation from likelihood theory. We will use $\mathcal{L}(\theta \mid x)$ as the likelihood function of the parameter vector $\theta$, given the data $x$, and will denote the natural logarithm of this expression as the log-likelihood function, $\log(\mathcal{L}(\theta \mid x))$. In the following material, we will make use of the maximized log-likelihood, i.e., the numerical value of $\log(\mathcal{L}(\theta \mid x))$ evaluated at the maximum likelihood estimates (MLEs). Of course, the MLEs are defined as those values of $\hat{\theta}$ that maximize the log-likelihood function, given the data $x$.

## 2.1 The Distance or Discrepancy Between Two Models

Well over a century ago measures were derived for assessing the "distance" between two models or probability distributions. Most relevant here is Boltzmann's (1877) concept of generalized entropy (see Section 2.6) in physics and thermody-

Ludwig Eduard Boltzmann, 1844–1906, one of the most famous scientists of his time, made incredible contributions in theoretical physics. He received his doctorate in 1866; most of his work was done in Austria, but he spent some years in Germany. He became full professor of mathematical physics at the University of Graz, Austria, at the age of 25. His mathematical expression for entropy was of fundamental importance throughout many areas of science. The negative of Boltzmann's entropy is a measure of "information" derived over half a century later by Kullback and Leibler. J. Bronowski wrote that Boltzmann was "an irascible, extraordinary man, an early follower of Darwin, quarrelsome and delightful, and everything that a human should be." Several books chronicle the life of this great figure of science, including Cohen and Thirring (1973) and Broda (1983), and his collected technical papers appear in Hasenöhrl (1909).

namics (see Akaike 1985 for a brief review). Shannon (1948) employed entropy in his famous treatise on communication theory. Kullback and Leibler (1951) derived an information measure that happened to be the negative of Boltzmann's entropy, now referred to as the Kullback–Leibler (K-L) distance. The motivation for Kullback and Leibler's work was to provide a rigorous definition of "information" in relation to Fisher's "sufficient statistics." The K-L distance has also been called the K-L discrepancy, divergence, information, and number—we will treat these terms as synonyms, but tend to use *distance* or *information* in the material to follow.

The Kullback–Liebler distance can be conceptualized as a directed "distance" between two models, say $f$ and $g$ (Kullback 1959). Strictly speaking, this is a measure of "discrepancy"; it is not a simple distance because the measure from $f$ to $g$ is not the same as the measure from $g$ to $f$—it is a directed, or oriented, distance (Fig. 2.2). The K-L distance is perhaps the most fundamental of all in-

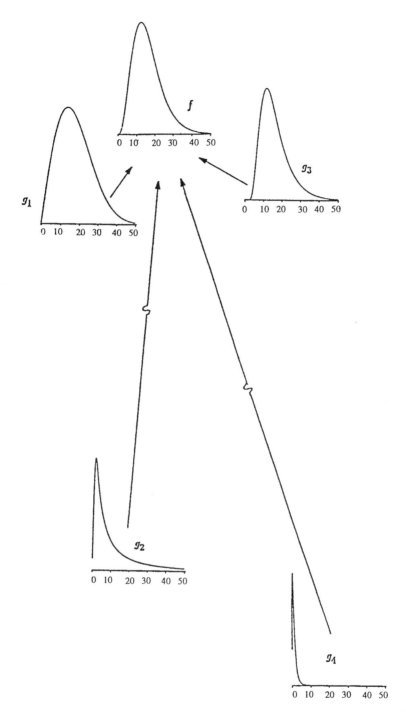

FIGURE 2.2. The Kullback–Liebler discrepancy $I(f, g_i)$ is a directed distance from the various candidate models $g_i$ to $f$. Knowing the K-L distance would allow one to find the approximating model that is *closest* to the model $f$, the "best approximating model" ("best" meaning the best of the 4 models considered). Here, $f$ is gamma (4, 4) and the 4 approximating models are $g_1$ = Weibull (2, 20), $g_2$ = lognormal (2, 2), $g_3$ = inverse Gaussian (16, 64), and $g_4$ = F distribution (4, 10).

formation measures in the sense of being derived from minimal assumptions and its additivity property. The K-L distance is an extension of Shannon's concept of information (Hobson and Cheng 1973, Soofi 1994) and is sometimes called a "relative entropy." The K-L distance between models is a *fundamental quantity* in science and information theory (see Akaike 1983) and is the logical basis for model selection as defined by Akaike. In the heuristics given here, we will assume that the models in question are continuous probability distributions denoted by $f$ and $g$. Biologists are familiar with the normal, log-normal, gamma, and other continuous distributions. We will, of course, not limit ourselves to these common, simple types.

## 2.1.1  $f$, Truth, Full Reality, and "True Models"

In the initial discussion, we let $f$ represent merely the model to be measured against $g$—and we want simply the distance between two models, $f$ and $g$. Further, to illustrate the concept of the K-L distance we will let $f$ be of a simple form and have only 2 parameters. This simplicity will motivate the concept of K-L distances and temporarily avoid, for tutorial reasons, the real-world intricacies implicit in $f$ when used in a model selection context.

In the remainder of the book we will want a more general, conceptual view of $f$, and we will use it to reflect truth or full reality. Here, reality is rarely (if ever) a model; rather it reflects the complex biological (and measuring or sampling) process that generated the observed data $x$. For this reason, we will not usually explicitly parametrize the complex function $f$ because, as it represents full reality (truth), it might not even have parameters in a sense that would be analogous to $\theta$ in a modeling framework. In fact, thinking that truth is parametrized is itself a type of (artificial) model-based conceptualization. Sometimes it is useful to think of $f$ as full reality and let it have (conceptually) an infinite number of parameters (see Section 1.2.4). This "crutch" of infinite-dimensionality at least keeps the concept of reality even though it is in some unattainable perspective. Thus, $f(x)$ represents full truth, and might be conceptually based on a very large number of parameters (of a type we may have not even properly conceived) that give rise to a set of data $x$. Finally, we will see how this conceptualization of reality ($f$) collapses into a constant and that AIC does not rest on the specific existence of a "true" model.

When Monte Carlo methods are used to generate data, then in a sense there is a "true" model, and this is sometimes useful in examining the operating characteristics of various estimation and model selection methods. In this unique (artificial) case, there is a model ($f$) with known functional form and known parameters; we will use the term *generating model* for such Monte Carlo studies.

## 2.1.2  $g$, Approximating Models

We will let $g$ be the approximating model being compared to (measured against) $f$. We use $x$ to denote the data being modeled and $\theta$ to denote the parameters in the approximating model $g$. In later sections we will use $g(x)$ as an approximating model, whose parameters must be estimated from these data (in fact, we will make this explicit using the notation $g(x \mid \theta)$, read as "the approximating model $g$ for data $x$ given the parameters $\theta$). If the parameters of the model $g$ have been estimated, using ML or LS methods, we will denote this by $g(x \mid \hat{\theta})$. Generally, in any real-world problem, the model $g(x \mid \theta)$ is a function of sample data (often multivariate), and the number of parameters ($\theta$) in $g$ might often be of high dimensionality. Finally, we will want to consider a set of approximating models as candidates for the representation of the data; this set of models is denoted by $\{g_i(x \mid \theta), i = 1, \ldots, R\}$. Often, approximating models might be some of those reviewed in Section 1.3.1.

We make frequent use of the same model structure in terms of the likelihood function, $\mathcal{L}(\theta \mid x)$. Here the conditioning has been reversed, and this function is read as "the likelihood of the parameters $\theta$, given the data $x$." This likelihood function and the log-likelihood function are useful in statistical inference, while the model $g(x \mid \theta)$ is useful in computing probabilities of events, given knowledge of the model parameters.

## 2.1.3 The Kullback–Liebler Distance (or Information)

We begin without any issues of estimation of model parameters and deal with very simple expressions for the models $f$ and $g$, assuming that their parameters are known. The K-L distance between the models $f$ and $g$ is defined for continuous functions as the (usually multidimensional) integral

$$I(f, g) = \int f(x) \log \left( \frac{f(x)}{g(x \mid \theta)} \right) dx,$$

where log denotes the natural logarithm. Kullback and Leibler (1951) developed this quantity from "information theory." Thus they used the notation **$I(f, g)$ as it relates to the "information" lost when $g$ is used to approximate $f$.** Of course, we seek an approximating model that loses as little information as possible; this is equivalent to minimizing $I(f, g)$ over $g$. The function $f$ is considered to be given (fixed), and only $g$ varies over a space of models indexed by $\theta$. An equivalent interpretation of minimizing $I(f, g)$ is finding an approximating model that is the "shortest distance" away from truth. We will use both interpretations throughout this book, as both seem useful. Similarly, Cover and Thomas (1991) note that the K-L distance is ". . . a measure of the inefficiency of assuming that the distribution is $g$ when the true distribution is $f$."

The expression for the K-L distance in the case of discrete distributions such as the Poisson, binomial, or multinomial is

$$I(f, g) = \sum_{i=1}^{k} p_i \log \left( \frac{p_i}{\pi_i} \right).$$

Here, there are $k$ possible outcomes of the underlying random variable; the true probability of the $i$th outcome is given by $p_i$, while $\pi_1, \ldots, \pi_k$ constitute the approximating probability distribution (i.e., the approximating model). In the discrete case, we have $0 < p_i < 1$, $0 < \pi_i < 1$, and $\sum p_i = \sum \pi_i = 1$. Hence, here

$f$ and $g$ correspond to $p$ and $\pi$, respectively. In the following material, we will generally think of K-L information in the continuous case and use the notation $f$ and $g$ for simplicity.

At a heuristic level, "information" is defined as $-\log_e(f(x))$ for some continuous probability density function or $-\log_e(p_i)$ for the discrete case. Kullback–Liebler information is a type of "cross entropy," a further generalization. In either the continuous or discrete representation, the right-hand side is an expected value (i.e., $\int f(x)(\cdot)dx$ for the continuous case or $\sum_{i=1}^{k} p_i(\cdot)$ for the discrete case) of the logarithm of the ratio of the two distributions ($f$ and $g$) or two discrete probabilities ($p_i$ and $\pi_i$). In the continuous case one can think of this as an average (with respect to $f$) of $\log_e(f/g)$, and in the discrete case it is an average (with respect to the $p_i$) of the logarithm of the ratio ($p_i/\pi_i$). The foundations of these expressions are both deep and fundamental (see Boltzmann 1877, Kullback and Leibler 1951, or contemporary books on information theory).

The K-L distance ($I(f, g)$) is always positive except when the two distributions $f$ and $g$ are identical (i.e., $I(f, g) = 0$ if and only if $f(x) = g(x)$ everywhere). More detail and extended notation will be introduced in Chapter 6; here we will employ a simple notation and use it to imply considerable generality in the sample data ($x$) and the multivariate functions $f$ and $g$.

An example will illustrate the K-L distances ($I(f, g_i)$). Let $f$ be a gamma distribution with 2 parameters ($\alpha = 4, \beta = 4$). Then consider 4 approximating models $g_i$, each with 2 parameters (see below): Weibull, lognormal, inverse Gaussian, and the F distribution. Details on these simple probability models can be found in Johnson and Kotz (1970). The particular parameter values used for the four $g_i$ are not material here, except to stress that they are assumed known, not estimated. "Which of these parametrized distributions is the *closest* to $f$?" is answered by computing the K-L distance between each $g_i$ and $f$ (Fig. 2.2). These are as follows:

|  | Approximating model | $I(f, g_i)$ | Rank |
|---|---|---|---|
| $g_1$ | Weibull distribution ($\alpha = 2, \beta = 20$) | 0.04620 | 1 |
| $g_2$ | lognormal distribution ($\theta = 2, \sigma^2 = 2$) | 0.67235 | 3 |
| $g_3$ | inverse Gaussian ($\alpha = 16, \beta = 64$) | 0.06008 | 2 |
| $g_4$ | F distribution ($\alpha = 4, \beta = 10$) | 5.74555 | 4 |

Here, the Weibull distribution is closest (loses the least information) to $f$, followed by the inverse Gaussian. The lognormal distribution is a poor third, while the F distribution is relatively far from the gamma distribution, $f$ (see Fig. 2.3).

Further utility of the K-L distance can be illustrated by asking which of the approximating models $g_i$ might be closest to $f$ when the parameters of $g_i$ are allowed to vary (i.e., what parameter values make each $g_i$ optimally close to $f$?). Following a computer search of the parameter space for the Weibull, we found that the *best* Weibull had parameters $\alpha = 2.120$ and $\beta = 18.112$ and a K-L distance of 0.02009; this is somewhat closer than the original parametrization (0.04620, above). Using the same approach, the best lognormal model had parameters $\theta = 2.642$ and $\sigma^2 = 0.2838$ and a K-L distance of 0.02195, while the best inverse Gaussian model had parameters $\alpha = 16$ and $\beta = 48$ with a K-L distance of

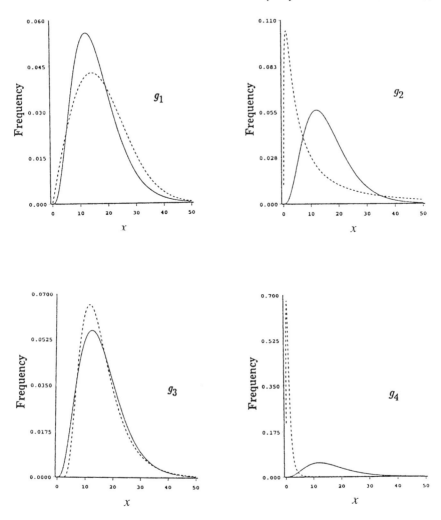

FIGURE 2.3. Plots of $f$ (= gamma (4, 4), solid line) against each of the 4 approximating models $g_i$ (dashed lines) as a function of $x$. Here, $g_1$ = Weibull (2, 20), $g_2$ = lognormal (2, 2), $g_3$ = inverse Gaussian (16, 64), and $g_4$ = F distribution (4, 10). Only in the most simple cases can plots such as these be used to judge closeness between models. Model $f$ is the same in all 4 graphs, it is merely scaled differently to allow the $g_i(x)$ to be plotted on the same graph.

0.03726, and the approximately best F distribution had parameters $\alpha \approx 300$, $\beta = 0.767$ and a K-L distance of approximately 1.486 (the K-L distance is not sensitive to $\alpha$ in this case, but is quite difficult to evaluate numerically). Thus, K-L distance indicates that the best Weibull is closer to $f$ than is the best lognormal (Fig. 2.4). Note that the formal calculation of K-L distance requires knowing the true distribution $f$ as well as all the model parameters (i.e., parameter estimation has not yet been addressed).

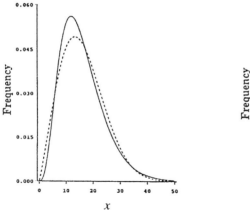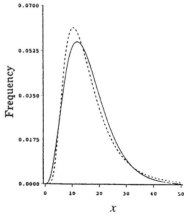

FIGURE 2.4. Plots of $f$ (= gamma $(4, 4)$) against the best Weibull (left) and lognormal models. The Weibull model that was closest to $f$ had parameters $(2.120, 18.112)$ with K-L distance $= 0.02009$, while the best lognormal had parameters $(2.642, 0.2838)$ with K-L distance $= 0.02195$. Compare these optimally-parametrized models with those in Fig. 2.3 (top).

These are *directed* distances; in the first Weibull example, $I(f, g_1) = 0.04620$, while $I(g_1, f) = 0.05552$ (in fact we would rarely be interested in $I(g_1, f)$, as this is the information lost when $f$ is used to approximate $g$!). The point here is that these are directed or oriented distances and $I(f, g_1) \neq I(g_1, f)$, *nor should they be equal, because the roles of truth and model are not interchangeable.*

These are all univariate functions; thus one could merely plot them on the same scale and visually compare each $g_i$ to $f$; however, this graphical method will work only in the simplest cases. In addition, if two approximating distributions are fairly close to $f$, it might be difficult to decide which is better by only visual inspection. Values of the K-L distance are not based on only the mean and variance of the distributions; rather, the distributions in their entirety are the subject of comparison.

The F-distribution ($\alpha = 4, \beta = 10$) provided a relatively poor approximation to the gamma distribution with ($\alpha = 4, \beta = 4$). Even the best 2-parameter F distribution remains a relatively poor approximation (K-L distance $= 1.486$). However, in general, adding more parameters will result in a closer approximation (e.g., the classic use of the Fourier series in the physical sciences or Wel's (1975) elephant-fitting problem). If we allow the addition of a third parameter ($\lambda$) in the F distribution (the noncentral F distribution), we find that the best model ($\alpha = 1.322$, $\beta = 43.308$, and $\lambda = 18.856$) has a K-L distance of only $0.001097$; this is better than any of the other 2-parameter candidate models (Fig. 2.5). Closeness of approximation can always be increased by adding more parameters to the candidate model. When we consider *estimation* of parameters and the associated uncertainty, then the principle of parsimony must be addressed (see Section 2.3), or over-fitted models will be problematic.

2.1 The Distance or Discrepancy Between Two Models  41

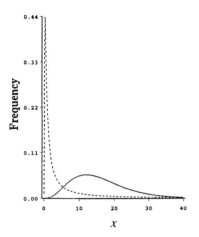
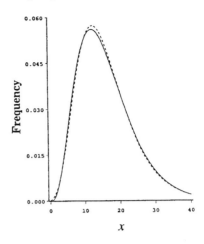

FIGURE 2.5. Plots of $f$ (= gamma (4, 4)) against the best 2-parameter F distribution (left) and the best 3-parameter (noncentral) F distribution. The best 2-parameter model was a poor approximation to $f$ (K-L distance = 1.486), while the best 3-parameter model is an excellent approximation (parameters 1.322, 43.308, 18.856) with K-L distance = 0.001097. Approximating models with increasing numbers of parameters typically are closer to $f$ than approximating models with fewer parameters.

## 2.1.4 Truth, $f$, Drops Out as a Constant

The material above makes it obvious that both $f$ and $g$ (and their parameters) must be known to compute the K-L distance between these two models. We see that this requirement is diminished, as $I(f, g)$ can be written equivalently as

$$I(f, g) = \int f(x) \log(f(x)) dx - \int f(x) \log(g(x \mid \theta)) dx.$$

Note that each of the two terms on the right of the above expression is a statistical expectation with respect to $f$ (truth). Thus, the K-L distance (above) can be expressed as a difference between two expectations,

$$I(f, g) = E_f \left[ \log(f(x)) \right] - E_f [\log(g(x \mid \theta))],$$

each with respect to the true distribution $f$. This last expression provides easy insights into the derivation of AIC. The important point is that the K-L distance $I(f, g)$ is a measure of the directed "distance" between the probability models $f$ and $g$ (Fig. 2.2).

The first expectation $E_f[\log(f(x))]$ is a constant that depends only on the unknown true distribution, and it is clearly not known (i.e., we do not know $f(x)$ in actual data analysis). Therefore, treating this unknown term as a constant, only a measure of *relative* directed distance is possible (Bozdogan 1987, Kapur and Kesavan 1992:155) in practice. Clearly, if one computed the second expectation, $E_f[\log(g(x \mid \theta))]$, one could estimate $I(f, g)$, up to a constant $C$ (namely $E_f[\log(f(x))]$),

$$I(f, g) = C - E_f[\log(g(x \mid \theta))],$$

or

$$I(f, g) - C = -E_f[\log(g(x \mid \theta))].$$

The term $(I(f, g) - C)$ is a *relative* directed distance between the two models $f$ and $g$, if one could compute or estimate $E_f[\log(g(x \mid \theta))]$. Thus, $E_f[\log(g(x \mid \theta))]$ becomes the quantity of interest. From the preceding example, where $f$ is gamma (4,4), then $\int f(x) \log(f(x)) dx = 3.40970$, and this term is constant across the models being compared,

$$I(f, g) - 3.40970 = -E_f[\log(g(x \mid \theta))].$$

The *relative* distances between the gamma (4, 4) model and the four approximating models are shown below:

|  | Approximating model | Relative distance $I(f, g_i) - C$ | Rank |
|---|---|---|---|
| $g_1$ | Weibull distribution ($\alpha = 2, \beta = 20$) | 3.45591 | 1 |
| $g_2$ | lognormal distribution ($\theta = 2, \sigma^2 = 2$) | 4.08205 | 3 |
| $g_3$ | inverse Gaussian ($\alpha = 16, \beta = 64$) | 3.46978 | 2 |
| $g_4$ | F distribution ($\alpha = 4, \beta = 10$) | 9.15525 | 4. |

Note that the ranking of "closeness" of the four candidate models to $f$ is preserved even though only relative distances are used.

The calculation of the two components of K-L distance (above) is in effect based on a sample size of 1. If the sample size were 100, then each component would be 100 times larger, and the difference between the two components would also be 100 times larger. For example, if $n = 100$, then $\int f(x) \log(f(x)) dx = 3.40970 \times 100 = 340.970$ and $E_f[\log(g_1(x \mid \theta))]$ (the Weibull) $= 3.45591 \times 100 = 345.591$. Thus, the difference between the two components of K-L distance would be 4.620; the *relative* difference is large when sample size is large.

Typically, as in the example above, the analyst would postulate several *a priori* candidate models $g_i(x \mid \theta)$ and want to select the *best* among these as a basis for data analysis and inference. Definition of "best" will involve the principle of parsimony and the related concept of a "best approximating model." In data analysis, the parameters in the various candidate models are not known and must be estimated from the empirical data. This represents an important distinction from the material above, as one usually has only models with estimated parameters, denoted by $g_i(x \mid \hat{\theta})$. In this case, one needs *estimates* of the relative directed distances between the unknown, true distribution $f(x)$ that generated the data and the various candidate models $g_i(x \mid \hat{\theta})$. Then, knowing the estimated relative distance from each $g_i(x)$ to $f(x)$, we select the candidate model that is estimated to be "closest" to truth for inference (Fig. 2.2). That is, we select the model with the smallest *relative* distance. Alternatively, we select an approximating model that loses the least information about truth. The conceptual model $f$ becomes a constant term, and nothing need be assumed about $f$, as the constant is the same across the candidate models and is irrelevant for comparison. (Similarly, it is interesting to note that the log-likelihood function also involves an additive constant that is the same across models; this term is known, but generally ignored, as it is often difficult to compute.) In practice, we can obtain only an approximate *estimator* of the relative distance from each approximating model $g_i(x \mid \hat{\theta})$ to $f(x)$ using AIC.

## 2.2 Akaike's Information Criterion

This section presents a heuristic introduction to the derivation of AIC; a more technical account is given in Section 2.3, while complete details on the conceptual and mathematical statistics of the derivation are provided in Chapter 6, Section 6.2. Akaike's (1973) seminal paper proposed the use of the Kullback–Liebler (K-L) distance as a fundamental basis for model selection. However, K-L distance cannot be computed without full knowledge of both $f(x)$, the form of the proposed model, $g(x \mid \theta)$, and a defined parameter space $\Theta$ that is the set of all possible values for $\theta$. Then the specific value of the parameter $\theta$ needed in the model can be computed given $f(x)$ and the functional form of the model. There is a specific value of $\theta$ defining the model $g(x \mid \theta)$ as one member of a family of models indexed by $\theta \in \Theta$. For well-conceived structural models there is a unique value of $\theta$ that minimizes K-L distance $I(f, g)$, $g \equiv g(x \mid \theta)$. This minimizing value of $\theta$ depends on truth $f$, the model $g$ through its structure, and the sample space, i.e., the structure and nature of the data that can be collected. In this sense there is a "true" value of $\theta$ underling ML estimation, but that value of $\theta$ depends on the assumed model, hence can vary by model. For the "true" $\theta$ under model $g$, K-L is minimized. This property of the model $g(x \mid \theta)$ as the minimizer of K-L, over all $\theta \in \Theta$, is an important feature involved in the derivation of AIC.

In data analysis the model parameters must be estimated, and there is usually substantial uncertainty in this estimation. Models based on estimated parameters, hence on $\hat{\theta}$ not $\theta$, represent a major distinction from the case where model parameters would be known. This distinction affects how we must use K-L distance as a basis for model selection. The difference between having $\theta$ (we do not) and having $\hat{\theta}$ (we do) is quite important and basically causes us to change our model selection criterion to that of minimizing *expected* estimated K-L distance rather than minimizing known K-L distance (over the set of $R$ models considered).

As a motivation of what AIC is, we continue here to use the *concept* of selecting a model based on minimizing the estimated Kullback–Liebler distance

$$\hat{I}(f, g) = \int f(x) \log \left( \frac{f(x)}{g(x \mid \hat{\theta}(y))} \right) dx. \tag{2.1}$$

In formula (2.1) it is important to realize that the variable of integration, $x$, does not represent the data. Rather, the data are represented above by "$y$." In fact, properly thought of, $x$ has been integrated out of $I(f, g)$, and what remains as $I(f, g)$ is really a function of $\theta$ given the structural form of the model, $g$, and the influence of truth, $f$. That is, $x$ is gone from $I(f, g)$, hence gone from $\hat{I}(f, g)$. Thus, the theoretical K-L distance between $f$ and $g$ does not involve any data. There is, however, of necessity an underlying defined sample space from which will arise a data point $y$. The data, $y$, as a random variable come from the same sample space as does the integrand, $x$. As conceptual random variables, $x$ and $y$ have the same probability distribution, $f$.

The data arise from the study at hand and can then be used to compute the MLE of $\theta$ given a model. Truth is not required to compute $\hat{\theta}$; but the concept of truth is required to interpret $\theta$ and $\hat{\theta}$. Given the data, hence $\hat{\theta}$, we would then (if we could) estimate the K-L distance using formula (2.1). We cannot actually compute formula (2.1) as our target criterion for selecting a best model. Moreover, we should not want to use (2.1) as our model selection criterion because having had to estimate $\theta$, we have slightly changed the character of the problem. Now we would be dealing with an estimate, $\hat{I}(f, g)$, of true distance, not true distance itself. Thus even if our model form were truth (i.e., $g = f$ held), our baseline value to judge, on average, a perfect-fitting model should not be 0. The value of 0 would occur only if $\hat{\theta}(y)$ exactly equaled $\theta$. When this does not happen, which is for all, or almost all, samples, we would find $\hat{I}(g, g) > 0$, even though $I(g, g) = 0$.

In the context of repeated sampling properties of an inference procedure, as one often-used guide to statistical inference we expect our estimated K-L to have on average the positive value

$$E_{\hat{\theta}}[\hat{I}(f, g)] = \int f(y) \left[ \int f(x) \log\left(\frac{f(x)}{g(x \mid \hat{\theta}(y))}\right) dx \right] dy. \quad (2.2)$$

Because for model structure $g$, $\theta$ minimizes true K-L distance, we are assured that

$$E_{\hat{\theta}}[\hat{I}(f, g)] > I(f, g).$$

Thus, because estimated K-L distance will be systematically greater than true K-L distance we should readjust our conceptual idea of optimal approximation by the fitted model to be not the minimizing of $I(f, g)$ over models (we cannot do this anyway), but rather the minimizing on average of the the slightly larger target criterion given by formula (2.2).

To translate concepts into methodology we break both formulae (2.1) and (2.2) into their two parts (as in the Section 2.1.4 discussion of this matter):

$$\hat{I}(f, g) = \int f(x) \log(f(x)) dx - \int f(x) \log(g(x \mid \hat{\theta}(y))) dx$$
$$= \text{Constant} - E_x \left[ \log(g(x \mid \hat{\theta}(y))) \right] \quad (2.3)$$

and

$$E_{\hat{\theta}}[\hat{I}(f, g)] = \int f(y) \left[ \int f(x) \log(f(x)) dx \right] dy$$
$$- \int f(y) \left[ \int f(x) \log(g(x \mid \hat{\theta}(y))) dx \right] dy$$
$$= \left[ \int f(y) dy \right] \left[ \int f(x) \log(f(x)) dx \right]$$

## 2.2 Akaike's Information Criterion

$$-\int f(y) \left[ \int f(x) \log(g(x \mid \hat{\theta}(y))) dx \right] dy$$

$$= \text{Constant} - E_y E_x \left[ \log(g(x \mid \hat{\theta}(y))) \right]. \tag{2.4}$$

If we were to base K-L model selection on formula (2.3); we would have to compute, at least approximately, the estimated relative K-L distance, $E_x[\log(g(x \mid \hat{\theta}(y)))]$. In fact, this cannot be done: We can neither compute the term $E_x[\log(g(x \mid \hat{\theta}(y)))]$ nor any estimate of it (this technical matter is touched on again in Section 2.3, and is fully developed in Chapter 6). Above, we have given a conceptual argument that formula (2.3) is not the correct basis for K-L information-theoretic model selection when $\theta$ must be estimated. That argument lead to formula (2.4), hence to the realization that the necessary basis for applied K-L information theoretic model selection is the quantity

$$E_y E_x[\log(g(x \mid \hat{\theta}(y)))] \tag{2.5}$$

that appears in formula (2.4). The quantity, given above, i.e., in (2.5), is related to K-L distance by

$$E_y E_x \left[ \log(g(x \mid \hat{\theta}(y))) \right] = \text{Constant} - E_{\hat{\theta}}[\hat{I}(f, g)].$$

Thus $E_y E_x[\log(g(x \mid \hat{\theta}(y)))]$ is the relative expected K-L distance; it depends on truth $f$, model structure $g$, and true parameter value $\theta$ given $f$ and $g$, and it can be estimated based on the data and the model.

Consequently, **we can determine a method to select the model $g_i$ that on average minimizes, over the set of models $g_1, \ldots, g_R$, a very relevant expected K-L distance**. The objective function, to be maximized, is given by formula (2.5). The model $g_i(x \mid \theta)$ that maximizes formula (2.5) is the K-L best model when $\theta$ must be estimated, and ML theory is used to get $\hat{\theta}$. In applications we fit each model to the data (getting $\hat{\theta}$, by model) and then get estimated values of our criterion (2.5). For large sample sizes, the maxima of both $E_y E_x[\log(g(x \mid \hat{\theta}(y)))]$ and $E_x\left[\log(g(x \mid \theta))\right]$ have a ratio of almost 1, so AIC will then essentially be selecting **on average** the K-L best model from the set of models considered.

Akaike (1973, 1985, 1994) first showed that the critical issue for getting an applied K-L model selection method was to estimate

$$E_y E_x[\log(g(x \mid \hat{\theta}(y)))].$$

This expression involves what looks like the log of the probability model for the data. That is, the conceptual $\log(g(x \mid \hat{\theta}(y)))$ bears a strong resemblance to the actual log-likelihood, $\log(\mathcal{L}(\hat{\theta}(y) \mid y)) \equiv \log(g(y \mid \hat{\theta}(y)))$. Moreover, because all we have directly available to estimate this target model selection criterion is our log-likelihood (for each model), the temptation is to just estimate $E_y E_x[\log(g(x \mid \hat{\theta}(y)))]$ by the maximized $\log(\mathcal{L}(\hat{\theta}(y) \mid y))$ for each model.

However, Akaike (1973) showed first that the maximized log-likelihood is biased upward as an estimator of the model selection target criterion (2.5). Second, he showed that under certain conditions (these conditions are important, but quite technical), that this bias is approximately equal to $K$, the number of estimable parameters in the approximating model. Thus, an approximately unbiased estimator of

$$T = \mathrm{E}_y \mathrm{E}_x[\log(g(x \mid \hat{\theta}(y)))]$$

for large samples and "good" models, is

$$\hat{T} = \log(\mathcal{L}(\hat{\theta} \mid y)) - K \qquad (2.6)$$

(where the notation $\hat{\theta} \equiv \hat{\theta}(y)$). This result is equivalent to

$$\log(\mathcal{L}(\hat{\theta} \mid y)) - K = \text{Constant} - \hat{\mathrm{E}}_{\hat{\theta}}[\hat{I}(f, g)],$$

or

$$-\log(\mathcal{L}(\hat{\theta} \mid y)) + K = \text{estimated relative expected K-L distance}.$$

The bias correction term ($K$) above is a special case of a more general result derived by Takeuchi (1976) and described in Chapter 6. If the approximating models in the candidate set are poor (far from $f$), then Takeuchi's information criterion (TIC) is an alternative if sample size is large. AIC is a special case of TIC, and as such, AIC is a parsimonious approach to the estimation of relative expected K-L distance (see Section 2.8). **Akaike's finding of a relation between the relative K-L distance and the maximized log-likelihood has allowed major practical and theoretical advances in model selection and the analysis of complex data sets** (see Stone 1982, Bozdogan 1987, and deLeeuw 1992).

Akaike (1973) then defined "**an** *information criterion*" (AIC) by multiplying $\hat{T}$ (i.e., formula 2.6) by $-2$ ("taking historical reasons into account") to get

$$\text{AIC} = -2\log(\mathcal{L}(\hat{\theta} \mid y)) + 2K.$$

This has became known as "*Akaike's information criterion*," or AIC. Here it is important to note that AIC has a strong theoretical underpinning, based on information theory and Kullback–Liebler information within a realistic data analysis philosophy that no model is true; rather, truth as $f$ is far more complex than any model used. Akaike's inferential breakthrough was realizing that a predictive expectation version of the log-likelihood could (as one approach) be used to estimate the relative expected K-L distance between the approximating model and the true generating mechanism. Thus, rather than having a simple measure of the directed distance between two models (i.e., the K-L distance), one has instead an *estimate* of the expected relative, directed distance between the fitted model and the unknown true mechanism (perhaps of infinite dimension) that actually generated the observed data. Because the expectation of the logarithm of $f(x)$ drops out as a constant, independent of the data, AIC is defined without specific reference to a "true model" (Akaike 1985:13). The constant term ($\mathrm{E}_f[\log(f(x))]$) is the same across models, and therefore has no relevance to the relative K-L distances being estimated and compared across approximating models. Thus, one should select the model that yields the smallest value of AIC (or, equivalently, the model with the smallest estimate of $-2T$) because this model is estimated to be "closest" to the unknown reality that generated the data, from among the candidate models considered. This seems a very natural, simple concept; select the fitted approximating model that is estimated, on average, to be closest to the unknown $f$.

$I(f, g)$ can be made smaller by adding more known (not estimated) parameters in the approximating model $g$. Thus, for a fixed data set, the further addition of parameters in a model $g_i$ will allow it to be closer to $f$. However, when these parameters must be estimated (rather than being known or "given"), further uncertainty is added to the *estimation* of the relative K-L distance. At some point, the addition of still more estimated parameters will have the opposite from desired effect (i.e., reduce $E_{\hat{\theta}}[\hat{I}(f, g)]$), and the estimate of the relative K-L distance will increase because of "noise" in estimated parameters that are not really needed to achieve a good model. This can be seen by examination of the information criterion being minimized,

$$\text{AIC} = -2\log(\mathcal{L}(\hat{\theta} \mid y)) + 2K,$$

where the first term tends to decrease as more parameters are added to the approximating model, while the second term ($2K$) gets larger as more parameters are added to the approximating model. This is the tradeoff between bias and variance or the tradeoff between under-fitting and over-fitting that is fundamental to the principle of parsimony (see Section 1.4.2).

Usually, AIC is positive; however, it can be shifted by any additive constant and a shift can somtimes result in negative values of AIC. Computing AIC from regression statistics (see Section 1.4.2) often results in negative AIC values (see Fig. 1.3). The model with the smallest AIC value is estimated to be "closest" to truth and is the best approximation for the information in the data, relative to the other models considered (in the sense of the principle of parsimony). Perhaps none of the models in the set are good, but AIC attempts to select the best approximating model of those in the candidate set. Thus, every effort must be made to assure that the set of models is well founded.

Let candidate models $g_1$, $g_2$, $g_3$, and $g_4$ have AIC values of 3,400, 3,560, 3,380, and 3,415, respectively. Then one would select model $g_3$ as the best single model as the basis for inference because $g_3$ has the smallest AIC value. Because these values are on a relative (additive) scale, one could subtract, say, 3,000 from each AIC value and have the following rescaled AIC values: 400, 560, 380, and 415. Of course, such rescaling does not change the ranks of the models, nor the pairwise differences in the AIC values. In our work, we have seen minimum AIC values that range from large negative numbers to as high as 340,000. It is not the absolute size of the AIC value, it is the relative values, and particularly the differences between AIC values, that are important.

Because AIC is on an relative scale, we routinely recommend computing (and presenting in publications) the **AIC differences** (rather than the actual AIC values),

$$\Delta_i = \text{AIC}_i - \min \text{AIC} \doteq E_{\hat{\theta}}[\hat{I}(f, g_i)] - \min E_{\hat{\theta}}[\hat{I}(f, g_i)],$$

over all candidate models in the set. Such differences estimate the relative expected K-L differences between $f$ and $g_i(x \mid \theta)$. These $\Delta_i$ values are easy to interpret and allow a quick comparison and ranking of candidate models and are also useful in computing Akaike weights (Section 4.2).

The larger $\Delta_i$ is, the less plausible is the fitted model $g_i(x \mid \hat{\theta})$ as being the K-L best model for samples such as the data one has. As a rough rule of thumb, models for which $\Delta_i \leq 2$ have substantial support and should receive consideration in making inferences. Models having $\Delta_i$ of about 4 to 7 have considerably less support, while models with $\Delta_i > 10$ have either essentially no support, and might be omitted from further consideration, or at least those models fail to explain some substantial explainable variation in the data. If observations are not independent but are assumed to be independent, then these simple guidelines cannot be expected to hold.

The best (i.e., the model with the minimum AIC value) of the four hypothetical models in the above example, model $g_3$, has $\Delta_3 = 3{,}380 - 3{,}380 = 0$. Models $g_1$, $g_2$, and $g_4$ have $\Delta_i$ values of 20, 180, and 35, respectively. This simple rescaling to a minimum relative AIC value of zero makes comparisons between the best model and other candidate models easy. In the example, it is then readily seen that fitted model $g_3$ is the best, followed in order, by $g_1$, $g_4$, and finally $g_2$.

The interpretation of the $\Delta_i$ is not well studied for nonnested models, and the whole subject of these differences in AIC values and their interpretation needs further study (e.g., Linhart 1988). However, we can say with considerable confidence that in real data analysis with several models, or more, and large sample size (say $n > 10 \times K$ for the biggest model) a model having $\Delta_i = 35$, such as model $g_4$, would be a very, very poor approximating model for the data at hand.

The material to this point has been based on likelihood theory, as it is a very general approach. In the special case of least squares (LS) estimation with normally distributed errors, and apart from an arbitrary additive constant, AIC can be expressed as

$$\mathrm{AIC} = n \log(\hat{\sigma}^2) + 2K,$$

where

$$\hat{\sigma}^2 = \frac{\sum \hat{\epsilon}_i^2}{n} \text{ (the MLE of } \sigma^2\text{)},$$

and $\hat{\epsilon}_i$ are the estimated residuals for a particular candidate model. A common mistake with LS model fitting, when computing AIC, is to take the estimate of $\sigma^2$ from the computer output, instead of computing the ML estimate above. Also, for LS model fitting, $K$ is the total number of estimated parameters, including the intercept and $\sigma^2$. The value of $K$ is sometimes determined incorrectly because either $\beta_0$ (the intercept) or $\sigma^2$ (or both) is mistakenly ignored in determining $K$. Thus, AIC is easy to compute from the results of LS estimation in the case of linear models and is now included in the output of many software packages for regression analysis.

## 2.3 Akaike's Predictive Expected Log-Likelihood

This section provides further insights into Akaike's (1973) result that an approximately unbiased estimator of the relative K-L distance for large samples and good models is $\log(\mathcal{L}(\hat{\theta} \mid \text{data})) - K$. The argument that we should select a model to minimize K-L distance (initially ignoring parameter estimation) leads us to focus on the quantity $E_x[\log(g(x \mid \theta))]$. A logical next step would be to plug $\hat{\theta}(y)$ into this function. Conceptually we can do so, but still $E_x[\log(g(x \mid \hat{\theta}(y)))]$ (which appears in formula 2.3) cannot be computed in reality.

Progress can be made because we can consider finding an approximation to $E_x[\log(g(x \mid \hat{\theta}(y)))]$. A second-order Taylor series expansion of $\log(g(x \mid \hat{\theta}(y)))$ about $\theta$ leads to an interesting quadratic approximation. We remind the reader that the parameter, $\theta$, is actually a $K \times 1$ vector, and the MLE, $\hat{\theta}$, has a $K \times K$ sampling variance–covariance matrix denoted by $\Sigma$. If the model is "close" to truth, then the following result is a decent approximation:

$$E_x[\log(g(x \mid \hat{\theta}(y)))] \doteq E_{\hat{\theta}}\left[\log(\mathcal{L}(\hat{\theta} \mid \text{data}))\right] - \frac{1}{2}K - \frac{1}{2}[\hat{\theta} - \theta]\Sigma^{-1}[\hat{\theta} - \theta]' \quad (2.7)$$

(Section 6.2 gives the derivation of a more general result). The log-likelihood, $\log(\mathcal{L}(\hat{\theta} \mid \text{data}))$, is an unbiased estimator of $E_{\hat{\theta}}\left[\log(\mathcal{L}(\hat{\theta} \mid \text{data}))\right]$ and of course we know $K$. There are several ways to compute a good estimate of the unknown variance–covariance matrix, $\Sigma$. Thus we can get to

$$\hat{E}_x[\log(g(x \mid \hat{\theta}(y)))] = \log(\mathcal{L}(\hat{\theta} \mid \text{data})) - \frac{1}{2}K - \frac{1}{2}[\hat{\theta} - \theta]\hat{\Sigma}^{-1}[\hat{\theta} - \theta]'.$$

However, we fall short of our goal of a computable estimator of the left-hand side of formula (2.7): We do not know $\theta$, and it is no use reusing $\hat{\theta}$ in the quadratic term because then that term vanishes, thus producing a very biased estimator ($\equiv 0$). However, if we replace the quadratic term by its expected value (approximately $K$ because that quadratic is approximately a central chi-square random variable on $K$ df) we can obtain a criterion that is computable. This is equivalent to taking the expectation of both sides of formula (2.7), hence using

$$E_y E_x[\log(g(x \mid \hat{\theta}(y)))] \doteq E_{\hat{\theta}}\left[\log(\mathcal{L}(\hat{\theta} \mid \text{data}))\right] - K \quad (2.8)$$

(the *data* are $y$, and expectation with respect to $\hat{\theta}$, $E_{\hat{\theta}}$ is the same as $E_y$).

Akaike, in defining and deriving AIC, sometimes uses a particular predictive expectation for the log-likelihood function of the data and unknown parameters; this has advantages and properties that are still not well recognized in the literature. Deeper insights into the derivation of AIC are given in Akaike (1973, 1987:319, 1992), Bozdogan (1987), Sakamoto (1991), deLeeuw (1992), and in Chapter 6.

This alternative path to AIC again shows us that we have to adopt as our criterion for selecting the best fitted model, the maximization of $E_y E_x[\log(g(x \mid \hat{\theta}(y)))]$, the expected relative K-L distance. Hence we are really implementing formula (2.4) as our model selection criterion; this is forced on us because we must estimate $\theta$. Given this realization, it seems of interest to consider what $\log(g(x \mid \hat{\theta}(y)))$ might mean; in fact, Akaike called it a predictive likelihood.

Deriving AIC is sometimes given in terms of a statistical expectation based on a different, independent sample, $x$, as well as data, $y$. Then one considers the conceptual predictive model $g(x \mid \hat{\theta}(y))$ (used for making predictions about a new sample, $x$, that could be collected, based on actual data, $y$). The predictive log-likelihood is $\log(g(x \mid \hat{\theta}(y)))$, with expected value $E_y E_x[\log(g(x \mid \hat{\theta}(y)))]$, useful as a criterion for model selection. It is this expectation over a second conceptual, independent "data set" that provides AIC with a cross validation property (see Tong 1994, Stone 1977).

This perception of AIC as based on an expectation over a different, independent hypothetical data set was one of Akaike's major contributions and represents a rather deep issue, even for those with a very good background in mathematical statistics (cf. Chapter 6). Symbolically, Akaike's expected predictive log-likelihood can be given as

$$E_p[\log(\mathcal{L}(\hat{\theta}))] = E_y E_x[\log(\mathcal{L}(\hat{\theta}(y) \mid x))];$$

$E_p$ is used to denote Akaike's predictive expectation. Thus, $E_y E_x[\log(\mathcal{L}(\hat{\theta}(y) \mid x))]$ is an alternative expression for the target criterion to get the K-L best-fitted model. AIC provides an estimate of the predictive expectation for each fitted model $g_i$ (Fig. 2.2).

For the second term, $2K$, in AIC it has been shown that one-half of this (the first $K$) is due to the asymptotic bias correction in the maximized log-likelihood. The other half (the second $K$) is the result of his predictive expectation with respect to $\theta$ (note the form of formula 2.7) (Akaike 1974). This is related to the cross validation property for independent and identically distributed samples (Stone 1977 and Stoica et al. 1986). Golub et al. (1979) show that AIC asymptotically coincides with generalized cross validation in subset regression (also see review by Atilgan 1996). Considering Akaike's predictive expectation representation of AIC and the independent conceptual data set $x$, the cross validation aspect of AIC is reasonably apparent. Akaike's papers provide the statistical formalism for AIC (also see Bozdogan 1987, Sakamoto 1991, Shibata 1983, 1989, and Chapter 6).

The fact that AIC is an estimate only of **relative** expected K-L distance is almost unimportant. It is the fact that AIC **is only an estimate** of these relative distances from each model $g_i$ to $f$ that is less than ideal. It is important to recognize that there is usually substantial uncertainty as to the best model for a given data set. After all, these are stochastic biological processes, often with relatively high levels of uncertainty.

In as much as a statistical model can provide insight into the underlying biological process, it is important to try to determine as accurately as possible the basic underlying *structure* of the model that fits well the data. "Let the data speak" is of interest to both biologists and statisticians in objectively learning from empirical data. The data then help determine the proper complexity (order) of the approximating model used for inference and help determine what effects or factors are justified. We must admit that if much more data were available, then further effects could probably be found and supported. "Truth" is elusive; model selection tells us what inferences the data support, not what full reality might be.

## 2.4 Important Refinements to AIC

### 2.4.1 A Second-Order AIC

Akaike derived an estimator of the K-L information quantity. However, AIC may perform poorly if there are too many parameters in relation to the size of the sample (Sugiura 1978, Sakamoto et al. 1986). Sugiura (1978) derived a second-order variant of AIC that he called c-AIC. Hurvich and Tsai (1989) further studied this small-sample (second-order) bias adjustment, which led to a criterion that is called $AIC_c$:

$$AIC_c = -2\log(\mathcal{L}(\hat{\theta})) + 2K\left(\frac{n}{n-K-1}\right),$$

where the penalty term is multiplied by the correction factor $n/(n-K-1)$. This can be rewritten as

$$AIC_c = -2\log(\mathcal{L}(\hat{\theta})) + 2K + \frac{2K(K+1)}{n-K-1},$$

or, equivalently,

$$AIC_c = AIC + \frac{2K(K+1)}{n-K-1},$$

where $n$ is sample size (also see Sugiura 1978). $AIC_c$ merely has an additional bias correction term. If $n$ is large with respect to $K$, then the second-order correction is negligible and AIC should perform well. Findley (1985) noted that the study of the bias correction is of interest in itself; the exact small-sample bias correction term varies by type of model (e.g., normal, exponential, Poisson). Bedrick and Tsai (1994) provide a further refinement, but it is more difficult to compute (also see Hurvich and Tsai 1991 and 1995a and b, and Hurvich et al. 1990). While $AIC_c$ was derived under Gaussian assumptions for linear models (fixed effects), Burnham et al. (1994) found this second-order approximation to the K-L distance to be useful in product multinomial models. Generally, we advocate the use of $AIC_c$ when the ratio $n/K$ is small (say $< 40$). In reaching a decision about the use of AIC vs. $AIC_c$, one must use the value of $K$ for the highest-dimensioned model in the set of candidates. If the ratio $n/K$ is sufficiently large, then AIC and $AIC_c$ are similar and will tend to select the same model. One should use either AIC or $AIC_c$ consistently in a given analysis, rather than mixing the two criteria. Few software packages provide $AIC_c$ values, but these can easily be computed by hand. **Unless the sample size is large with respect to the number of estimated parameters, use of $AIC_c$ is recommended.**

## 2.4.2 Modification to AIC for Overdispersed Count Data

In general, if the random variable $n$ represents a random count under some simple distribution (e.g., Poisson or binomial), it has a known expectation, $\mu(\theta)$, and a known theoretical variance function, $\sigma^2(\theta)$ ($\theta$ still is unknown). In a model of overdispersed data the expectation of $n$ is not changed, but the variance model must be generalized, for example using a multiplicative factor, e.g., $\gamma(\theta)\sigma^2(\theta)$. The form of the factor $\gamma(\theta)$ can be partly determined by theoretical considerations and can be complex (see, e.g., McCullagh and Nelder 1989). Overdispersion factors typically are small, ranging from just above 1 to perhaps 3 or 4, if the model structure is correct and overdispersion is due to small violations of assumptions such as independence and parameter homogeneity over individuals. Hence, a first approximation for dealing with overdispersion is to use a simple constant $c$ in place of $\gamma(\theta)$.

Count data have been known not to conform to simple variance assumptions based on binomial or multinomial distributions (e.g. Bartlett 1936, Fisher 1949, Armitage 1957, and Finney 1971). There are a number of statistical models for count data (e.g., Poisson, binomial, negative binomial, multinomial). In these, the sampling variance is theoretically determined, by assumption (e.g., for the Poisson model, $\text{var}(n) = E(n)$). If the sampling variance exceeds the theoretical (model based) variance, the situation is called "overdispersion." Our focus here is on a lack of independence in the data leading to overdispersion, or "extra-binomial variation." Eberhardt (1978) provides a clear review of these issues in the biological sciences. For example, Canada geese (*Branta* spp.) frequently mate for life, and the pair behaves almost as an individual, rather than as two independent "trials." The young of some species continue to live with the parents for a period of time, which can also cause a lack of independence of individual responses. Further reasons for overdispersion in biological systems include species whose members exist in schools or flocks. Members of such populations can be expected to have positive correlations among individuals within the group; such dependence causes overdispersion. A different type of overdispersion stems from parameter heterogeneity, that is, individuals having unique parameters rather than the same parameter (such as survival probability) applying to all individuals.

The estimators of model parameters often remain unbiased in the presence of overdispersion, but the model-based, theoretical variances overestimate precision (McCullagh and Nelder 1989). To properly cope with overdispersion one needs to model the overdispersion and then use generalized likelihood inference methods. Quasi-likelihood (Wedderburn 1974) theory is a basis for the analysis of overdispersed data (also see Williams 1982, McCullagh and Pregibon 1985, Moore 1987, and McCullagh and Nelder 1989). Hurvich and Tsai (1995b) provide information on the use of $AIC_c$ with overdispersed data.

Cox and Snell (1989) discuss modeling of count data and note that the first useful approximation is based on a single variance inflation factor ($c$), which can be estimated from the goodness-of-fit chi-square statistic ($\chi^2$) of the global model and its degrees of freedom,

$$\hat{c} = \chi^2/\text{df}.$$

The variance inflation factor should be estimated from the global model. Cox and Snell (1989) assert that the simple approach of a constant variance inflation factor should often be adequate, as opposed to the much more arduous task of seeking a detailed model for the $\gamma(\theta)$. In a study of these competing approaches on five data sets, Liang and McCullagh (1993) found that modeling overdispersion was clearly better than use of a single $\hat{c}$ in only one of five cases examined.

Given $\hat{c}$, empirical estimates of sampling variances ($\text{var}_e(\hat{\theta}_i)$) and covariances ($\text{cov}_e(\hat{\theta}_i, \hat{\theta}_j)$) can be computed by multiplying the estimates of the theoretical (model-based) variances and covariances by $\hat{c}$ (a technique that has long been used; see e.g., Finney 1971). These empirical measures of variation (i.e., $\hat{c} \cdot \widehat{\text{var}}_t(\hat{\theta}_i)$) must be treated as having the degrees of freedom used to compute $\hat{c}$ for purposes of setting confidence limits (or testing hypotheses). Generally, quasi-likelihood adjustments (i.e., use of $\hat{c} > 1$) are made only if some reasonable lack of fit has been found (for example, if the observed significance level $P \geq 0.15$ or $0.25$) and the degrees of freedom $\geq 10$, as rough guidelines.

We might expect $c > 1$ with real data but would not expect $c$ to exceed about 4 if model structure is acceptable and only overdispersion is affecting $c$ (see Eberhardt 1978). Substantially larger values of $c$ (say, 6–10) are usually caused partly by a model structure that is inadequate; that is, the fitted model does not actually represent all the explainable variation in the data. Quasi-likelihood methods of variance inflation are most appropriate only after a reasonable structural adequacy of the model has been achieved. The estimate of $c$ should be computed only for the global model; one should not make and use separate estimates of this variance inflation factor for each of the candidate models in the set. The issue of the structural adequacy of the model is at the very heart of good data analysis (i.e., the reliable identification of the structure vs. residual variation in the data). Patterns in the goodness-of-fit statistics (Pearson $\chi^2$ or G-statistics) might be an indication of structural problems with the model. Of course, the biology of the organism in question should provide clues as to the existence of overdispersion; one should not rely only on statistical considerations in this matter.

Principles of quasi-likelihood suggest simple modifications to AIC and $\text{AIC}_c$; we denote these modifications as (Lebreton et al. 1992)

$$\text{QAIC} = -\left[2\log(\mathcal{L}(\hat{\theta}))/\hat{c}\right] + 2K,$$

and

$$\text{QAIC}_c = -\left[2\log(\mathcal{L}(\hat{\theta}))/\hat{c}\right] + 2K + \frac{2K(K+1)}{n-K-1},$$

$$= \text{QAIC} + \frac{2K(K+1)}{n-K-1}.$$

Of course, when no overdispersion exists, $c = 1$, the formulae for QAIC and $\text{QAIC}_c$ reduce to AIC and $\text{AIC}_c$, respectively.

One must be careful when using some standard software packages (e.g., SAS GENMOD), as they were developed some time ago under a hypothesis testing mode (i.e., adjusting $\chi^2$ test statistics by $\hat{c}$ to obtain $F$-tests). In some cases, a separate estimate of $c$ is made for each model, and variances and covariances are multiplied by this model-specific estimate of the variance inflation factor. Some software packages compute an estimate of $c$ for every model, thus making the correct use of model selection criteria tricky unless one is careful. Instead, we recommend that the global model be used as a basis for the estimation of a single variance inflation factor, $c$. Then the empirical log-likelihood for each of the candidate models is divided by $\hat{c}$ and QAIC or QAIC$_c$ computed and used for model selection. The estimated variances and covariances should also be adjusted using $\hat{c}$ from the global model, unless there are few degrees of freedom left; then one might estimate $c$ from the selected model.

Some commercial software computes AIC, while AIC$_c$ is rarely available, and no general software package computes QAIC or QAIC$_c$. In almost all cases, AIC, AIC$_c$, QAIC, and QAIC$_c$ can be computed easily by hand from the material that is output from standard computer packages (either likelihood or least squares estimation). In general, we recommend using this extended "information-theoretic criterion" for count data, and we will use QAIC$_c$ in some of the practical examples in Chapter 3. Of course, often the overdispersion parameter is near 1, negating the need for quasi-likelihood adjustments, and just as often the ratio $n/K$ is large, negating the need for the additional penalty term in AIC$_c$. AIC, AIC$_c$, and QAIC$_c$ are all estimates of the relative K-L information. We often use the generic term "AIC" to mean any of these criteria.

## 2.5 A Useful Analogy

In some ways, selection of a best approximating model is analogous to auto racing or other similar contests. The goal of such a race is to identify the best (fastest) car/driver combination, and the data represent results from a major race (e.g., the Indianapolis 500 in the USA, the 24 Heures du Mans in France). Only a relatively few car/driver combinations "qualify," based on pre-race trials (e.g., 33 cars at Indianapolis)—this is like the set of candidate models (i.e., only certain models "qualify," based on the science of the situation). It would be chaotic if all car/driver combinations with an interest could enter the race, just as it makes little sense to include a very large number of models in the set of candidates (and risk Freedman's paradox). Cars that do not qualify do not win, even though they might indeed have been the best (fastest) had they not failed to qualify. Similarly, models, either good or bad, not in the set of candidates remain out of consideration.

At the end of the race the results provide a ranking ("placing") of each car/driver combination, from first to last. Furthermore, if a quantitative index of quality is available (e.g., elapsed time for each finisher), then a further "calibration" can be considered. Clearly, the primary interest is in "who won the race" or "which was the first"—this is like the model with the minimum AIC value. This answers the question, "Which is best in the race"; the results could differ for another (future) race or another data set, but these are, as yet, unavailable to us.

## 2.5 A Useful Analogy

Some (secondary) interest exists in the question "Who was in second place?" and in particular, was second place only thousandths of a second behind the winner or 5 minutes behind? The race time results provide answers to these questions, as do the $\Delta$-AIC values in model selection. In the first case, the best inference might be that the first two cars are essentially tied and that neither is appreciably better than the other (still, the size of the purse certainly favors the first-place winner!), while in the second case, the inference probably favors a single car/driver combination as the clear best (with a 5 minute lead at the finish). The finishing times provide insights into the third and fourth finishers, etc. In trying to understand the performance of car/driver combinations, one has considerable information from both the rankings and their finishing times, analogous to the AIC values (both the ranks and the $\Delta_i$ values). In later sections we will see how the $\Delta_i$ can be used to estimate model selection probabilities, and these will provide additional insights. Note that the absolute time of the winner is of little interest because of temperature differences, track conditions, and other variables; only the *relative* times for a given race are of critical interest. Similarly, the absolute values of AIC are also of little interest, as they reflect sample size, among other things.

The winner of the race is clearly the best for the particular race. If one wants to make a broader inference concerning races for an entire year, then results (i.e., ranks) from several races can be pooled or weighted. Similarly, statistical inferences beyond a single observed data set can sometimes be broadened by some type of model averaging using, for example, the nonparametric bootstrap (details in Chapters 4 and 5) and the incorporation of model selection uncertainty in estimators of precision.

The race result might not always select the best car/driver combination, as the fastest qualifying car/driver may have had bad luck (e.g., crash or engine failure) and finished well back from the leader (if at all). Similarly, in model selection one has only one realization of the stochastic process and an *estimated* relative distance as the basis for the selection of a best approximating model (a winner). If the same race is held again with the same drivers, the winner and order of finishers are likely to change somewhat. Similarly, if a new sample of data could be obtained the model ranks will likely change somewhat.

To carry the analogy a bit further, data dredging would be equivalent to watching a race as cars dropped out and others came to the lead. Then one continually shifts the bet and predicted winner, based on the car/driver in the lead at any point in time (i.e., an unfair advantage). In this case, the final prediction would surely be improved, but the rules of play have certainly been altered! Alternatively, the definition of winning might not be established prior to the initiation of the race. Only after the race are the rules decided (e.g., based, in part, on who they think "ought" to win). Then, one might question the applicability of this specific prediction to other races. Indeed, we recommend "new rules" when data dredging has been done. That is, if a particular result was found following data dredging, then this should be fully admitted and discussed in resulting publication. We believe in fully examining the data for all the information and insights they might provide. However, the sequence leading to data dredging should be revealed and results following should be discussed in this light.

Many realize that there is considerable variation in cars and drivers from race to race and track to track. Similarly, many are comfortable with the fact that there is often considerable sampling variation (uncertainty) associated with an estimate of a parameter from data set to data set. Similarly, if other samples (races) could be taken, the estimated best model (car/driver) might also vary from sample to sample (or race to race). Both components of sampling variation and model selection uncertainty should ideally be incorporated into measures of precision.

## 2.6 Some History

Akaike (1973) considered AIC and its information-theoretic foundations "... a natural extension of the classical maximum likelihood principle." Interestingly, Fisher (1936) anticipated such an advance over 60 years ago when he wrote,

> "... an even wider type of inductive argument may some day be developed, which shall discuss methods of assigning from the data the functional form of the population."

This comment was quite insightful; of course, we might expect this from R. A. Fisher! Akaike was perhaps kind to consider AIC an extension of classical ML theory; he might just as well have said that classical likelihood theory was a special application of the more general information theory. In fact, Kullback believed in the importance of information theory as a unifying principle in statistics.

### 2.6.1 The G-Statistic and K-L Information

For discrete count data for $k$ mutually exclusive categories there is a close relationship between the G-statistic for goodness-of-fit testing and the K-L distance. The G-statistic is usually written as

$$G = 2 \sum_{j=1}^{k} O_j \log\left(\frac{O_j}{E_j}\right),$$

where $O_j$ is the observed count and $E_j$ is the expectation under some fitted model. Under mild conditions, G is asymptotically distributed as chi-squared under the null hypothesis that the model is an adequate fit to the discrete data. Such G-statistics are additive, whereas the more traditional Pearson's goodness-of-fit test statistic

$$\text{Pearson} = \sum_{j=1}^{k} \left((O_j - E_j)^2 / E_j\right)$$

is not. The K-L distance for discrete data is written as

$$I(f, g) = \sum_{i=1}^{k} p_i \log\left(\frac{p_i}{\pi_i}\right)$$

and is almost identical in form to the G-statistic.

Given a sample of count data, $n_1, \ldots, n_k$ ($n = \sum n_i$), let $p_j = n_j/n$ correspond to the observed relative frequencies. Denote the estimated expected probabilities under the approximating model by $\hat{\pi}_j(\theta)$; thus $n\hat{\pi}_j(\theta) = E_j$. In the discrete case, we have $0 < p_i < 1$, $0 < \pi_i < 1$, and these quantities each sum to 1, as do their estimators. Then $I(\hat{f}, \hat{g})$ can be rewritten as

$$\sum_{j=1}^{k} (n_j/n) \log\left(\frac{n_j/n}{E_j/n}\right).$$

Now K-L distance between these (estimated) distributions can be written as

$$\frac{1}{n} \sum_{j=1}^{k} (n_j) \log\left(\frac{n_j}{E_j}\right),$$

or

$$\frac{1}{n} \sum_{j=1}^{k} O_j \log\left(\frac{O_j}{E_j}\right).$$

Thus, the G-statistic and K-L information differ by a constant multiplier of $2n$, i.e., in this context, $G = 2n \cdot I(\hat{f}, \hat{g})$. Similar relationships exist between K-L information expectations of likelihood ratio statistics for continuous data (G is a likelihood ratio test (LRT) for discrete data). Thus, the LRT is fundamentally related to the K-L distance.

## 2.6.2 Further Insights

Much of the research on model selection has been in regression and time series models, with some work being done in log-linear and classical multivariate (e.g., factor analysis) models. Bozdogan (1987) provides a recent review of the theory and some extensions. However, the number of published papers that critically examine the performance of AIC-selected models is quite limited. One serious problem with the statistical literature as regards the evaluation of AIC has been the use of Monte Carlo methods using only very simple generating models with a few large effects and no smaller, tapering effects. Furthermore, these Monte Carlo studies usually have the wrong objective, namely, to evaluate how often AIC selects the simple generating model—we believe that this misses the point entirely with respect to real data analysis. Such evaluations are often done even without regard for sample size (and often use AIC when $AIC_c$ should have been used).

In Monte Carlo studies it would be useful to generate data from a much more realistic model with several big effects and a series of smaller, tapering effects (Speed and Yu 1993). Then interest is refocused onto the selection of a good approximating model and its statistical properties, rather than trying to select the simple, artificial model used to generate the data. AIC attempts to select a best approximating model for the data at hand; if (as with reality) the "true model" is at all complex, its use would be poor for inference, even if it existed and its functional form (but not parameter values) were known (e.g., Sakamoto et al. 1986). This counterintuitive result occurs because the (limited) data would have to be used to estimate all the unknown parameters in the "true model," which would likely result in a substantial loss of precision (see Fig. 1.3B).

AIC reformulates the problem explicitly as a problem of *approximation* of the true structure (probably infinite-dimensional, at least in the biological sciences) by a *model*. Model selection then becomes a simple function minimization, where AIC (or more properly K-L information loss) is the criterion to be minimized. AIC selection is objective and represents a very different paradigm to that of hypothesis testing and is free from the arbitrary $\alpha$ levels, the multiple testing problem, and the fact that some candidate models might not be nested. The problem of what model to use is inherently not a hypothesis testing problem (Akaike 1974). However, the fact that AIC allows a simple comparison of models does not justify the comparison of all possible models (Akaike 1985 and Section 1.3.3). If one had 10 variables, then there are 1,024 possible models, even if interactions and squared or cubed terms are excluded. If sample size is $n \leq 1,000$, over-fitting the data is almost a certainty. It is simply not sensible to consider such a large number of models, because a model that over-fits the data will almost surely result, and the science of the problem has been lost. *Even in a very exploratory analysis it seems poor practice to consider all possible models; surely, some science can be brought to bear on such an unthinking approach* (otherwise, the scientist is superfluous and the work could be done by a technician).

### 2.6.3 Entropy

Akaike's (1977) term *"entropy maximization principle"* comes from the fact that the negative of K-L information is Boltzmann's entropy (in fact, K-L information has been called negative entropy or "negentropy"). Conceptually,

$$\text{Boltzmann's entropy} = -\log\left(\frac{f(x)}{g(x)}\right).$$

Then,

$$-\text{Boltzmann's entropy} = \log\left(\frac{f(x)}{g(x)}\right),$$

and

$$\begin{aligned}
\text{K-L} &= E_f(-\text{Boltzmann's entropy}) \\
&= E_f\left(\log\left(\frac{f(x)}{g(x)}\right)\right), \\
&= \int f(x) \log\left(\frac{f(x)}{g(x)}\right) dx.
\end{aligned}$$

Thus, minimizing the K-L distance is equivalent to maximizing the entropy; hence the name *maximum entropy principle* (see Jaynes 1957, Akaike 1983a, 1985 and

Bozdogan 1987, Jessop 1995 for further historical insights). However, maximizing entropy is subject to a constraint—the model of the information in the data. A good model contains the information in the data, leaving only "noise." It is the noise (entropy or uncertainty) that is maximized under the concept of the entropy maximization principle (Section 1.2.4). Minimizing K-L information then results in an approximating model that loses a minimum amount of information in the data. Entropy maximization results in a model that maximizes the uncertainty, leaving only information (the model) "maximally" justified by the data. The concepts are equivalent, but minimizing K-L distance (or information loss) certainly seems the more direct approach.

The K-L information is *averaged* entropy, hence the expectation with respect to $f$. While the theory of entropy is a large subject by itself, readers here can think of entropy as nearly synonymous with uncertainty.

Boltzmann derived the fundamental theorem that

**entropy is proportional to log(probability).**

Entropy, information, and probability are thus linked, allowing probabilities to be multiplicative, while information and entropies are additive. (This result was also derived by Shannon 1948). Fritz Hasenöhrl, a student of Boltzmann's, Boltzmann's successor at Vienna University, and a famous theoretical physicist himself, noted that this result "... is one of the most profound, most beautiful theorems of theoretical physics, indeed all of science." Further information concerning Boltzmann appears in Brush (1965, 1966), while interesting insights into Akaike's career are found in Findley and Parzen (1995).

## 2.6.4  A Summary

The principle of parsimony provides a conceptual guide to model selection, while K-L information provides an objective criterion, based on a deep theoretical justification. AIC and $AIC_c$ provide a practical method for model selection and associated data analysis and are estimates of relative K-L information. AIC, $AIC_c$, and TIC represent an extension of classical likelihood theory, are applicable across a very wide range of scientific questions, and AIC and $AIC_c$ are quite simple to use in practice.

## 2.7  Further Comments

Data analysis involves the question, "How complex a model will the data support?" and the proper tradeoff between bias and variance. The estimation of expected K-L distance is a natural and simple way to view model selection; given a good set of candidate models, select that model where information loss is minimized. Proper model selection is reflected in good achieved confidence interval coverage for the estimators in the model; otherwise, perhaps too much bias has been accepted in the tradeoff to gain precision, giving a false sense of high precision. This represents the worst inferential situation—a highly precise but quite biased estimate. These ideas have had a long history in statistical thinking.

### 2.7.1  A Heuristic Interpretation

After Akaike's elegant derivation of AIC, people noticed a heuristic interpretation that was both interesting and sometimes misleading. The first term in AIC,

$$\text{AIC} = -2\log(\mathcal{L}(\hat{\theta}\mid x)) + 2K,$$

is a measure of lack of model fit, while the second term ($2K$) can be interpreted as a "penalty" for increasing the size of the model (the penalty enforces parsimony in the number of parameters). This heuristic explanation does not do justice to the much deeper theoretical basis for AIC (i.e., the link with K-L distance and information theory). The heuristic interpretation led some statisticians to consider "alternative" penalty terms, and this has not always been productive (see Section 6.2). The so-called penalty term in AIC is not arbitrary; rather, it has a basis in information theory and is the result of Akaike's predictive expectation. [Note that of course, had Akaike defined $\text{AIC} = -\log(\mathcal{L}(\hat{\theta}\mid x)) + K$, the minimization would be unchanged; some authors use this expression, but we will use AIC as Akaike defined it.]

The heuristic view of the components of AIC clearly shows a bias versus variance tradeoff and insight into how the principle of parsimony is met by using AIC (see Gooijer et al. 1985:316). Still, we recommend viewing AIC as an estimate of the relative Kullback–Liebler distance between model pairs (i.e., each $g_i$ vs. $f$). Minimizing this relative, directed distance provides an estimated best approximating model for that particular data set (i.e., the *closest* approximating model to $f$). The relative Kullback–Liebler distance is the link between information theory and the log-likelihood function that is a critical element in AIC model selection.

### 2.7.2  Interpreting Differences Among AIC Values

Akaike's information criterion (AIC) and other information-theoretic methods can be used to rank the candidate models from best to worst. Often data do not support only one model as clearly best for data analysis. Instead, suppose three models are essentially tied for best, while another, larger, set of models is clearly not appropriate (either under- or over-fit). Such virtual "ties" for the best approximating model must be carefully considered and admitted. Poskitt and Tremayne (1987) discuss a "portfolio of models" that deserve final consideration. Chatfield (1995b) notes that there may be more than one model that is to be regarded as "useful." The inability to ferret out a single best model is not a defect of AIC or any other selection criterion; rather, it is an indication that the data are simply inadequate to reach such a strong inference. That is, the data are ambivalent concerning some effect or parametrization or structure.

## 2.7 Further Comments

It is perfectly reasonable that several models would serve nearly equally well in approximating a set of data. Inference must admit that there are sometimes competing models and the data do not support selecting only one. Using the principle of parsimony, if several models fit the data equally well, the one with the fewest parameters might be preferred; however, some consideration should be given to the other (few) competing models that are essentially tied as the best approximating model. Here the science of the matter should be fully considered. The issue of competing models is especially relevant in including model selection uncertainty into estimators of precision and model averaging (Chapter 4).

The use of the information-theoretic criteria vs. hypothesis testing in model selection can be quite different, and this is an important issue to understand. These differences can be illustrated by considering a set of nested candidate models, each successive model differing by one parameter. Model $M_i$ is the null model with $i$ parameters, and model $M_{i+j}$ is the alternative with $i+j$ parameters. Model $i$ is nested within model $i+j$; thus likelihood ratio tests (LRT) can be used to compare the null model with any of the alternative models $M_{i+j}$, where $j \geq 1$. Thus, if model $M_i$ has 12 parameters, then model $M_{i+1}$ has 13, model $M_{i+2}$ has 14, and so on.

This concept of a set of nested models is useful in illustrating some differences between AIC versus LRT for model selection. First, assume that the AIC value for each of the models is exactly the same; thus no model in the set has more support than any other model. Second, in each case we let the null hypothesis be model $M_i$ and assume that it is an adequate model for the data. Then, we entertain a set of alternative hypotheses, models $M_{i+j}$; these are each hypothesized to offer a "significantly" better explanation of the data. That is, $M_i$ (the null) is tested individually against the $j \geq 1$ alternative models in the set. The first test statistic ($M_i$ vs. $M_{i+1}$) here is assumed to be distributed as $\chi^2$ with 1 df, while the second test statistic ($M_i$ vs. $M_{i+2}$) has an assumed $\chi^2$ distribution with 2 df, and so on. The following relations will be useful:

$$\text{AIC}_i = -2\log(\mathcal{L}_i) + 2i,$$
$$\text{AIC}_{i+j} = -2\log(\mathcal{L}_{i+j}) + 2(i+j),$$
$$\text{LRT} = -2\big(\log(\mathcal{L}_i) - \log(\mathcal{L}_{i+j})\big) \text{ with } j \text{ df.}$$

Then, in general,

$$\text{LRT} = \text{AIC}_i - \text{AIC}_{i+j} + 2j.$$

Now, for illustration of a point about the difference between LRTs and AIC in model selection, assume

$$\text{AIC}_i \equiv \text{AIC}_{i+j}.$$

If this boundary condition were to occur (where K-L–based selection is indifferent to the model), then we would have,

$$\text{LRT} = 2j \text{ on } j \text{ degrees of freedom.}$$

TABLE 2.1. Summary of P-values (i.e., Prob$\{\chi^2 \geq 2\mathrm{df} = 2j\}$) for likelihood ratio tests between two nested models where the two corresponding AIC values are equal, but the number of estimable parameters differs by $j$ (after Sakamoto et al. 1986).

| $j$ | $\chi^2$ | P |
|---|---|---|
| 1 | 2 | 0.157 |
| 2 | 4 | 0.135 |
| 3 | 6 | 0.112 |
| 4 | 8 | 0.092 |
| 5 | 10 | 0.075 |
| 6 | 12 | 0.062 |
| 7 | 14 | 0.051 |
| 8 | 16 | 0.042 |
| 9 | 18 | 0.035 |
| 10 | 20 | 0.029 |
| 15 | 30 | 0.012 |
| 20 | 40 | 0.005 |
| 25 | 50 | 0.005 |
| 30 | 60 | 0.001 |

Now, a difference of 1 df between $M_i$ vs. $M_{i+1}$ corresponds to a $\chi^2$ value of 2 with 1 df, and a P-value of 0.157 (Table 2.1). Similarly, a difference of 4 df ($j = 4$) between $M_i$ vs. $M_{i+4}$ corresponds to a $\chi^2$ value of 8 and a P value of 0.092. If the df $\leq$ about 7 (assuming $\alpha = 0.05$), then hypothesis testing methods support the null model ($M_i$) over any of the alternative models ($M_{i+1}$, $M_{i+2}$, $M_{i+3}$, ...) (Table 2.1). This result is in contrast with AIC-selection, where in this example all the models are supported equally.

Test results change in this scenario when there are $> j = 8$ additional parameters in the alternative model (Table 2.1). Here, the null model ($M_i$) is rejected with increasing strength, as the alternative model has an increasing number of parameters. For example, the likelihood ratio test of $M_i$ vs. $M_{i+10}$ has 10 df, $\chi^2 = 20$, and P = 0.029. More striking is the test of $M_i$ vs. $M_{i+30}$, which has 30 df, $\chi^2 = 60$, and P = 0.001, even though the AIC value is the same for all the models (the null and the various alternatives). In these cases (i.e., $> 8$ parameters difference between the null and alternative model), the testing method indicates increasingly strong support of the models with many parameters and strong rejection of the simple, null model $M_i$ (see Sakamoto 1991 and Sakamoto and Akaike 1978:196 for additional insights on this issue).

More extreme differences between the two approaches can be shown by letting $\text{AIC}_i = \text{AIC}_{i+j} - x$ for $x$ in the range of about 0 to 4. It is convenient to work with the $\Delta_{i+j}$ values; then relative to the selected model, $\Delta$ for model $M_{i+j}$ is $x$. If $x = 4$, the choice of model $M_i$ is compelling in the context of nested models, as judged by AIC. For comparison, the LRT statistic is $2j - x$. Let $x = 4$ and $j = 20$; then the LRT statistic is 36 on 20 df and P = 0.0154. Most would take this P-value as compelling evidence for the use of model $M_{i+j}$. Thus, AIC can clearly support the simple model $M_i$, while LRT can clearly support model $M_{i+j}$ with 20 additional parameters. The solution to this dilemma is entirely a matter of which the model selection approach has a sound theoretical basis: Information criteria based on K-L information does; likelihood ratio testing does not.

Those individuals holding the belief that the results of hypothesis tests represent a "gold standard" will be surprised at the information in Table 2.1 and may even believe that AIC "loses power" as the difference in parameters between models increases beyond about 7. [Note: The concept of "power" has no utility in the information-theoretic approach because it is not a "test" in any way.] Akaike (1974) noted, "The use of a fixed level of significance for the comparison of models with various numbers of parameters is wrong, since it does not take into account the increase of the variability of the estimates when the number of parameters increased." The $\alpha$-level should be related to sample size and the degrees of freedom if hypothesis testing is to be somehow used as a basis for model selection (see Akaike 1974). However, the $\alpha$-level is usually kept fixed, regardless of sample size or degrees of freedom, in the hypothesis testing approach. This practice of keeping the $\alpha$-level constant corresponds to asymptotically inconsistent results from hypothesis testing. For example, if the null hypothesis is true and $\alpha$ is fixed (at, say, 0.05), then even as the degrees of freedom approach $\infty$ we still have a 0.05 probability of rejecting the null hypothesis, even with near infinite sample size. The inconsistency is that statistical procedures in this simple context should converge on truth with probability 1 as $n \to \infty$.

AIC provides a ranking of the models; thus the analyst can determine which model is best, which are essentially tied for best, and which models are clearly in an inferior class (and perhaps some that are in an intermediate class). These ranks are, of course, estimates based on the data. Still, the rankings are quite useful (cf. Section 2.5 and Sakamoto et al. 1986:84) and suggest that primary inference be developed using the model for which AIC is minimized or the small number of models where there is an essential tie for the minimum AIC (i.e., within about 1 or 2 to perhaps as many as 3 or 4 AIC units from the minimum for nested models successively differing by one parameter). In the context of a string of nested models, when there is a single model that is clearly superior (say, the next best model is > 7–10 AIC units from the minimum) there is little model selection uncertainty, and the theoretical standard errors can be used (e.g., Flather's data in Sections 1.2.2 and 2.9). When the results of model selection are less clear, then methods described in Chapter 4 can be considered. AIC allows a ranking of models and the identification of models that are nearly equally useful versus those that are clearly poor explanations for the data at hand. Hypothesis testing provides no way to rank models, even for models that are nested.

## 2.7.3 Nonnested Models

A substantial advantage in using AIC is that it is valid for nonnested models (e.g., Table 2.2). Of course, traditional likelihood ratio tests are defined only for nested models, and this represents a substantial limitation in the use of hypothesis testing in model selection. The ranking of models using AIC helps clarify the importance of modeling (Akaike 1973: 173); for example, some models for a particular data set are simply poor and should not be used for inference.

A well thought out global model (where applicable) is very important, and substantial prior knowledge is required during the entire survey or experiment, including the clear statement of the question to be addressed and the collection of the data. This prior knowledge is then carefully input into the development of the set of candidate models (Section 1.4.3). Without this background science, the entire investigation should probably be considered only very preliminary.

### 2.7.4 Model Selection Uncertainty

One must keep in mind that there is often considerable uncertainty in the selection of a particular model as the "best" approximating model. The observed data are conceptualized as random variables; their values would be different if another, independent set were available. It is this "sampling variability" that results in uncertain statistical inference from the particular data set being analyzed. While we would like to make inferences that would be robust to other (hypothetical) data sets, our ability to do so is still quite limited, even with procedures such as AIC, with its cross validation properties, and with independent and identically distributed sample data. Various computer-intensive resampling methods will further improve our assessment of the uncertainty of our inferences, but it remains important to understand that proper model selection is accompanied by a substantial amount of uncertainty. The bootstrap technique can effectively allow insights into model uncertainty; this and other similar issues are the subject of Chapters 4 and 5.

### 2.7.5 AIC When Different Data Sets Are to Be Compared

Models can be compared using AIC only when they have been fitted to exactly the same set of data (this applies also to likelihood ratio tests). For example, if nonlinear regression model A is fitted to a data set with $n = 140$ observations, one cannot validly compare it with model B when 7 outliers have been deleted, leaving only $n = 133$. Furthermore, AIC cannot be used to compare models where the data are ungrouped in one case (model U) and grouped (e.g., grouped into histograms classes) in another (model G).

### 2.7.6 Order Not Important in Computing AIC Values

The order in which AIC is computed over the set of models is clearly not important. Often, one may want to compute AIC starting with the global model and proceed to simpler models with fewer parameters. Others may wish to start with the simple models and work up to the more general models with many parameters; this might be best if numerical problems are encountered when fitting some models. The order is irrelevant here, as opposed to the various hypothesis testing approaches where the order may be both arbitrary and the results quite dependent on the choice of order (e.g., stepup (forward) vs. stepdown (backward) testing).

Hypothesis testing is commonly used in the early phases of exploratory analysis to iteratively seek model structure and understanding. Here, one might start with 5–8 models, compute various test statistics for each, and note that several of the better models each have a gender effect. Thus, additional models are derived to include a gender effect, and more tests are conducted. Then the analyst notes that several of these models have a trend in time for some parameter set; thus more models with this effect are derived, and so on. While this procedure violates several theoretical aspects of hypothesis testing, it is commonly used, and if the results are treated only as alternative hypotheses for a more confirmatory study to be conducted later, might be an admissible practice. We suggest that the information-theoretic criteria might serve better as a data dredging tool; at least key assumptions upon which these criteria are based are not terribly violated. Data dredging using an information-theoretic criterion instead of some form of test statistic eliminates statistical inferences as P-values from being made, but one must still worry about over-fitting and spurious effects. The ranking of alternative models (the $\Delta_i$ values) might be useful in preliminary work. While we do not condone the use of AIC in data dredging, we suggest that it might be a more useful tool than hypothesis testing in the analysis of very preliminary data where little *a priori* knowledge is available.

## 2.7.7 Hypothesis Testing Is Still Important

A priori hypothesis testing plays an important role when a formal experiment (i.e., treatment and control groups being formally contrasted in a replicated design with random assignment) has been done and specific *a priori* alternative hypotheses have been identified. In these cases, there is a very large body of statistical theory on testing of treatment effects in such experimental data. We certainly acknowledge the value of traditional testing approaches to the analysis of these *experimental* data. Still, the primary emphasis should be on the size of the treatment effects and their precision; too often we find a statement regarding "significance" while the treatment and control means are not even presented. Akaike (1981) suggests that the "multiple comparison" of several treatment means should be viewed as a model selection problem, rather than resorting to one of the many testing methods that have been developed. Here, *a priori* considerations would be brought to bear on the issue and a set of candidate models derived, letting AIC values aid in sorting out differences in treatment means — a refocusing on parameter estimation, instead of on testing.

In observational studies, where randomization and replication are not achievable, we believe that "data analysis" should be largely viewed as a problem in model selection and associated parameter estimation. This seems especially the case where nuisance parameters are encountered in the model, such as the recapture or resighting probabilities in capture–recapture or band recovery studies. Here, it is not always clear what either the null or the alternative hypothesis should be in a hypothesis testing framework. In addition, often hypotheses that are tested are naive, as Johnson (1995) points out with such simplicity (e.g., is there any reason to test formally hypotheses such as "$\mathcal{H}_0$: the number of robins is the same in cities A and B"? — of course not!). One should merely assume that the number is different and proceed to estimate the magnitude of the difference: an estimation problem, not a hypothesis testing problem.

Finally, we advise that the theories underlying the information-theoretic approaches and hypothesis testing are fundamentally quite different. AIC is not a "test" in any sense, and there are no associated concepts such as test power or $\alpha$-levels; statistical hypothesis testing represents a very different paradigm. The results of model selection under the two approaches might happen to be similar with simple problems and a large amount of data; however, in more complex situations, with many candidate models and less data, the results of the two approaches can be quite different (see Section 3.5). It is critical to bear in mind that there is a theoretical basis to information-theoretic approaches to model selection criteria.

## 2.8 Comparisons with Other Criteria

Under the frequentist paradigm for model selection, one generally has three main approaches: (I) optimization of some selection criteria, (II) tests of hypotheses, and (III) ad hoc methods. One has a further classification under (I): (1) criteria based on some form of mean squared error (e.g., Mallows's $C_p$, Mallows 1973) or mean squared prediction error (e.g., PRESS, Allen 1970), (2) criteria that are estimates of K-L information or distance (e.g., TIC and the special cases AIC and $AIC_c$), and (3) criteria that are consistent estimators of $K$, the dimension of the "true model" (e.g., BIC). We will explore (2) and (3) in the following material.

### 2.8.1 Information Criteria That Are Estimates of K-L Information

AIC, $AIC_c$, and $QAIC_c$ are estimates of the relative K-L distance between truth $f(x)$ and the approximating model $g(x)$. These criteria were derived based on the concept that truth is very complex and that no "true model" exists (or at least, it was immaterial to the argument). Thus, one could only *approximate* truth with a model, say $g(x)$. Given a good set of candidate models for the data, one could estimate which approximating model was best (among those candidates considered, given the data and their sample size). Linhart and Zucchini (1986) speak of "approximating families" of models. Hurvich and Tsai (1994) explain that these criteria select the best finite-dimensional approximating model in large samples when truth is infinite-dimensional. The basis for these criteria seem reasonable in the biological sciences.

If one had sample sizes that were quite large, there have been other criteria derived that might offer advantages in model selection and inference (e.g., TIC in Chapter 6). These criteria specifically allow for "misspecification" of the ap-

## 2.8 Comparisons with Other Criteria

proximating models: the fact that the set of candidate models does not include $f(x)$, or any model very close to $f(x)$. Here we will note 4 criteria, even though their operating properties have received little attention in the published statistical literature (but see Konishi and Kitagawa, 1996).

Takeuchi (1976) provides a general derivation from K-L information to AIC. An intermediate result indicated a selection criterion useful when the candidate models were not particularly close approximations to $f$. He derived TIC (Takeuchi's information criterion) for model selection that has a more general bias adjustment term to allow $-2\log_e(\mathcal{L})$ to be adjusted to be an asymptotically unbiased estimate of relative K-L,

$$\text{TIC} = -2\log(\mathcal{L}) + 2 \cdot \text{tr}\big(J(\theta)I(\theta)^{-1}\big).$$

The matrices $J(\theta)$ and $I(\theta)$ involve first and second mixed partial derivatives of the log-likelihood function, and "tr" denotes the matrix trace function. AIC is only an approximation to TIC, where $\text{tr}\big(J(\theta)I(\theta)^{-1}\big) \doteq K$. The approximation is excellent when the approximating model is quite "good" and becomes poor when the approximating model is poor. One might consider always using TIC and worry less about the adequacy of the models in the set of candidates. This consideration involves two issues that are problematic. First, one must always worry about the quality of the set of approximating models being considered; this is not something to shortcut. Second, using the expanded bias adjustment term in TIC involves estimation of the elements of the matrices $J(\theta)$ and $I(\theta)$ (details provided in Chapter 6). Shibata (in prep.) notes that estimation error of these two matrices can cause instability of the results of model selection (note that the matrices are of dimension $K \times K$). If overdispersion is found in count data, then the log-likelihood could be divided by an estimated variance inflation factor, given QTIC. In most practical situations, AIC and $\text{AIC}_c$ are very useful approximations to relative K-L information.

Linhart and Zucchini (1986) proposed a further generalization, and Amari (1993) proposed a network information criterion (NIC) potentially useful in training samples in neural network models. Shibata (in prep.) provides technical details on these advanced methods, while Konishi and Kitagawa (1996) suggest even more general criteria for model selection and provide further insights into AIC and TIC and their derivation. Shibata (1989) developed a complicated criterion, based on the theory of penalized likelihoods. His method has been called RIC for "regularized information criterion." We will not explore these methods, as they would take us too far afield from our stated objectives and they do not have the direct link with information theory and the estimation of relative K-L distance. However, we note that almost no work has been done to evaluate the utility of these extensions in applied problems. Surely, the use of these criteria must be reserved for problems where the sample size is quite large and good estimates of the elements of the matrices ($I(\theta)$ and $J(\theta)$) in the bias adjustment term are available.

[Mallows's $C_p$ (Mallows 1973, 1995) statistic is well known for variable selection, but limited to LS regression problems with normal errors. However, $C_p$ lacks any direct link to K-L information. Atilgan (1996) provides a relationship between AIC and Mallows's $C_p$, shows that under some conditions AIC selection behaves like minimum mean squared error selection, and notes that AIC and $C_p$ are somewhat equivalent criteria. When the usual multiple linear regression assumptions hold, the two criteria seem to select the same model and rank the contending models in the same order, but they are not equivalent. We have not found a small-sample version of $C_p$ that would be useful when the sample size is small compared to the number of regressor variables (like $AIC_c$) (see Fujikoshi and Satoh 1997). Ronchetti and Staudte (1994) provide a robust version of $C_p$ (also see Sommer and Huggins 1996). Of course, adjusted $R^2$ has been used in classical multiple linear regression analysis.]

## 2.8.2 Criteria That Are Consistent for K

This section deals with a class of criteria used in model selection that are "consistent" or "dimension consistent" and with how these criteria differ from those that are estimates of Kullback–Liebler information. Several criteria have been developed, based on the assumptions that an exactly "true model" exists, that it is one of the candidate models being considered, and that the model selection goal is to select the *true* model. Implicit is the assumption that truth is of fairly low dimension (i.e., $K = 1-5$ or so) and that $K$ is fixed as sample size increases. Here, the criteria are derived to provide a consistent estimator of the order or dimension $(K)$ of this "true model," and the probability of selecting this "true model" approaches 1 as sample size increases. Bozdogan (1987) provides a nice review of many of the "dimension consistent" criterion. The best known of the "dimension consistent criteria" was derived by Schwarz (1978) in a Bayesian context and is termed BIC for Bayesian information criterion (or occasionally SIC for Schwarz's information criterion); it is simply

$$\text{BIC} = -2\log(\mathcal{L}) + K \cdot \log(n).$$

BIC arises from a Bayesian viewpoint with equal priors on each model and very vague priors on the parameters, given the model. The assumed purpose of the BIC-selected model was often simple prediction; as opposed to scientific understanding of the process or system under study. BIC is not an estimator of relative K-L.

Rissanen (1989) proposed a criterion he called minimum description length (MDL), based on coding theory, another branch of information theory (see Yu 1996 for a recent review). While the derivation and its justification are difficult to follow without a strong background in coding theory, his criterion are equivalent to BIC. Hannan and Quinn (1979) derived a criterion (HQ) for model selection whereby the penalty term was

$$c \cdot \log(\log(n)),$$

where $n$ is sample size and $c$ is a constant $> 2$ (see Bozdogan 1987:359). This criterion, while often cited, seems to have seen little use in practice. Bozdogan (1987) proposed a criterion he called CAICF (C denoting "consistent" and F denoting the use of the Fisher information matrix),

$$\text{CAICF} = -2\log(\mathcal{L}) + K\{\log(n) + 2\} + \log|I(\hat{\theta})|,$$

where $\log|I(\hat{\theta})|$ is the natural logarithm of the determinant of the estimated Fisher information matrix. He has recently advanced a somewhat similar criterion based on a notion of complexity (ICOMP, Bozdogan 1988). Neither CAICF or ICOMP are invariant to 1-to-1 transformations of the parameters, and this feature would seem to limit their application. AIC, $\text{AIC}_c$, QAIC, and TIC are invariant to 1-to-1 transformations.

We question (deny, actually) the concept of a simple "true model" in the biological sciences (see the Preface) and would surely think it unlikely that even if a "true model" existed, it might be included in the set of candidate models! If an investigator knew that a true model existed and that it was in the set of candidate models, would not he know which one it was? We see little utility in these criteria in the biological, social, or medical sciences, although they have seen frequent application. Relatively few people seem to be aware of the differences in the basis and assumptions for these dimension-consistent criteria relative to criteria that are estimates of K-L information. The dimension-consistent criteria are directed at a very different problem than criteria that are estimates of K-L.

People have often (mis) used Monte Carlo methods to study the various criteria, and this has been the source of confusion in some cases (such as in Rosenblum 1994). In Monte Carlo studies, one *knows* the generating model and often considers it to be "truth." The generating model is nearly always quite simple, and it is included in the set of candidate models. In the analysis of the simulated data, attention is (mistakenly) focused on what criterion most often finds this true model (e.g., Bozdogan 1987, Fujikoshi and Satoh 1997, Ibrahim and Chen 1997). Under this objective, we would suggest the use of the dimension-consistent criteria in this artificial situation, especially if the order of the true model was quite low (e.g., $K = 3\text{–}5$), or the residual variation ($\sigma^2$) was quite small, or the sample size was quite large. However, this contrived situation is far from that confronted in the analysis of empirical data in the biological sciences. Monte Carlo studies to evaluate model selection approaches to the analysis of real data must employ generating models with a range of tapering effect sizes and substantial complexity. Such evaluations should then focus on selection of a best approximating model and ranking of the candidate models; the notion that the true (in this case, the generating) model is in the set should be discarded.

Research into the dimension-consistent criteria has often used a generating model with only a few large effects. More realistic models employing a range of tapering effects have been avoided. In addition, the basis for the dimension-consistent criteria assumes that the true model remains fixed as sample size approaches infinity. In biological systems increased sample size stems from the addition of new geographic field sites or laboratories, the inclusion of additional years, and the inclusion of new animals with genetic variation over individuals. Thus, as substantial increases in sample size are achieved, the number of factors in the model also increases. The "true model" does not remain fixed as $n \to \infty$. We have found that the dimension-consistent criteria perform poorly in open population capture–recapture models even in the case where $K$ is small, but the parameters reflect a range of effect sizes (Anderson et al. 1998).

Notwithstanding our objections above, the sample sizes required to achieve the benefits of dimension-consistent estimation of model order ($K$) are often very, very large by any usual standard. In the examples we have studied (that have substantial residual variances), we have seen the need for sample sizes in the thousands or much more before the consistent criteria begin to point to the "true model" with a high probability. In cases where the sample size was very large, say 100,000, one might merely examine the ratios $\hat{\theta}/\widehat{\text{se}}(\hat{\theta})$ to decide on the parametrization, with little regard for the principle of parsimony (given the assumption that the true model is being sought, and it *is* in the set of candidates). It should be emphasized that these dimension-consistent criteria are not linked directly to K-L information and are "information-theoretic" only in the weakest sense. Instead, their motivation veered to consistent estimation of the order ($K$) of the supposed "true model" by employing alternative penalty terms (but see Section 2.7.1).

When sample size is less than very large for realistic sorts of biological data, these dimension-consistent criteria tend to select under-fitted models with the attendant large bias, overestimated precision, and associated problems in inference. Umbach and Wilcox (1996:1341) present the results of Monte Carlo simulations conducted under the BIC-type assumptions. For sample size up to 100,000, AIC performed better than BIC in terms of the selected set coinciding with the "correct" set. The two criteria were tied for sample size $= 125,000$. However, even at that large sample size, BIC only selected the "correct" set in 79% of the cases; this is still far from selecting the correct set with probability $= 1$. While these criteria might be useful in some of the physical sciences and engineering, we suspect that they have relatively little utility in the biological and social sciences or medicine. Findley (1985) notes that "... consistency can be an undesirable property in the context of selecting a model."

### 2.8.3 Contrasts

As Reschenhofer (1996) notes, regarding criteria that are estimates of relative K-L information vs. criteria that are dimension consistent, they "... are often employed in the same situations, which is in contrast to the fact that they have been designed to answer different questions" (also see Potshcher 1991, Hurvich and Tsai 1995a and 1996, and Anderson and Burnham 1999b). In the biological and social sciences and medicine, we argue that the AIC-type criteria (e.g., AIC, $AIC_c$, QAIC, $QAIC_c$, and TIC) are reasonable for the analysis of empirical data. The dimension-consistent criteria (e.g., BIC, MDL, HQ, CAICF, and ICOMP) might find use in some physical sciences where a simple true model might exist and where sample size is quite large (perhaps thousands or tens of thousands, or more). Still, we question the likelihood that this true model would be in the set of candidate models. Even in cases where a simple true model exists and it is contained in the set of candidates, AIC might frequently have better inferential properties than the dimension-consistent criteria.

Still other, somewhat similar criteria have been derived (see Sclove 1987, 1994 and Stoica et al. 1986 for recent reviews). A large number of other methods have appeared including the lasso (Tibshirani 1996), the little bootstrap (Breiman 1992), the nonnegative garrote (Breiman 1995), predictive least quasi-deviance (Qian et al. 1996), various Bayesian methods (e.g., Ibrahim and Chen 1997) including the use of Gibbs sampling (George and McCulloch 1993). Some of these approaches seem somewhat ad hoc, while others are difficult to understand, interpret, or compute. Often the methods lack generality; for example, several are applicable only to regression-type models. We will not pursue these methods here, as they take us too far from our objectives.

In summary, we recommend the class of information-theoretic criteria that are estimates of relative K-L information such as AIC, $AIC_c$ for general use in the selection of a parsimonious approximating model for statistical inference. If count data are found to be overdispersed, then QAIC and $QAIC_c$ are useful. If large samples are available, then TIC might offer an improvement over AIC or $AIC_c$. However, our limited investigations suggest that the simpler criteria perform as well as TIC in cases we examined (Chapter 6). We cannot recommend the dimension-consistent criteria for the analysis of real data.

## 2.9 Return to Flather's Models

We now extend the example in Chapter 1 where 9 models for the species-accumulation curve for data from Indiana and Ohio were analyzed by Flather (1992, 1996). The simple computation of AIC was done by hand from the regression output from program NLIN in SAS (SAS Institute, Inc. 1985). In this case, apart from a constant,

$$\text{AIC} = n \cdot \log(\hat{\sigma}^2) + 2K,$$

where $\hat{\sigma}^2 = \text{RSS}/n$ and $K$ is the number of regression parameters plus 1 (for $\sigma^2$). AIC values for the 9 models are given in Table 2.2. The last model is clearly the best approximating model for these data. Values of $\Delta_i = \text{AIC}_i - \min \text{AIC} = \text{AIC}_i + 585.48$ are also given and allow the nonoptimal values to be more easily interpreted. Here, the second- and third-best models are quickly identified (corresponding to $\Delta_i$ values of 163.40 and 542.63, respectively); however, these $\Delta$ values are very large, and the inference here is that the final model is clearly the best of the candidate models considered for these specific data. This conclusion seems to be born out by Flather (1992), as he also selected this model based on a careful analysis of residuals for each of the 9 models and Mallows's $C_p$. The remaining question is whether a still better model might have been postulated with 6 or 7 parameters and increased structure. Of course, information criteria only attempt to select the best model from the candidate models available; if a better model exists, but is not offered as a candidate, then the information-theoretic approach cannot be expected to identify this new model.

TABLE 2.2. Summary of nine *a priori* models of avian species-accumulation curves from the Breeding Bird Survey (from Flather 1992 and 1996). Models are shown, including the number of parameters ($K$), AIC values, and $\Delta_i = \text{AIC}_i - \min$ AIC values for the Indian–Ohio Major Land Resource Area. AIC is computed for each model; the order is not relevant. Here the models are shown in order according to the number of parameters ($K$). However, this is only a convenience. This elaborates on the example in Table 1.1.

| Model | Number of parameters[a] | AIC value | $\Delta_i$ |
|---|---|---|---|
| $ax^b$ | 3 | 227.64 | 813.12 |
| $a + b\log(x)$ | 3 | 91.56 | 677.04 |
| $a\big(x/(b+x)\big)$ | 3 | 350.40 | 935.88 |
| $a(1 - e^{-bx})$ | 3 | 529.17 | 1114.65 |
| $a - bc^x$ | 4 | 223.53 | 809.01 |
| $(a + bx)/(1 + cx)$ | 4 | 57.53 | 643.01 |
| $a(1 - e^{-bx})^c$ | 4 | -42.85 | 542.63 |
| $a\big(1 - [1 + (x/c)^d]^{-b}\big)$ | 5 | -422.08 | 163.40 |
| $a[1 - e^{-(b(x-c))^d}]$ | 5 | -585.48 | 0 |

[a] $K$ is the number of parameters in the regression model plus 1 for $\sigma^2$.

## 2.10 Summary

The research problem must be carefully stated, followed by careful planning, proper sampling, and an adequate data gathering program. Then, the science of the matter, experience, and expertise are used to define an *a priori* set of candidate models. These are philosophical issues that must receive increased attention in the future. The basis for the information-theoretic approach to model selection is Kullback–Liebler information,

$$I(f, g) = \int f(x) \log\left(\frac{f(x)}{g(x \mid \theta)}\right) dx.$$

$I(f, g)$ is the "information" lost when the model $g$ is used to approximate full reality, or truth, $f$. An equivalent interpretation of $I(f, g)$ is a "distance" between the approximating model $g$ and full truth or reality, $f$. Under either interpretation, we seek to find a candidate model that minimizes $I(f, g)$ over the candidate models. This is a conceptually simple, yet powerful, approach.

Clearly, $I(f, g)$ cannot be used directly, because it requires knowledge of full truth, or reality. Akaike (1973), in a landmark paper, provided a way to *estimate* relative $I(f, g)$, based on the empirical log-likelihood function. He found that the maximized log-likelihood value was a biased estimate of the relative Kullback–Liebler information and that under certain conditions, this bias was approximately equal to $K$, the number of estimable parameters in the approximating model, $g$. His method, *Akaike's information criterion* (AIC), allowed model selection to be firmly based on a fundamental theory and opened to door to further conceptual work. AIC is an approximation to relative Kullback–Liebler information (or distance); better approximations (AIC$_c$) were offered by Sugiura (1978) and Hurvich and Tsai (1989 and several subsequent papers). Takeuchi (1976) derived an asymptotically unbiased estimator of Kullback–Liebler information that applies in general (i.e., without the special conditions underlying the direct derivation of AIC), but his method (TIC for Takeuchi's information criterion) requires large sample sizes to estimate elements of $K \times K$ matrices in the bias-adjustment term applied to $\log_e(\mathcal{L})$. Still, TIC is an important conceptual advance. Thus, investigators working in applied data analysis have several powerful methods for selecting a model for making inferences from data to the population or process of interest.

Following Akaike's finding that the maximized log-likelihood value was a positively biased estimator of relative Kullback–Liebler information, we can summarize three main approaches to adjusting for this bias (the bias adjustment term is subtracted from the maximized log-likelihood),

| Criterion | Bias adjustment term |
|---|---|
| AIC | $K$ |
| AIC$_c$ | $K + \frac{K(K+1)}{n-K-1}$ |
| TIC | $\text{tr}\left(J(\theta)I(\theta)^{-1}\right)$ |

When sample size ($n$) is small compared to $K$, AIC$_c$ is an improved approximation (2nd order) relative to AIC. Both criteria are easy to compute, quite effective in many applications, and we recommend their use. Neither AIC nor AIC$_c$ is asymptotically unbiased in general; instead, they are good approximations when there are "good" models in the candidate set.

TIC allows for substantial model misspecification; perhaps none of the candidate models are very good. In this case, AIC and AIC$_c$ may not perform as well as TIC (especially if sample size is quite large). TIC is much more complicated to compute because its bias-adjustment term involves the estimation of $K \times K$ matrices of first and second partial derivatives, $J(\theta)$ and $I(\theta)$, and the inversion of the matrix $I(\theta)$. The benefit of TIC is that one achieves asymptotic unbiasedness for K-L model selection. The disadvantage with TIC is that the estimate of $\text{tr}(J(\theta)I(\theta)^{-1})$ may be, in a sense, over-fit (unstable) with small or moderate sample sizes. In such cases, the bias adjustment terms $K$ or $K + \frac{K(K+1)}{n-K-1}$, corresponding to AIC and AIC$_c$, represent a *parsimonious* approach to bias correction! [In fact, if $f$ was assumed to be in the set of candidate models, then for that model $\text{tr}(J(\theta)I(\theta)^{-1}) \equiv K$. If the set of candidate models included a very good, or even just good, model, then $\text{tr}(J(\theta)I(\theta)^{-1})$ is approximately $K$ for those models.]

The maximized log-likelihood for poor models will be much less than that for good models. Recall that the first term in AIC (or AIC$_c$) measures lack of fit. Models that fit poorly have smaller maximized log-likelihood values, and AIC will be appropriately large. The bias correction in AIC or AIC$_c$ may be less than perfect (because TIC could not be used; probably because a very large sample was not available for analysis), but this is a relatively minor issue for models that are poor. For models that are "good," TIC reduces to AIC.

In practice, one need not assume that the "true model" is in the set of candidates (although this is sometimes mistakenly stated in the technical literature). Similarly, one often sees that AIC is an asymptotically unbiased estimate of relative K-L; this is not true without assuming that $f$ is in the set of candidate models. Takeuchi's TIC is an asymptotically unbiased estimate of relative K-L. AIC and $AIC_c$ are approximations, given certain conditions, but are very useful in many practical situations.

The principle of parsimony provides a philosophical basis for model selection, K-L information provides an objective target based on deep theory, and AIC, $AIC_c$, or TIC provides an estimator of relative K-L information. Objective model selection may be rigorously based on these principles. In any event, we recommend presentation of $\log(\mathcal{L}(\hat{\theta}))$, $K$, the appropriate information criterion (AIC, $AIC_c$, or TIC), and $\Delta_i$ for various models considered in research papers. Additional theory (Chapter 4) allows further insights into model selection uncertainty, based on the $\Delta_i$ and other considerations.

# 3
# Practical Use of the Information-Theoretic Approach

Model building and data analysis in the biological sciences somewhat presupposes that the person has some advanced education in the quantitative sciences, and statistics in particular. This requirement also implies that a person has substantial knowledge of statistical hypothesis-testing approaches. Such people, including ourselves over the past several years, often find it difficult to understand the information-theoretic approach, only because it is conceptually so very different from the testing approach that is so familiar. Relatively speaking, the concepts and practical use of the information-theoretic approach are much simpler than those of statistical hypothesis testing, and very much simpler than some of the various Bayesian approaches to data analysis (e.g., Laud and Ibrahim 1995 and Carlin and Chib 1995).

While the derivation of AIC (Chapter 6) lies deep in the theory of mathematical statistics, its application is quite simple. The initial example is a simple multiple linear regression model of cement hardening and is a classic example in the model-selection literature. The remaining examples in this chapter focus on more complex data sets and models. These examples will provide insights into real-world complexities and illustrate the ease and general applicability of AIC in model selection. Several of these examples are followed in later chapters as additional concepts and methods are provided. Several examples deal with survival models, as that has been one of our research interests.

## 3.1 Computation and Interpretation of AIC Values

Given a model, likelihood inference provides a quantitative assessment of the "strength of evidence" in the data regarding the plausible values of the parameters in the model (Royall 1997). Given a well-developed set of *a priori* candidate models, information-theoretic methods provide a quantitative assessment of the "strength of evidence" in the data regarding the plausibility of which model is "best." For example, AIC can be computed for each of the models in the set

$$\text{AIC} = -2\log(\mathcal{L}(\hat{\theta} \mid x)) + 2K,$$

where $\log(\mathcal{L}(\hat{\theta} \mid x))$ is the value of the log-likelihood function at its maximum (i.e., evaluated at the MLEs) and $K$ is the total number of estimable parameters in the particular model. Such computation can be done easily and objectively; the order in which AIC is computed for the various candidate models is immaterial. Usually, the investigator does not examine the parameter estimates or associated measures of precision during the computation of AIC for each model. The model with the lowest AIC value can be selected for inference (additional considerations are found in Chapters 4 and 5). Model selection is thus a matter of function minimization (see Sakamoto et al. 1986, Sakamoto and Akaike 1978), just as ML estimation involves function maximization. AIC can be computed and interpreted without the aid of subjective judgment (e.g., $\alpha$-levels or Bayesian priors) once a set of candidate models has been derived. **If the ratio $n/K$ is small, say $< 40$, then $\text{AIC}_c$ should be employed.** Similarly, if there is evidence of overdispersion in some types of models of count data, then QAIC or $\text{QAIC}_c$ should be used. If a large sample is available for analysis, then TIC might be considered for model selection. Generally, we encourage the use of $\Delta_i$ values (using AIC, $\text{AIC}_c$, $\text{QAIC}_c$, or TIC) in the publication of results because the AIC values (or $\text{AIC}_c$, $\text{QAIC}_c$, or TIC values) themselves are on a relative scale.

It is important to note that to compute and interpret $\text{AIC}_c$ (or $\text{QAIC}_c$) in model comparison, all models must have the same response variable (say, $y$). In linear regression, for example, one cannot directly compare a model fitted as $y = \beta_0 + \beta_1 x + \epsilon$ with one fitted as $\log(y) = \beta_0 + \beta_1 x + \epsilon$ using information-theoretic methods. Instead, one must fit the second model as $y = \exp(\beta_0 + \beta_1 x) + \epsilon$; this allows valid use of information-theoretic methods to compare the models. One must have all models based directly on the distribution of $y$, hence $g(y \mid \underline{\theta})$; the parameter vector can enter in any form at all. For valid model comparability under K-L model selection one can transform the parameters (and predictor variables) over models, but one must maintain the same form of the response variable over models to be compared.

*Akaike (1981b) believed that the most important contribution of his general approach was the clarification of the importance of modeling and the need for substantial, prior information on the system being studied.* A well-founded global model is often important, as it incorporates the *a priori* scientific and biological knowledge that seems relevant. The set of candidate models can then be defined, usually as special cases of the global model. Then the analyst turns to the data in order to make valid inferences. These steps tend to prevent data dredging, which often leads to over-fitting and its associated problems of inference.

At some early point in the analysis of count data, the goodness-of-fit of the global model should be assessed using standard methods. Overdispersion can be an issue in analyzing count data, where often something about the variation is assumed by the simple model used. Overdispersion can be assessed by computing $\hat{c}$ from a chi-squared goodness-of-fit test of the global model *and* by careful consideration of the design and sampling protocol (e.g., were the data from samples of litter mates?). If $\hat{c}$ is in the range 1–3 and there is some biological reason to suggest some overdispersion, then one might use QAIC or QAIC$_c$ and proceed. Lacking some reason to suspect overdispersion or if $\hat{c} > 4$, then some structural failure may be suggested, and this should prompt one to spend more time developing the set of candidate models.

If, after proper attention to the *a priori* considerations, the global model still fits poorly, then information-theoretic methods will only select the best of the set of poor-fitting models. This undesirable situation probably reflects on the poor science that went into the modeling and definition of the set of candidate models. Lack of fit of the global model should be a flag warning that still more consideration must be given to the modeling, based on an understanding of the questions being asked and the design of the data collection. Perhaps the effort must be classed as exploratory and very tentative; this would allow some data dredging, leading perhaps to some tentative models and suggestive conclusions. Treated as the results of a pilot study, then new data could be collected and the analysis could proceed in a more confirmatory fashion using the techniques we outline in this book.

Often, LS is used for the analysis of continuous data, especially when the errors can be assumed to be normally distributed. Essentially all statistical software packages for LS estimation provide the unbiased estimate of $\sigma^2$, and this can be mapped easily into the MLE of $\hat{\sigma}^2 = \text{RSS}/n$ (Section 1.3.2), and from this, one can compute the equivalent value of the maximized log-likelihood. The correct specification of $K$ can be tricky and error-prone. Here, $K =$ the number of slope (regression) parameters $+$ one for the intercept $+$ one for the estimate of $\sigma^2$ (thus $K = p + 2$, where $p$ is the number of slope parameters). Equivalently, if one defines $p$ to be the total number of regression parameters, including the intercept ($\beta_0$), then $K = p + 1$. Computer programs for likelihood methods nearly always provide the value of the log-likelihood at its maximum. AIC or AIC$_c$ can be easily computed by hand from standard output of LS programs, assuming that $K$ and $\hat{\sigma}^2$ can be computed correctly. While many software packages currently print AIC, relatively few print the value of AIC$_c$, and this is a limitation (see Example 1 below, where AIC performs poorly, as the ratio $n/K$ is small). In such cases, one should compute the additional term $(2K(K+1))/(n-K-1)$ by hand and add it to the AIC value printed.

## 3.2 Example 1—Cement Hardening Data

The first example is a small set of data on variables thought to be related to the heat evolved during the hardening of Portland cement (Woods et al. 1932:635–649). While not a biological example, these data represent a simple use of multiple linear regression analysis (see Section 1.3.2) and will serve well as a first example. This data set (the "Hald data") has been used by various authors (e.g., Hald 1952:635–649, Seber 1977, Daniel and Wood 1971, Draper and Smith 1981:294–342 and 629–673, Stone and Brooks 1990, George and McCulloch 1993, Hjorth 1994:31–33, Ronchetti and Staudte 1994, Laud and Ibrahim 1996, and Sommer and Huggins 1996) and will illustrate a variety of important points. The data include 4 regressor variables and have a sample size of 13 (Table 3.1). The regressor variables (in percent of the weight) are $x_1$ = calcium aluminate ($3CaO \cdot Al_2O_3$), $x_2$ = tricalcium silicate ($3CaO \cdot SiO_2$), $x_3$ = tetracalcium alumino ferrite ($4CaO \cdot Al_2O_3 \cdot Fe_2O_3$), and $x_4$ = dicalcium silicate ($2CaO \cdot SiO_2$), while the response variable is $y$ = total calories given off during hardening per gram of cement after 180 days. Daniel and Wood (1971) provide further details on these data for the interested reader. *"What approximating model to use?"* is the focus of this example.

The small size of the sample necessitates the use of $AIC_c$ (Section 2.4.1); however, we will present comparable values for AIC in this example. We will use an obvious notation for denoting what variables are in each candidate model. That is, if variables $x_1$ and $x_3$ are in a particular model, we denote this as model {13}; each model has an intercept ($\beta_0$).

TABLE 3.1. Cement hardening data from Woods et al. (1932). Four regressor variables (in percent by weight) [$x_1$ = calcium aluminate ($3CaO \cdot Al_2O_3$), $x_2$ = tricalcium silicate ($3CaO \cdot SiO_2$), $x_3$ = tetracalcium alumino ferrite ($4CaO \cdot Al_2O_3 \cdot Fe_2O_3$), $x_4$ = dicalcium silicate ($2CaO \cdot SiO_2$)], are used to predict the dependent variable, $y$ = calories of heat evolved per gram of cement after 180 days of hardening.

| $x_1$ | $x_2$ | $x_3$ | $x_4$ | $y$ |
|---|---|---|---|---|
| 7 | 26 | 6 | 60 | 78.5 |
| 1 | 29 | 15 | 52 | 74.3 |
| 11 | 56 | 8 | 20 | 104.3 |
| 11 | 31 | 8 | 47 | 87.6 |
| 7 | 52 | 6 | 33 | 95.9 |
| 11 | 55 | 9 | 22 | 109.2 |
| 3 | 71 | 17 | 6 | 102.7 |
| 1 | 31 | 22 | 44 | 72.5 |
| 2 | 54 | 18 | 22 | 93.1 |
| 21 | 47 | 4 | 26 | 115.9 |
| 1 | 40 | 23 | 34 | 83.8 |
| 11 | 66 | 9 | 12 | 113.3 |
| 10 | 68 | 8 | 12 | 109.4 |

## 3.2.1 Set of Candidate Models

Because only 4 variables are available, the temptation is to consider all possible models ($2^4 - 1 = 15$) involving at least one of the regressor variables. In view of the small sample size we will consider this example as largely exploratory, and lacking any personal knowledge concerning the physics or chemistry of cement hardening, we will consider the full set of models, including the global model, {1234} with $K = 6$ parameters. While we generally advise against consideration of all possible models of the $x_i$ (but no interactions or powers of the predictor variables), this approach will allow some comparisons with others in the published literature (e.g., Draper and Smith 1981, Hjorth 1994, and Hoeting and Ibrahim 1996). We note, however, that the 4 models with only a single variable might have been excluded on *a priori* grounds because cement involves a mixture of at least two compounds. We will extend this example in Chapter 4 to examine the issue of model selection uncertainty.

## 3.2.2 Some Results and Comparisons

The use of $AIC_c$ suggests model {12} as the best approximating model for these

TABLE 3.2. Summary of 15 models for the cement hardening data, including the total number of estimable parameters ($K$), the ML estimated mean squared error ($\hat{\sigma}^2$), and $\Delta_i$ values for both AIC and $AIC_c$. Models are ordered in terms of $\Delta_i$ for $AIC_c$.

| Model | $K$ | $\hat{\sigma}^2$ | $\Delta_i$ − AIC | $\Delta_i$ − $AIC_c$ |
|---|---|---|---|---|
| {12}[1] | 4 | 4.45 | 0.4346 | 0.0000 |
| {124} | 5 | 3.69 | 0.0000 | 3.1368 |
| {123} | 5 | 3.70 | 0.0352 | 3.1720 |
| {14} | 4 | 5.75 | 3.7665 | 3.3318 |
| {134} | 5 | 3.91 | 0.7528 | 3.8897 |
| {234} | 5 | 5.68 | 5.6072 | 8.7440 |
| {1234} | 6 | 3.68 | 1.9647 | 10.5301 |
| {34} | 4 | 13.52 | 14.8811 | 14.4465 |
| {23} | 4 | 31.96 | 26.0652 | 25.6306 |
| {4} | 3 | 67.99 | 33.8785 | 31.1106 |
| {2} | 3 | 69.72 | 34.2052 | 31.4372 |
| {24} | 4 | 66.84 | 35.6568 | 35.2222 |
| {1} | 3 | 97.37 | 38.5471 | 35.7791 |
| {13} | 4 | 94.39 | 40.1435 | 39.7089 |
| {3} | 3 | 149.18 | 44.0939 | 41.3259 |

[1] Here, $\log(\mathcal{L}) = -n/2 \cdot \log_e(\hat{\sigma}^2) = -9.7039$, $AIC = -2\log_e(\mathcal{L}) + 2K = 27.4078$, and $AIC_c = AIC + \frac{2K(K+1)}{n-K-1} = 32.4078$.

data (Table 3.2). The estimated regression coefficients in the selected model are

$$\hat{E}(y) = 52.6 + 1.468(x_1) + 0.662(x_2),$$

where the estimated standard errors of the 3 estimated regression parameters (*given this model*) are 2.286, 0.121, and 0.046, respectively (this result is in agreement with Hald 1952). The adjusted $R^2 = 0.974$ and the MLE $\hat{\sigma} = 2.11$ for the AIC$_c$-selected model. The second-best model is {124}, but it is 3.14 AIC$_c$ units from the best model (Table 3.2). Other candidate models are ranked, and clearly many of the models represent poor approximations to these (scant) data (at least the models in Table 3.2 with $\Delta_i$ values $> 8$). Note the differences in $\Delta_i$ and associated rankings between AIC vs. AIC$_c$ in Table 3.2; clearly, AIC$_c$ is to be preferred over AIC, because the ratio $n/K$ ($= 13/6$) is only 2.2 for the global model (model {1234}).

Using a type of cross-validation criterion ($Q_{cv}$), Hjorth (1994:33) selected model {124} with $K = 5$ for these data. Here, his result is,

$$\hat{E}(y) = 71.6 + 1.452(x_1) + 0.416(x_2) - 0.236(x_4),$$

where the estimated standard errors are 14.142, 0.117, 0.186, and 0.173, respectively. Model {124} has an adjusted $R^2 = 0.976$ and $\hat{\sigma} = 1.921$. Draper and Smith (1981:325–327) used cross validation and the PRESS (Allen 1970) selection criterion, which is quite similar to $Q_{cv}$, and also selected model {124}. Note, had AIC been used, ignoring the ratio $n/K \doteq 2$, model {124} would have been selected (Table 3.2); AIC$_c$ should be used if this ratio is small (i.e., $< 40$).

Is there any basis to say that AIC$_c$ selected a better approximating model than Hjorth's cross-validation procedure or AIC or the PRESS criterion? This is difficult to answer conclusively because truth is not known here. However, the regression coefficient on $x_4$ is not "significant" under the traditional hypothesis testing scenario ($t = 1.36$, 9 df) and the estimated standard error on the regression coefficient for $x_2$ increased by a factor of 4 from 0.046 to 0.186 compared to model {12}. The adjusted $R^2$ statistics for Hjorth's selected model is 0.976 (vs. 0.974), but it has one additional parameter. The correlation coefficient between $x_1$ and $x_3$ was $-0.824$, while the correlation between $x_2$ and $x_4$ was $-0.973$. Just on the basis of this latter correlation it seems unwise to allow both $x_2$ and $x_4$ in the same model (if $n$ were 3,000 instead of only 13, perhaps there would be more support for including both $x_2$ and $x_4$). While not completely compelling, it would seem that AIC$_c$ has selected the better parsimonious model in this case. An additional, negative consideration is the computer-intensive nature of Hjorth's cross-validation algorithm ($Q_{cv}$) compared to the information-theoretic approach. With more reasonable sample sizes or with more models to consider, the cross-validation approaches may often become computationally too "costly."

Draper and Smith (1981) used Mallows's $C_p$ statistic and also selected model {12}, in agreement with AIC$_c$ (this might be fortuitous, because no small sample version of $C_p$ is available). They further point out that $\sum_{j=1}^{4} x_{ij} =$ a constant (approximately 98%) for any $i$; thus the $X'X$ matrix for model {1234} is theoretically singular. Small rounding errors were eventually introduced as the percentage data were expressed as integers, leaving the $X'X$ matrix barely nonsingular. At best, model {1234} would be a poor model for the analysis of these data. They also warn against the unthinking use of all possible regressions and present a detailed analysis of forward, backward, and stepwise approaches, based on tests of hypotheses and arbitrary $\alpha$ levels. Draper and Smith (1981) also used the stepwise procedure (with $\alpha = 0.15$), which resulted in model {12}, after starting at step 1 with $x_4$, eventually dropping it, and retaining only $x_1$ and $x_2$. This represents an improvement over routines that merely add new variables, without looking to see whether a particular variable has become redundant. Draper and Smith (1981) provide a good discussion of the various older model selection alternatives and offer some useful recommendations (but do not discuss any of the information-theoretic approaches). They provide an intensive analysis of the cement data over several chapters and include detailed computer output in 2 large appendices.

Another analysis approach involves computation of the principal components on the (centered) $X'X$ matrix and examination of the correlation matrix for the 4 explanatory variables (see Draper and Smith 1981:327–332, Stone and Brooks 1990). The principal component eigenvalues here are 2.23570, 1.57607, 0.18661, and 0.00162. Approximately 95.3% of the total variance is contained in the first 2 eigenvectors, while 99.96% is in the first 3 eigenvectors. These results certainly suggest that the global model over-fits these data (i.e., 4 regressors are redundant). In addition, it might suggest that 2 regressors will suffice (given $n = 13$). Critical interpretation of the percent eigenvalues requires some judgment and subjectivity. Furthermore, relatively few biologists are familiar with the concept of eigenvalues and eigenvectors. *We believe that the investigators should understand the methods leading to the results of their work; this is sometimes difficult with some advanced methods.* Such understanding seems relatively easy with the information-theoretic approaches.

One could ask whether there is a need for model selection when there are only 4 regressor variables (i.e., why not merely take the global model with 6 parameters and use it for inference?). This simple strategy is often very poor, as we illustrate here. First, note that this global model has $\Delta_i = 10.5301$, relative to model {12}, and is therefore a poor approximation to the meager data available. The estimates of parameters for the global model {1234} are

$$\hat{E}(\hat{y}) = 62.4 + 1.551(x_1) + 0.510(x_2) + 0.102(x_3) - 0.144(x_4),$$

where the estimated standard errors, given this model, are 70.071, 0.745, 0.728, 0.755, and 0.709, respectively. These standard errors are large because the $X'X$ is nearly singular (the percent coefficients of variation for $\hat{\beta}_0$, $\hat{\beta}_1$, and $\hat{\beta}_2$ were 4.3, 8.2, and 6.9 under model {12}, compared to 112.3, 48.0, and 142.7, respectively, under model {1234}. Only the regression coefficient for $x_1$ might be judged as "significant" in a hypothesis testing sense, and the model is clearly over-fit (see Fig. 1.3B). Model {1234} has an adjusted $R^2 = 0.974$ and $\hat{\sigma} = 1.918$. Surely a parsimonious model, such as {12}, would better serve the analyst in this case.

Loss of precision is expected in using an over-fit global model; however, there is also a nonnegligible probability that even the sign of the estimated parameter may be incorrect in such cases. It seems somewhat compelling to withhold judgment if the information (data) is inadequate for reliable inference on a parameter or effect, as the estimate might be very misleading.

If all the regressor variables are mutually orthogonal (uncorrelated), model selection is not quite as critical, and the global model with $K = 6$ might not be so bad. Orthognality arises in controlled experiments where the factors and levels are *designed* to be orthogonal. In observational studies, there is a high probability that some of the regressor variables will be mutually quite dependent. Rigorous experimental methods were just being developed during the time these data were taken (about 1930). Had such design methods been widely available and the importance of replication understood, then it would have been possible to break the unwanted correlations among the $x$ variables and establish cause and effect.

### 3.2.3   A Summary

In summary, the simple approach of using $AIC_c$ appears to have given a good parsimonious model as the basis for inference from these data. The use of $AIC_c$ sharpens the inference about which parsimonious model to use, relative to AIC. A priori information could have resulted in fewer candidate models and generally strengthened the process (note, Hald (1952) first presented only an analysis of $x_1$ and $x_2$ and presented the analysis of the 2 additional variables several pages later). It seems likely that models with only a single variable might have been excluded from serious consideration based on what must have been known about cement in the late 1920s. Similarly, we suspect that Woods et al. (1932) had knowledge of the negative relationship between $x_2$ and $x_4$; after all, model {14} was their second best model. $AIC_c$ avoided use of both $x_2$ and $x_4$ in the same model (where the correlation was $-0.973$) and the over-parametrized global model. An important feature of the information-theoretic approach is that it provides a ranking of alternative models, allowing some inferences to be made about other models that might also be useful. In addition, the rankings suggest some models that remain very poor (e.g., models {24}, {1}, {13}, and {3} for the cement data). The importance of carefully defining a small set of candidate models, based on the objective and what is known about the problem, cannot be overemphasized.

An investigator with, say, 10 explanatory variables cannot expect to learn much from his data and a multiple linear regression analysis unless there is some substantial supporting science that can be used to help narrow the number of models to consider. In this example, there would be 1,024 models (many more if transformations or interaction terms were allowed), and over-fitting would surely be a serious risk. The analysis, by whatever method, should probably be considered exploratory and the results used to design further data gathering leading to a more confirmatory analysis, based on some *a priori* considerations.

## 3.3 Example 2—Time Distribution of an Insecticide Added to a Simulated Ecosystem

This example concerns the addition of the insecticide DURSBAN® to a laboratory system that simulates a pond of water. The original work was done by Smith (1966) and his colleagues; our main reference for this example was Blau and Neely (1975), but also see Carpenter (1990) for a simplified Bayesian analysis of these data.

Blau and Neely note (1975) that the determination of the ultimate fate and distribution of this introduced chemical into an ecosystem is an important environmental issue. They go on to mention that "... a true mathematical model describing each step of the process would be extremely complex. It is important, however, to try to find a suitable model to identify the most important chemical, physical, and biological phenomena taking place and to predict the long-term environmental consequences." This view of modeling is consistent with Akaike's and the one recommended here. This example is used because it rests on a system of first-order differential equations whose parameters, given a model, are estimated by least squares. Such results can easily be used to compute AIC values to aid in selection of a parsimonious approximating model.

The active ingredient of DURSBAN® is 0,0-Diethyl 0-(3,5,6-trichloro-2-pyridyl) phosphorothioate and was labeled with radioactive carbon 14 in the pyridyl ring and added at a level of 1 mg/6 gal in a 10-gallon glass jar (see Fig. 3.1). This aquarium contained 2 in. of soil (13.3% organic matter), plants (salvinia, anacharis,

FIGURE 3.1. Glass aquarium used in the studies of DURSBAN® (from Smith 1966).

TABLE 3.3. Distribution of radioactive carbon in DURSBAN® in a simulated ecosystem (from Blau and Neely 1975).

|  | Percent radioactivity | | |
|---|---|---|---|
| Time after DURSBAN® addition (hours) | Fish $C$ | Soil & Plants $B$ | Water $A$ |
| 0 | 0 | 0 | 100 |
| 1.5 | 15.2 | 35.2 | 49.7 |
| 3.0 | 19.0 | 46.0 | 28.3 |
| 4.0 | 19.3 | 56.0 | 24.5 |
| 6.0 | 20.7 | 61.0 | 18.3 |
| 8.0 | 23.0 | 60.5 | 17.0 |
| 10.0 | 24.2 | 59.3 | 18.2 |
| 24.0 | 21.2 | 51.5 | 26.5 |
| 48.0 | 23.0 | 38.3 | 34.5 |
| 72.0 | 22.7 | 38.3 | 39.5 |
| 96.0 | 20.5 | 36.3 | 43.0 |
| 120.0 | 17.3 | 38.3 | 44.5 |

milfoil, and water cucumber), and 45 goldfish. Samples of the various components were analyzed for radioactivity at 12 different time periods, following the addition of DURSBAN®. Three samples at each time period yielded a sample size ($n$) of 36. The data (Table 3.3) are in percentages from the crude radioactivity measurements (Blau and Neely 1975:150). The authors of the study assumed that the model residuals were normally distributed, with zero means and a constant standard deviation of 1% (we take this to mean the actual measurement error of the instrument used).

### 3.3.1 Set of Candidate Models

Blau and Neely (1975) had a great deal of knowledge about this system, and they exploited this in *a priori* model building. They began by postulating that an equilibrium exists between DURSBAN® in the water ($A$), soil and plant components ($B$) and a direct uptake of the chemical by the fish ($C$). This led to their Model 1 (Fig. 3.2) and was represented by a system of differential equations, where the rate parameters to be estimated are denoted by $k_i$,

$$dx_A(t)/dt = -k_1 x_A(t) + k_2 x_B(t) - k_3 x_C(t),$$
$$dx_B(t)/dt = k_1 x_A(t) - k_2 x_B(t),$$
$$dx_C(t)/dt = k_3 x_A(t),$$

with initial conditions $x_A(0) = 100$, $x_B(0) = 0$, and $x_C(0) = 0$. This is a type of compartment model (Brown and Rothery 1993) and is often used in some fields. Blau and Neely (1975) used $x_A(t)$, $x_B(t)$, and $x_C(t)$ as the percentages at time ($t$) of $A$, $B$, and $C$, respectively, with the restriction that,

$$x_A(t) + x_B(t) + x_C(t) = 100.$$

3.3 Time Distribution of an Insecticide Added to a Simulated Ecosystem 85

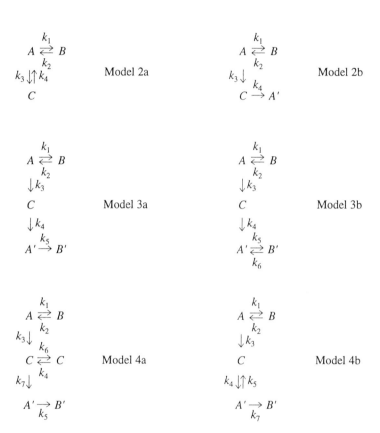

FIGURE 3.2. Summary of models used by Blau and Neely (1975) for the data on DURSBAN® in a simulated pond ecosystem.

They used nonlinear least squares to estimate model parameters (the $k_i$ and $\sigma^2$), and their analytic methods were quite sophisticated. The parameter estimates for this model were $\hat{k}_1 = 0.510$, $\hat{k}_2 = 0.800$, $\hat{k}_3 = 0.00930$, and $\hat{\sigma}^2 = 149.278$ (this is their residual sum of squares (RSS) divided by $n$ to obtain the MLE of $\sigma^2$); thus, $K = 4$ for this model.

Blau and Neely (1975) built 6 other models, each based on their knowledge of the system, but also based on examination of the residuals from prior models (there are some inconsistencies here that we were unable to resolve; thus we will use the

86     3. Practical Use of the Information-Theoretic Approach

material from their paper). While some data dredging was evident, their main derivation of additional models seemed to stem primarily from their knowledge of the processes. They were well aware of the principle of parsimony and included a very nice discussion of LS and ML methods and their relationships in an introductory part of their paper. They computed goodness-of-fit tests and separated "pure error" from the remaining residual terms. Model selection was accomplished by using statistical hypothesis tests (likelihood ratio tests) and examining the RSS. They found Model 4a (see Fig. 3.2) to be the best and also found some support for Model 4b.

### 3.3.2   Some Results

Analysis of these data under an information-theoretic paradigm is simple, given Blau and Neely's (1975) table II, as they provide values for $K-1$ and RSS for each of their 7 models. Due to the relationships between LS estimation and ML theory (see Section 1.3.2),

$$\log(\mathcal{L}(\hat{\underline{k}}\hat{\sigma}^2 \mid data)) = -n/2 \cdot \log(\hat{\sigma}^2),$$

where $\hat{\sigma}^2 = $ RSS$/n$. Then,

$$\text{AIC} = -2 \cdot \log(\mathcal{L}(\hat{\underline{k}}\hat{\sigma}^2 \mid data)) + 2K$$

and

$$\text{AIC}_c = \text{AIC} + \left((2K(K+1))/(n-K-1)\right).$$

These computations were done by hand on a simple calculator and took approximately 20 minutes. The results of this extended analysis are shown in Table 3.4 and suggest that Model 4a is the best to use for inference, in agreement with Blau and Neely (1975). Only Model 4b is a competitor, but it has a $\Delta_i$ value of 7.611 and seems relatively implausible for these data.

Carpenter (1990) used these data and 7 models under a simplified Bayesian analysis with equal Bayesian prior probabilities on the models but with no prior

TABLE 3.4. Summary of model-selection statistics (the first 3 columns taken from Blau and Neely 1975). Statistics for the AIC$_c$-selected model are shown in bold.

| Model | K | RSS | $\log(\mathcal{L}(\hat{\underline{k}}\hat{\sigma}^2 \mid data))$ | AIC | AIC$_c$ | $\Delta_i - $ AIC$_c$ |
|---|---|---|---|---|---|---|
| 1  | 4 | 5374  | −90.105 | 188.209 | 189.499 | 150.626 |
| 2a | 5 | 1964  | −71.986 | 153.972 | 155.972 | 117.099 |
| 2b | 5 | 848   | −56.869 | 123.737 | 125.737 | 86.864  |
| 3a | 6 | 208.3 | −31.598 | 75.196  | 78.094  | 39.221  |
| 3b | 7 | 207.9 | −31.563 | 77.127  | 81.127  | 42.254  |
| **4a** | **8** | **58.6** | **−8.770** | **33.540** | **38.873** | **0.0** |
| 4b | 7 | 79.4  | −14.238 | 42.475  | 46.475  | 7.602   |

probabilities specified on the model parameters in that same semi-Bayesian context. He also concluded that Model 4a was the best, with Model 4b being a poor second. In this example, $K$ ranged from only 4 to 8; thus the various methods might be expected to be in somewhat close agreement. This example illustrates that it is often easy to perform a reanalysis of data on complex systems, based on information provided in published papers.

## 3.4 Example 3—Nestling Starlings

We generated a set of Monte Carlo data to illustrate many of the points discussed with a much more complicated example of an experimental setting. Thus, in a sense, the generating model is "truth"; we will accept this bit of unrealism for the moment, but mitigate it by including many parameters ($K = 34$) and a wide variety of tapering treatment effects. In addition, we will choose a global model that has 4 fewer parameters than the generating model; thus the generating model is not in the set of candidate models. Furthermore, this example contains many so-called nuisance parameters (sampling probabilities). This is the only example in Chapter 3 where "truth" is known, and some interesting insights can be gained from this knowledge. The essential question is what parsimonious, approximating model can be used for data analysis that will lead to valid inference about the structure of the system, its parameters, and the effects of the treatment.

### 3.4.1 Experimental Scenario

We generated data to mimic the experiment presented by Stromborg et al. (1988) (also see Burnham et al. 1987:343–348). The hypothetical research question relates to the survival effects of an organophosphate pesticide administered to nestling European starlings (*Sturnus vulgaris*). We assume for illustration that a simple field experiment is designed using artificial nest boxes placed on a 5,000 ha island. Fledgling birds are assumed not to leave the island during the summer and early fall months when the experiment is conducted (geographic closure). Nest boxes are monitored during the nesting season to determine the date of hatching. All nestlings are leg-banded with uniquely numbered bands 16 days following hatching, and half of those nestlings are randomly assigned to a treatment group and the remaining birds assigned to a control group. In total, we will assume that 600 nestling starlings are banded and returned to the nest box (i.e., the number of starlings originally released in each group is 300). All nest boxes contain 4 young birds (thus 2 treatment and 2 control), and we assume these to be of nearly uniform size and age and that once fledged, they move about and behave independently. Starlings in the treatment group receive an oral dose of pesticide mixed in corn oil. Birds in the control groups are given pure corn oil under otherwise very similar conditions. Colored leg bands provide a unique identification for each starling and therefore its group membership, on each weekly resighting occasion. Data collection will be assumed to begin after a 4-day period following dosage, and for simplicity, we assume that no birds die due to handling effects following marking but before resighting efforts begin a week later. Surviving starlings are potentially resighted during the following 9 weeks; sampling covers the entire island and is done on each Friday for 9 weeks. Thus, the data are collected on 10 occasions; occasion 1 is the initial marking and release period, followed by 9 resighting occasions.

The pesticide is hypothesized to affect conditional survival probability (the parameters of interest) and resighting probabilities (the nuisance parameters); however, the pesticide industry's position is that only minor survival effects are likely, while environmental groups suspect that there are substantial acute (short-term) and chronic (long-term) effects on survival probabilities and worry that the resighting probabilities might also be affected by the treatment. Thus, the set of candidate models might span the range of the controversy. In practice, of course, one might design the experiment to include several "lots" of starlings, released at different, independent locations (islands), and these data would be the basis for empirical estimates of treatment effect and precision (see Burnham et al. 1987 for a discussion of experiments of this general type). Here we will focus on an example of the model selection issue and not on optimal design.

### 3.4.2 Monte Carlo Data

Monte Carlo data were generated using the following relationships for conditional survival probability ($\phi$) and resighting probability ($p$) for treatment ($t$) and control ($c$) groups at week $i$,

$$\phi_{ti} = \phi_{ci} - (0.1)(0.9)^{i-1} \text{ for } i = 1, \ldots, 9 \text{ and}$$
$$p_{ti} = p_{ci} - (0.1)(0.8)^{i-2} \text{ for } i = 2, \ldots, 10,$$

using program RELEASE (Burnham et al. 1987). These relationships allow a smooth temporal tapering of effect size due to the treatment in both conditional survival and resighting probabilities. That is, each week the effect of the pesticide is diminished. We used the initial per-week survival and resighting probabilities for the control group as 0.9 and 0.8, respectively. Conditional survival and resighting probabilities for the control group did not differ by week (i.e., $\phi_{ci} \equiv \phi_c \equiv 0.9$ and $p_{ci} \equiv p_c \equiv 0.8$). The data are given in Table 3.5 for each treatment and control group.

### 3.4.3 Set of Candidate Models

Define $\phi_{vi}$ as the conditional probability of survival for treatment group $v$ ($v = t$ for treatment and $c$ for control) from week $i$ to $i + 1$ ($i = 1$ to 9) and $p_{vi}$ as the conditional probability of resighting for treatment group $v$ at week $i$ (for $i = 2$ to 10). The set of models that seem reasonable might include one with no treatment effects ($M_0$), a model for an acute effect only on the first survival probability ($M_{1\phi}$), and a model for an acute effect on both the first survival probability and the first

### 3.4 Example 3—Nestling Starlings

TABLE 3.5. Summary of the starling data as the matrix $m_{vij}$, where $v$ = treatment or control group, $i$ = week of release ($i = 1, \ldots, 9$), and $j$ = week of resighting ($j = 2, \ldots, 10$). The data given for each group ($v$) are the number of starlings first captured in week $j$ after last being released at time $i$. $R_i$ = the number of birds released at week $i$; note that all of those released in weeks $2, \ldots, 9$ were merely rereleased. Each row ($i$) plus the term $\left(R(i) - \sum_j m_{ij}\right)$ is modeled as a multinomial distribution with sample size $R(i)$.

| Week | $R(i)$ | Observed Recaptures for Treatment Group $m(i, j)$ | | | | | | |
|---|---|---|---|---|---|---|---|---|
| | | $j = 2$ | 3 | 4 | 5 | 6 | 7 | 8 | 9 |
| 1 | 300 | 158 | 43 | 15 | 5 | 0 | 0 | 0 | 0 |
| 2 | 158 | | 82 | 23 | 7 | 1 | 1 | 0 | 0 |
| 3 | 125 | | | 69 | 17 | 6 | 1 | 0 | 0 |
| 4 | 107 | | | | 76 | 8 | 2 | 0 | 0 |
| 5 | 105 | | | | | 67 | 20 | 3 | 0 |
| 6 | 82 | | | | | | 57 | 14 | 1 |
| 7 | 81 | | | | | | | 53 | 12 |
| 8 | 70 | | | | | | | | 46 |

| Week | $R(i)$ | Observed Recaptures for Control Group $m(i, j)$ | | | | | | |
|---|---|---|---|---|---|---|---|---|
| | | $j = 2$ | 3 | 4 | 5 | 6 | 7 | 8 | 9 |
| 1 | 300 | 210 | 38 | 5 | 1 | 0 | 0 | 0 | 0 |
| 2 | 210 | | 157 | 20 | 8 | 2 | 0 | 0 | 0 |
| 3 | 195 | | | 138 | 24 | 2 | 1 | 0 | 0 |
| 4 | 163 | | | | 112 | 24 | 2 | 0 | 0 |
| 5 | 145 | | | | | 111 | 16 | 6 | 0 |
| 6 | 139 | | | | | | 105 | 16 | 4 |
| 7 | 124 | | | | | | | 93 | 12 |
| 8 | 115 | | | | | | | | 89 |

resighting probability (denote this by $p_2$, as it occurs at week 2) (model $M_{2p}$). This initial line of *a priori* consideration leads to three models:

| Model | Parametrization |
|---|---|
| $M_0$ | All $\phi_{ti} = \phi_{ci}$ and all $p_{ti} = p_{ci}$ (no treatment effect) |
| $M_{1\phi}$ | $M_0$, except $\phi_{t1} \neq \phi_{c1}$ (an acute effect on $\phi_1$) |
| $M_{2p}$ | $M_{1\phi}$, except $p_{t2} \neq p_{c2}$ (acute effects on $\phi_1$ and $p_2$) |

Chronic effects might arise from starlings that are in poor health due to effects of the pesticide; these starlings might be more susceptible to predation (this would be revealed in lessened survival during the summer period) or might be less active in foraging (this might be revealed in differing probabilities of resighting compared to the control starlings, as sampling is done during the summer period). Chronic effects, if they exist, might be reduced with time. That is, one might expect chronic effects to diminish over time, relative to the starlings in the control group. Agreement is reached, based on biological evidence, that chronic effects, if they exist, should not last beyond the 7th week.

Define $S_i = \phi_{ti}/\phi_{ci}$ for $i = 1$ to 7 as the measure of treatment effect on conditional survival probability, compared to the control group. (Starlings in the control group will experience some mortality as the summer progresses; here the interest is in any *additional* mortality incurred by starlings caused by the pesticide treatment.) The parameters $S_i$ ($i = 1, 2, \ldots, 7$) are 0.889, 0.911, 0.929, 0.943, 0.954, 0.964, and 0.971, respectively. With dampened chronic effects, one expects $S_2 < S_3 < S_4 < \cdots < S_7 < 1$, as can be seen from the parameters above (of course, the unconstrained *estimates* of these parameters, based on some approximating model, might not follow these inequalities). Here, it seems reasonable to consider the presence of chronic effects only as additional impacts to the hypothesized acute effects. Thus, several models of chronic effects on both conditional survival and resighting probabilities are defined and might be included in the set of candidate models:

| Model | Parametrization |
|---|---|
| $M_{2\phi}$ | $M_{2p}$, except $\phi_{t2} \neq \phi_{c2}$ (chronic effect on $\phi_2$) |
| $M_{3p}$ | $M_{2\phi}$, except $p_{t3} \neq p_{c3}$ (chronic effect on $p_3$) |
| $M_{3\phi}$ | $M_{3p}$, except $\phi_{t3} \neq \phi_{c3}$ (more chronic effects) |
| $M_{4p}$ | $M_{3\phi}$, except $p_{t4} \neq p_{c4}$ (more chronic effects) |
| $\vdots$ | |
| $M_{7\phi}$ | All $\phi_{vi}$ and $p_{vi}$ differ by treatment group for 7 weeks |

This last candidate model ($M_{7\phi}$) allows chronic treatment effects on both conditional survival and resighting probabilities up through the 7th sampling week, in addition to the acute treatment effects on $\phi_{t1}$ and $p_{t2}$. This model will serve as our global model and has 30 parameters. The treatment effect extends through the 9th week; thus, the generating model is not in the set of candidate models and has more parameters than the global model (34 vs. 30).

Model $M_0$ has 17 parameters, while model $M_{7\phi}$ has 30 parameters. The simplest model would have a constant survival and resighting probability for each group ($M_{\phi,p}$) and thus no treatment or week effects on either conditional survival or resighting probabilities. This model would have only 2 parameters ($\phi$ and $p$). Alternatively, a 4-parameter model could allow the time-constant parameters to differ by treatment group ($\phi_t$, $\phi_c$, $p_t$, and $p_c$). Considering the relatively large sample size in this example, these models seem to be too simple and unlikely to be useful based on initial biological information, and we might well exclude these from the set of candidate models. Models without biological support should not be included in the set of candidate models. However, as an example, we will include these simple models for consideration and note that these models might well be viewed as more viable models if the initial sample size released were 60 instead of 600.

The effective sample size in these product multinomial models is the number of starlings released (or rereleased) at each week. [The effective sample size in these product multinomial models is a complicated issue, but we will not divert attention to this matter here, except to say that here we used $n = \sum R_i$ in the context of $AIC_c$. Technical notes on this subject may be obtained from KPB.] In this example, $n = 2{,}583$ releases (a resighting is equivalent to being "recaptured and rereleased"). Because 600 starlings (300 in each group) were released at week 1 (the nest boxes), the remaining 1,983 starlings were resighted at least once. Because of the large effective sample size, the use of $AIC_c$ is unnecessary; however, if one chose always to use $AIC_c$ in place of AIC, no problems would be encountered because $AIC_c$ and AIC converge as $n/K$ gets large.

A statistician on the research team suggests adding several models of the possible tapering treatment effects on conditional survival or resighting probabilities. This is suggested both to conserve the number of parameters (recognizing the bias–variance tradeoff, Fig. 1.2) and to gain additional insights concerning possible long-term, chronic treatment effects. Models employing a type of sine transformation on the parameters ($\phi_{vi}$ and $p_{vi}$) will be used here. In this transformation, the parameter ($\theta$, representing either $\phi$ or $p$, assumed to be between 0 and 1) to be modeled as a function of an external covariate (e.g., $X$) is replaced by the expression $(\sin(\alpha + \beta X) + 1)/2$. The new parameters $\alpha$ and $\beta$ are the intercept and slope parameters, respectively, in the covariate model. The transformation utilizes one half of a sine wave to model increasing or decreasing sigmoid-like functions and is an example of a link function in generalized linear models. In particular, submodel $M_{\sin \phi_t}$ and submodel $M_{\sin p_t}$ were defined for the dynamics of starlings in the treatment group:

$$M_{\sin \phi_t} \quad \sin(\phi) = \alpha + \beta(\text{week}),$$
$$M_{\sin p_t} \quad \sin(p) = \alpha' + \beta'(\text{week}).$$

These submodels each have only 2 parameters (intercepts $\alpha$ and $\alpha'$ and slopes $\beta$ and $\beta'$) and assume that $\sin(\phi_t)$ or $\sin(p_t)$ is a linear function of week (e.g., conditional survival of starlings in the treatment group will gradually increase as the summer period progresses, eventually approximating that of starlings in the control group).

These above two submodels for the treatment group can be crossed with four submodels below for the control group:

$M_{\phi_{ci}}$    $\phi$ is allowed to differ for each week; hence $(i = 1, \ldots, 8)$.
$M_{\phi_c}$    $\phi$ is assumed constant across weeks.
$M_{p_{ci}}$    $p$ is allowed to differ for each week; hence $(i = 2, \ldots, 9)$.
$M_{p_c}$    $p$ is assumed constant across weeks.

For example, a model can be developed using $M_{\sin \phi_t}$ for conditional survival of the treatment group and model $M_{\phi_c}$ for the conditional survival of the control group. This part of the model has 3 parameters: $\alpha$, $\beta$, $\phi_c$, plus the parametrization of the resighting probabilities. Thus, one could consider model $M_{\sin p_t}$ for the treatment group and model $M_{p_{ci}}$ for the control group as one parametrization for

TABLE 3.6. Summary of Akaike's information criterion (AIC) and associated statistics for 24 candidate models for the analysis of the simulated data on nestling starlings dosed with a pesticide. (All values are scaled by the additive constant $-4,467.779$; thus $\Delta_i = 0$ for the best model.)

| Model | No. Parameters | $\Delta_i$ |
|---|---|---|
| $M_{7\phi}$ (global) | 30 | 27.63 |
| $M_{7p}$ | 29 | 25.84 |
| $M_{6\phi}$ | 28 | 23.87 |
| $M_{6p}$ | 27 | 22.11 |
| $M_{5\phi}$ | 26 | 23.90 |
| $M_{5p}$ | 25 | 24.15 |
| $M_{4\phi}$ | 24 | 22.42 |
| $M_{4p}$ | 23 | 21.25 |
| $M_{3\phi}$ | 22 | 21.85 |
| $M_{3p}$ | 21 | 24.84 |
| $M_{2\phi}$ | 20 | 34.03 |
| $M_{2p}$ | 19 | 49.24 |
| $M_{1\phi}$ | 18 | 55.71 |
| $M_0$ | 17 | 64.82 |
| $M_{\sin\phi_t,\phi_{ci},\sin p_t,p_{ci}}$ | 21 | 17.89 |
| $M_{\sin\phi_t,\phi_{ci},\sin pt,pc}$ | 14 | 7.47 |
| $M_{\sin\phi_t,\phi_c,\sin p_t,p_{ci}}$ | 14 | 11.58 |
| $\mathbf{M_{\sin\phi_t,\phi_c,\sin p_t,p_c}}$ | **6** | **0.0** |
| $M_{\sin\phi_t,\phi_{ci},p_{ti},p_{ci}}$ | 28 | 20.85 |
| $M_{\sin\phi_t,\phi_{ci},p_{ti},p_c}$ | 21 | 10.43 |
| $M_{\sin\phi_t,\phi_c,p_t,p_{ci}}$ | 13 | 16.92 |
| $M_{\sin\phi_t,\phi_c,pt,pc}$ | 5 | 5.34 |
| $M_{\phi_t,\phi_c,p_t,p_c}$ | 4 | 302.70 |
| $M_{\phi,p}$ | 2 | 356.13 |

the resighting probabilities. This would add the parameters $\alpha'$, $\beta'$, $p_{c2}$, $p_{c3}$, ..., $p_{c10}$, for a total of $K = 14$ parameters. As an illustration, we consider a rich mixture of candidate models in Table 3.6 (a set of 24 candidate models). If this were a real situation, still other *a priori* models might be introduced and carefully supported with biological reason. If this experiment was based on only 60 nestlings, then several simple models should be included in the set and high-dimensional models would be deleted. This set of 24 candidate models will serve as a first example where there is some substantial complexity.

### 3.4.4 Data Analysis Results

As one would expect with simulated data, they fit the model used for their generation; $M_{9\phi}$ ($\chi^2 = 35.5$, 36 df, P = 0.49). [A large literature on goodness-of-fit testing in this class of models exists (e.g., Burnham et al. 1987 and Pollock et al. 1990); we will not pursue such tests here.] These data were simulated such that no overdispersion was present, and an estimate of the overdispersion factor $c$ could be computed under the generating model from the results of the goodness-of-fit test, $\hat{c} = \chi^2/\text{df} = 35.5/36 \doteq 1$. The global model, $M_{7\phi}$, has fewer parameters than the generating model, but also fits these data well ($\chi^2 = 35.4$, 30 df, P $= 0.23$). The value of $\hat{c}$ for the global model was 1.18, reflecting no overdispersion in this case, but some lack of fit (which is known to be true in this instance); after all, it, too, is only a model of "truth." One could consider a quasi-likelihood inflation of the variances and covariances of the estimates from the selected model by multiplying these by 1.18 (or the standard errors by the square root, 1.086). In particular, one might consider using the modifications to AIC given in Section 2.5 (i.e., QAIC $= -2\log(\mathcal{L})/1.18 + 2K$)—we will mention these issues at a later point. The critical information needed for selection of a parsimonious model is shown in Table 3.6.

The model with the minimum AIC value was $M_{\sin\phi_t.\phi_c.\sin p_t.p_c}$ with $K = 6$ parameters ($\alpha$, $\beta$, $\alpha'$, $\beta'$, $\phi_c$, and $p_c$). Using estimates of these 6 parameters one can derive MLEs of the survival and resighting parameters of interest; the MLEs for the treatment survival probabilities were:

| $i$ | $\phi_{ti}$ | $\hat{\phi}_{ti}$ | $\widehat{\text{se}}(\hat{\phi}_{ti})$ |
|---|---|---|---|
| 1 | 0.800 | 0.796 | 0.021 |
| 2 | 0.810 | 0.810 | 0.016 |
| 3 | 0.819 | 0.824 | 0.014 |
| 4 | 0.827 | 0.838 | 0.014 |
| 5 | 0.834 | 0.851 | 0.160 |
| 6 | 0.841 | 0.864 | 0.019 |
| 7 | 0.847 | 0.876 | 0.022 |
| 8 | 0.852 | 0.887 | 0.026 |
| 9 | 0.857 | 0.898 | 0.029 |

The survival parameter for the control group was 0.90, and its MLE from the selected model was 0.893 ($\widehat{\text{se}} = 0.008$). These estimates are reasonably close to the parameter values, and one can correctly infer the diminishing, negative effect of the treatment on weekly survival probabilities. On a technical note, the 9 estimates of survival probability for the treatment group (above) were derived from the MLEs of $\alpha$ and $\beta$ in the submodel $\sin(\phi_{ti}) = \alpha + \beta(\text{week } i)$.

Model $M_{\sin\phi_t.\phi_c.\sin p_t.p_c}$ had the lowest AIC value (4,467.78, $\Delta_i = 0$); the AIC value is large because the sample size is large (Section 2.1.4). Here, the sine model estimates the acute and chronic effects of the treatment on both the conditional survival and resighting probabilities for birds in the treatment group. The conditional survival and resighting probabilities for birds in the control group were constant over weeks in this model, but differed from those in the treatment group. The AIC-selected model captures the main structure of the generated process. Figure 3.3 illustrates the similarities among the true values, the estimates from the global model ($M_{7\phi}$), and the estimates from the AIC-selected model in terms of the treatment effect, $1 - S_i$.

94     3. Practical Use of the Information-Theoretic Approach

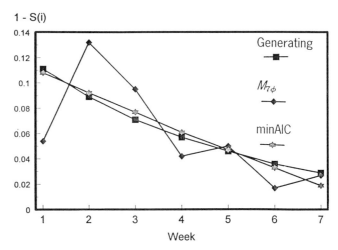

FIGURE 3.3. Treatment effect $(1 - S_i,$ for $i = 1, \ldots,$ 7th week) for the starling data from the generating models $(M_{9\phi})$ with 34 parameters, compared with estimates of these parameters from the global model $(M_{7\phi})$ with 30 parameters and the AIC-selected model with 6 parameters.

Part of the reason that this analysis was successful was the *a priori* reasoning that led to *modeling* the treatment effects, rather than trying to estimate the week-specific treatment effects (i.e., the $S_i$) individually. Such modeling allowed substantial insight into the tapering, chronic effects in this case. Note: The two simplest models ($M_{\phi_t,\phi_c,p_t,p_c}$ with $K = 4$ and $M_{\phi,p}$ with $K = 2$) were not at all plausible ($\Delta_i = 302.70$ and 356.13, respectively); recall that these models would not normally have been considered in a well-designed experiment, as they lacked any reasonable biological support, given the large sample size involved. Of course, had sample size been very small, then these models might have been more reasonable to include in the set of candidates.

If sample size is small, one must realize that relatively little information is probably contained in the data (unless the effect size if very substantial), and the data may provide few insights of much interest or use. People routinely err by building models that are far too complex for the (often meager) data at hand. They do not realize how little structure can be reliably supported by small amounts of data that are typically "noisy." Some experience is required before analysts get a feeling for modeling based on sample size and what is known about the science of the problem of interest.

### 3.4.5   *Further Insights into the First 14 Nested Models*

We now examine further the results had the set of candidate models included just the first 14 models in Table 3.6. Substantial theory (e.g., the estimators exist in closed form) and software (program RELEASE, Burnham et al. 1987) exist for this sequence of nested models, allowing the illustration of a number of deeper points. First, we must notice that these 14 models are clearly inferior to the models hypothesizing tapering treatment effects (a diminishing linear treatment effect embedded in a sine link function) for birds in the treatment group (e.g., the best model of the 14, model $M_{4p}$, is 21.25 AIC units above the selected model and has 23 parameters, compared to only 6 parameters for the AIC-selected model). Again, this points to the importance of a good set of candidate models. Second, many smaller, chronic effects could not be identified by model $M_{4p}$ (i.e., the relative treatment effects on survival in the later time periods, $S_4$, $S_5$, and $S_6$); however, the $\Delta_i$ values provide clues that at the very least, models $H_{4\phi}$ (therefore, $S_4$) and $M_{6p}$ (therefore, $S_5$) are also somewhat supported by the data (Table 3.6). These models have AIC values within 1.17 and 0.86, respectively, of model $M_{4p}$. In fact, models $M_{3p}$ through model $M_{6\phi}$ have fairly similar AIC values (Table 3.6 and Fig. 3.4). Unless the data uniquely support a particular model, we should not take the resulting model as *the* answer for the issue at hand; just the best that the particular data set can provide. Perhaps more than one model should be considered for inference from the 14 models (Chapters 4 and 5).

The program RELEASE (Burnham et al. 1987) allows approximate expected values of estimators and theoretical standard errors to be computed easily for models in this class (i.e., the 14 appearing at the top of Table 3.6). These results allow insight into why the more minor, chronic effects were not identified by model

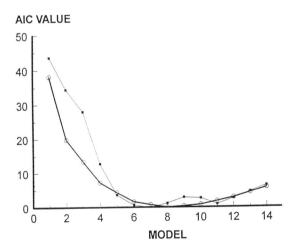

FIGURE 3.4. Estimated theoretical (heavy line) and sample $\Delta_i$ values for the 14 nested models used for the starling experiment. The estimated ($n = 50,000$ Monte Carlo reps) theoretical, expected AIC values (shown as open circles) are minimized ($\Delta_i = 0$) at model $M_{5p}$, while the realized AIC value from the sample data is minimized at model $M_{4p}$ (see Table 3.6). Generally there is good agreement between the theoretical and sample values, here plotted as $\Delta_i$ values).

$M_{4p}$ (the model estimated to be the best among the 14):

| $i$ | $1 - E(\hat{S}_i)$ | $\widehat{se}(1 - \hat{S}_i)$ | $(1 - E(\hat{S}_i))/\widehat{se}(1 - \hat{S}_i)$ |
|---|---|---|---|
| 4 | 0.057 | 0.053 | 1.08 |
| 5 | 0.046 | 0.055 | 0.84 |
| 6 | 0.036 | 0.057 | 0.63 |

The expected treatment effect size (i.e., $1 - E(\hat{S})$) was small (near 0), while the standard errors were of a similar magnitude or larger, as shown in the final two columns above. The larger effects (i.e., $S_1$ and $S_2$) are relatively easy to identify; however, at some point, the effect size is too small to detect directly with confidence from the information contained in the finite sample. Still, if one had only the first 14 models and had used AIC to select model $M_{4p}$, inference from the data in this example would have been fairly reasonable, but hardly optimal. The acute and larger chronic effects would have been convincingly identified. Comparison of AIC values for models $M_{4\phi}$ ($K = 24$) and $M_{6p}$ ($K = 28$) would have provided reasonable evidence for some extended, chronic treatment effects. Still, having to estimate 23–28 parameters would lead to imprecise estimators, compared to those under the best model ($M_{\sin \phi_t, \phi_c, \sin p_t, p_c}$). AIC, AIC$_c$, and QAIC$_c$ are fundamental criteria that provide a basis for a unified approach to the statistical analysis of empirical data in the biological sciences. Further details concerning this class of models are provided by Anderson et al. (1994), Burnham et al. (1994), Burnham et al. (1995a and b), and Anderson et al. (1998).

### 3.4.6 Hypothesis Testing and Information-Theoretic Approaches Have Different Selection Frequencies

At this point it is illustrative to examine briefly how information-theoretic selection compares to traditional approaches based on statistical hypothesis testing. Thus, Monte Carlo methods were employed to generate 50,000 independent samples (data sets) using the same methods as were used to generate the original set of simulated data on nestling starlings. That is, model $M_{9\phi}$ and the numbers released and all parameter values were identical to those used to generate the first set of data. Six methods were used to select a model for inference: The first 3 methods involve well-known selection methods based on hypothesis testing (stepup or forward selection, stepdown or backward selection, and stepwise selection), each using $\alpha = 0.05$. Three information-theoretic methods were also used on each of the data sets: AIC, AIC$_c$, and QAIC$_c$ (using $\hat{c}$ as a variance inflation factor, estimated for each simulated data set). The results (Table 3.7 and Fig. 3.5) show substantial differences among model-selection frequencies for the various methods.

Stepup selection, on average, selects model $M_{2\phi}$ with 20 parameters (the average was 20.3 parameters selected). These results are similar to the stepwise approach, which selects, on average, model $M_{3p}$ with 21 parameters (here the average number of parameters was 20.6). Given that data were simulated under model $M_{9\phi}$ with

TABLE 3.7. Selection percentages for six selection methods, based on 50,000 Monte Carlo repetitions. The hypothesis testing approaches use $\alpha = 0.05$. The data sets were generated under model $M_{9\phi}$ with 34 parameters, which was parametrized to reflect a tapering treatment effect on both conditional survival and resighting probabilities for the treatment group.

|  | Model | Hypothesis Testing | | | Information-Theoretic | | |
| --- | --- | --- | --- | --- | --- | --- | --- |
|  |  | Stepup | Stepdown | Stepwise | AIC | $AIC_c$ | $QAIC_c$ |
| 1 | $M_0$ | 0.6 | 0.0 | 0.0 | 0.0 | 0.0 | 0.0 |
| 2 | $M_{1\phi}$ | 17.4 | 0.0 | 13.9 | 0.0 | 0.0 | 0.0 |
| 3 | $M_{2p}$ | 14.8 | 0.3 | 13.5 | 0.2 | 0.2 | 0.2 |
| 4 | $M_{2\phi}$ | 26.8 | 1.5 | 26.3 | 1.2 | 1.3 | 1.4 |
| 5 | $M_{3p}$ | 16.3 | 2.8 | 16.9 | 2.7 | 2.8 | 2.6 |
| 6 | $M_{3\phi}$ | 14.6 | 6.9 | 16.1 | 6.8 | 7.4 | 7.1 |
| 7 | $M_{4p}$ | 5.9 | 7.5 | 7.1 | 8.5 | 9.0 | 8.3 |
| 8 | $M_{4\phi}$ | 2.7 | 11.9 | 3.8 | 13.5 | 14.0 | 13.1 |
| 9 | $M_{5p}$ | 0.8 | 10.3 | 1.3 | 12.0 | 12.3 | 11.4 |
| 10 | $M_{5\phi}$ | 0.2 | 13.3 | 0.7 | 14.3 | 14.1 | 13.5 |
| 11 | $M_{6p}$ | 0.0 | 10.9 | 0.2 | 11.3 | 10.9 | 10.6 |
| 12 | $M_{6\phi}$ | 0.0 | 12.9 | 0.2 | 11.3 | 11.1 | 11.3 |
| 13 | $M_{7p}$ | 0.0 | 10.2 | 0.1 | 8.8 | 8.1 | 8.9 |
| 14 | $M_{7\phi}$ | 0.0 | 11.5 | 0.1 | 9.5 | 8.8 | 11.5 |

34 parameters, these methods seem to select relatively simple models that miss most of the chronic treatment effects. Stepdown testing resulted, on average, in model $M_{6p}$ (mean 26.9 parameters) and resulted in quite different model selection frequencies than the other hypothesis testing approaches. Of course, the selection frequencies would differ substantially if a different (arbitrary) $\alpha$ level (say, 0.15 or 0.01) had been chosen or had the treatment impacts differed (i.e., a different model used to generate the data). The practical utility of hypothesis testing procedures is of limited value in model identification (Akaike 1981:722).

In this example, AIC selection averaged 25.4 parameters (approximately model $M_{5p}$). Both AIC and the stepdown testing did reasonably well at detecting the larger chronic effects. The differences between AIC, $AIC_c$, and $QAIC_c$ are trivial, as one would expect from the large sample sizes used in the example (Table 3.7 and Fig. 3.5). Even the use of $QAIC_c$ made relatively little difference in this example because the estimated variance inflation factor was near 1 ($\hat{c} = 1.18$).

While both AIC and likelihood ratio tests employ the maximized log-likelihood, their operating characteristics can be quite different, as illustrated in this simulated example. In addition, one must note that substantial uncertainty exists in model choice for all six approaches (Fig. 3.5). This makes the material in Chapter 4 particularly important, as this component of uncertainty should be incorporated into estimates of precision of the parameter estimators. In this example the statistical hypothesis testing approach is a poor alternative to selection based on estimating

98     3. Practical Use of the Information-Theoretic Approach

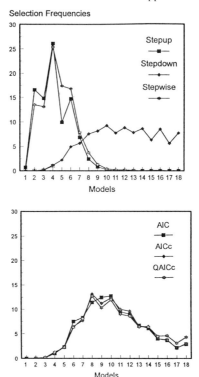

FIGURE 3.5. Model-selection probabilities for three hypothesis testing approaches (top) and three information-theoretic approaches based on 50,000 Monte Carlo repetitions of the starling data, generated under model $M_{9\phi}$.

the relative K-L information. In general, we recommend strongly against the use of hypothesis testing in model selection.

BIC selection in the face of tapering treatment effect size, a sample size of "only" 2,500 (but still quite less than $\infty$), and a generating model with 34 parameters that was not (quite) in the set of candidate models performed poorly, as theory would suggest. BIC selection most frequently chose model $M_{2\phi}$ (26.8%), followed by model $M_{1\phi}$ (24.5%), and model $M_{2p}$ (19.4%). BIC selection frequencies fell rapidly for models $M_{3p}$, $M_{3\phi}$, $M_{4p}$, and $M_{4\phi}$ (13.0, 9.8, 3.1, and 1.3, respectively). These selected models are substantially under-fit and would have poor confidence interval coverage. If the set of candidate models were to be expanded to include the generating model ($M_{9\phi}$), and as sample size increased, BIC should select model $M_{9\phi}$ with probability one. The initial number of nestlings required for BIC to select the generating model (its objective, where $K = 34$) with probability approaching one in this moderately complex example is approximately 108,000 birds (instead of the 600 used in this example).

Clearly, it would often be impossible to find and band 108,000 birds on one small island for a particular year; it might be quite unusual to have such a large number of nestlings present! Thus, to realize substantial increases in initial sample size, one must include other islands and conduct the study over several years. However, in so doing, other factors become important, and the conceptualization of truth must include obvious factors such as island and year, in addition to slightly less obvious factors such as technicians with differing resighting probabilities and islands with differing vegetation that also affects resighting probabilities. The "year" effect is not so much the actual calendar year, but a host of covariates (most unmeasured) that affect both survival and resighting probabilities in complex, nonlinear ways across time. Of course, there is individual heterogeneity that is substantial (e.g., weight, hatching data, growth rate, dispersal distance). Thus, the concept of truth, or full reality, is very complex. To think that such reality exists as an exactly true model is not useful; to think that such a true model is included in the set of candidate models seems absurd. The primary foundations of the dimension-consistent criteria do not apply in the biological sciences and medicine and the other "noisy" sciences. Reality is not fixed as sample size is increased by orders of magnitude in biological systems; rather, the target "true model" sought by BIC increases in size as $n$ increases. This simple fact is a violation of the assumptions that form the basis of BIC; however, this is allowed under the AIC-type criteria.

## *3.4.7 Further Insights Following Final Model Selection*

Selection of the best model and the relative ranking of all the candidate models is objective, given a set of candidate models, and can be implemented without the aid of subjective judgment. The formal data-based search for a best model is a key part of the analysis. In the example, model $M_{9\phi}$ was used to generate the data. Thus "truth" is known and serves here as a basis for comparisons. AIC does not try to select the model that generated these data; rather, it estimates which model is the "best approximating model" for analysis of these data in the sense of having the smallest K-L distance from approximating model to truth. Further information concerning the statistical properties of AIC-selected models in the $M_0, \ldots, M_{9\phi}$ class are given in Anderson et al. (1998).

The starling example illustrates an ideal *a priori* strategy; however, let us explore some potential realities *after* the analysis has been completed to this point. We select model $M_{\sin \phi_t, \phi_c, \sin p_t, p_c}$ as the best (with 6 parameters), but must also consider perhaps models

$M_{\sin \phi_t, \phi_c, p_t, p_c}$     ($\Delta_i = 5.34$ with only 5 parameters) and
$M_{\sin \phi_t, \phi_{ci}, \sin p_t, p_c}$     ($\Delta_i = 7.47$, with 14 parameters)

before making some final inferences (at this time, the analyst must address the variance component due to model uncertainty, Chapter 4). Model $M_{\sin \phi_t, \phi_{ci}, \sin p_t, p_c}$ is somewhat inconsistent—why would there be unrestricted weekly variation in conditional survival for the birds in the control group, but only smooth time trends in the treatment group? Perhaps this finding might lead to a thorough review of field methods in an effort to detect some anomaly? Perhaps this model should not have been in the set of candidate models considered, as it may be picking up random variation in the data? Note, too, that $\Delta_i$ is 7.47 and this model has 8 additional parameters over the AIC-selected model. It would seem that this model is a relatively poor one for these data, although it might play a role in estimating the variance component due to model-selection uncertainty.

After the analysis of the data to this point, suppose that one of the team members asks about models where there is no treatment effect on the resighting probabilities, just hypothesized treatment effects on conditional survival—either acute or chronic or both. Can new models reflecting these hypotheses be added to the set of candidate models and more AIC values computed? This question brings up several points. First, if this suggestion was made after examining the estimates of $p_{t2}, p_{t3}, p_{t4}, \ldots, p_{p7}$ vs. $p_{c_2}, p_{c3}, p_{c4}, \ldots, p_{c7}$, and noting that there seemed to be little difference between successive week-dependent pairs, then this is a form of data dredging, and any subsequent results should clearly detail the process by which the additional models were considered. *We encourage full investigation of the data to gain all possible insights; we only want investigators to reveal the extent of any data dredging that took place.* Second, if that suggestion was made on conceptual grounds rather than by studying the intermediate results, then the new class of models can be added to the list, AIC computed, the $\Delta_i$ values recomputed, and inferences made. However, in this second case, the team could be somewhat faulted for not considering the set of models more carefully in the first place.

### 3.4.8 Why Not Always Use the Global Model for Inference?

Some might argue that the global model (or another model with many parameters representing likely effects) should always be used for making inferences. After all, this model has been carefully defined by existing biological considerations and has all the effects thought to be reasonable. Why bend to the principle of parsimony and endure the various issues concerning model selection, selection uncertainty, etc.? We can illustrate problems with this approach using the starling data and the global model ($M_{7\phi}$) with 30 parameters. The key results are given here in detail for estimates of $S_i$ for $i = 1, \ldots, 7$,

| $i$ | $\hat{S}_i$ | $\widehat{se}(\hat{S}_i)$ | $CI_L$ | $CI_U$ |
|---|---|---|---|---|
| 1 | 0.946 | 0.044 | 0.858 | 1.033 |
| 2 | 0.868 | 0.052 | 0.766 | 0.969 |
| 3 | 0.905 | 0.058 | 0.792 | 1.017 |
| 4 | 0.958 | 0.057 | 0.847 | 1.069 |
| 5 | 0.950 | 0.047 | 0.859 | 1.042 |
| 6 | 0.983 | 0.051 | 0.883 | 1.084 |
| 7 | 0.973 | 0.059 | 0.858 | 1.088 |

The poor precision is illustrated by the upper confidence interval ($CL_U$), as 6 of the 7 include the value of 1 (i.e., no treatment effect; often such upper limits would be truncated at 1.0). The average coefficient of variation on the $\hat{\phi}_{ti}$ under model $M_{7\phi}$ is 4.75% vs. 2.08% under the AIC-selected model.

Attempts to select a properly parsimonious model for inference has its rewards; primarily an approximating model that has a reasonable tradeoff between bias and variance. The tradeoff between bias and variance is a byproduct of model selection where expected K-L information loss is minimized (i.e., select that model whose estimated distance to truth is the shortest). The routine reliance on the global model may have little bias, but will likely give estimates of model parameters that are unnecessarily imprecise (see Fig. 1.3B and Section 3.2.2), and this weakens the inferences made. In fact, the estimates fail to show the real patterns that can be validly inferred from these data, such as the smooth decrease in $\phi_{ti}$ and thus in $S_i$. Sometimes the global model might have 50, 100, or even 200 parameters, and this makes interpretation difficult. One cannot see patterns and structure, as there are so many parameters, most estimated with poor precision. Thus, some analysts have tried to make analyses of these estimated parameters in order to "see the forest for the trees." This has rarely been done correctly, as the estimators usually have substantial sampling correlations, making simple analysis results misleading. It is far better to embed the reduced model in the log-likelihood function and use the information-theoretic criteria to select a simple, *interpretable* approximation to the information in the data.

## 3.5 Example 4—Sage Grouse Survival

### 3.5.1 Introduction

Data from sage grouse (*Centrocercus urophasianus*) banded in North Park, Colorado, provide insights into hypothesis testing and information-theoretic criteria in data analysis. The example is taken from Zablan (1993), and additional details are found there. Here we will use data on subadult (birds less than 1 year old) and adult (birds more than 1 year old) male grouse banded on leks during the breeding season (first week of March through the third week of May), from 1973 through 1987. Sage grouse are hunted in the fall, and nearly all of the band recoveries were from hunters who shot and retrieved a banded bird and reported it to the Colorado Division of Wildlife. During this time 1,777 subadult and 1,847 adult males were banded, and the subsequent numbers of band recoveries were 312 and 270, respectively (Table 3.8). The basic theory for modeling and estimation for these types of sampling data is found in Brownie et al. (1985).

Two types of parameters are relevant here: $S_i$ is the conditional survival probability relating to the annual period between banding times $i$ to $i+1$, and $r_i$ is the conditional probability of a band from a bird being reported in year $i$, given that the bird died in year $i$. In the model building it is convenient to use $a$ as a subscript to denote age (subadult vs. adult) and $t$ to denote annual variation (e.g., $S_{a*t}$ denotes survival probabilities that vary by both age ($a$) and year ($t$); some models assume that $r_i$ is a constant, thus resulting in identifiability of the parameter $S_{15}$).

102    3. Practical Use of the Information-Theoretic Approach

TABLE 3.8. Summary of banding and recovery data for subadult (top) and adult male sage grouse banded in North Park, Colorado (from Zablan 1993).

| Year Banded | Number banded | Recoveries by hunting season |||||||||||||||
|---|---|---|---|---|---|---|---|---|---|---|---|---|---|---|---|
| | | 73 | 74 | 75 | 76 | 77 | 78 | 79 | 80 | 81 | 82 | 83 | 84 | 85 | 86 | 87 |
| 1973 | 80 | 6 | 4 | 6 | 1 | 0 | 1 | 0 | 0 | 0 | 0 | 0 | 0 | 0 | 0 | 0 |
| 1974 | 54 | | 6 | 5 | 2 | 1 | 0 | 0 | 0 | 0 | 0 | 0 | 0 | 0 | 0 | 0 |
| 1975 | 138 | | | 18 | 6 | 6 | 2 | 0 | 1 | 0 | 0 | 0 | 0 | 0 | 0 | 0 |
| 1976 | 120 | | | | 17 | 5 | 6 | 2 | 1 | 1 | 0 | 0 | 0 | 0 | 0 | 0 |
| 1977 | 183 | | | | | 20 | 9 | 6 | 2 | 1 | 1 | 0 | 0 | 0 | 0 | 0 |
| 1978 | 106 | | | | | | 14 | 4 | 3 | 1 | 0 | 0 | 0 | 0 | 0 | 0 |
| 1979 | 111 | | | | | | | 13 | 4 | 0 | 1 | 0 | 0 | 0 | 0 | 0 |
| 1980 | 127 | | | | | | | | 13 | 5 | 3 | 1 | 0 | 0 | 0 | 0 |
| 1981 | 110 | | | | | | | | | 13 | 5 | 4 | 0 | 0 | 0 | 0 |
| 1982 | 110 | | | | | | | | | | 7 | 1 | 3 | 1 | 1 | 0 |
| 1983 | 152 | | | | | | | | | | | 15 | 10 | 2 | 0 | 0 |
| 1984 | 102 | | | | | | | | | | | | 12 | 4 | 0 | 0 |
| 1985 | 163 | | | | | | | | | | | | | 16 | 4 | 1 |
| 1986 | 104 | | | | | | | | | | | | | | 5 | 2 |
| 1987 | 117 | | | | | | | | | | | | | | | 8 |
| | | 73 | 74 | 75 | 76 | 77 | 78 | 79 | 80 | 81 | 82 | 83 | 84 | 85 | 86 | 87 |
| 1973 | 99 | 7 | 4 | 1 | 0 | 1 | 0 | 0 | 0 | 0 | 0 | 0 | 0 | 0 | 0 | 0 |
| 1974 | 38 | | 8 | 5 | 1 | 0 | 0 | 0 | 0 | 0 | 0 | 0 | 0 | 0 | 0 | 0 |
| 1975 | 153 | | | 10 | 4 | 2 | 0 | 1 | 1 | 0 | 0 | 0 | 0 | 0 | 0 | 0 |
| 1976 | 114 | | | | 16 | 3 | 2 | 0 | 0 | 0 | 0 | 0 | 0 | 0 | 0 | 0 |
| 1977 | 123 | | | | | 12 | 3 | 2 | 3 | 0 | 0 | 0 | 0 | 0 | 0 | 0 |
| 1978 | 98 | | | | | | 10 | 9 | 3 | 0 | 0 | 0 | 0 | 0 | 0 | 0 |
| 1979 | 146 | | | | | | | 14 | 9 | 3 | 3 | 0 | 0 | 0 | 0 | 0 |
| 1980 | 173 | | | | | | | | 9 | 5 | 2 | 1 | 0 | 1 | 0 | 0 |
| 1991 | 190 | | | | | | | | | 16 | 5 | 2 | 0 | 1 | 0 | 0 |
| 1982 | 190 | | | | | | | | | | 19 | 6 | 2 | 1 | 0 | 1 |
| 1983 | 157 | | | | | | | | | | | 15 | 3 | 0 | 0 | 0 |
| 1984 | 92 | | | | | | | | | | | | 8 | 5 | 1 | 0 |
| 1985 | 88 | | | | | | | | | | | | | 10 | 1 | 0 |
| 1986 | 51 | | | | | | | | | | | | | | 8 | 1 |
| 1987 | 85 | | | | | | | | | | | | | | | 10 |

### 3.5.2  Set of Candidate Models

The biological objective of this study was to increase understanding of the survival process of sage grouse. Zablan (1993) used model $\{S_{a*t}, r_{a*t}\}$ as the global model, with 58 parameters. For the purpose of this particular example, the set of candidate models includes the following:

## 3.5 Example 4—Sage Grouse Survival

| Number/Model | | K | Comment |
|---|---|---|---|
| Models with $r$ constant: | | | |
| 1 | $S, r$ | 2 | Constant $S$ |
| 2 | $S_t, r$ | 16 | Year-dependent $S$ |
| 3 | $S_a, r$ | 3 | Age-dependent $S$ |
| 4 | $S_{a+t}, r$ | 17 | Age- and year-dependent $S$, no interaction |
| 5 | $S_{a*t}, r$ | 31 | Age- and year dependent $S$, interaction |
| Models with $r$ year-dependent ($t$): | | | |
| 6 | $S, r_t$ | 16 | Constant $S$ |
| 7 | $S_t, r_t$ | 29 | Year-dependent $S$ |
| 8 | $S_a, r_t$ | 17 | Age-dependent $S$ |
| 9 | $S_{a+t}, r_t$ | 30 | Age- and year dependent $S$, no interaction |
| 10 | $S_{a*t}, r_t$ | 44 | Age- and year-dependent $S$, interaction |
| Models with $r$ age-dependent ($a$): | | | |
| 11 | $S, r_a$ | 3 | Constant $S$ |
| 12 | $S_t, r_a$ | 17 | Year-dependent $S$ |
| 13 | $S_a, r_a$ | 4 | Age-dependent $S$ |
| 14 | $S_{a+t}, r_a$ | 18 | Age- and year-dependent $S$, no interaction |
| 15 | $S_{a*t}, r_a$ | 32 | Age- and year-dependent $S$, interaction |

Models with interaction terms (denoted by the $a*t$) allow each age class to have its own set of time-dependent parameters. The additive models (denoted by "+") exclude interaction terms; i.e., for model $S_{a+t}$ there is a constant difference between subadult and adult survival parameters, and the year-to-year estimates of survival probabilities for subadults and adults are parallel on a logit scale and separated by $\beta_0$ (see below). A logit transformation has been made on $S$, and age ($a$) and year (as a dummy variable, $t_i$) enter as a linear function,

$$S_{a+t} \text{ denotes logit}(S) = \beta_0 + \beta_1(a) + \beta_2(t_1) + \beta_3(t_2) + \cdots + \beta_{16}(t_{15}).$$

This approach is similar to logistic regression. However, this submodel is embedded in the log-likelihood function. Such models are often quite useful and can be biologically realistic. Models without interaction terms have fewer parameters; for example, model $S_{a+t}, r_a$ has 18 parameters, compared to 32 parameters for model $S_{a*t}, r_a$.

These 15 models plus models $\{S_a, r_{a*t}\}$ and $\{S_a, r_{a+t}\}$ and the global model $\{S_{a*t}, r_{a*t}\}$ seem like sound, initial choices; however, further biological considerations might lead one to exclude models with many parameters, in view of the relatively sparse data available (Table 3.8). We realize that this *a priori* set of models would ideally be fine-tuned in a real-world application. For example, if a long-term increase or decrease in survival was hypothesized, one might introduce submodels for survival such as

$$\text{logit}(S) = \beta_0 + \beta_1(a) + \beta_2(T) + \beta_3(a*T),$$

where $T$ indexes year as a continuous covariate $\{1, 2, \ldots, 15\}$. Alternatively, it was known that 1979 and 1984 had very severe winters; the survival probability in these years (say, $S_s$) could have been parametrized to differentiate them from the survival probability in more normal years ($S_n$). We assume that a great deal of thought has been put into the development of a set of candidate models.

A primary interest in banding studies is often to estimate survival probabilities and assess what external covariates might influence these. Thus, Zablan (1993) modeled sage grouse survival using 4 year-dependent environmental covariates (cov$_t$): winter precipitation ($wp$), winter temperature ($wt$), spring precipitation ($sp$), and spring temperature ($st$) (she provided operational definitions of these variables; we will not need to note the specific details here). Submodels with survival probabilities of the form

$$\text{logit}(S_t) = \beta_0 + \beta_1(\text{cov}_t) \quad \text{or} \quad \text{logit}(S_{t+a}) = \beta_0 + \beta_1(a) + \beta_2(\text{cov}_t)$$

could be constructed. Such submodels for survival have only 2 or 3 parameters (an intercept and 1 slope coefficient for the first submodel and an intercept, 1 age effect, and 1 slope coefficient for the second submodel), but also provide some insights into biological correlates, which themselves are time-dependent.

### 3.5.3 *Model Selection*

Zablan's analysis was done using the programs ESTIMATE and BROWNIE (Brownie et al. 1985) and SURVIV (White 1983). Zablan (1993) found that the global model $\{S_{a*t}, r_{a*t}\}$ fit the data well ($\chi^2 = 34.34$, 30 df, P = 0.268), and she computed a variance inflation factor from the global model as $\hat{c} = 34.34/30 = 1.14$. Her calculations are in agreement with ours (deviance of model $\{S_{a*t}, r_{a*t}\} = 87.85$, 80 df, $\hat{c} = 1.10$). There was little evidence of overdispersion; thus there was no compelling reason to use QAIC. The effective sample size for parameter estimation in these surveys is the sum of the number of birds banded and equaled 3,624 in this case. Zablan's global model had 58 parameters, giving the ratio $n/K = 3,624/58 = 62$; thus AIC could have been safely used instead of AIC$_c$. We used the program MARK (White and Burnham 1999) to compute MLEs of the parameters and their conditional covariance matrix, the maximized value of the log-likelihood function, AIC$_c$, and $\Delta_i$ for each of the 17 candidate models without a weather covariate and 4 models with one of the weather covariates.

AIC$_c$ selected model $\{S_a, r_a\}$ with 4 parameters (Table 3.9) among the models without a weather covariate on survival. This approximating model assumes that conditional survival and reporting probabilities are age-dependent (subadult vs. adult), but constant over years. Here, the ML estimate of adult survival probability was 0.407 ($\widehat{\text{se}} = 0.021$), while the estimated survival for subadults was higher, at 0.547 ($\widehat{\text{se}} = 0.055$). The respective percent coefficients of variation were 5.2 and 10.0. An inference here is that male subadult grouse survive at a higher rate than male adults; perhaps this reflects the cost of breeding and increased predation

## 3.5 Example 4—Sage Grouse Survival

TABLE 3.9. Candidate models for male sage grouse, number of estimable parameters ($K$), and $\Delta_i = \text{AIC}_{c(i)} - \min \text{AIC}_c$.

| Number | Model | K | AIC$_c$ | $\Delta_i$ |
|---|---|---|---|---|
| Without environmental covariates: | | | | |
| 1 | $S, r$ | 2 | 4435.13 | 4.41 |
| 2 | $S_t, r$ | 16 | 4442.30 | 11.57 |
| 3 | $S_a, r$ | 3 | 4436.20 | 5.47 |
| 4 | $S_{a+t}, r$ | 17 | 4441.76 | 11.04 |
| 5 | $S_{a*t}, r$ | 31 | 4461.11 | 30.38 |
| 6 | $S, r_t$ | 16 | 4441.94 | 11.21 |
| 7 | $S_t, r_t$ | 29 | 4447.71 | 16.98 |
| 8 | $S_a, r_t$ | 17 | 4443.22 | 12.50 |
| 9 | $S_{a+t}, r_t$ | 30 | 4447.84 | 17.12 |
| 10 | $S_{a*t}, r_t$ | 44 | 4466.17 | 35.44 |
| 11 | $S, r_a$ | 3 | 4435.44 | 4.72 |
| 12 | $S_t, r_a$ | 17 | 4443.26 | 12.53 |
| **13** | **$S_a, r_a$** | **4** | **4430.72** | **0** |
| 14 | $S_{a+t}, r_a$ | 18 | 4439.96 | 9.23 |
| 15 | $S_{a*t}, r_a$ | 32 | 4456.72 | 25.99 |
| 16 | $S_a, r_{a*t}$ | 32 | 4459.73 | 29.01 |
| 17 | $S_{a*t}, r_{a*t}$ | 58 | 4467.03 | 36.31 |
| With environmental covariates: | | | | |
| 18 | $S_{a+wp}, r_a$ | 5 | 4431.67 | 0.95 |
| 19 | $S_{a+wt}, r_a$ | 5 | 4432.68 | 1.96 |
| 20 | $S_{a+sp}, r_a$ | 5 | 4431.66 | 0.93 |
| 21 | $S_{a+st}, r_a$ | 5 | 4431.62 | 0.90 |

on breeding males. Estimated reporting probabilities (the $\hat{r}$) for first-year subadult birds were also different from adult birds (0.227 ($\widehat{se} = 0.031$) and 0.151 ($\widehat{se} = 0.008$), respectively). The use of the AIC$_c$-selected model does not indicate that there was no year-dependent variation in the parameters, only that this variation was relatively small in the sense of a bias–variance tradeoff and K-L information loss.

The estimated next-best model ($\Delta_i = 4.41$) without a covariate was model $\{S, r\}$ with only 2 parameters, while the third-best model ($\Delta_i = 4.72$) was $\{S, r_a\}$ with 3 parameters. There is relatively little structure revealed by these data; this is not surprising, as the data are somewhat sparse (Table 3.8). The AIC-selected model assumed that the survival and reporting probabilities varied by age class. Thus we considered 4 models where logit($S$) was a linear function of age (subadult vs. adult) and one of the weather covariates ($wp, wt, sp,$ or $st$), while retaining the age-specific reporting probability. The results were interesting and suggest that any of the weather covariate models are nearly tied with the AIC-selected model $\{S_a, r_a\}$ (Table 3.9).

### 3.5.4 Hypothesis Tests for Year-Dependent Survival Probabilities

Zablan (1993) computed a likelihood ratio test between models $\{S_a, r_{a*t}\}$ and $\{S_{a*t}, r_{a*t}\}$ (the global model) using the program BROWNIE and found strong evidence of year-dependent survival ($\chi^2 = 46.78$, 26 df, $P = 0.007$). The program MARK provides similar results ($\chi^2 = 46.03$, 26 df, P = 0.009). This test allowed a fairly general structure on the reporting probabilities and therefore seemed convincing and provided evidence that survival probabilities varied "significantly" by year. In fact, it might be argued that had a more simple structure been imposed on the reporting probabilities (e.g., $r_a$), the power of the test for year-dependent survival probabilities (i.e., $S$ vs. $S_t$) would have increased and the test result been even more "significant." However, contrary to this line of reasoning, the test of $\{S_a, r_a\}$ vs. $\{S_{a*t}, r_a\}$ gives $\chi^2 = 30.58$, 28 df, and P = 0.336. Still other testing strategies are possible, and it is not clear which might be deemed the best.

Given a believed year-dependence in annual survival probabilities, Zablan (1993) asked whether this variability was partially explained by one of the 4 covariates, with or without an age effect. However, she was unable to find a relationship between annual survival probabilities and any of the 4 covariates using likelihood ratio tests (the smallest P-value for the four covariates was 0.194). She used model $\{S_{a*t}, r_{a*t}\}$ (the global model) as a basis for inference.

### 3.5.5 Hypothesis Testing Versus AIC in Model Selection

An apparent paradox can be seen in the results for the male sage grouse data, and this allows us to further compare alternative paradigms of statistical hypothesis testing and AIC for model selection. The test between the 2 models $\{S_a, r_{a*t}\}$ and $\{S_{a*t}, r_{a*t}\}$ attempts to answer the questions, "Given the structure $\{r_{a*t}\}$ on the recovery probabilities, is there evidence that survival is also (i.e., in addition to age) year-dependent?" The answer provided is yes (P = 0.007 or 0.009). But this answer is seemingly in contrast to the inferences from the AIC-selected model, where there is no hint of time-dependence in either $S$ or $r$ (Table 3.9). The $\Delta_i$ values for models $\{S_a, r_{a*t}\}$ and $\{S_{a*t}, r_{a*t}\}$ are 28.068 and 30.400, respectively. AIC lends little support for a best approximating model that includes year-dependent survival or reporting probabilities.

The answer to this paradox is interesting and important to understand. The null hypothesis that $S_1 = S_2 = S_3 = \cdots = S_{14}$ for a given age class is *obviously* false, so why test it? This is not properly a hypothesis testing issue (see Johnson 1995 and Yoccoz 1991 for related issues). The test result is merely telling the investigator whether or not they have enough information (data, sample size) to show that the null hypothesis is false. *However, a significant test result does not relate directly to the issue of what approximating model is best to use for inference.* One model-selection strategy that has often been used in the past is to do likelihood ratio tests of each structural factor (e.g., $a, t, a+t, a*t$, for each of the parameters $S$ and $r$) and then use a model with all the factors that were "significant" at, say, $\alpha = 0.05$. However, there is no theory that would suggest that this strategy would

## 3.5 Example 4—Sage Grouse Survival 107

lead to a model with good inference properties (i.e., small bias, good precision, and achieved confidence interval coverage at the nominal level).

Clearly, one must also worry about the multiple testing problem here and the fact that many such tests would not be independent. Furthermore, the test statistics may not be chi-square distributed for nontrivial problems such as these, where sample sizes are far from asymptotic and, in particular, where many models contain nuisance parameters (the $r_i$). The choice of an $\alpha$-level is arbitrary as well. Many of the models in Table 3.9 are not nested; thus likelihood ratio tests are not possible between these model pairs. We note a certain lack of symmetry (this is again related to the $\alpha$-level) between the null and alternative hypotheses and how this might relate to selection of a "best approximating model" (see section 2.7.2). A very general and important problem here is how the test results are to be incorporated into building a good model for statistical inference. This problem becomes acute when there are many (say, > 8–10) candidate models. Using just the set of 17 models for the sage grouse data, one would have 136 potential likelihood ratio tests; however, some of these models were not nested, prohibiting a test between these pairs. With 136 (or even 36) test results there is no theory or unique way to decide what the best model should be and no rigorous, general way to rank the models (e.g., which model is second best? Is the second-best model virtually as good as the best, or substantially inferior?). Finally, what is to be done when test results are inconsistent, such as those found in Section 3.5.4?

The biological question regarding annual survival probabilities would better be stated as, "How much did annual survival vary during the years of study?" Has survival varied little over the 14 or 15 years, or has there been large variation in these annual parameters? Such questions are *estimation problems*, not ones of hypothesis testing. Here, the focus of inquiry should be on the amount of variation among the population parameters ($S_1, S_2, \ldots, S_{14}$) for each of the two age classes; we will denote this standard deviation among these parameters by $\sigma_s$. Of course, if we knew the parameters $S_i$, then

$$\hat{\sigma}_s = \sqrt{\sum_{i=1}^{14}(S_i - \bar{S})^2/13}.$$

We next ask why the AIC procedure did not pick up the "fact" that survival varied by year? The reason is simple; AIC attempts to select a parsimonious approximating model for the observed data. In the sense of K-L information loss or a tradeoff between bias and variance, it was poor practice to add some 54 additional parameters (the difference in parameters between models $\{S_a, r_a\}$ and $\{S_{a*t}, r_{a*t}\}$) or even 26 additional parameters (the difference between models $\{S_a, r_{a*t}\}$ and $\{S_{a*t}, r_{a*t}\}$) to model the variation in $S_i$ or $r_i$ (Table 3.9). Note that the difference in AIC values for model $\{S_a, r_a\}$ versus model $\{S_a, r_{a*t}\}$ is 29.01, suggesting that model $\{S_a, r_{a*t}\}$ is highly over-fit. Whether differences among survival probabilities are large enough to be included in a model is a *model-selection* problem, not one of hypothesis testing.

TABLE 3.10. MLEs of year-dependent survival probabilities under model $\{S_{a*t}, r_a\}$. The first 15 estimates relate to grouse banded as adults, while the second set of estimates relate to first-year survival of subadults. The model assumes that the subadults become adults the second year after banding and thus have the same year-dependent survival probabilities as birds banded as adults.

| Year($i$) | $\hat{S}_i$ | Standard Error | 95% Confidence Interval[a] | |
|---|---|---|---|---|
| | | | Lower | Upper |
| 1 | 0.462 | 0.128 | 0.238 | 0.702 |
| 2 | 0.500 | 0.092 | 0.327 | 0.673 |
| 3 | 0.357 | 0.073 | 0.230 | 0.508 |
| 4 | 0.412 | 0.074 | 0.277 | 0.561 |
| 5 | 0.464 | 0.073 | 0.328 | 0.606 |
| 6 | 0.507 | 0.069 | 0.375 | 0.639 |
| 7 | 0.465 | 0.066 | 0.340 | 0.595 |
| 8 | 0.357 | 0.062 | 0.246 | 0.486 |
| 9 | 0.397 | 0.063 | 0.282 | 0.524 |
| 10 | 0.340 | 0.061 | 0.233 | 0.466 |
| 11 | 0.321 | 0.063 | 0.212 | 0.455 |
| 12 | 0.358 | 0.073 | 0.231 | 0.509 |
| 13 | 0.171 | 0.071 | 0.071 | 0.355 |
| 14 | 0.339 | 0.133 | 0.138 | 0.621 |
| 15 | 0.549 | 0.129 | 0.305 | 0.771 |
| 1 | 0.725 | 0.114 | 0.462 | 0.891 |
| 2 | 0.629 | 0.152 | 0.321 | 0.859 |
| 3 | 0.524 | 0.106 | 0.323 | 0.717 |
| 4 | 0.528 | 0.112 | 0.316 | 0.731 |
| 5 | 0.566 | 0.093 | 0.383 | 0.732 |
| 6 | 0.446 | 0.120 | 0.237 | 0.677 |
| 7 | 0.386 | 0.117 | 0.193 | 0.623 |
| 8 | 0.513 | 0.110 | 0.307 | 0.715 |
| 9 | 0.497 | 0.118 | 0.282 | 0.713 |
| 10 | 0.615 | 0.111 | 0.389 | 0.801 |
| 11 | 0.547 | 0.101 | 0.351 | 0.729 |
| 12 | 0.368 | 0.121 | 0.173 | 0.618 |
| 13 | 0.440 | 0.107 | 0.251 | 0.649 |
| 14 | 0.744 | 0.104 | 0.498 | 0.895 |
| 15 | 0.695 | 0.111 | 0.450 | 0.864 |

[a] Based on a back transformation of the interval endpoints on a logit scale (Burnham et al. 1987:214).

Estimates of the 30 survival probabilities under model $\{S_{a*t}, r_a\}$ are given in Table 3.10. The average of the 15 estimates of adult survival was 0.400, nearly the same as that from the 4 parameter model selected by $AIC_c$ (0.407). However, the average percent coefficient of variation for each $\hat{S}_i$ was 20.4 for model $\{S_{a*t}, r_a\}$ compared to only 5.2 for $\hat{S}$ in the $AIC_c$-selected model. Thus, the AIC-selected model indicates that the best estimate of annual survival in a particular year is merely $\hat{S}$ (from model $\{S_a, r_a\}$).

The situation was similar for subadult survival; the average survival from model $\{S_{a*t}, r_a\}$ was 0.548 compared to 0.547 for the $AIC_c$-selected model. The respective average percent coefficients of variation were 20.6% and 10% for models $\{S_{a*t}, r_a\}$ and $\{S_a, r_a\}$. In summary, 54 (or even 26) additional parameters "cost too much" in terms of increased variability of the estimates (see Fig. 1.3B and Table 3.10) and reflect substantial over-fitting. The lack of precision illustrated in Table 3.9 for model $\{S_{a*t}, r_a\}$ was worse still when using model $\{S_{a*t}, r_{a*t}\}$; coefficients of variation were 35.6% for adult survival and 30.5% for subadult survival. The model suggested by the hypothesis testing approach had 58 parameters, while the $AIC_c$-selected model has only 4 parameters. This illustrates the cost of using overparametrized models, even though the results of hypothesis tests clearly show "significance" for year-dependent survival (and reporting) probabilities. Models $\{S_a, r_{a*t}\}$ and $\{S_{a*t}, r_{a*t}\}$ are very general for these data and lie far to the right of the bias-variance tradeoff region in Fig. 1.2. Zablan recognized the problems in using model $\{S_{a*t}, r_{a*t}\}$ and commented, "While significant differences were found between survival and recovery rates of males and of both age classes, and between years, survival estimates had unacceptably wide confidence intervals."

The researcher could use the $AIC_c$-selected model $\{S_a, r_a\}$ to obtain estimates of parameters and then proceed to obtain an estimate $\sigma_s$ for each of the two age classes, using model $\{S_{a*t}, r_{a*t}\}$, or $\{S_{a*t}, r_{a+t}\}$, if desired. It is not trivial to embed this parameter into the likelihood framework, allowing an ML estimate of $\sigma_s$; this becomes a "random effects" model. However, this is a problem in "variance components," and consistent estimates of $\sigma_s$ can be computed using, say, model $\{S_{a*t}, r_a\}$, following, for example, Anderson and Burnham (1976: 62–66); and Burnham et al. (1987:260–269). This approach is conceptually based on the simple partitioning of the total variance ($\text{var}(\hat{S}_i)$) into its two additive components: the variance in the population parameters $\sigma_s^2$ and the conditional sampling variance ($\text{var}(\hat{S}_i \mid model)$). The approach assumes that the true $S_i$ are independently and identically distributed random variables (in this case, these assumptions are weak and the effect somewhat innocent). Here, one has ML estimates of the sampling covariance matrix and can compute $\text{var}(\hat{S}_i)$ directly from the estimates $\hat{S}_i$; thus, by subtraction one can obtain an estimate of $\sigma_s^2$. Thus, inferences could be made from the AIC-selected model, hopefully after incorporating model-selection uncertainty.

The estimate of $\sigma_s$ would provide some insight into the variation in the survival parameters; this would be done in the context that one *knows* that the $S_i$ vary. Exact details of the optimal methodology would take us too far afield; however, some unpublished results seem exciting. For the adult data, $\hat{\sigma}_s = 0.0426$, 95% profile likelihood interval [0, 0.106], and cv = 10.8% on annual survival probability and for the subadult data $\hat{\sigma}_s = 0.0279$, 95% profile likelihood interval [0, 0.129], and cv = 4.9% on $S$. Thus, one can infer that the relative variation in the true, annual survival probabilities was fairly small (cv ≈ 5–10%). Thus, the large variation in the estimates of annual survival probabilities under model $\{S_{a*t}, r_a\}$ (Table 3.9) is due primarily to sampling variation, as the large estimated standard errors suggest.

Additional details, including shrinkage estimates of annual survival probabilities, will not be given here.

### 3.5.6 A Class of Intermediate Models

Ideally, the number of parameters in the various candidate models would not have large increments (see Section 2.7.2). In the grouse models, a submodel for the survival probabilities without year-dependent survival might have one or two (if age is included) parameters, while a model with year-dependent survival would have as many as 30 parameters (15 for each of the two age groups). Large differences in the number of parameters between certain candidate models are not ideal and one should consider intermediate models while deriving the set of candidate models. Zablan's (1993) various covariate models represent an example; here an intercept and slope parameter on one of the covariates would introduce 2 parameters (3 with age) instead of 15 (30 with age). In contrast with the hypothesis testing approach, AIC-selection showed the 4 weather covariate models to be essentially tied with the AIC-selected model $\{S_a, r_a\}$ (Table 3.9). These results suggest the importance of the weather covariates and provide a possible reason (hypothesis) as to *why* annual survival varies.

Hypothesis testing and AIC in model selection are fundamentally very different paradigms in model selection and in drawing inferences from a set of data. In the sage grouse example, $AIC_c$ tries to select a model that well approximates the information in the data. That selected model then provides estimates of the parameters $S_1, S_2, \ldots, S_{15}$ for each age group in the sense of K-L information loss (or a bias vs. variance tradeoff). That is, an estimate of average survival ($\hat{S}$) from model $\{S_a, r_a\}$ (i.e., 0.407 for adults) would be used to estimate, for example, $S_5$ for adult grouse (hence, $\hat{S}_5 = 0.407$, 95% confidence interval [0.368, 0.448]), and this estimate would have better inferential properties than using model $\{S_{a*t}, r_a\}$, whereby $S_5$ would be estimated using the year-specific estimator $\hat{S}_5$ (see Table 3.10, where this estimate is given as 0.464 with 95% confidence interval of 0.328 to 0.606). If inference about the conditional survival in the fifth year is made from the general model $\{S_{a*t}, r_{a*t}\}$, then the estimate is 0.336, and the precision is worse yet (95% confidence interval of [0.168, 0.561]). In the last two cases, the precision is relatively poor (e.g., compare Figs. 1.3B and C for further insights). We will revisit this example in Chapter 4, where model-selection uncertainty is quantified and the adequacy of the standard errors of estimates, conditional on model $\{S_a, r_a\}$, is assessed.

## 3.6 Example 5—Resource Utilization of *Anolis* Lizards

This example illustrates the use of information-theoretic criteria in the analysis of count data displayed as multidimensional contingency tables (see also Sakamoto 1982). Schoener (1970) studied resource utilization in species of lizards of the genus *Anolis* on several islands in the Lesser Antilles, in the Caribbean. Here we use his data collected on *Anolis grahami* and *A. opalinus* near Whitehouse, on Jamaica, as provided by Bishop et al. (1975). These data have been analyzed by Fienberg (1970), Bishop et al. (1975), and McCullagh and Nelder (1989) and Qian et al. (1996), and the reader is urged to compare the approaches given in these papers to that presented here.

In his general studies of species overlap, Schoener (1970) studied the two species of lizards in an area of trees and shrubs that had been cleared for grazing. The height ($< 5$ or $\geq 5$ ft) and diameter ($< 2$ or $\geq 2$ in) of the perch, insolation (sunny or shaded), and time of day (roughly early morning, midday and late afternoon) were recorded for each of the two species of lizard seen. Data were taken on each individual only once per "census"; data were not recorded if the lizard was disturbed, and the census route was varied considerably from one observation period to the next. The data of interest for this example can be summarized as a $2 \times 2 \times 2 \times 3 \times 2$ contingency table corresponding to height (H), diameter (D), insolation (I), time of day (T), and species (S), shown in Table 3.11 (from McCullagh and Nelder 1989:128–135). The data on 546 observations of lizards appear in the table with 48 cells.

## 3.6.1 Set of Candidate Models

Because these data were collected nearly 30 years ago and we have little expertise in lizard ecology and behavior, decisions concerning an *a priori* set of candidate models will necessarily be somewhat contrived. We will focus attention on modeling and model selection issues as an example and comment on several inference issues. In reading Schoener's (1970) paper and the literature he cited, it would

TABLE 3.11. Contingency table of site preference for two species of lizard, *Anolis grahami* and *A. opalinus* (denoted g and o, respectively) on the island of Jamaica (from Schoener 1970).

| | | | Time of Day (T); Species (S) | | | | | |
|---|---|---|---|---|---|---|---|---|
| Insolation | Diameter | Height | Early morning | | Midday | | Late afternoon | |
| (I) | (D) in | (H) ft | g | o | g | o | g | o |
| Sunny | $\leq 2$ | $< 5$ | 20 | 2 | 8 | 1 | 4 | 4 |
| | | $\geq 5$ | 13 | 0 | 8 | 0 | 12 | 0 |
| | $> 2$ | $< 5$ | 8 | 3 | 4 | 1 | 1 | 3 |
| | | $\geq 5$ | 6 | 0 | 0 | 0 | 1 | 1 |
| Shaded | $\leq 2$ | $< 5$ | 34 | 11 | 69 | 20 | 18 | 10 |
| | | $\geq 5$ | 31 | 5 | 55 | 4 | 13 | 3 |
| | $> 2$ | $< 5$ | 17 | 15 | 60 | 32 | 8 | 8 |
| | | $\geq 5$ | 12 | 1 | 21 | 5 | 4 | 4 |

seem that a model with all the main effects (i.e., H, D, I, T, and S) might serve as a starting point for models to be considered. Several second-order interactions might be suspected, e.g., H * T and H * I and I * T. If the two species are partitioning their resources, then models with H * S, D * S, I * S, and T * S included should be reasonable. As the study was designed and data were collected, it was probably evident that site occupancy was affected by several variables as well as some interactions. This might suggest that a model with all main effects and second-order interactions might be considered. Then issues remain concerning possible higher-order interactions. On biological grounds, it might seem reasonable to consider third-order terms such as H * D * I, H * D * T, and H * D * S; or further, to add H * I * S, H * T * S, and I * T * S. Finally, the second-order term D * T seems unlikely to be important; thus some models without this term were considered. We will use the short set of models in Table 3.12 for illustrative purposes here. Even in the late 1960s, T. W. Schoener and his colleagues, including S. E. Fienberg, could have developed a better set of *a priori* candidate models than ours.

### 3.6.2 Comments on Analytic Method

We used a loglinear model with Poisson errors following Agresti (1990: chapter 5 and pages 453–456), and the analysis was made conditional on $\sum n_j$. Specifically, we used the SAS program GENMOD (SAS 1985). The form of the log-likelihood can be expressed as

$$\log(\mathcal{L}) = \sum_{j=1}^{48} \left( n_j \cdot \log(\mu_j) - \mu_j \right) - \sum_{j=1}^{48} \log(n_j!),$$

where $n_j$ is the number of observations in cell $j$ with Poisson mean $\mu_j$. Then one has the log-linear model $\log(\mu_i) = X\beta$ as in analysis of variance. Thus, $\beta$ is the vector of effects and the grand mean. The final term in the log-likelihood is a constant; thus SAS GENMOD omits this term, and the resulting log-likelihood is positive. AIC is computed in the usual manner, even though the AIC values are scaled by $\sum_{j=1}^{48} \log(n_j!)$. Note that such arbitrary, additive constants are not present in the $\Delta_i$ values. Several software packages allow ML estimates from discrete data such as these and provide a number of relevant analysis options (see summary in Agresti 1990: 484–488), and some also print AIC values (e.g., SAS). However, we do not know of any packages that print $AIC_c$. Section 2.4.2 provides additional guidance if overdispersion is thought to exist in count data. We note Agresti's (1990) comment, "In practice, we learn more from *estimating* descriptive parameters than from *testing hypotheses* about their values."

### 3.6.3 Some Tentative Results

Clearly, a model with only main effects is not adequate for these data (see Table 3.12 for a summary of the candidate models along with $K$, log-likelihood, and differences, $\Delta_i$). A model with the 5 main effects and all the second-order interac-

TABLE 3.12. Summary of *a priori* models considered, the number of model parameters ($K$), log-likelihood, and $AIC_c$ for the lizard data (from Schoener 1970). The model with the minimum $AIC_c$ is shown in bold print.

| Model | $K$ | $\log(\mathcal{L})$ | $\Delta_i$ |
|---|---|---|---|
| All main effects, H D I T S | 7 | 1181.08 | 100.88 |
| All main effects and second-order interactions | 21 | 1181.86 | 1.32 |
| **Base[1] but drop DT** | **19** | **1180.52** | **0.00** |
| Base[1] plus HDI, HDT and HDS terms | 25 | 1182.97 | 7.10 |
| Base[1] plus HDI, HDS, HIT, HIS, HTS, and ITS | 30 | 1185.75 | 11.55 |
| Base[1] plus HIT, HIS, HTS, and ITS | 28 | 1185.48 | 8.09 |
| Base[1] plus HIS, HTS, and ITS | 26 | 1184.01 | 7.02 |
| Base[1] plus HIT, HIS, HTS, and ITS, but drop DT | 26 | 1184.15 | 6.75 |
| Base[1] plus HIT, HIS, and ITS, but drop DT | 24 | 1183.29 | 4.46 |
| Base[1] plus HIT and HIS, but drop DT | 22 | 1182.18 | 2.69 |

[1] "Base" is a model with the 5 main effects plus all second-order interaction terms.

tions (called the "base" model, for convenience) is actually quite good ($\Delta_i = 1.32$ in the set of models considered). Various models including a few third-order interactions did nothing to improve upon this ("base") model ($\Delta_i$ varied from 7.02 to 11.55). Deleting the second-order term D ∗ T provided one model that was also relatively good (Table 3.12). It would seem that site occupancy is related to all the main effects measured and that many important second-order interactions are suggested. Some support might exist for a few third-order interactions, but this becomes much more speculative. The data are unable to provide definitive results; instead, we are left with 3 to 4 models that offer somewhat equal utility. Note that if another model were to be added, the $\Delta_i$ values would need to be recomputed (but not the AIC values). Such values are always with respect to the minimum AIC model, given a set of candidate models, and inferences derived from the data via a set of models are effectively conditional on the set of models considered. Strict experimentation might be expected to provide additional insights into the issue of resource utilization in these lizard species.

The best model of those considered here was a model with the 5 main effects and all second-order interaction terms except D ∗ T. This model had $K = 19$ parameters and fit the data well ($\chi^2 = 23.76$, 29 df, P = 0.741), with no evidence of overdispersion. Four models have $\Delta_i$ values that are < 5, while 4 more models have $\Delta_i$ values less than 8; this suggests substantial model uncertainty. We will return to this example in Chapter 4 after several methods for dealing with model selection uncertainty have been introduced.

If our original interest had been only on differences in resource use by the two species, we could examine a model with all 5 main effects (H, D, I, T, S) vs. a similar model without a species effect (H, D, I, T). Here, there might have been only 2 candidate models considered. The difference $\Delta_i$ for these two models is 163.70, indicating that the two species were using their resources differently. Alternatively, we could make a similar comparison of two models, but also include the relevant second-order interactions,

H, D, I, T, S, H*D, H*I, H*T, H*S, D*I, D*T, D*S, I*T, I*S, T*S

vs.

H, D, I, T, H*D, H*I, H*T, D*I, D*T, I*T.

Here, the difference $\Delta_i$ is 84.63 and again indicates that the two species are utilizing their habitat differently. Other alternative analysis might be pursued if more were known about the study design and field protocol.

Bishop et al. (1975) performed a number of hypothesis tests to build a model, and this resulted in a model with all the main effects plus the second-order interaction terms $H * D$, $H * S$, $D * S$, $T * S$ and the third-order term $H * D * S$ ($K = 13$); we find that here $\Delta_i = 7.26$ (relative to the model with the minimum AIC value, above). If this study is considered to be merely exploratory, then one might consider a much wider class of candidate models, and many models with "small" $\Delta_i$ could be found. However, the number of possible models would be very large (perhaps 2,000, depending on what rules might be applied concerning the presence of lower-order effects if higher-order effects are included in the model). Even with powerful computing equipment, an exhaustive study of all models for this $2 \times 2 \times 2 \times 3 \times 2$ table is nearly prohibitive (see Agresti 1990: 215). This illustrates again the importance of a set of good *a priori* models, even if the study is somewhat exploratory. Of course, data dredging could be very effective in finding a model that "fits" *these* data. However, the goals in data analysis usually stress an inference about the population or process, not merely data description. Thus, the results from intensive data dredging should be viewed as tenuous.

The rigorous analysis of multidimensional contingency tables remains problematic because of the large number of possible models. Gokhale and Kullback (1978:19) suggest that conclusions drawn from contingency tables should be only exploratory. While a good *a priori* set of models seems essential, it may be difficult to forecast higher-order interactions in many situations. Several other problematic issues are associated with contingency table analysis. Other analytic approaches exist (Manly et al. 1993), and alternative model formulations besides a log link and Poisson errors can be considered (see Santer and Duffy 1989 for additional information).

## 3.7 Summary

The purpose of the "analysis" of empirical data is not to attempt to find the "true model"—not at all. Instead, we wish to find a "best approximating model," based on the data, and then develop statistical inferences from this model. In some sense, the model is *the* inference from the data available. We search, then, not for a "true model," but rather for a *parsimonious model* giving an accurate approximation to the interpretable information in the data at hand. Data analysis involves the question, "What level of model complexity will the data support?" and both under-

Solomon Kullback was born in 1907 in Brooklyn, New York. He graduated from the City College of New York in 1927, received an M.A. degree in mathematics in 1929, and completed a Ph.D. in mathematics at the George Washington University in 1934. Kully, as he was known to all who knew him, had two major careers: one in the Defense Department (1930–1962) and the other in the Department of Statistics at George Washington University (1962–1972). He was chairman of the Statistics Department from 1964–72. Much of his professional life was spent in the National Security Agency, and most of his work during this time is still classified. Clearly, most of his studies on information theory were done during this time. Many of his results up to 1958 were published in his 1959 book "*Information Theory and Statistics*." Additional details on Kullback may be found in Greenhouse (1994) and Anonymous (1997).

and over-fitting are to be avoided. Larger data sets tend to support more complex models, and the selection of the "size" of the model represents a tradeoff between bias and variance. Proper parsimony must reflect good achieved confidence interval coverage as an integrating metric. We view the "analysis" as thoughtful *a priori* model building, model selection, parameter estimation, and quantification of uncertainty. Then, inference is based on this 4-part package and its underlying philosophy. Information-theoretic approaches are based on a fundamental measure, the Kullback–Leibler distance. AIC and $AIC_c$ are estimates of the relative K-L distance and are simple to compute and interpret, but flexible enough to be used in a very wide variety of practical situations in the biological sciences (see Burnham et al. 1996 for a comprehensive example dealing with the Northern Spotted Owl, *Strix occidentalis carina*, including an exhaustive assessment of goodness-of-fit).

Data from the simulated starling experiment and the banded sage grouse population both illustrate moderate complexity. In both examples, inference based on the global model would have been (needlessly) poor, inefficient, and difficult to

Richard Arthur Leibler was born in Chicago, Illinois, on March 18, 1914. He received bachelor's and master's degrees in mathematics from Northwestern University and a Ph.D. in mathematics from the University of Illinois and Purdue University (1939). After serving in the Navy during WWII, he was a member of the Institute for Advanced Study and a member of the von Neumann Computer Project 1946–48. From 1948–1980 he worked for the National Security Agency (1948–58 and 1977–80) and the Communications Research Division of the Institute for Defense Analysis (1958-77). He is the president of Data Handling Inc., a consulting firm for the Intelligence Community (1980–present).

interpret. The global model for the sage grouse data had 58 parameters, and the resulting MLEs had wide confidence intervals (Table 3.10) and dozens of substantial sampling correlations (e.g., $\widehat{\text{corr}}(\hat{S}_i, \hat{S}_{i+1})$ and $\widehat{\text{corr}}(\hat{S}_i, \hat{r}_i)$; see Brownie et al. 1985), making it difficult to examine patterns, trends, and associations (e.g., are age-specific survival probabilities linked across years? are there declines in survival probabilities over years?).

People often attempt to perform some further, separate analysis (e.g., multiple regression) of the estimates from a very general model in an effort to understand the structure of the process and gain insight into its behavior. However, such external analyses are not easily done correctly; the estimates for the starling and grouse data have a multinomial, not normal, variance structure, the variances are unequal (not constant); and successive estimates are dependent (not independent). When the analyst feels a need for such further "external" analysis of the parameter estimates of a fitted model, it is clear that a properly parsimonious, easily interpretable model has not been achieved, and hence the analysis has partially failed.

We recommend carefully developing a set of candidate models to explore the science of the issue (e.g., embed additive submodels for recovery probabilities into the log-likelihood using an appropriate link function) and obtain the MLEs under these models. Then one can focus on model selection to identify a properly parsimonious model(s) (the models in the set of candidates that are "close" to truth in the K-L information sense) that will serve as a basis for inference. Model-selection uncertainty should be addressed (Chapter 4) to obtain good estimates of precision of the model parameters. In contrast, if only the high dimensional global model is employed and estimates of parameters obtained by ML or LS, then the purpose of the analysis is virtually defeated, because a parsimonious interpretation of results may be impossible, patterns often cannot be found, and estimates are very imprecise. Zablan (1993) used a global model with 58 parameters in her analysis of the sage grouse data and correctly observed, "... survival estimates had unacceptably wide confidence intervals."

Statistical analysis of empirical data should not be just number crunching, given only a set of data (the numbers). The cement hardening data (Section 3.2) have too frequently been analyzed without examining the collection and treatment of the data—the important *a priori* considerations that we have stressed here. What was known about the chemistry of cement hardening (if not in the 1930s, then at least by the 1960s)? For example, can cement be expected to harden well with only a single ingredient? If not, this might have put the 4 single-variable models out of consideration. Given the chemical composition of variables $x_2$ and $x_4$ ($3CaO \cdot SiO_2$ vs. $2CaO \cdot SiO_2$), is it surprising that these variables are highly correlated ($r = -0.973$)? The situation involving variables $x_1$ and $x_3$ is similar. Surely, these conditions were known *a priori* and could have affected the models in the set to be considered. How many analysts have known, or bothered to find out, that $\sum_{j=1}^{4} x_{ij}$ = a constant for each of the 13 observations and therefore excluded the 4-variable model from consideration, based on these *a priori* grounds? Instead, unthinking approaches have been the *modus operandi*, and "all possible models" have frequently been tried. "Let the computer find out" is a poor strategy for researchers who do not bother to think clearly about the problem of interest and its scientific setting. *The sterile analysis of "just the numbers" will continue to be a poor strategy for progress in the sciences.*

The presentation of results in scientific publications should detail the logic in arriving at a set of candidate models and present and discuss at least the $\log(\mathcal{L})$ values, $K$, AIC, and $\Delta_i$ values for the better models and the corresponding models themselves. Such information allows the merits of each model to be compared to the best model. Material in Chapter 4 will allow additional insights that should often be included in published results. If data dredging followed the formal analysis, this activity should be clearly noted and the insights from these activities provided. We would not encourage the word "*significant*" be used in publication of results, as this word is so tied to statistical null hypothesis testing, the arbitrary $\alpha$ level, and the resulting $P$ values.

# 4
# Model-Selection Uncertainty with Examples

## 4.1 Introduction

The understanding of model-selection uncertainty requires that one consider the process that generates the sample data we observe. For a given field, laboratory, or computer simulation study, data are observed on some process or system. If a second, independent, data set could be observed on the same process or system under nearly identical conditions, the new data set would differ somewhat from the first. Clearly, both data sets would contain information about the process, but the information would likely be slightly different, by chance. An obvious goal of data analysis is to make an inference about the process based on the data observed. A less obvious goal of data analysis is to make inferences about the process that are not overly specific with respect to the (single) data set observed. That is, we would like our inferences to be robust, with respect to the particular data set observed, in such a way that we tend to avoid problems associated with over-fitting (overinterpreting) the limited data we have. Thus, we would like some ability to make inferences about the process as if a large number of other data sets were also available. The interpretation of a confidence interval is similar; i.e., in repeated samples from the process, 95% of the data sets will generate a confidence interval that includes the true parameter value. This idea extends to the idea of generating a confidence (sub) set of the models considered such that with high relative frequency, over samples, that set of models contains the actual K-L best model of the set of models considered, while being as small a subset as possible (analogous to short confidence intervals).

With only a single data set, one could use $AIC_c$ and select the best model for inference. However, *if* several other independent data sets were available, would the same model be selected? The answer is, perhaps it would be; but generally there would be variation in the selected model from data set to data set, just as there would be variation in parameter estimates over data sets, given that the same model is used for analysis. The fact that other data sets might suggest the use of other models leads us to model-selection uncertainty and hence another variance component that should be included in measures of precision of parameter estimates.

If analysts select a model using $AIC_c$ (or using some other criterion such as cross validation) and make estimates of the sampling variance of an estimated parameter in that model, they invariably do so conditionally on the selected model. The estimated precision will then likely be overestimated, because the variance component due to model-selection uncertainty has been omitted. The standard errors we compute conditionally on the model will be too small, confidence intervals will be too narrow, and achieved coverage will be below the nominal level. Chatfield (1995b) reviews this issue in some detail; also see Rencher and Pun (1980), Chow (1981), Hurvich and Tsai (1990), Pötscher (1991), Goutis and Casella (1995), and Kabaila (1995).

Based on simulation studies, we are usually surprised by how much variation there is in selecting a parsimonious model for a given problem. It is demonstrably the case that in many real-world problems there is substantial model-selection uncertainty. We will return to the cement data used in Section 3.2 to illustrate model-selection uncertainty. Here, 13 observations were made on 4 variables (the data set is shown in Table 3.1) and the $AIC_c$-selected model {12} (i.e., with independent variables $x_1$ and $x_2$) with $K = 4$ parameters ($\beta_0, \beta_1, \beta_2, \sigma^2$). To estimate model-selection uncertainty for this data-analysis problem we generated 10,000 bootstrap samples from these data. The parameters, in each of the 15 models shown in Table 3.2, were estimated and $AIC_c$ was computed for each bootstrap sample. The following summary shows the model-selection frequencies applying $AIC_c$ (models not shown had zero selections) to each of the 10,000 bootstrap samples. Also shown are $\Delta_i$ values from the original data (from Table 3.2):

| Model | K | Sel. Freq. | $\Delta - AIC_c$ |
|---|---|---|---|
| {12} | 4 | 0.5338 | 0.0000 |
| {124} | 5 | 0.0124 | 3.1368 |
| {123} | 5 | 0.1120 | 3.1720 |
| {14} | 4 | 0.2140 | 3.3318 |
| {134} | 5 | 0.0136 | 3.8897 |
| {234} | 5 | 0.0766 | 8.7440 |
| {1234} | 6 | 0.0337 | 10.5301 |
| {34} | 4 | 0.0039 | 14.4465 |

As might be expected with such a small sample size, the selection frequencies varied substantially, and model {12} was selected as the best in only about 53% of the bootstrapped samples. Model {14} was selected 21% of the time; recall that the simple correlation between variables $x_2$ and $x_4$ was $r = -0.973$; thus it is a quite reasonable result that models {12} and {14} are somewhat aliased. Other models had much lower selection frequencies in this example.

AIC model selection is most often thought of as a way to select just the best model. However, this information-theory-based approach is more general than this simplistic concept of model-selection. Given a set of models (specified independent of the sample data before being fit to that data) we can make inferences based on the entire set of models, not just based conditionally on the selected best model. Part of this inference process involves ranking the fitted models from best to worst and then going a step further and calibrating the relative plausibility of each fitted model ($M_i$) by a weight of evidence ($w_i$) relative to the selected best model. Then using the full set of models, and associated information such as conditional sampling variances and the model weights, we can produce unconditional inferences over the entire set of models, such as unconditional sampling variances or model-averaged parameter estimates. Thus model-selection uncertainty is a substantial subject in its own right, well beyond just the issue of what is the best model.

The set of issues addressed in this chapter are twofold. First, how can model-selection uncertainty be quantified? Second, how can estimators of parameters and estimates of their precision incorporate model-selection uncertainty? More research is required to develop and understand general methods, but we provide several approaches that seem useful. Attempts to make estimates of precision without being conditioned on a specific model will likely be a research area for some years to come.

## 4.2 Methods for Assessing Model-Selection Uncertainty

This section presents a variety of methods that can be used to (1) measure the uncertainty associated with model-selection, either as regards what is the actual best model, or as regards uncertainty about selected variables; and (2) provide measures of unconditional precision (e.g., sampling variance, standard errors, and confidence intervals) for parameter estimators, rather than just using the usual measures of sampling uncertainty conditional on a selected model. Also, there are estimation and inference methods presented here that arise specifically because several models are considered (i.e., model averaging for parameter estimation, and a confidence set on "good" models) that do not arise when only one model is considered. Some of these ideas are well developed from the Bayesian perspective (especially model averaging; see, e.g., Madigan and Raftery 1994, Draper 1995, Raftery 1996; Newman 1997 provides an application), but do not seem to have been adapted yet into applied frequentist inferences. The theoretical basis for these approaches and ideas is not given here; rather, this section just presents a variety of practical methods. Justification of these methods is either in the statistical literature, or in sections of Chapter 6 (or in both places). These methods are in need of more evaluation (some more so than others) to understand their properties and limitations, but these

## 4.2 Methods for Assessing Model-Selection Uncertainty

methods are sound enough to present. The purpose of this section is to present in one place various methods for measuring the uncertainty of inferences made after formal, objective, information-theory-based model-selection. The application of these methods is illustrated in the rest of this chapter (i.e., after Section 4.2) and in Chapter 5.

There are three general approaches to assessing model-selection uncertainty: (1) theoretical studies, mostly using Monte Carlo simulation methods; (2) the bootstrap applied to a given set of data; and (3) utilizing the set of AIC differences (i.e., $\Delta_i$) from the set of models fit to data. Monte Carlo and bootstrap methods are both computer intensive. Useful insights can be obtained about model-selection and associated uncertainties by extensive Monte Carlo simulations of model-selection; however, use of the bootstrap and $\Delta_i$ values apply directly to a single data set; hence they represent our focus here. The bootstrap is a type of Monte Carlo method applied case by case based on realized data (Efron and Tibshirani 1993, Mooney and Duval 1993). We assume that the reader has sufficient understanding of these ideas and methods, so we do not need to elaborate much, except for a little about the bootstrap. The bootstrap has enormous potential for assessing model-selection uncertainty. However, its computer-intensive nature will hinder its use. We believe that at least 1,000 bootstrap reps are needed in any application, preferably 10,000, for reliable results, and thus currently it could take days of computer time to apply the bootstrap to complex data-analysis cases. In contrast, the third method (i.e., use of $\Delta_i$ values) is easily computable and merits more development.

Monte Carlo investigations generate data from a stated generating model and 10,000 to 100,000 independent samples of generated data. Then we apply model-selection to each sample and summarize resultant relative frequencies of models selected and other information of interest, such as variation of the $\Delta_i$ and unconditional variances of parameter estimators over models. For results to apply fully to the K-L model-selection paradigm envisioned for real data, the generating model (which is truth, $f$, in the simulation study) should be complex and not contained in the set of approximating models, $M_1, \ldots, M_R$ (we now use this notation to denote the models, rather than the notation $g_i$ or $g_i(x \mid \theta)$). Many simulation studies on model-selection do not at all meet these conditions (e.g., Wang et al. 1996). Rather, they are far too simplistic because (1) a simple generating model is used (so no tapering effects and small $K$), (2) the set of models considered contains the generating model (i.e., contains "truth"), and (3) the model-selection goal is usually to select the generating model (hence to select "truth"). None of these features are realistic of real data-analysis problems; hence we discount the results of such simulation studies as appropriate guides to real-world model-selection issues.

The fundamental idea of the model-based sampling theory approach to statistical inference is that the data arise as a sample from some conceptual probability distribution, $f$, and hence the uncertainties of our inferences can be measured if we can estimate $f$. There are ways to construct a nonparametric estimator of (in essence) $f$ from the sample data. The most fundamental idea of the bootstrap method is that we compute measures of our inference uncertainty from that estimated sampling distribution of $f$. However, in practical application, the bootstrap means using some form of resampling with replacement from the actual data, $\underline{x}$, to generate $B$ bootstrap samples, $\underline{x}^*$. Often, the data (sample) consist of $n$ independent units, and it then suffices to take a simple random sample of size $n$, with replacement, from the $n$ units of data, to get one bootstrap sample (i.e., "rep"). However, the nature of the correct bootstrap data resampling can be more complex for more complex data structures.

The set of $B$ bootstrap samples is a proxy for a set of $B$ independent real samples from $f$ (in reality we have only one actual sample of data). Properties expected from replicate real samples are inferred from the bootstrap samples by analyzing each bootstrap sample exactly as we first analyzed the real data sample. From the set of results of sample size $B$ we measure our inference uncertainties from sample to (conceptual) population. The bootstrap can work well for large sample sizes ($n$), but may not be reliable for small $n$ (say 5, 10, or even 20), regardless of how many bootstrap samples, $B$, are used.

A word about notation. Any quantity $Q$ computed from the real data, and then also from a bootstrap sample, is denoted simply by $Q^*$, or by $Q_b^*$ to indicate that it came from the $b$th bootstrap sample. Also, let $R$ denote the number of models fit in the model-selection process; hence we have models $M_1$ to $M_R$ to consider. Also, we will refer herein to "AIC," but the methods apply equally well to $AIC_c$, QAIC, $QAIC_c$, and TIC.

## 4.2.1 AIC Differences and a Confidence Set on the K-L Best Model

The difference between the AIC values for two models is meaningful; it is an estimate of the difference between the K-L distance for the two models. Hence, we focus on $\Delta_i$ values, where for the $i$th model, $\Delta_i = AIC_i - \min AIC$, where min AIC is the smallest AIC value. The model producing min AIC is our (estimated to be) expected K-L best model in the set of $R$ models. We can order the $\Delta_i$ from smallest to largest, and the same ordering of the models puts a rank ordering on those models as regards how good they are as an approximation to the expected K-L best model. For example, with seven models we might have the ordered $\Delta_i$ as 0, 1.2, 1.9, 3.5, 4.1, 5.8, and 7.3. An important question is how big of a difference matters? This should be asked in the sense of when is a model not to be considered competitive with the selected best model as plausibly the actual K-L best model in the set of models fit to the data. The question has no unambiguous answer; it is like asking how far away from an MLE, $\hat{\theta}$, an alternative value of $\theta$ must be (assuming that the model is a good model) before we would say that alternative $\theta$ is unlikely as "truth." This question ought to be answered with a confidence (or credibility) interval on $\theta$ based on $\hat{\theta}$ and its estimation uncertainty. A conventionally accepted answer here is that $\theta$ is unlikely as truth if it is further away than $\pm 2\,\widehat{se}(\hat{\theta})$ (there are, of course, reasons for using such a procedure).

In the same spirit, but with less certainty of belief, a useful rule of thumb can be given for AIC model-selection as to when $\Delta_i$ is "big." In a sampling theory context, $\Delta_i$ can be considered big if it is at, or beyond, the 95th percentile of the sampling distribution of a statistic analogous to $\theta - \hat{\theta}$. Thus we can envision a useful basis for a confidence set on models (elaborated on below). First, however, we must note that such a cutoff point is dependent (perhaps strongly) on the assumption of independent observations, and it is weakly dependent on sample size and on certain characteristics of the selection situation in the same sense as the $t$-distribution is approximated by the standard normal distribution, and a sample mean is approximately normally distributed.

The most common context for model-selection is that in which we have several models to select from (say, at least five), most of these models are "nested" within a global model, nested sequences occur, and preferably we also have a large sample size. In this context, both some theory and simulation suggest that if a model has $\Delta_i \geq 4$, we may justifiably think that it is not a highly plausible model for the data (again, this is only a first approximation as a useful rule of thumb). The value of 4 needs to be bigger for small sample sizes, or when there are models for which $K/n$ is too large. Also, as regards a theoretical basis, this value of 4 is obtained from a too simplistic situation wherein one model is the true generating model and it is nested in a sequence of increasingly more general alternative models (that have unneeded parameters). This simple context is useful in that it allows analytical and numerical evaluation to determine such an initial rule of thumb.

In realistic situations there will exist a best approximating model (we will not know which model that is) and other models that are increasingly poorer approximations due only to the finite amount of data available. Based on simulations, and the idea of using the 95th percentile of a sampling distribution (applied to $\Delta$), in this latter situation one might better think of a range of 4–7 as the useful cutoff point for $\Delta_i$ to define, in effect, a 95% confidence set of models. That is, such a subset of the models considered would include the actual K-L best model in 95% of all samples. At the other end of the selection spectrum, selecting from only two models with no nesting of one in the other, a simple rule of thumb about a "cutoff" $\Delta_i$ value may not exist from a frequentist sampling viewpoint. However, we are willing to risk the assertion that in any situation if a model has $\Delta_i > 10$ this is strong evidence that this model is not competitive as the K-L best model.

## 4.2.2 Likelihood of a Model and Akaike Weights

We will return to the matter of a critical value for a confidence set on plausible models in terms of a $\Delta_i$ value. First, however, we note that what is needed is to better quantify the plausibility of each model as being the actual K-L best model. This can be done by extending the concept of the likelihood of the parameters given both the data and model, i.e., $\mathcal{L}(\underline{\theta} \mid \underline{x}, M_i) = g_i(\underline{x} \mid \underline{\theta})$, to a concept of the likelihood of the model given the data, hence $\mathcal{L}(M_i \mid \underline{x})$ ("$\propto$" means "is proportional to"):

$$\mathcal{L}(M_i \mid \underline{x}) \propto \exp(-\tfrac{1}{2}\Delta_i) = C\mathcal{L}(\hat{\underline{\theta}} \mid \underline{x}, M_i)e^{-K}; \tag{4.1}$$

Hirotugu Akaike was born in 1927 in Fujinomiya-shi, Shizuoka-jen, in Japan. He received B.S. and D.S. degrees in mathematics from the University of Tokyo in 1952 and 1961, respectively. He worked at the Institute of Statistical Mathematics for over 30 years, becoming its Director General in 1982. He has received many awards, prizes, and honors for his work in theoretical and applied statistics (de Leeuw 1992, Parzen 1994). The 3-volume set, "*Proceedings of the First US/Japan Conference on the Frontiers of Statistical Modeling: An Informational Approach*" (Bozdogan 1994) commemorated Professor Hirotugu Akaike's 65th birthday. Bozdogan (1994) records that the idea of a connection between the Kullback–Leibler discrepancy and the empirical log-likelihood function occurred to Akaike on the morning of March 16, 1971, as he was taking a seat on a commuter train.

"$C$" is an arbitrary constant. Akaike (see, e.g., Akaike 1983b) advocates the above $\exp(-\frac{1}{2}\Delta_i)$ as being the relative likelihood of the model (after allowing for MLE of parameters based on the same data, $\underline{x}$, as show up in $\mathcal{L}(\hat{\theta} \mid \underline{x}, M_i)$). To better interpret the relative likelihood, (4.1), of a model given the data and the set of $R$ models, we normalize the $\mathcal{L}(M_i \mid \underline{x})$ to be a set of positive "Akaike weights," $w_i$, summing to 1:

$$w_i = \frac{\exp(-\frac{1}{2}\Delta_i)}{\sum_{r=1}^{R} \exp(-\frac{1}{2}\Delta_r)}. \qquad (4.2)$$

This idea of the likelihood of the model given the data, and hence these model weights, has been suggested for many years by Akaike (e.g., Akaike 1978b, 1979, 1980, 1981b and 1983b; also see Bozdogan 1987 and Kishino 1991) and has been researched some by Buckland et al. (1997). These model weights seemed not to have a name, so we call them Akaike weights. This name will herein apply also when using $AIC_c$, QAIC, $QAIC_c$, and even TIC. A given $w_i$ is considered as the weight of evidence in favor of model $i$ as being the actual K-L best model for the situation at hand *given* that one of the $R$ models must be the K-L best model of that set of $R$ models. We could just compute the relative likelihood of model pairs as $\mathcal{L}(M_i \mid \underline{x})/\mathcal{L}(M_j \mid \underline{x})$; this ratio will be identical to the ratio $w_i/w_j$. Hence, given that there are only $R$ models and one of them must be best in this set of models, it is convenient to normalize the relative likelihoods to sum to 1.

For the estimated K-L best model (taken to be model $M_k$), $\Delta_k = 0$; hence, for that model $\exp(-\frac{1}{2}\Delta_k) = 1$. The odds for the $i$th model actually being the K-L best model are thus $\exp(-\frac{1}{2}\Delta_i)$ to 1, or just the "ratio" $\exp(-\frac{1}{2}\Delta_i)$. It is convenient to express such odds as the set of Akaike weights. The bigger a $\Delta_i$ is, the smaller the $w_i$, and the less plausible is model $i$ as being the actual K-L best model for $f$ based on the design and sample size used. Computing the sampling distribution of $\Delta_p$ by Monte Carlo or bootstrap methods gives us a way to interpret the observed $\Delta_i$; the Akaike weights give us a second way to calibrate (interpret) the $\Delta_i$ values. These weights also have other uses and interpretations that are given below.

For the example above of seven $\Delta_i$ values as 0, 1.2, 1.9, 3.5, 4.1, 5.8, and 7.3, the relative likelihoods of the models are proportional to the values (in order) 1, 0.54881, 0.38674, 0.17377, 0.12873, 0.05502, and 0.02599. Normalized to Akaike weights, $w_i$, we have (in order) 0.431, 0.237, 0.167, 0.075, 0.056, 0.024, and 0.010. As weight of evidence for each model we can see that the selected best model is not convincingly best; it is only about twice as likely as the next best model. This weak support for the best model suggests that we should expect to see a lot of variation in the selected best model from sample to sample if we could, in this situation, draw multiple independent samples.

Relative frequencies for model $i$ being selected as the best model would be similar to the Akaike weights, but would not be identical. There is no reason, nor need, for the data-based weights of evidence (as the set of $w_i$) to be the same as the sampling relative frequencies at which the models are selected by AIC as being best. In general, likelihood provides a better measure of data-based weight of evidence about parameter values given a model and data (see, e.g., Royall 1997), and we think that this concept (i.e., evidence for the best model is best represented by the likelihood of a model) rightly extends to evidence about a best model given an *a priori* set of models.

We further elaborate on the interpretation of the Akaike weights as being conceptually different from the sampling-theory-based relative frequencies of model-selection. It has has been noted in the literature (e.g., Akaike 1981a, 1994, Bozdogan 1987) that there is a Bayesian basis for interpreting the Akaike weight $w_i$ as being the probability that model $M_i$ is the expected K-L best model given the data (for convenience we usually drop this "expected" distinction and just think of the K-L best model). Once we have accepted formula (4.1) as being the likelihood of model $M_i$ given the data, $\mathcal{L}(M_i \mid \underline{x})$, then we can compute the posterior probability that model $M_i$ is the K-L best model if we are willing to specify prior probabilities on the models. That is, we first must specify an *a priori* probability distribution, $\tau_1, \ldots, \tau_R$, which provides our belief that fitted model $M_i$ will be the

K-L best model for the purpose of analyzing the data. These probabilities, $\tau_i$, must be specified independently of (basically, prior to) fitting any models to the data.

In the absence of any prior information about which of these models might be the K-L best model for the data at hand, we are compelled by a certain aspect of information theory itself (see Jaynes 1957, Jessop 1995) to set the $\tau_i$ all equal, hence to use $\tau_i \equiv 1/R$. In fact, doing so depends on all $R$ of the models being on an equal footing to be selected as the K-L best model; this may not always be the case (which opens the door to having unequal prior probabilities). The issue here is one of possible partial model redundancy (this subject is deferred to Section 4.2.9) if the set of models considered is not carefully defined for the purpose of the analysis. Given no model redundancy, the concept of what $\tau_i$ must mean seems clear: It is our prior state of information, or lack thereof, hence belief that model $M_i$ fitted to the data provides the K-L best model for this sampling situation, hence for the data at hand. This is a deceptively complex issue, as it relates to ideas of both models as best approximations to truth and to expected model-fitting tradeoff of bias versus sampling variances.

To us it seems impossible to have any real prior basis for an informative differential assessment of the $\tau_i$ (other than on how the models might be structurally interrelated or partially redundant). Using the maximum entropy principle of Jaynes (1957), we should take the $\tau_i$ to represent maximal uncertainty about all unknown aspects of the probability distribution represented by the $\tau_i$. Thus we determine the $\tau_i$ that maximize the entropy, $-\sum \tau_i \log(\tau_i)$, subject to constraints that express whatever information (in the colloquial sense) we have about the distribution. In the "no information" case the only constraint we have is that $\sum \tau_i = 1$ (plus the essential $0 < \tau_i < 1$). The maximum entropy (hence maximum uncertainty) prior is then $\tau_i \equiv 1/R$. [It takes us too far afield to delve into the aspects of information theory underlying the maximum entropy principle. This principle is fundamentally tied to both Boltzmann's entropy and to information theory and can be used to justify noninformative Bayesian priors—when they exist. The interested reader is referred to Kapur and Kesavan 1992, or the less technical Jessop 1995.]

Given any set of prior probabilities (the $\tau_i$), generalized Akaike weights are given by

$$w_i = \frac{\mathcal{L}(M_i \mid \underline{x})\tau_i}{\sum_{r=1}^{R} \mathcal{L}(M_r \mid \underline{x})\tau_r}. \tag{4.3}$$

There may be occasions to use unequal prior probabilities, and hence use (4.3) rather than its special case (4.2). However, in general, by Akaike weights we mean (4.2), and hence all $\tau_i = 1/R$.

The computation (via formula 4.3) and use of a probability $w_i$ of model $M_i$ being the K-L best-fitted model for the sampling situation, hence data, at hand is not a true Bayesian approach. A Bayesian approach to model-selection requires both the prior $\tau_i$ on the model and a prior probability distribution on the parameters $\underline{\theta}$ in model $M_i$ for each model. Then the derivation of posterior results requires a lot of integration (usually only achievable by Markov chain Monte Carlo methods). Persons wishing to learn the Bayesian approach to model-selection can start with the following sources: Raftery et al. (1993), Madigan and Raftery (1994), Carlin and Chib (1995), Chatfield (1995b), Draper (1995), Kass and Raftery (1995), Hoeting and Ibrahim (1996), and Raftery (1996a, 1996b).

A brief comparison is given here of what we mean by the prior probabilities, $\tau_i$, in (4.3) under this information-theoretic approach to model-selection versus what seems to be meant by the prior probabilities of models in the Bayesian approach. The Bayesian approach generally assumes that one of the models in the set of $R$ models is true (or it is assumed that some weighted average of these $R$ models is true if model averaging is to be used). Hence, $\tau_i$ is then one's prior degree of belief that model $M_i$ is the true model, or the degree of "correctness" of the model (see, e.g., Newman 1997). Under the information-theoretic approach we do not assume that truth, $f$, is in the set of models, and $\tau_1, \ldots, \tau_R$ is a probability distribution of our prior information (or lack there of) about which of the $R$ models is the K-L best model for the data. Information theory itself (Kapur and Kesavan 1992) then justifies determination of the $\tau_i$, generally as $\tau_i \equiv 1/R$. For data analysis we believe that the issue cannot be which model structure is truth, because none of the models considered are truth. Rather the issue is, Which model, *when fit to the data* (i.e., when $\underline{\theta}$ is *estimated*), is the best model for purposes of representing the (finite) information in the data.

## 4.2.3 *More Options for a Confidence Set for the K-L Best Model*

There exists a concept of a confidence set for the K-L best model based on the data, just as there is a confidence interval for a parameter based on a model and data. For a 95% confidence set on the actual K-L best model, a rational (but not unique) approach is to sum the Akaike weights from largest to smallest until that sum is just $\geq 0.95$; the corresponding subset of models is a type of confidence set on the K-L best model. In this example (assuming that we have indexed the models as 1 to 7 in order of decreasing weights), the confidence set is models {1, 2, 3, 4, 5}, which has sum of weights $= 0.966$. In using this approach to a confidence set of models we are interpreting the Akaike weights as posterior probabilities (i.e., given the data and the set of a prior models) that model $i$ is the K-L best model.

There is another approach to developing a confidence set of models based on the idea of a $\Delta_i$ being a random variable with a sampling distribution. In particular, let index value $k$ correspond to the actual expected K-L best model in the set. There always is a K-L best model in the set of models (ignoring that ties might occur). It is thus model $M_k$ that we should use for the data analysis; we just do not happen to know *a priori* the value of $k$. Then the $\Delta_i$ of conceptual interest is

$$\Delta_p = \text{AIC}_k - \min \text{AIC}. \tag{4.4}$$

This unobservable random variable ($\Delta_p$) is analogous to $\theta - \hat{\theta}$, which can often be used (after normalization by $\widehat{\text{se}}(\hat{\theta})$) as a pivotal value for construction of a confidence interval on $\theta$. A pivotal quantity is one whose sampling distribution is independent of any unknown parameters—a $t$-distributed pivotal, for example.

The "$p$" in the $\Delta$ defined by (4.4) denotes that this $\Delta$ is a conceptual pivotal value rather than an actual $\Delta_i$ that we can compute from real data.

It is not exact to consider $\Delta_p = \text{AIC}_k - \min \text{AIC}$ as a pivotal quantity, but it seems a useful approximation in some contexts. The context it seems useful in is one of complex truth; tapering effect sizes; many models, some being good approximations to truth, with full truth not in the set of models used; and a lot of nested sequences of models (like the starling experiment example in Chapter 3). Monte Carlo studies on the above $\Delta_p$ can be done; we have done lots of this and results support the conclusion that in this context, the sampling distribution of this $\Delta_p$ has substantial stability and the 95th percentile of the sampling distribution of $\Delta_p$ is generally much less that 10, and in fact generally less than 7 (often closer to 4 in simple situations). This means that an alternative rule of thumb for an approximate 95% confidence set on the K-L best model is the subset of all models $M_i$ having $\Delta_i \leq$ some value that is roughly 4 to 7. In fact, the $\Delta$ value to use, for when a model is not competitive as a candidate for the K-L best model, is variable, but is probably somewhere between 2 and 10 in almost any situation. Thus, a $\Delta_i$ of 2 is not large, while a $\Delta_i = 10$ is strong evidence against model $M_i$ being the K-L best model in the set of models considered, if sample size is not small. These guidelines, rough as they are, are useful.

We recapitulate this interpretation of evidence from the $\Delta_i$ when observations are independent and sample sizes are large: For any model with $\Delta \leq 2$ there is no credible evidence that the model should be ruled out as being the actual K-L best model for the population of all possible samples. For a model with $2 < \Delta \leq 4$ there is weak evidence that the model is not the K-L best model. If a model has $4 < \Delta \leq 7$ there is definite evidence that the model is not the K-L best model, and if $7 < \Delta \leq 10$, there is strong evidence that the model is not the K-L best model. Finally, if $\Delta > 10$, there is very strong evidence that the model is not the K-L best model. The reader should not take these guidelines as inviolate, as there are situations to which they do not apply well (such as when there is a small sample size or dependent observations). We had these guidelines well in mind when we encountered similar guidelines for the Bayes factor. The Bayes factor is a Bayesian-based ratio for the relative data-based likelihood of one model versus another model, but without considering any priors on the set of models (Berger and Pericchi 1996, Raftery 1996a); it is somewhat analogous to $\exp(-\frac{1}{2}\Delta_i)$. Raftery (1996a:252, 1996b:165) presents a similar scale for interpretation of 2 log(Bayes factor) as evidence for the simpler model being considered.

A third reasonable basis for a confidence set on models is motivated by likelihood-based inference (see e.g., Edwards 1992, Azzalini 1996, Royall 1997), hence is analogous to a profile likelihood interval on a parameter given a model. Here we would simply agree on some value of the relative likelihood of model $i$ versus the estimated K-L best model $k$ as being a cutoff point for a set of relatively more plausible models. Thus our confidence set of models is all models for which the ratio $\mathcal{L}(M_i \mid \underline{x})/\mathcal{L}(M_k \mid \underline{x})$ is "small" (such as $< 1/8$; see e.g., Royall 1997:89–90). This criterion translates exactly into being the same as the set of all models for which $\Delta_i \leq$ some fixed cutoff point. However, now there is no direct sampling theory interpretation required and no necessary appeal to the idea of the selected subset of models including the K-L best model with a preset, known, long-run inclusion relative frequency such as 95%.

We have presented three approaches to finding a confidence set on models: (1) base it directly the Akaike weights, interpreted as probabilities of each model being the actual best model, given the data; (2) use a cutoff $\Delta_i$ motivated by the idea of the sampling distribution of the approximate pivotal, $\Delta_p$ (using, say, the 95th percentile of this distribution as the cutoff $\Delta$); or (3), think in terms of relative likelihood and hence (for $k$ indexing the selected AIC best model) use a cutoff value of $\Delta$ for which $\mathcal{L}(M_i \mid \underline{x})/\mathcal{L}(M_k \mid \underline{x}) \equiv \exp(-\frac{1}{2}\Delta_i)$ is small, say 0.135 ($\Delta_i = 4$), 0.082 ($\Delta_i = 5$), or 0.050 ($\Delta_i = 6$).

The use of intervals, or testing procedures, based purely on relative likelihood is soundly supported by statistical theory (cf. Berger and Wolpert 1984, Edwards 1992, Azzalini 1996, Royall 1997), but rarely taught or used. Rather, most users of statistics have been taught to think of confidence intervals in terms of coverage probability; hence they might feel more at home with methods (1) and (2), both of which are motivated by the sampling theory idea of a 95% confidence interval on a parameter. More needs to be known about the properties of these three methods to construct a confidence set of models before we would be comfortable recommending just one approach.

### 4.2.4 $\Delta_i$, Model-Selection Probabilities and the Bootstrap

For a given set of data we can estimate the sampling distribution of model-selection frequencies and the distribution of $\Delta_p = \text{AIC}_k - \min \text{AIC}$ (i.e., formula 4.4) using the bootstrap method. In this method the role of the actual (unknown) K-L best model is played by the model selected as best from the data analysis; denote that model by model $M_k$. For each bootstrap sample we fit each of the $R$ models, compute all $R$ of the $\text{AIC}_i^*$, and then find the single $\Delta_p^* = \text{AIC}_k^* - \min \text{AIC}^*$; $k$ does not change over bootstrap reps. That is, $\text{AIC}_k^*$ is always the AIC value, from the given bootstrap sample, for model $k$, which is the selected AIC best model for the data. The model producing min $\text{AIC}^*$ varies by bootstrap sample. However, it will often be model $k$ that is the best model in a bootstrap sample, in which case $\Delta_p^* = 0$. When it is not model $k$ that produces min $\text{AIC}^*$, then $\Delta_p^* > 0$.

The $B$ bootstrap reps provide $B$ values of $\Delta_p^*$ that are independent, conditional on the data. The percentiles of the empirical probability distribution function of $\Delta_p^*$ provide the estimate of the percentiles of the sampling distribution of $\Delta_p$, and hence provide a basis for a confidence set on the K-L best model for the actual data. For a $(1 - \alpha)100\%$ confidence set on the K-L best model, order the $\Delta_{p,(b)}^*$ (smallest to largest) and find the $\Delta_{p,(b)}^*$ value for $b = [(1 - \alpha)B]$. For the actual data analysis results, the subset of the $R$ models $M_i$ having $\Delta_i \leq \Delta_{[(1-\alpha)B]}^*$ is the desired confidence set. For reliable results on the upper tail percentiles of $\Delta_p$, $B$ needs to be 10,000.

Other information can be gained from these bootstrap results about model-selection uncertainty, in particular, the frequency of selection of each of the $R$

models. Let $b_i$ be the number of reps in which model $i$ was selected as the K-L best model. Then an estimator of the relative frequency of model-selection in the given situation is $\hat{\pi}_i = b_i/B$. These estimated selection probabilities are useful for assessing how much sampling variation there is in the selection of the best model: They directly quantify model-selection uncertainty. These estimated selection probabilities are similar to, but not identical in meaning to, the Akaike weights, which also quantify strength of evidence about model-selection uncertainty.

Also, for each bootstrap rep we can compute the Akaike weights,

$$w_i^* = \frac{\exp(-\frac{1}{2}\Delta_i^*)}{\sum_{r=1}^R \exp(-\frac{1}{2}\Delta_r^*)},$$

and then average these over the $B$ reps to get $\bar{w}_i^*$. Comparison of the $w_i$, $\bar{w}_i^*$, and $\hat{\pi}_i$ is informative as to the coherence of these methods, each of which provides information about the sampling uncertainty in model-selection. The theoretical measure of model-selection sampling uncertainty is the set of true, unknown selection probabilities, $\pi_1 \ldots, \pi_R$. Either the $\hat{\pi}_i$ (from the bootstrap) or the Akaike weights, $w_i$ (we often prefer the latter because they do not require computer-intensive calculations and they relate more directly to strength of evidence based on the data at hand), may be taken as the estimated inference uncertainty about model-selection. Note that in this usage of the term "model" we mean the structural form of the model (such as which variables are included vs. excluded) without consideration of the specific parameter values required in each model. Parameter-estimation uncertainty is conceptually separable from (but influenced by) model-selection uncertainty.

### 4.2.5 Concepts of Parameter-Estimation and Model-Selection Uncertainty

Statistics should emphasize estimation of parameters and associated measures of estimator uncertainty. Given a correct model (most theory assumes $g = f$), an MLE is reliable, and we can compute a reliable estimate of its sampling variance, and a reliable confidence interval (such as a profile likelihood interval; see Royall 1997). If the model is selected entirely independently of the data at hand, and is a good approximating model, and if $n$ is large, then the estimated sampling variance is essentially unbiased, and any appropriate confidence interval will essentially achieve its nominal coverage. This would be the case if we used only one model, decided on *a priori*, and it was a good model, $g$, of the data generated under truth, $f$. However, even when we do objective, data-based model-selection (which we are advocating here), the selection process is expected to introduce an added component of sampling uncertainty into any estimated parameter; hence classical theoretical sampling variances are too small: They are conditional on the model and do not reflect model-selection uncertainty. One result is that conditional confidence intervals can be expected to have less than nominal coverage.

## 4.2 Methods for Assessing Model-Selection Uncertainty

Consider a scalar parameter $\theta$, which may be used in all or only some of the models considered, but is in the selected model, and therein has unknown value $\theta_i$ given model $M_i$. Here, the subscript $i$ denotes the model used to estimate $\theta$ with the understanding that this parameter means the same thing in all models in which it appears. There is a conceptual true value of $\theta$ for the given study. However, the value of $\theta$ that we would infer, in the sense of $\mathrm{E}_f(\hat{\theta}_i \mid M_i) = \theta_i$ (for large sample size) from model $g_i$ applied to the data, may vary somewhat by model. Given model $M_i$, the MLE, $\hat{\theta}_i$, has a conditional sampling distribution, and hence a conditional sampling variance $\mathrm{var}(\hat{\theta}_i \mid M_i)$. We mean the notation $\mathrm{var}(\hat{\theta}_i \mid M_i)$ to be functionally and numerically identical to $\mathrm{var}(\hat{\theta}_i \mid \theta_i)$. The latter notation is more traditional; we use the former notation when we want to emphasize the importance of assuming the model in its totality when in fact other models are also being considered.

There is a concept of the true value of $\theta$: It is the value of $\theta$ we would compute based on knowing truth, $f$, even though $\theta$ need not literally appear in $f$. To the extent a model, $M_i$, is wrong (i.e., $g_i \neq f$), $\theta_i$ may not equal $\theta$ when this K-L best value is determined for $\theta$ under assumed model $M_i$. That is, even when data are generated by $f$, if those data are interpreted under model $g_i$, we will infer (for large $n$) that the value of $\theta$ is $\theta_i$. For a good model this possible "bias" (i.e., $\theta_i - \theta$) is not of great concern, because it will be dominated by the conditional sampling standard error of $\hat{\theta}_i$ (in essence, this domination is one feature of a "good" model). The bias $\theta_i - \theta$ induces one source of model-selection uncertainty into $\hat{\theta}$; that is, this bias varies over models in an unknown manner. In many situations the model, as such, means something to us, and we will then take $\hat{\theta}_i$ derived only from the selected model, $g_i$, as the most meaningful estimator of $\theta$. This is what has been commonly done, and seems sensible, so much so that an alternative called model averaging (below) seems at first strange, but is an alternative to getting an estimator of $\theta$ based on multiple models. Model averaging arises in a natural way when we consider the unconditional sampling variance of $\hat{\theta}_i$.

Another source of model-selection uncertainty arises because we risk getting a negatively biased estimator of conditional sampling variance, $\mathrm{var}(\hat{\theta}_i \mid M_i)$, if we over-fit the data when we must use an estimate of sampling variance that depends on residuals from the fitted model. To think about this idea, we need to note two situations. The first occurs when, given the model structure, there is a known theoretical sampling variance, such as for example under Poisson, binomial, multinomial (includes contingency-table-based-models), or negative binomial models, where, given the model structure, we can infer the sampling variance theoretically. Under-estimation of sampling variance due to structural over-fitting does not seem to be a serious problem in cases where such a theoretical sampling variance is known. In particular, this is true if we also use a variance inflation factor, $\hat{c}$, applied for all models in the set of $R$ models (so selection is not an issue as regards $\hat{c}$) to adjust for any modest structural lack of fit of the global model.

The second situation occurs often in regression models where the residual sampling variance, $\sigma^2$, is functionally unrelated to the model structure and there is no true replication. Then $\sigma^2$ must be estimated only from residuals to the fitted model.

In this case there is neither true replication nor a theoretical basis to infer $\sigma^2$, such as there is in models for count data. If we over-fit the structural component of the model to the data, we will get $\hat{\sigma}^2$ biased low, and hence estimated sampling standard errors of any $\hat{\theta}_i$ will be biased low (there is likely to be a compensating increase in the factor $(\hat{\theta}_i - \theta)^2$). True replication at the level of the regressor values can eliminate this problem, but often we do not have such true replication.

The ideas of classical sampling theory can be used to derive the theoretical sampling variance of $\hat{\theta}$ resulting from the two-stage process of (1) model-selection, then (2) using $\hat{\theta} \equiv \hat{\theta}_i$ given that model $M_i$ was selected. Imagine this process carried out many times, $m$, each time on an independent sample. For sample $j$ we get $\hat{\theta}_j$ as our estimator of $\theta$. This conceptual $\hat{\theta}_j$ comes from the selected model in repetition $j$, but we do not need, hence avoid using, a doubly indexed notation (such as $\hat{\theta}_{i,j}$) to denote both rep $j$ and selected model $i$ given rep $j$.

The estimated unconditional sampling variance, $\widehat{\text{var}}(\hat{\theta})$, from these $m$ replicates would be $\sum(\hat{\theta}_j - \overline{\hat{\theta}})^2/(m-1)$; $\overline{\hat{\theta}}$ is the simple average of all $m$ estimates (hence the $\hat{\theta}_j$ have been averaged over selected models). This variance estimator represents the total variation in the set of $m$ values of $\hat{\theta}$; hence both within and between model variation is included. This set of $m$ values of $\hat{\theta}$ can be partitioned into $R$ subsets, one for each model wherein the $i$th subset contains all the $\hat{\theta}$'s computed under selected model $i$. Then one can compute from the $i$th subset of the $\hat{\theta}$ values an estimate of the conditional sampling variance of $\hat{\theta}$ when model $M_i$ was selected. Formal mathematics along this line of partitioning the above $\widehat{\text{var}}(\hat{\theta})$ into $R$ components and taking expectations to get a theoretical unconditional sampling variance gives the result for $\text{var}(\hat{\theta})$ as a weighted combination of conditional variances, $\text{var}(\hat{\theta}_i \mid M_i)$ plus a term for variation among $\theta_1, \ldots, \theta_R$. The weights involved are the model-selection probabilities. Relevant formulae are given in the next section, but first we mention one more issue.

The above heuristics were presented as if the parameter of interest appeared in every model. However, a given parameter may appear only in some of the models. In this case the basis for unconditional inference about that parameter can be made based on just those models in which that parameter appears. An example is variable selection in linear regression, say $y$ versus $p$ regressors, $x_1, \ldots, x_p$ (plus an intercept). There are $2^p$ possible models, but each regressor appears in only half of these models (i.e., in $2^{p-1}$ models). Thus if regressor variable $x_j$, hence parameter $\beta_j$, is in the selected AIC best model, we must restrict ourselves to just that subset of models that contains $\beta_j$ in order to directly estimate the unconditional sampling variance of $\hat{\beta}_j$. All the above (and below) considerations about conditional and unconditional variances with regard to a particular parameter can be interpreted to apply to just the subset of models that include the parameter.

We have emphasized models as approximations to truth. Thus, a model being "wrong" is technically called model misspecification (see White 1994), and the usual theoretical sampling variances of MLEs, $\text{var}(\hat{\theta}_i \mid M_i)$, may be wrong, but only trivially so if the model is a good approximation to truth. There is theory that gives the correct conditional (on the model) sampling variance of $\hat{\theta}_i$ in the event of model misspecification (Chapter 6 gives some of this theory). However, the correct estimator of var($\hat{\theta}_i \mid M_i$) is then so much more complex and variable (a type of instability) that it generally seems better to use the theoretical estimator supplied by the usual model-specific information matrix (which assumes that the model is correct). This simplified approach seems especially defensible when done in conjunction with sound model-selection procedures intended to minimize both serious over- and under-fitting. We believe AIC is suitable for this selection purpose and that the only additional consideration is thus to get reliable unconditional sampling variances (and confidence intervals) for MLEs after model-selection.

## 4.2.6 Including Model-Selection Uncertainty in Estimator Sampling Variance

We continue to assume that the scalar parameter $\theta$ is in common to all models considered. This will often be the case for our full set of *a priori* specific models, and is always the case if our objective is prediction with the fitted model, such as interpolation or extrapolation with a generalized linear model. Alternatively, if our focus is on a model structural parameter that appears only in a subset of our full set of models, then we must restrict ourselves to that subset in order to make the sort of inferences considered here about the parameter of interest. In the latter case we simply consider the relevant subset as the full set of models under consideration.

In repeated (conceptual) samples there is a probability $\pi_i$ of selecting each model. Presentation of a defensible way to augment $\widehat{\text{var}}(\hat{\theta}_i \mid M_i)$ with model-selection uncertainty involves the idea of model averaging in that we must define a model averaged parameter value, $\theta_a$, as

$$\theta_a = \sum_{i=1}^{R} \pi_i \theta_i, \tag{4.5}$$

and an estimator as

$$\hat{\theta}_a = \sum_{i=1}^{R} \hat{\pi}_i \hat{\theta}_i. \tag{4.6}$$

In some theory development we use $\pi_i$ rather than $\hat{\pi}_i$, but we still use notation such as $\hat{\theta}_a$.

It is important to realize that $\theta_a$ is not the same as $\theta$, which is computable from absolute truth $f$. Under classical sampling theory the $\hat{\theta}$ ($\equiv \hat{\theta}_i$ for some selected model $M_i$ that varies by sample), arrived at in the two-stage process of model-selection and then parameter estimation given the model, is by definition an unbiased estimator of $\theta_a$ as given by (4.5). Therefore, the unconditional sampling variance of $\hat{\theta}$ is to be measured with respect to $\theta_a$. Any remaining bias, $\theta_a - \theta$, in $\hat{\theta}_a$ cannot be measured or allowed for in model-selection uncertainty. However, part of the intent of having a good set of models and sound model-selection is to render this bias negligible with respect to the unconditional se($\hat{\theta}$).

134   4. Model-Selection Uncertainty with Examples

The theoretical, unconditional sampling variance of the estimator of $\theta$ is given by

$$\text{var}(\hat{\theta}) = \sum_{i=1}^{R} \pi_i \left[ \text{var}(\hat{\theta}_i \mid M_i) + (\theta_i - \theta_a)^2 \right]. \tag{4.7}$$

This result follows directly from frequentist sampling theory. It is noted in Section 4.2.3 that if we had $m$ independent samples and then applied model-selection to each sample to get $\hat{\theta}_j$, $j = 1, \ldots, m$, then an estimator of $\widehat{\text{var}}(\hat{\theta})$ would be

$$\widehat{\text{var}}(\hat{\theta}) = \sum (\hat{\theta}_j - \bar{\hat{\theta}})^2 \big/ (m - 1).$$

Here, $j$ indexes the sample that $\hat{\theta}_j$ came from (whatever the model used), whereas $i$ indexes that $\hat{\theta}_i$ arose from model $i$ (whatever the sample was). This notation allows us to focus on different aspects of the model-selection problem without a notation so complex that it hinders understanding of concepts. Letting $m$ become infinite, the above estimator of $\text{var}(\hat{\theta})$ converges to the theoretical unconditional sampling variance of $\hat{\theta}$. By first grouping the set of $m$ different $\hat{\theta}$ values by model and then taking the needed limit as $m \to \infty$ we get formula (4.7).

The quantity $\text{var}(\hat{\theta}_i \mid M_i) + (\theta_i - \theta_a)^2$ is just the mean square error of $\hat{\theta}_i$ given model $i$. Thus, in one sense the unconditional variance of $\hat{\theta}$ is just an average mean square error. Specifically,

$$E[(\hat{\theta}_i - \theta_a)^2 \mid M_i] = \text{var}(\hat{\theta}_i \mid M_i) + (\theta_i - \theta_a)^2,$$

and we recommend thinking of the above quantity as the sampling variance of $\hat{\theta}_i$, given model $i$, when $\hat{\theta}_i$ is being used as an estimator of $\theta_a$. The incorporation of model-selection uncertainty into the variance of $\hat{\theta}_i$ requires some new thinking like this. The matter arises again when we must consider a type of covariance, $E[(\hat{\theta}_i - \theta_a)(\hat{\theta}_j - \theta_a) \mid M_i]$, that also allows for model-selection uncertainty.

One might think to estimate the augmented sampling variance of $\hat{\theta}_i$ by $\widehat{\text{var}}(\hat{\theta}_i \mid M_i) + (\hat{\theta}_i - \hat{\theta}_a)^2$. Such an estimator is not supported by any theory and is likely to be both biased (it could be bias-corrected) and quite variable; however, we have not investigated this possible estimator. Rather, to get an estimator of $\text{var}(\hat{\theta})$ we could plug estimated values into (4.7) to get $\widehat{\text{var}}(\hat{\theta}) = \sum \hat{\pi}_i [\widehat{\text{var}}(\hat{\theta}_i \mid M_i) + (\hat{\theta}_i - \hat{\theta}_a)^2]$. Ignoring that the $\pi_i$ and $\text{var}(\hat{\theta}_i \mid M_i)$ are estimated (they are not the major source of estimation variation in this variance estimator), we can evaluate $E(\widehat{\text{var}}(\hat{\theta}))$ to bias-correct $\widehat{\text{var}}(\hat{\theta})$. The result involves the sampling variance, $\text{var}(\hat{\theta}_a)$, of the model averaged estimator and is

$$E(\widehat{\text{var}}(\hat{\theta})) = \text{var}(\hat{\theta}) + \sum \pi_i \, \text{var}(\hat{\theta}_i \mid M_i) - \text{var}(\hat{\theta}_a),$$

which leads to

$$\text{var}(\hat{\theta}) = \text{var}(\hat{\theta}_a) + \sum \pi_i E(\hat{\theta}_i - \hat{\theta}_a)^2. \tag{4.8}$$

It seems that we cannot avoid estimating $\theta_a$ and $\text{var}(\hat{\theta}_a)$ even though it is $\hat{\theta}$ and $\widehat{\text{var}}(\hat{\theta})$ that we are seeking. First, note that efforts to evaluate $\sum \pi_i E(\hat{\theta}_i - \hat{\theta}_a)^2$

are circular, thus useless. Anyway, at worst we would only need to estimate this quantity without (much) bias, which we can clearly do. Second, (4.8) shows us that as one might expect, if our goal is to estimate $\theta_a$, then the model averaged $\hat{\theta}_a$ is to be preferred to $\hat{\theta}_i$ because it will have a smaller sampling variance. However, given that our goal is to estimate $\theta$, there is no theoretical basis to claim that $\hat{\theta}_a$ is the superior estimator as compared to $\hat{\theta} \equiv \hat{\theta}_i$.

From Buckland et al. (1997) we will take the needed $\text{var}(\hat{\theta}_a)$ as

$$\text{var}(\hat{\theta}_a) = \left[ \sum_{i=1}^{R} \pi_i \sqrt{\text{var}(\hat{\theta}_i \mid M_i) + (\theta_i - \theta_a)^2} \right]^2, \quad (4.9)$$

with the estimator as

$$\widehat{\text{var}}(\hat{\theta}_a) = \left[ \sum_{i=1}^{R} \hat{\pi}_i \sqrt{\widehat{\text{var}}(\hat{\theta}_i \mid M_i) + (\hat{\theta}_i - \hat{\theta}_a)^2} \right]^2 \quad (4.10)$$

(derivation of formula 4.7 is given below in this subsection). Formula (4.9) does entail an assumption of perfect pairwise correlation, $\rho_{ih}$, of $\hat{\theta}_i - \theta_a$ and $\hat{\theta}_h - \theta_a$ for all $i \neq h$ (both $i$ and $h$ index models). Such pairwise correlation of $\rho_{ih} = 1$ is unlikely; however, it will be high. The choice of a value of $\rho_{ih} = 1$ is conservative in that $\text{var}(\hat{\theta}_a)$ computed from formula (4.9) will tend to be too large if this assumption is in error. Also, by just plugging estimators into (4.10) a further upward bias to (4.9) results. Thus from (4.8) the use of $\widehat{\text{var}}(\hat{\theta}) = \widehat{\text{var}}(\hat{\theta}_a) + \sum \pi_i (\hat{\theta}_i - \hat{\theta}_a)^2$ with $\widehat{\text{var}}(\hat{\theta}_a)$ from (4.10) seems to risk too much positive bias. Hence, we are now suggesting just using $\widehat{\text{var}}(\hat{\theta}) = \widehat{\text{var}}(\hat{\theta}_a)$ from (4.10).

All simulations we have done so far, in various contexts, have supported use of this estimator:

$$\widehat{\text{var}}(\hat{\theta}) = \left[ \sum_{i=1}^{R} \hat{\pi}_i \sqrt{\widehat{\text{var}}(\hat{\theta}_i \mid M_i) + (\hat{\theta}_i - \hat{\theta}_a)^2} \right]^2. \quad (4.11)$$

These simulations have used $\theta$ (i.e., truth), not $\theta_a$, as the target for confidence interval coverage, and this may be another reason that the dual usage of (4.10) and (4.11) is acceptable. Improved estimation of unconditional sampling variances under model-selection may be possible. However, our objective is to give practical solutions to some problems under model-selection with the expectation that improvements will be further explored.

The $\hat{\pi}_i$ in (4.11) (and the equivalent (4.10)) will usually be taken as the Akaike weights, $w_i$, of (4.2). In general, $w_i \neq \pi_i$; rather, $w_i$ can be considered to approximate $\pi_i$, but (4.11) seems robust to slightly imprecise values of the weights. Alternatively, one can use the bootstrap estimates, $\hat{\pi}_i = b_i / B$; however, given that one has bootstrap samples, the analytical formulae above are not needed.

As a final part of this section we give some details of the derivation of (4.9) in a more restricted context than was used in Buckland et al. (1997). Specifically, we do not assume that the $R$ models are randomly selected from all possible models. Rather, we just condition on the set of $R$ models that have been provided; hence, inferences are conditional on just this set of models. Formally, each $\hat{\theta}_i$ is considered as an estimator of $\theta_a$, and it is this conceptualization that is critical to getting a variance formula that includes model-selection uncertainty.

Ignoring that the $\pi_i$ in (4.9) need to be estimated, the variance of $\hat{\theta}_a$ can be expressed as

$$\mathrm{var}(\hat{\theta}_a) = \sum_{i=1}^{R} (\pi_i)^2 \mathrm{E}[(\hat{\theta}_i - \theta_a)^2 \mid M_i]$$
$$+ \sum_{i=1}^{R} \sum_{\substack{h=1 \\ h \neq i}}^{R} \pi_i \pi_h \left[ \mathrm{E}(\hat{\theta}_i - \theta_a)(\hat{\theta}_h - \theta_a) \mid M_i, M_h) \right].$$

From above we know that

$$\mathrm{E}[(\hat{\theta}_i - \theta_a)^2 \mid M_i] = \mathrm{var}(\hat{\theta}_i \mid M_i) + (\theta_i - \theta_a)^2.$$

In order to coherently allow for model-selection uncertainty and to be consistent with the definition of a correlation, we must interpret the covariance term in this expression for $\mathrm{var}(\hat{\theta}_a)$ as

$$\mathrm{E}(\hat{\theta}_i - \theta_a)(\hat{\theta}_h - \theta_a) \mid M_i, M_h) = \rho_{ih} \sqrt{\mathrm{E}[(\hat{\theta}_i - \theta_a)^2 \mid M_i] \mathrm{E}[(\hat{\theta}_h - \theta_a)^2 \mid M_h]};$$

hence,

$$\mathrm{E}(\hat{\theta}_i - \theta_a)(\hat{\theta}_h - \theta_a) \mid M_i, M_h)$$
$$= \rho_{ih} \sqrt{[\mathrm{var}(\hat{\theta}_i \mid M_i) + (\theta_i - \theta_a)^2][\mathrm{var}(\hat{\theta}_h \mid M_h) + (\theta_h - \theta_a)^2]}.$$

Thus we have

$$\mathrm{var}(\hat{\theta}_a) = \sum_{i=1}^{R} (\pi_i)^2 \left[ \mathrm{var}(\hat{\theta}_i \mid M_i) + (\theta_i - \theta_a)^2 \right]$$
$$+ \sum_{i=1}^{R} \sum_{\substack{h=1 \\ h \neq i}}^{R} \pi_i \pi_h \rho_{ih} \sqrt{[\mathrm{var}(\hat{\theta}_i \mid M_i) + (\theta_i - \theta_a)^2][\mathrm{var}(\hat{\theta}_h \mid M_h) + (\theta_h - \theta_a)^2]}.$$

We have no basis to estimate the across-model correlation of $\hat{\theta}_i - \theta_a$ with $\hat{\theta}_h - \theta_a$ (other than the bootstrap, but then we do not need theory for $\mathrm{var}(\hat{\theta}_a)$). The above simplifies if we assume that all $\rho_{ih} = \rho$:

$$\mathrm{var}(\hat{\theta}_a) = (1 - \rho) \left[ \sum_{i=1}^{R} (\pi_i)^2 \left[ \mathrm{var}(\hat{\theta}_i \mid M_i) + (\theta_i - \theta_a)^2 \right] \right]$$
$$+ \rho \left[ \sum_{i=1}^{R} \pi_i \sqrt{\mathrm{var}(\hat{\theta}_i \mid M_i) + (\theta_i - \theta_a)^2} \right]^2. \quad (4.12)$$

From (4.12), if we further assume $\rho = 1$, then we get (4.9):

$$\text{var}(\hat{\theta}_a) = \left[ \sum_{i=1}^{R} \pi_i \sqrt{\text{var}(\hat{\theta}_i \mid M_i) + (\hat{\theta}_i - \hat{\theta}_a)^2} \right]^2.$$

### 4.2.7 Unconditional Confidence Intervals

The matter of a $(1 - \alpha)100\%$ unconditional confidence interval is now considered. We have two general approaches: the bootstrap (see, e.g., Buckland et al. 1997), or analytical formulae based on analysis results from just the one data set. The analytical approach requires less computing; hence we start with it.

The simplest such interval is given by the endpoints $\hat{\theta}_i \pm z_{1-\alpha/2} \, \widehat{\text{se}}(\hat{\theta}_i)$, where $\widehat{\text{se}}(\hat{\theta}_i) = \sqrt{\widehat{\text{var}}(\hat{\theta}_i)}$. One substitutes the model-averaged $\hat{\theta}_a$ for $\hat{\theta}_i$ if that is the estimator used. A common form used and recommended as a conditional interval is $\hat{\theta}_i \pm t_{\text{df}, 1-\alpha/2} \, \widehat{\text{se}}(\hat{\theta}_i \mid M_i)$. When there is no model-selection, or it is ignored, it is clear what the degrees of freedom (df) are for the $t$-distribution here. For formula (4.11) it is not clear what the degrees of freedom should be. Note, however, that we are focusing on situations where sample size is large enough that the normal approximation will be applicable. These simple forms are based on the assumption that $\hat{\theta}_i$ has a normal sampling distribution.

We will hazard a suggestion here; it has not been evaluated in this context, but a similar procedure worked in a different context. If for each fitted model we have degrees of freedom $\text{df}_i$ for the estimator $\widehat{\text{var}}(\hat{\theta}_i \mid M_i)$, then for generally small degrees of freedom one might try using the interval $\hat{\theta}_i \pm z_{1-\alpha/2} \, \widehat{\text{ase}}(\hat{\theta}_i)$, where the adjusted standard error estimator is

$$\widehat{\text{ase}}(\hat{\theta}_a) = \sum_{i=1}^{R} \hat{\pi}_i \sqrt{\left(\frac{t_{\text{df}_i, 1-\alpha/2}}{z_{1-\alpha/2}}\right)^2 \widehat{\text{var}}(\hat{\theta}_i \mid M_i) + (\hat{\theta}_i - \hat{\theta}_a)^2}.$$

In cases where $\hat{\theta}_i \pm z_{1-\alpha/2} \, \widehat{\text{se}}(\hat{\theta}_i)$ is not justified by a normal sampling distribution (as judged by the conditional distribution of $\hat{\theta}_i$), intervals with improved coverage can be based on a transformation of $\hat{\theta}_i$ if a suitable transformation is known. Log and logit transforms are commonly used, often implicitly in the context of general linear models. In fact, in general linear models the vector parameter $\underline{\theta}$ will be linked to the likelihood by a set of transformations, $\underline{\theta} = W(\underline{\beta})$. Then it is $\underline{\beta}$ that is directly estimated, and it is often the case that the simple normal-based confidence limits on components of $\underline{\beta}$ can be reliably used. An interval constructed from a component of $\hat{\underline{\beta}}$ and its unconditional sampling variance (formula 4.9 applies) can be back-transformed to an interval on the corresponding component of $\underline{\theta}$.

The above methods are justified asymptotically, or if a normal sampling distribution applies to $\hat{\theta}$. However, "asymptotically" means for some suitable large sample size $n$. We do not know when to trust that $n$ is suitably large in nonlinear and nonnormal random variation models. A general alternative when there is no model-selection is the profile likelihood interval approach. We suggest here an adaptation of that approach that widens the likelihood interval to account for model-selection uncertainty.

Let the vector parameter $\underline{\theta}$ be partitioned into the component of interest, $\theta$, and the rest of the parameters, denoted here by $\underline{\gamma}$. Then the profile likelihood, as a function of $\theta_i$ (the subscript denotes the model used) for model $M_i$ is given by

$$\mathcal{PL}(\theta_i \mid \underline{x}, M_i) = \max_{\underline{\gamma}_i \mid \theta_i} \left[ \mathcal{L}(\theta_i, \underline{\gamma}_i \mid \underline{x}, M_i) \right];$$

almost always $\mathcal{PL}(\theta_i \mid \underline{x}, M_i)$ has to be computed numerically. We define a profile deviance as

$$\mathcal{PD}(\theta_i) = 2 \left[ \mathcal{PL}(\hat{\theta}_i \mid \underline{x}, M_i) - \mathcal{PL}(\theta_i \mid \underline{x}, M_i) \right]. \quad (4.13)$$

The large sample profile likelihood interval ignoring model-selection uncertainty is the set of $\theta_i$ that satisfy the condition $\mathcal{PD}(\theta_i) \leq \chi^2_{1,1-\alpha}$. Here, $\chi^2_{1,1-\alpha}$ is the upper $1 - \alpha$ percentile of the central chi-squared distribution on 1 df. This interval is approximately a $(1 - \alpha)100\%$ confidence interval.

We propose an interval that is a version of (4.13) adjusted (widened) for model-selection uncertainty: the set of all $\theta_i$ that satisfy

$$\mathcal{PD}(\theta_i) \leq \left[ \frac{\widehat{\text{var}}(\hat{\theta})}{\widehat{\text{var}}(\hat{\theta}_i \mid M_i)} \right] \chi^2_{1,1-\alpha}. \quad (4.14)$$

It suffices to solve (numerically) this inequality for the confidence interval endpoints, $\hat{\theta}_{i,L}$ and $\hat{\theta}_{i,U}$. In the event we are not doing model averaging, it seems logical to use the resultant confidence interval from (4.14).

If we decide to use the model-averaged $\hat{\theta}_a$ of formula (4.6), then a logical interval is also of the model-averaged form. For each model $i$, compute $\hat{\theta}_{i,L}$ and $\hat{\theta}_{i,U}$ from (4.14) and then use

$$\hat{\theta}_L = \sum_{i=1}^{R} \hat{\pi}_i \hat{\theta}_{i,L} \text{ and } \hat{\theta}_U = \sum_{i=1}^{R} \hat{\pi}_i \hat{\theta}_{i,U}. \quad (4.15)$$

Generally, one would use Akaike weights, hence $\hat{\pi}_i \equiv w_i$ from (4.2).

All of the above was assuming that the parameter of interest occurred in each of the $R$ models in the full set of models. Often this will not be the case. Rather, there will be a subset of size $Q < R$ of the models in which the parameter $\theta$ occurs. In this event we suggest applying all the above theory to just that subset of $Q$ models. The $R$-$Q$ models in which $\theta$ does not appear seem totally uninformative about the value of $\theta$; hence they cannot play a direct role in inference about $\theta$.

In the case that $Q = 1$ ($\theta$ is unique to one model in the set of $R$ models), none of the above results can be used. In this case it seems that there may not be a direct way to include model-selection uncertainty into the uncertainty about the value of $\theta$. An approach we can envision here is to adjust upward the conditional sampling variance estimator, $\widehat{\text{var}}(\hat{\theta} \mid M_i)$, by some variance inflation factor. What the variance inflation factor would be is not clear because we have looked at the variance inflation factor

$$\frac{\widehat{\mathrm{var}}(\hat{\theta})}{\widehat{\mathrm{var}}(\hat{\theta}_i \mid M_i)},$$

and found it can vary greatly by parameter. Thus, estimation of a variance inflation factor for $\hat{\theta}$ based on a different parameter in a subset of different models (from the one model $\theta$ appears in) seems very problematic. Fundamentally, it is not clear that we should inflate the conditional sampling variance of a parameter unique to just one model in the set of models. Perhaps all we can, and should, do in this case is note the uncertainty about whether that model is likely to be the K-L best model in the full set of models and use the model-specific conditional sampling variance for that $\hat{\theta}$. Confidence intervals for $\theta$ are then constructed based on just the one model in which $\theta$ appears (e.g., profile likelihood interval, or other parametric methods such as $\hat{\theta} \pm z_{1-\alpha/2} \, \widehat{\mathrm{se}}(\hat{\theta})$).

Bootstrap construction of an unconditional confidence interval on a parameter in the selected model is not fundamentally different from such bootstrap-based interval construction without model-selection. The latter is much discussed in the statistical literature (see, e.g., Efron and Tibshirani 1993, Mooney and Duval 1993, Hjorth 1994).

First one generates a large number, $B$, of bootstrap samples from the data and applies model-selection to each bootstrap sample. All $R$ models in the original set are fit to each bootstrap sample, and one of these models will be selected as best. Only the estimated parameters from that selected best model are kept in an output set of parameter estimates for each bootstrap rep (plus the index of the selected model for each rep). Hence if a parameter $\theta$ is not in the selected model $M_i$ for bootstrap rep $b$, the value of $\hat{\theta}_b^*$ is missing for bootstrap sample $b$ (the subscript $b$ and $\hat{\theta}_b^*$ denote that $\hat{\theta}^*$ is from the $b$th bootstrap sample; the model used to get this $\hat{\theta}^*$ varies over bootstrap samples). For any parameter in common over all models, there will be $B$ values of $\hat{\theta}_b^*$ in the output data set. In either case the variation in the output set of (not missing) values of $\hat{\theta}_b^*$, $b = 1, \ldots, m$ ($\leq B$), reflects both model-selection variation in $\hat{\theta}$ and within-model sampling variation of $\hat{\theta}$ given a model ($\theta$ may not appear in some models; hence if one of those models is selected in a bootstrap sample we have a missing $\hat{\theta}_b^*$ value).

As noted in Section 4.2.2, the model-selection frequencies can be estimated from these bootstrap results; but our focus here is on unconditional estimator uncertainty and confidence intervals (Efron and Tibshirani 1993). The average of all $m$ values of $\hat{\theta}_b^*$, $\overline{\hat{\theta}^*}$ is an estimator of the model-averaged parameter $\theta_a$. Hence the empirical variance of the set of $m$ values of $\hat{\theta}_b^*$,

$$\mathrm{var}(\hat{\theta}^*) = \sum (\hat{\theta}_b^* - \overline{\hat{\theta}^*})^2 \, / \, (m-1),$$

is the bootstrap estimator of the unconditional sampling variance of $\hat{\theta}$ ($\equiv \hat{\theta}_i$ for the parameter $\theta$ estimated from the selected best model, $i$). Given this bootstrap $\mathrm{var}(\hat{\theta}^*) = \widehat{\mathrm{var}}(\hat{\theta})$, the simplest confidence interval is $\hat{\theta} \pm z_{1-\alpha/2} \, \widehat{\mathrm{se}}(\hat{\theta})$. However, such an interval fails to make full use of the value of the bootstrap method in finding upper and lower interval estimates that allow for a nonnormal sampling distribution for $\hat{\theta}$ under model-selection.

The most simple, direct bootstrap-based confidence interval on $\theta$ is the percentile interval. Order the bootstrap values $\hat{\theta}_b^*$ from smallest to largest and denote these ordered values by $\hat{\theta}_{(b)}^*$, $b = 1, \ldots, m$. For a $(1-\alpha)100\%$ confidence interval select the $\alpha/2$ lower and $1-\alpha/2$ upper percentiles of these ordered bootstrap estimates as $\hat{\theta}_L$ and $\hat{\theta}_U$. These percentiles may not occur at integer values of $b$, but if $m$ is large, it suffices to use $\hat{\theta}_L = \hat{\theta}_{(l)}^*$ and $\hat{\theta}_U = \hat{\theta}_{(u)}^*$, where $l = [m \cdot \frac{\alpha}{2}]$ and $u = [m \cdot (1-\frac{\alpha}{2})]$. More complex, but possibly better, unconditional intervals after model-selection based on the bootstrap are considered by Shao (1996) for regression problems.

Note that $B$ needs to be at least several hundred for the bootstrap method to even begin to work, and we recommend 10,000 (and at least use $B = 1,000$). If the parameter of interest is in every model, then $m = B$, which is user selected. If the parameter is not in every model, $m$ is random, and $B$ may need to be made larger to assure that a sufficient sample size, $m$, of relevant bootstrap samples is obtained.

## 4.2.8 Uncertainty of Variable Selection

Sometimes in data analysis the focus is considered to be on the variables to include versus exclude in the selected model. In fact, variable selection is often expressed as the focus of model-selection for linear regression models. The above provides methods for model-selection uncertainty as such and for other needs such as unconditional sampling variances on partial regression coefficients in the selected best model. However, one additional issue arises now: Quantify the evidence for the importance of each variable, or set of variables, as well as just for the best model.

For example, there might 10 models considered based on combinations of a number of regressor variables. Assume that the selected best model includes $x_1$ and has an Akaike weight of only 0.3. There is a lot of model-selection uncertainty here, and hence there would seem to be only weak evidence for the importance of variable $x_1$ based on the selected best model. But one must consider the Akaike weights of all other models that include $x_1$ in order to quantify the importance of $x_1$. It might be that all models that exclude $x_1$ have very low Akaike weights; that situation would suggest that $x_1$ is a very important predictor. The measure of this importance is to sum the Akaike weights (or the bootstrap $\hat{\pi}_i$) over the subset of models that include variable $x_1$. This idea is applicable in general to model-selection whenever it is equated to variable selection, for linear or nonlinear models of any type.

To help make this idea clear, we consider below the hypothetical example of three regressors, $x_1$, $x_2$, and $x_3$, and a search for the best of the eight possible models of the simple linear regression type: $y = \beta_0 + \beta_1 x_1 + \beta_2 x_2 + \beta_3 x_3 + \epsilon$. The possible combinations of regressors that define the eight possible models are shown below, along with hypothetical Akaike weights, $w_i$ (a 1 denotes that $x_i$ is in the model; otherwise, it is excluded):

| $x_1$ | $x_2$ | $x_3$ | $w_i$ |
|---|---|---|---|
| 0 | 0 | 0 | 0.00 |
| 1 | 0 | 0 | 0.10 |
| 0 | 1 | 0 | 0.01 |
| 0 | 0 | 1 | 0.05 |
| 1 | 1 | 0 | 0.04 |
| 1 | 0 | 1 | 0.50 |
| 0 | 1 | 1 | 0.15 |
| 1 | 1 | 1 | 0.15 |

While the selected best model has weight of only 0.5, i.e., a probability of 0.5 of being the actual K-L best model here, the sum of the weights for variable $x_1$ is 0.79. This is the correct evidence of the importance of this variable, across the models considered. Variable $x_2$ was not included in the selected best model; do we therefore think that it is of zero importance? No; its relative weight of evidence support is 0.35. Finally, the sum of the Akaike weights for independent variable $x_3$ is 0.85. Thus the evidence for the importance of variable $x_3$ is substantially more that just the weight of evidence for the best model. We can order the three independent variables in this example by their estimated importance: $x_3, x_1, x_2$ with importance weights of 0.85, 0.79, and 0.35. As with other methods recommended here, we see that we are able to use model-selection to go well beyond just noting the best model from a set of models.

This idea extends to subsets of variables. For example, we can judge the importance of a pair of variables, as a pair, by the sum of the Akaike weights of all models that include the pair of variables. For the pair $x_1$ & $x_2$, the weight of evidence for the importance of this pair is 0.19. For pair $x_2$ & $x_3$, the weight of evidence for importance is 0.23, while for the pair $x_1$ & $x_3$, the weight of evidence is 0.65 (compared to 0.5 for the selected model as such).

To summarize, in many contexts the AIC selected best model will include some variables and exclude others. Yet this inclusion or exclusion by itself does not distinguish differential evidence for the importance of a variable in the model. The model weights, $w_i$ or $\hat{\pi}_i$, summed over all models that include a given variable provide a better weight of evidence for the importance of that variable in the context of the set of models considered.

### 4.2.9 Model Redundancy

Consider a set of three models, except that models 2 and 3 are identical because a mistake was made in setting up the problem on the computer. Thus, models 2 and 3 are 100% redundant in the set of models; the model set should be only models 1 and 2. Assume $\Delta_1 = 0$ and $\Delta_2 = \Delta_3 = 4$. For the redundant set of three models we get, using formula (4.1), $\mathcal{L}(M_1)/\mathcal{L}(M_2) = \mathcal{L}(M_1)/\mathcal{L}(M_3) = 7.4$. Similarly, for the correct set of two models, $\mathcal{L}(M_1)/\mathcal{L}(M_2) = 7.4$. The unfortunate model redundancy has not affected the $\Delta_i$ nor the likelihood ratios of models. However, the (normalized) Akaike weights (formula 4.2) are affected: For the set of two models, $w_1 = 0.881$ and $w_2 = 0.119$; whereas for the set with model redundancy, $w_1 = 0.787$ and $w_2 = w_3 = 0.106$. Note that for either model set we still have $w_1/w_2 = 7.4 (= w_1/w_3)$: Likelihood ratios are not affected by model redundancy.

The difference between a $w_1$ of 0.881 and 0.787 is not large. However, our point is that this, clearly erroneous, model redundancy in the three-model set has affected the Akaike weights. The weights for the model set with a redundant model included are not correct because the value $\Delta_2$ shows up twice (one time "disguised" as $\Delta_3$). The effect on the weights is not dramatic here, but they are wrong, and this could affect (presumably adversely) calculations using the $w_i$ (as $\hat{\pi}_i$), as for example in model averaging (formula 4.6) and variance calculations (formula 4.11).

If the model redundancy was recognized, and we wanted to retain it, we could correct the situation by considering the set of models as having two subsets: Model $M_1$ is one subset; a second subset contains models $M_2$ and $M_3$. Given that we know that models $M_2$ and $M_3$ are 100% redundant, we allocate our prior beliefs, about which model is the expected K-L best model, as 1/2 to each subset, and the 1/2 is further divided equally for each model in a subset. Thus, $\tau_1 = 0.5$, $\tau_2 = 0.25$, and $\tau_3 = 0.25$. Now we use formula (4.3) to compute correct Akaike weights for the set of three models; thus

$$w_1 \propto 1.0 \cdot \frac{1}{2}, \quad w_2 \propto 0.135335 \cdot \frac{1}{4}, \quad w_2 \propto 0.135335 \cdot \frac{1}{4}.$$

The normalized (to add to 1) weights are 0.8808, 0.0596, and 0.0596. Now the sum of the weights for models $M_2$ and $M_3$ correctly add up to what they ought to be, so formulae such as (4.6) and (4.11) will produce correct results when applied to the redundant set of three models.

This hypothetical example presumably would not occur, but it serves to introduce the concept, and issue, of model redundancy in the set of models considered. It is possible to have actual model redundancy if one is not careful in constructing the set of models considered. For example, in analysis of distance sampling data (Buckland et al. 1993), the program DISTANCE (Laake et al. 1994) can consider, in effect, the full set of models structured into two or more subsets. Different subsets of models may be specified in such a way that they have one key model in common, to which adjustment terms are applied to get a sequence of models. The same key model can be used with different types of adjustment terms. Schematically, we can have this situation: The full set of models is given as two subsets of models, $\{M_1, M_2, M_3, M_4\}$ and $\{M_1, M_5, M_6, M_7\}$. If the full set of models is considered as just 8 different models, then the redundancy of model $M_1$ is not being recognized. The situation can easily be rectified if it is recognized: Either label the models as 1 to 7, or compute the $w_i$ from the $\Delta_i$, for the models label as 1 to 8, based on differential priors, $\tau_i$, as $\{\frac{1}{14}, \frac{1}{7}, \frac{1}{7}, \frac{1}{7}\}$ and $\{\frac{1}{14}, \frac{1}{7}, \frac{1}{7}, \frac{1}{7}\}$. Might there be more partial model redundancy in the models $M_1$ to $M_7$? That is not clear to us, but we think that it is not a problem.

To further illustrate model redundancy we consider some models for capture–recapture data, obtained on $k$ capture occasions, to estimate animal population size. The parameters of such models are population size ($N$) and capture probabilities (denoted by $p$), by occasion, animal, or by both factors. One possible type of model is model $M_b$ under which there are only two different capture probabilities: for first capture or for recapture. This model is for the case where animals have a behavioral response to first capture, but no other factors affect capture probability ($K = 3$).

A different model ($M_t$) allows capture probabilities to vary, but only by occasion; so we have $p_1, p_2, \ldots, p_k$ ($K = k + 1$). Thus, we can have two very different models. However, the model under which capture probabilities can vary by time allows for many submodels ($2^k - k$ possible models, including the most general case). Some example (sub) models are

$$M_{t1} : p_1 = p_2, \text{ other } p_i \text{ all differ } (K = k),$$
$$M_{t2} : p_1 = p_2 = p_3, \text{ other } p_i \text{ all differ } (K = k - 1),$$
$$M_{t3} : \text{all } p_i = p \ (K = 2),$$
$$M_{t4} : \text{all } p_i \text{ are different } (K = k + 1).$$

If we now take as our model the set $\{M_{t1}, M_{t2}, M_{t3}, M_{t4}, M_b\}$ (so these are in order, as models 1 to 5), we have model redundancy that if ignored could cause problems. If we were to get the $\Delta_i$, in order 1 to 5, as $\{15, 10, 16, 20, 0\}$, then model redundancy becomes irrelevant because $M_b$ is overwhelmingly the best model: The usual Akaike weight for that model is here $w_5 = 0.992$). We do claim, however, that the correct weights here should be based on model priors as $\{\frac{1}{8}, \frac{1}{8}, \frac{1}{8}, \frac{1}{8}, \frac{1}{2}\}$ not $\{\frac{1}{5}, \frac{1}{5}, \frac{1}{5}, \frac{1}{5}, \frac{1}{5}\}$. If the $\Delta_i$ are $\{2, 0, 1, 5, 20\}$, then again model redundancy is irrelevant (redundancy is only between model 5 and the others; there is no redundancy in models 1 to 4 if model 5 is ignored). But if the result is $\{2, 2, 2, 2, 0\}$, then model redundancy matters a lot as regards the proper $w_i$. For wrong priors $\{\frac{1}{5}, \frac{1}{5}, \frac{1}{5}, \frac{1}{5}, \frac{1}{5}\}$, $w_5 = 0.40$, but under correct priors $\{\frac{1}{8}, \frac{1}{8}, \frac{1}{8}, \frac{1}{8}, \frac{1}{2}\}$, $w_5 = 0.73$.

By adding submodels of the general time-specific model to our set of models, we dilute the absolute strength of evidence for model $M_b$ as measured by Akaike weights; and we must use such absolute weights in certain formulae (e.g., model averaging). In as much as these added models deal only with time variation in capture probabilities, they are all of a type (hence, redundant as regards their evidence *against* model $M_b$), so they unfairly "gang up" against model $M_b$, which is a totally different type of model.

The appropriateness of unequal priors if submodels of the general time model, $M_t$, are included is justified here on a theoretical basis. It is well documented in the capture–recapture literature that there is no practical advantage, as regards estimating $N$, of considering constrained versions of the general time-specific model. Thus, the original set of two models, $\{M_t, M_b\}$, should not be augmented as above. Hence, in the last example we should really have only these two models, and they have $\Delta_1$ and $\Delta_2$ as $\{2, 0\}$. Now, for model $M_b$, $w_2 = 0.73$. A key point here is that when we did have model redundancy, the use of the unequal priors did produce the correct Akaike weights. Thus, we think that model redundancy can be coped with analytically by appropriate modification of the otherwise equal model priors, $\tau_i$.

Even more important than accepting model redundancy, and therefore modifying model priors, is to construct the set of models to be considered so that there is no model redundancy. As the above example illustrates, all suitable knowledge about the correct formulation and use of models for the problem at hand should be utilized in defining the *a priori* set of models to consider. Another point worth repeating is that neither the $\Delta_i$ nor the relative likelihoods of the models will be affected by model redundancy. Thus confidence sets on models based on all models with $\Delta_l$ less than some cutoff value may be the safest type to use. Our ideas on the cutoff value to use can be obtained from the distribution of $\Delta_p$, but only for situations with no model redundancy. The recommendations already made on this matter were so developed.

The concept and issue of model redundancy was brought to our attention by S. T. Buckland (personal communication); the above ideas are our own. Professor Buckland (personal communication) suggested that the bootstrap would automatically be a solution to model redundancy as long as for a given bootstrap sample it is forced to select one best model. This seems reasonable; but we still perceive a need for analytical formulae, and we now think that the analytical solution to model redundancy lies in construction of unequal model priors. However, then we must be able to recognize model redundancy in our set of models. If we can do that (we can and should), we think that redundancy can be eliminated. If model redundancy operates at a more subtle level than considered here, the bootstrap would have an advantage. We are currently disinclined to think that there will be a model redundancy problem as regards Akaike weights as long as the set of models considered is carefully constructed. (More research on the issue would be helpful.)

## 4.2.10  *Recommendations*

If data analysis relies on model-selection, then inferences should acknowledge model-selection uncertainty. If the goal is to get the best estimates of a set of parameters in common to all models (this includes prediction), model averaging is recommended. If the models have definite, and differing, interpretations as regards understanding relationships among variables, and it is such understanding that is sought, then one wants to identify the best model and make inferences based on that model. Hence, reported parameter estimates should then be from the selected model (not model averaged values). However, even when selecting a best model, also note the competing models, as ranked by their Akaike weights. Restricting detailed comparisons to the models in a 90% confidence set on models should often suffice. If a single model is not strongly supported, and competing models give alternative inferences, this should be reported. It may occur that the basic inference(s) will be the same from all good models. However, this is not always the case, and then inference based on a single best model may not be sound if support for even the best model is weak.

We recommend that investigators compute and report unconditional measures of precision based on formula (4.11) when inference is based on a best model, unless the Akaike weight $w_i$ for the selected model is large. For an unconditional confidence interval, often the form $\hat{\theta} \pm 2\,\widehat{\text{se}}(\hat{\theta})$ will suffice, or an interval of this type back-transformed from a function of $\hat{\theta}$ such as occurs via the link function in general linear models. If such a simple interval has clear deficiencies, or in general if the computation can be done, use inflated profile likelihood intervals based on formulae (4.14) and (4.13).

If interest is really just on some parameters in common to all models, then we recommend using model-averaged parameter estimates from formula (4.6) taking for $\hat{\pi}_i$ the Akaike weight $w_i$. The sampling variance estimate to use is then formula (4.10). Again, often the form $\hat{\theta} \pm 2\,\widehat{\text{se}}(\hat{\theta})$ will suffice for a confidence interval. A logical alternative is to use weighted averaged confidence interval endpoints (formula 4.15); however, we recommend more study of this procedure.

We think that these analytical procedures can suffice, so we would not initially resort to use of the bootstrap to evaluate model-selection uncertainty. The bootstrap can produce robust estimates of unconditional sampling variances and confidence intervals, as by the percentile confidence intervals. The bootstrap provides direct, robust estimates of model-selection probabilities, $\pi_i$, but we have no reason now to think use of bootstrap estimates of model-selection probabilities rather than use of the Akaike weights will lead to superior unconditional sampling variances or model-averaged parameter estimators. The primary purpose of the bootstrap is to assess uncertainty about inferences; therefore, we recommend that the point estimates used be the actual MLEs from the selected model (not the bootstrap means). In analyses that are very complex, where there may be no suitable analytical or numerical estimators of conditional (on model) sampling variances, the bootstrap could be used to get conditional and unconditional measures of precision. We recommend that more bootstrap samples be used than is commonly the case; use 10,000 for really reliable results, but even 400 would be better than no assessment of model-selection uncertainty (no assessment has often been the default).

Be mindful of possible model redundancy. A carefully thought out set of *a priori* models should eliminate model redundancy problems and is a central part of a sound strategy for obtaining reliable inferences.

The theory here applies if the set of models is *a priori* to the data analysis. If any models considered have been included after some analyses, because said model(s) are suggested by the data, then theoretical results (such as variance formulae) might fail to properly apply (in principle the bootstrap can still be used). Even for such data-driven model-selection strategies we recommend assessing model-selection uncertainty rather than ignoring the matter.

## 4.3 Examples

### 4.3.1 Cement Data

We return to the cement data of Section 3.2 to compare bootstrap estimates of model-selection frequencies ($\pi_i$), $\Delta_i$ values, Akaike weights ($w_i$), and unconditional estimation of sampling variances. These quantities are summarized in Table

TABLE 4.1. Bootstrap selection probabilities, $\hat{\pi}_i$, for the models of the cement data used in Section 3.2; $B = 10,000$ bootstrap reps were used; also shown are AIC differences $\Delta_i$ and derived Akaike weights computed from the data.

| Model | K | $\hat{\pi}_i$ | $\Delta_i$ | $w_i$ |
|---|---|---|---|---|
| {12} | 4 | **0.5338** | **0.0000** | **0.5670** |
| {124} | 5 | 0.0124 | 3.1368 | 0.1182 |
| {123} | 5 | 0.1120 | 3.1720 | 0.1161 |
| {14} | 4 | 0.2140 | 3.3318 | 0.1072 |
| {134} | 5 | 0.0136 | 3.8897 | 0.0811 |
| {234} | 5 | 0.0766 | 8.7440 | 0.0072 |
| {1234} | 6 | 0.0337 | 10.5301 | 0.0029 |
| {34} | 4 | 0.0039 | 14.4465 | 0.0004 |

4.1; the $AIC_c$-selected model is shown there in bold. The remaining seven models are not shown in Table 4.1 because they were never selected in the 10,000 bootstrap samples (also, they have virtually zero Akaike weights). The three simple approaches shown in Table 4.1 provide useful insights into model-selection uncertainty for this very small ($n = 13$) data set. Clearly, model {12} is indicated as the best by all approaches. However, substantial model-selection uncertainty is evident because that best model has an Akaike weight of only 0.57 and a bootstrap selection probability of 0.53. All three approaches cast substantial doubt concerning the utility of the final three or four models in Table 4.1. Model {34} is particularly unsupported, with $\hat{\pi}_i < 0.004$ and $w_i = 0.0004$.

Evidence for the importance of each variable can be obtained by using the bootstrap and tallying the percentage of times that each variable occurred in the $AIC_c$ selected model (Section 4.2.6). For the 10,000 bootstrap samples, $x_1$ occurred in 93% of the models, followed by $x_2$ (76%), $x_3$ (23%), and $x_4$ (36%). Again, this simple approach indicates the importance of $x_1$ and $x_2$ relative to $x_3$ and $x_4$. Similar evidence can be obtained by summing the Akaike weights over those models with a particular variable present. Using that simple approach, the relative support of the 4 variables is as follows: $x_1$ (99%), $x_2$ (81%), $x_3$ (21%), and $x_4$ (32%). Considering the small sample size ($n = 13$), the bootstrap and Akaike weights seem to give similar results.

Using the idea of the pivotal $\Delta_p$ (Section 4.2.1, and formula 4.3) to obtain the bootstrap distribution as $\Delta_p^*$, we find that an approximate 90% confidence set occurs for $\Delta_i < 8.75$, while a 95% set is achieved if $\Delta_i < 13.8$. These bootstrap based percentile values of $\Delta_i$ are quite extreme here because the sample size in this example is so small ($n = 13$).

Using the bootstrap selection frequencies ($\hat{\pi}_i$), models {12}, {14}, and {123} represent an approximate 86% confidence set, while adding model {234} reflects an approximate 94% confidence set of models. Using Akaike weights ($w_i$), an approximate 90% confidence set includes models {12}, {124}, {123}, and {14}. The $\Delta_i$ values suggest that the final 3 models in Table 4.1 have little utility. These types

TABLE 4.2. Some analysis results for the cement data of Section 3.2; $\hat{Y}_o$ is a predicted expected response based on the fitted model (see text for $x_i$ values used); conditional-on-model measures of precision are given for $\hat{Y}_o$; $\overline{Y}$ denotes a model-averaged predicted value; and $\hat{Y}_o - \overline{Y}$ is the estimated bias in using a given model to estimate $Y_o$.

| Model | K | $\hat{Y}_o$ | $\widehat{se}(\hat{Y}_o \mid M_i)$ | $\widehat{var}(\hat{Y}_o \mid M_i)$ | $(\hat{Y}_o - \overline{Y}_{\text{bootstrap}})^2$ | $(\hat{Y}_o - \overline{Y})^2$ |
|---|---|---|---|---|---|---|
| {12}   | 4 | 100.4 | 0.732 | 0.536  | 4.264  | 1.503   |
| {124}  | 5 | 102.2 | 1.539 | 2.368  | 0.070  | 0.329   |
| {123}  | 5 | 100.5 | 0.709 | 0.503  | 3.861  | 1.268   |
| {14}   | 4 | 105.2 | 0.923 | 0.852  | 7.480  | 12.773  |
| {134}  | 5 | 105.2 | 0.802 | 0.643  | 7.480  | 12.773  |
| {234}  | 5 | 111.9 | 2.220 | 4.928  | 89.019 | 105.555 |
| {1234} | 6 | 101.6 | 5.291 | 27.995 | 0.748  | 0.001   |
| {34}   | 4 | 104.8 | 1.404 | 1.971  | 5.452  | 10.074. |

of ranking and calibration measures have not been available under a hypothesis testing approach or cross validation.

We now illustrate the computation of unconditional estimates of precision, first for a parameter in common to all models. What if one wanted to predict the value $\hat{E}(Y_o)$, denoted for simplicity as $\hat{Y}_o$, given the values $x_1 = 10$, $x_2 = 50$, $x_3 = 10$, and $x_4 = 20$ (cf. Table 3.1)? The prediction under each of the eight models of Table 4.1 is shown in Table 4.2; we used PROC REG (SAS 1985) in SAS to easily compute predicted values and their conditional standard errors, $\widehat{se}(\hat{Y}_o \mid M_i)$. Clearly, $\hat{Y}_o$ is high for model {234}, relative to the other models. The estimated standard error for model {1234} is very high, as might be expected because the X matrix is nearly singular. Both of these models have relatively little support, as reflected by the small relative weights, so the predicted value under these fitted models is of little credibility.

The predicted value for the $AIC_c$-selected model is 100.4 with an estimated conditional standard error of 0.73. However, this measure of precision is an underestimate because the variance component due to model-selection uncertainty has not been incorporated. Model averaging (formula 4.3) results in a predicted value of 102.5 using the bootstrap estimated weights ($\hat{\pi}_i$) and 101.6 using the Akaike weights ($w_i$). Using formula (4.10) the corresponding estimated unconditional standard errors are 3.0 using the bootstrap based weights and 1.9 using the Akaike weights (formula 4.10 can be computed here based on results in Tables 4.1 and 4.2). These unconditional standard errors are substantially larger than the conditional standard error of 0.73. In Monte Carlo studies we have done we find that the unconditional standard errors better reflect the actual precision of the predicted value, and conditional confidence interval coverage is often quite near the nominal level (Chapter 5).

Study of the final three columns in Table 4.2 above shows that the variation in the model-specific predictions (i.e., the $\hat{Y}_o$) from the weighted mean (i.e., $(\hat{Y}_o - \overline{Y}_{\text{bootstrap}})^2$ or $(\hat{Y}_o - \overline{Y})^2$) is substantial relative to the estimated variation, conditional

148    4. Model-Selection Uncertainty with Examples

on the model (i.e., the $\widehat{\text{var}}(\hat{Y}_o \mid M_i)$). Models {124} and {1234} are exceptions because they over-fit the data (i.e., more parameters than are needed). The Akaike weights are relatively easy to compute compared to the effort required to obtain the bootstrap estimates, $\hat{\pi}_i$; $w_i$ seem preferred for this reason, in this example. That is, we perceive no "payoff" here from the bootstrap-based results, compared to the Akaike-weight-based results, that compensates for the computational cost of the bootstrap (we do not claim that the bootstrap based results are any worse, just not better).

The investigator has the choice as to whether to use the predicted value from the $\text{AIC}_c$-selected model (100.4) or a model-averaged prediction (102.5 for the bootstrap weights, $\hat{\pi}_i$, or 101.6 for the Akaike weights, $w_i$). In this example, the differences in predicted values are small relative to the unconditional standard errors (3.0 for the bootstrap and 1.9 for Akaike weights); thus here the choice of weights makes no great difference. However, there is considerable model uncertainty associated with this data set, and we would suggest the use of model-averaged predictions (when prediction is the objective), based on the Akaike weights. Thus, we would use 101.6 as the predicted value with an unconditional standard error of 1.9. If the $\text{AIC}_c$-selected model was much more strongly supported by the data, then we might suggest use of the prediction based on that (best) model (i.e., $\hat{Y}_o = 100.4$) combined with use of the estimate of the unconditional standard error (1.9), based on the Akaike weights.

The selected model includes only regressor variables $x_1$ and $x_2$. For that model the estimated partial regression coefficients and their conditional standard errors are $\hat{\beta}_1 = 1.4683$ (conditional $\widehat{\text{se}} = 0.1213$) and $\hat{\beta}_2 = 0.6623$ (conditional $\widehat{\text{se}} = 0.0459$). Each of these parameters appears in eight models. To compute the estimate of unconditional sampling variation for $\hat{\beta}_1$ we first find each model containing $\beta_1$, its estimate and conditional variance in that model, and the model's Akaike weight:

| Model  | $\hat{\beta}_1$ | $\widehat{\text{se}}(\hat{\beta}_1 \mid M_i)$ | $w_i$  |
|--------|--------|--------|--------|
| {12}   | 1.4683 | 0.1213 | 0.5670 |
| {124}  | 1.4519 | 0.1170 | 0.1182 |
| {123}  | 1.6959 | 0.2046 | 0.1161 |
| {14}   | 1.4400 | 0.1384 | 0.1072 |
| {134}  | 1.0519 | 0.2237 | 0.0811 |
| {1234} | 1.5511 | 0.7448 | 0.0029 |
| {1}    | 1.8687 | 0.5264 | 0.0000 |
| {13}   | 2.3125 | 0.9598 | 0.0000 |

The first step is to renormalize the $w_i$ so they sum to 1 for this subset of models. Here that sum is 0.9925 before renormalizing, so we will not display the renormalized $w_i$, but they are the weights to use in applying formulae (4.6) and (4.11). The model-averaged estimate of $\beta_1$ is 1.4561 (from 4.3). Now apply (4.11) as

$$\widehat{\text{var}}(\hat{\theta}) = \left[ \sum_{i=1}^{8} w_i \sqrt{\widehat{\text{var}}(\hat{\theta}_i \mid M_i) + (\hat{\theta}_i - \hat{\theta}_a)^2} \right]^2.$$

For example, the first term in the needed sum is $0.069646 = 0.5713 \times \sqrt{(0.1213)^2 + (0.0122)^2}$. Completing the calculation, we get $\widehat{\text{var}}(\hat{\theta}) = (0.1755)^2$, or an estimated unconditional standard error on $\hat{\beta}_1$ of 0.1755, compared to the conditional standard error given the selected model of 0.1213.

For the same calculations applied for $\hat{\beta}_2$ we start with

| Model  | $\hat{\beta}_2$ | $\widehat{\text{se}}(\hat{\beta}_2 \mid M_i)$ | $w_i$  |
|--------|---------|---------|--------|
| {12}   | 0.6623  | 0.0459  | 0.6988 |
| {124}  | 0.4161  | 0.1856  | 0.1457 |
| {123}  | 0.6569  | 0.0442  | 0.1431 |
| {1234} | 0.5102  | 0.7238  | 0.0035 |
| {234}  | −0.9234 | 0.2619  | 0.0089 |
| {23}   | 0.7313  | 0.1207  | 0.0000 |
| {24}   | 0.3109  | 0.7486  | 0.0000 |
| {2}    | 0.7891  | 0.1684  | 0.0000 |

When all 16 models are considered, the Akaike weights for just the eight models above add to 0.8114. However, to compute results relevant to just these eight models we must renormalize the relevant Akaike weights to add to 1. Those renormalization Akaike weights are what are given above. The model-averaged estimator of $\beta_2$ is 0.6110, and the unconditional estimated standard error of $\hat{\beta}_2$ is 0.1206 (compared to the conditional standard error of 0.0459). It is important here to compute and use unconditional standard errors in all inferences after data-based model-selection. Note also that (to be conservative) confidence intervals on $\beta_1$ and $\beta_2$, using results from model {12}, should be constructed based on a $t$-statistic with 10 df ($t_{10, 0.975} = 2.228$ for a two-sided 95% confidence interval). Such intervals here will still be bounded well away from 0; for example, the 95% confidence interval for $\beta_2$ is 0.39 to 0.93.

We generated 10,000 bootstrap samples of these data and applied $AIC_c$ selection to all 16 models fit to each bootstrap sample. Then as per Section 4.2.7 (and common belief about the bootstrap) it should be acceptable to estimate the unconditional standard error of an estimated partial regression coefficient based on the standard deviation of the set of realized estimates, such as $\hat{\beta}^*_{1,b}$, over all bootstrap samples, $b$, wherein the selected model included variable $x_1$. The results are given below, along with the average value of the parameter estimate over all relevant bootstrap samples:

|           | bootstrap results | |
|-----------|---------|----------|
| parameter | average | st. error |
| $\beta_1$ | 1.461  | 0.760 |
| $\beta_2$ | 0.453  | 0.958 |
| $\beta_3$ | −0.420 | 1.750 |
| $\beta_4$ | −0.875 | 1.237 |

From the selected model, {12}, we get $\hat{\beta}_1 = 1.47$ and $\hat{\beta}_2 = 0.66$ with estimated unconditional standard errors of 0.18 and 0.12, respectively, as determined by analytical methods using Akaike weights.

Based on the above, and other comparisons not given, we conclude that the bootstrap failed here when all 16 models were allowed to be considered. As noted in Section 3.2 the full design matrix, $X$, for this regression example is essentially singular: The first three eigenvalues (in a principal components analysis on $X$) sum to 99.96% of the total of all four eigenvalues; the first two eigenvalues sum to 95.3% of the total. Also, the pairwise correlation of $x_2$ and $x_4$ is $r = -0.973$. This information, to us, strongly justifies (virtually forces) one to drop model {1234} from consideration and to drop all other models in which both $x_2$ and $x_4$ appear. Thus without any model fitting we can, and should, reduce the 16 possible models to 12 by eliminating models {24}, {124}, {234}, and {1234}. These sorts of considerations should be done routinely and do not compromise an *a priori* (as opposed to exploratory) model-selection strategy.

With the reduced set of 12 models we computed the $\Delta_i$, the $w_i$, and ran 10,000 new bootstrap reps (to get $\hat{\pi}_i$ and bootstrap estimates, $\hat{\theta}_b^*$) getting the results below (models not shown were never selected in the bootstrap samples):

| Model | K | $\hat{\pi}_i$ | $\Delta_i$ | $w_i$ |
|---|---|---|---|---|
| {12} | 4 | 0.5804 | 0.0000 | 0.6504 |
| {123} | 5 | 0.1315 | 3.1720 | 0.1332 |
| {14} | 4 | 0.2340 | 3.3318 | 0.1229 |
| {134} | 5 | 0.0465 | 3.8897 | 0.0930 |
| {34} | 4 | 0.0076 | 14.4465 | 0.0005 |

Applying here the method of Section 4.2.8 based on the Akaike weights, we get the support for the four variables as follows: $x_1$ (0.9995), $x_2$ (0.7836), $x_3$ (0.2267), and $x_4$ (0.2164). Using the methods of Section 4.2.5, especially formula (4.11) with the above Akaike weights, we computed unconditional standard errors as $\widehat{se}(\hat{\beta}_1) = 0.18$ (for $\hat{\beta}_1 = 1.47$, $\widehat{se}(\hat{\beta}_1 \mid M_{\{12\}}) = 0.12$) and $\widehat{se}(\hat{\beta}_2) = 0.046$ (for $\hat{\beta}_2 = 0.66$, $\widehat{se}(\hat{\beta}_2 \mid M_{\{12\}}) = 0.046$). The bootstrap estimates of unconditional standard errors are 0.34 and 0.047 for $\hat{\beta}_1$ and $\hat{\beta}_2$, respectively.

The two different methods (analytical-Akaike weights, and bootstrap) now agree for $\widehat{se}(\hat{\beta}_2)$ but not for $\widehat{se}(\hat{\beta}_1)$. The resolution of this discrepancy hinges on two items. First, the correlation in the data of $x_1$ and $x_3$ is $r = -0.82$; second, the sample size is only $n = 13$. As a result, fitted models {123} and {134} are very unstable over bootstrap samples as regards the estimate of $\beta_1$. For example, the sampling standard deviation (this estimates $se(\hat{\beta}_1 \mid M_{\{12\}})$) of the 1,315 bootstrap values of $\hat{\beta}_{1,b}$ that resulted when model {123} was selected by $AIC_c$ was 0.65 (the average of the bootstrap estimates $\hat{\beta}_{1,b}$ was 1.78). The theory-based estimate is $\widehat{se}(\hat{\beta}_1 \mid M_{\{12\}})) = 0.12$.

There are several points we wish to make with this example. Results are sensitive to having demonstrably poor models in the set of models considered; thus it is very important to exclude models that are *a priori* poor. The analytical method (vs. the bootstrap method) of assessing unconditional standard errors seems more stable, as regards having or excluding poor models from the set of models considered. In fact, the bootstrap approach failed when all 16 models were (erroneously) used.

However, the analytical approach seemed reasonable even with all 16 models considered (the results were more precise when only the 12 models were used). With the reduced set of models the bootstrap results are still suspect, but now only because sample size is so small ($n = 13$). Monte Carlo evaluation and comparison of both methods is needed before definitive statements about reliability will be possible.

### 4.3.2  Anolis Lizards in Jamaica

Schoener's (1970) data on resource utilization of 2 species of *Anolis* lizards in Jamaica were used as an example in Section 3.6. Here, we extend the example to illustrate the issue of model-selection uncertainty. First, we compute the Akaike weights; these weights suggest that only models 2, 3, 10, and perhaps 9 in Table 4.3 are supported by these data. Models 2, 3, and 10 comprise a rough 90% confidence set for the K-L best model here. The sum of the weights for these three models is 0.8894; contrast this to their being in the subset of models wherein $\Delta_i \leq 2.69$. If we include model 9 in our confidence set of models, then $\Delta_i \leq 4.46$, and we expect that this is about a 95% confidence set. The alternative measure of confidence in this set of four models is the sum of its Akaike weights, which is 0.9428. These are approximate, but useful, guidelines and procedures and allow us to say that there is little support from the data for the other 6 models.

### 4.3.3  Simulated Starling Experiment

The interpretation of the 24 models for the experimental starling data (Table 3.6) can be sharpened by examining the Akaike weights. Here, the the weight for

TABLE 4.3. Summary of *a priori* models considered, $\Delta_i$, and the Akaike weights for the lizard data (from Schoener 1970). The model with the minimum $AIC_c$ is shown in bold print.

| | Model | $\Delta_i$ | $w_i$ |
|---|---|---|---|
| 1. | All main effects, H D I T S | 100.88 | 0.0000 |
| 2. | All main effects and second-order interactions | 1.32 | 0.2586 |
| 3. | **Base[1] but drop DT** | **0.00** | **0.5004** |
| 4. | Base[1] plus HDI, HDT, and HDS terms | 7.10 | 0.0144 |
| 5. | Base[1] plus HDI, HDS, HIT, HIS, HTS, and ITS | 11.55 | 0.0016 |
| 6. | Base[1] plus HIT, HIS, HTS, and ITS | 8.09 | 0.0088 |
| 7. | Base[1] plus HIS, HTS, and ITS | 7.02 | 0.0150 |
| 8. | Base[1] plus HIT, HIS, HTS, and ITS, but drop DT | 6.75 | 0.0171 |
| 9. | Base[1] plus HIT, HIS, and ITS, but drop DT | 4.46 | 0.0538 |
| 10. | Base[1] plus HIT and HIS, but drop DT | 2.69 | 0.1304 |

[1] "Base" is a model with the 5 main effects plus all second-order interaction terms.

the AIC-selected model ($M_{\sin\phi_t.\phi_c.\sin p_t.p_c}$) is 0.906, while the second-best model ($M_{\sin\phi_t.\phi_c.p_t.p_c}$) has a weight of 0.063, and the third-best model ($M_{\sin\phi_t.\phi_{ci}.\sin p_t.p_c}$) has a weight of 0.022. The sum of the weights for the 21 remaining models is less than 0.01. In this case, one is left with strong support for the best model, with fairly limited support for the second-best model. Here, there is little need to attempt model averaging or bootstrapping to gain further robustness in inferences from these data. Rather, the use of conditional standard errors given the best model will likely suffice. Note: Bootstrapping in this example would be very, very difficult. Software development would be a very formidable task, and computer time on a Pentium PC would likely take more than a week. Thus, the Akaike weights provide a distinct advantage in complex problems such as this simulated starling experiment.

If only the first 14 models (Table 3.6) had been defined *a priori*, the inference concerning which model to use would have been far less clear. First, the best of these 14 models is over 25 units from the AIC-selected model, but this would not have been known. Second, seven models have AIC values within 4 units of the best of the 14. Thus, some additional steps would be necessary to incorporate model-selection uncertainty into inference for these experimental data for an analysis based on just the first 14 models.

### 4.3.4 Sage Grouse in North Park

Further insights into the sage grouse data (Section 3.5) can be obtained by computing Akaike weights for the first 17 models in Table 3.9 (ignoring here models 18–21). In this case, the weight for the AIC-selected model $\{S_a, r_a\}$ is 0.773, while the second-best model $\{S, r\}$ has a weight of only 0.085. The third- and fourth-best models had weights of 0.073 and 0.050, while the weights for the other models were nearly zero (the sum of the Akaike weights for the 13 remaining models was $< 0.02$). The annual variation in conditional survival probabilities was small (temporal process variation $\hat{\sigma}_s = 0.0426$ for adults and 0.0279 for subadults); thus model $\{S_a, r_a\}$ seems reasonable. The other three contending models all had $K$ less than that of the AIC-selected model; thus, conditional sampling variances from those models were smaller than from the AIC-selected model. In addition, these three models had small Akaike weights. These considerations lead to some trust in the conditional sampling variances from the best model as a good reflection of the precision of the parameter estimates. In this example, inference can probably be concentrated on model $\{S_a, r_a\}$ without a pressing concern about there being a substantial variance component due to model-selection uncertainty.

### 4.3.5 Sakamoto et al.'s (1986) Simulated Data

Here we return to the example used in Section 1.4.2 from Sakamoto et al.'s (1986) book. Ten data sets (each with $n = 21$) were generated from the simple model

$$y = e^{(x-0.3)^2} - 1 + \epsilon.$$

They used the simple polynomials of degree 0 to 5 as the set of candidate models to visually illustrate the concepts of under- and over-fitting (see Fig. 1.3). The averages of the ten $\Delta_i$ values, by model, are 24.88, 11.78, 0, 1.33, 2.34, and 4.27 for polynomials of degree 0, 1,..., 5, respectively, when those are the only models considered.

Because the sample size ($n = 21$) was small in relation to the dimension of the largest model in the set, $AIC_c$ should have been used for the analysis of data in this example. The averages of the ten $\Delta_i$ values when using $AIC_c$ are 23.05, 10.69, 0, 2.83, 5.84, and 10.38 for polynomials of degree 0, 1,..., 5, respectively, when those are the only models used. When the nonstandard special quadratic model is added, the $AIC_c$ and average $\Delta_i$ and derived average $w_i$ values are

| Model | K | $AIC_c$ | $\Delta_i$ | $w_i$ |
|---|---|---|---|---|
| Mean | 2 | −6.58 | 25.93 | 0.00 |
| Linear | 3 | −18.94 | 13.57 | 0.00 |
| Quadratic | 4 | −29.93 | 2.88 | 0.18 |
| Cubic | 5 | −26.80 | 5.71 | 0.04 |
| 4th-order | 6 | −23.79 | 8.72 | 0.01 |
| 5th-order | 7 | −19.26 | 13.25 | 0.00 |
| $\min f(x) = 0.3$ | 2 | −32.51 | 0.0 | 0.77 |

Clearly, the model based the on additional (hopefully *a priori*) information that $f(x)$ is 0 at $x = 0.3$ is the best of the set (compare these quantitative results with the plots in Fig. 1.3). The Akaike weights ($w_i$) more clearly sharpen the inference and suggest that only the quadratic model (with $K = 4$) is a competitor for these simulated data.

### 4.3.6  Pine Wood Data

We consider here an example of only two simple linear regression models, neither one nested in the other. This example has been used by Carlin and Chib (1995) on Bayesian model choice using Markov chain Monte Carlo methods. The data also appear elsewhere, such as in Efron (1984). The data (see Table 4.4) can be considered a trivariate response vector $(y, x, z)'$ for sample size $n = 42$. Variable $y$ is the measured strength of a piece of pine wood, $x$ is the measured density of that wood, and $z$ is the measured density after adjustment for the measured resin content of the wood. The scientific question is which of $x$ or $z$ is a better predictor of the wood strength, $y$, based on a linear model, either $y = a + bx + \epsilon$ or $y = c + dz + \delta$ ($\epsilon$ or $\delta$ are random residuals from the expected linear model structure). Residuals are taken to be normally distributed and homogeneous under either model. Scientifically, it is thought that wood density adjusted for resin content should be a better predictor of wood strength, but it takes more time and cost to measure variable $z$.

Simple linear regression can be used to find the MLE of $\sigma^2$ ($\hat{\sigma}^2$ = residual sum of squares divided by $n$) for each model. This gives us $\hat{\sigma}^2 = 109,589$ for model $M_x$ (i.e., $y$ vs. $x$) and $\hat{\sigma}^2 = 73,011$ for model $M_z$; both models have $K = 3$. The

154    4. Model-Selection Uncertainty with Examples

TABLE 4.4. Pine wood wood strength data y, wood density x, and wood density adjusted for resin content, z (from Carlin and Chib (1995)); $n = 42$.

| y | x | z | y | x | z | y | x | z |
|---|---|---|---|---|---|---|---|---|
| 3040 | 29.2 | 25.4 | 2250 | 27.5 | 23.8 | 1670 | 22.1 | 21.3 |
| 2470 | 24.7 | 22.2 | 2650 | 25.6 | 25.3 | 3310 | 29.2 | 28.5 |
| 3610 | 32.3 | 32.2 | 4970 | 34.5 | 34.2 | 3450 | 30.1 | 29.2 |
| 3480 | 31.3 | 31.0 | 2620 | 26.2 | 25.7 | 3600 | 31.4 | 31.4 |
| 3810 | 31.5 | 30.9 | 2900 | 26.7 | 26.4 | 2850 | 26.7 | 25.9 |
| 2330 | 24.5 | 23.9 | 1670 | 21.1 | 20.0 | 1590 | 22.1 | 21.4 |
| 1800 | 19.9 | 19.2 | 2540 | 24.1 | 23.9 | 3770 | 30.3 | 29.8 |
| 3110 | 27.3 | 27.2 | 3840 | 30.7 | 30.7 | 3850 | 32.0 | 30.6 |
| 3160 | 27.1 | 26.3 | 3800 | 32.7 | 32.6 | 2480 | 23.2 | 22.6 |
| 2310 | 24.0 | 23.9 | 4600 | 32.6 | 32.5 | 3570 | 30.3 | 30.3 |
| 4360 | 33.8 | 33.2 | 1900 | 22.1 | 20.8 | 2620 | 29.9 | 23.8 |
| 1880 | 21.5 | 21.0 | 2530 | 25.3 | 23.1 | 1890 | 20.8 | 18.4 |
| 3670 | 32.2 | 29.0 | 2920 | 30.8 | 29.8 | 3030 | 33.2 | 29.4 |
| 1740 | 22.5 | 22.0 | 4990 | 38.9 | 38.1 | 3030 | 28.2 | 28.2 |

$AIC_c$ values are 493.97 for model $M_x$ and 476.96 for model $M_z$. The latter model being the estimated K-L best model, we select z as the best predictor. The two $\Delta_i$ are 0 and 17.01 for models $M_z$ and $M_x$, respectfully. The corresponding Akaike weights are 0.9998 and 0.0002.

The context in which we developed guidelines about interpreting $\Delta_i$ is one having more complexity and more models than here. Therein, a $\Delta_i$ of 17 would be considered overwhelming evidence for the superiority of model $M_z$. But the matter of interpretation of the strength of evidence is uncertain here. Therefore, for this model-selection problem we recommend applying the bootstrap as well as using the above analytical results. The bootstrap is quite feasible here, much more so than in more complex model-selection situations.

Based on 10,000 bootstrap samples of the data in Table 4.4 followed by AIC model-selection we obtained results as follows. The bootstrap sampling distribution of $\Delta_p^*$ gives us estimated percentiles for the sampling distribution of $\Delta_p$, as 1.29 (95 percentile), 2.56 (96 percentile), 4.13 (97 percentile), 9.84 (99 percentile), and 17 is at about percentile 99.85. The Akaike weights are thus consistent here with the bootstrap sampling distribution of $\Delta_p^*$, which gives us a basis to interpret $\Delta_i$ as regards the plausibility that model $M_i$ is actually the K-L best model for the data. However, when we tabulate the model-selection relative frequencies from the bootstrap we find model $M_z$ selected in 93.8% of the 10,000 bootstrap reps. This is still strong evidence in favor of variable z as the better predictor.

It is clear that we select, based on strong evidence, the model structure $E(y) = c + dz$ as the better model. The estimates of the structural parameters of this model, and their conditional standard errors, are $\hat{c} = -1917.6$ ($\widehat{se} = 252.9$), $\hat{d} = 183.3$ ($\widehat{se} = 9.3$). We do not have a way we would consider reliable to compute unconditional standard errors when a parameter is unique to a single model (which $d$ definitely is). However, when the evidence is strongly in favor of the selected model, such as here, it is reasonable to act as if we considered only that model, hence act as if that model would always be the one fit to such data, in which case conditional standard errors apply. So here we accept use of the conditional standard errors as a measure of estimator precision. As a rule of thumb we will hazard the suggestion that if the selected model has Akaike weight $\geq 0.95$, it is acceptable to use the conditional standard errors. The exact value (i.e., 0.95) is not critical; the concept is that if (and only if) the data support the selected model strongly enough ($w_i \geq 0.9$ would probably also be a safe rule of thumb), then conditional and unconditional standard errors will be nearly the same. In a case like this if one is bothered by the issue here of using conditional inference after data-based model-selection, a modest simulation study can be done to explore that issue, as well as other matters.

Monte Carlo simulation methods are very useful to explore model-selection issues (and much more will be done with this idea in Chapter 5). We will introduce one use of simulation here, namely generating simulated "data" that closely mimic the apparent nature of the real data. This allows us to explore model-selection in a case like these pine wood data when we know what model generated the data. Generally, we would be against such simple simulations as having relevant applicability to AIC model-selection issues. However, here the issue is clearly one of just deciding between two linear models, so more complex data simulation models than used below do not seem needed.

We proceed by considering $(y, x, z)'$ as a trivariate normal random variable with mean vector $\mu$ and variance–covariance matrix $\Sigma = DCD$ where the diagonal matrix $D$ has as its diagonal the marginal standard deviations of $y$, $x$, and $z$, and $C$ is the matrix of correlations

$$\begin{bmatrix} 1 & \rho_{yx} & \rho_{yz} \\ \rho_{xy} & 1 & \rho_{xz} \\ \rho_{yz} & \rho_{zx} & 1 \end{bmatrix}.$$

From the data we obtain $\hat{\mu}' = (2992, 27.86, 26.79)$; the estimated marginal standard deviations are $894.60 \, (= \hat{\sigma}_y)$, $4.4946$ and $4.6475$; and the correlation estimates are $\hat{\rho}_{yx} = 0.9272$, $\hat{\rho}_{yz} = 0.9521$, and $\hat{\rho}_{xz} = 0.9584$. To generate a simulated observation mimicking the data we generate three independent standard normal random variables (i.e., normal(0,1)), say $\underline{v}$, then compute

$$(y, x, z)' = \underline{\mu} + DC^{0.5}\underline{v} \tag{4.16}$$

for some parameter choices "near" the estimated parameters needed in (4.16). There are many software packages that will find the needed "square root" of matrix $C$ (we used MATLAB, Anonymous 1994).

The best model here is the one that has the smaller residual standard error of $y$, given the predictor. Those true residual variances are $\sigma_{y|x}^2 = \sigma_y^2(1 - \rho_{yx}^2)$ and $\sigma_{y|z}^2 = \sigma_y^2(1 - \rho_{yz}^2)$. Therefore, in the simulation, the best variable for predicting $y$ is the one with the biggest correlation coefficient with $y$ ($z$, here). What we cannot determine without simulation is performance aspects of the model-selection method.

We can tell from theory that only the values of $\rho_{yx}$, $\rho_{yz}$, and $\rho_{xz}$ affect model-selection performance, including the distribution of $\Delta_p$ and $\Delta_i$ values, and selection frequencies, hence Akaike weights. Hence, parameter values for $\mu$ and $D$ are irrelevant to that aspect of the problem (we might be more concerned about values for $\mu$ and $D$ if we wanted a realistic evaluation of model-selection bias on parameter estimators). Therefore, in the simulations here it sufficed to set $\mu = 0$ and $D = 1$. Using these values results in $a = c = 0$ and $b = \rho_{yx}$, and $d = \rho_{yz}$, and knowing these as truth, we can infer relative degrees of model-selection bias that might occur.

We generated 10,000 simulated observations using (4.16) for the three correlation coefficients being at their estimated values (as truth) and then did AIC model selection. To look some at the sensitivity of results to correlation coefficients, this process was repeated for four more sets of correlations with the same value of $\rho_{yz} - \rho_{yx} = 0.025$ as in the real data and a final case wherein all three true correlation coefficients used in the simulation were set to 0.95. Our primary objective was to determine the relative frequency of model-selection as $\pi_z$ (this is without loss of generality as $\pi_z + \pi_x = 1$) and the expected Akaike weight $E(w_z)$ (also without loss of generality as $E(w_x) + E(w_z) = 1$) and the 95th and 99th percentiles of $\Delta_p$ (denoted below by $\Delta_{p,0.95}$ and $\Delta_{p,0.99}$). Results are given below by assumed sets of correlation coefficients (estimated proportions have coefficients of variation of about 1%; the estimated $\Delta$ have coefficients of variation more like 2.5%—and this for 10,000 reps):

| $\rho_{yx}$ | $\rho_{yz}$ | $\rho_{xz}$ | $\pi_z$ | $E(w_z)$ | $\Delta_{p,0.95}$ | $\Delta_{p,0.99}$ |
|---|---|---|---|---|---|---|
| 0.927 | 0.952 | 0.958 | 0.97 | 0.96 | 0.0 | 4.5 |
| 0.927 | 0.952 | 0.900 | 0.90 | 0.89 | 4.6 | 12.8 |
| 0.927 | 0.952 | 0.980 | 0.99 | 0.99 | 0.0 | 0.0 |
| 0.900 | 0.925 | 0.958 | 0.92 | 0.90 | 1.8 | 7.1 |
| 0.900 | 0.925 | 0.900 | 0.84 | 0.83 | 7.6 | 15.8 |
| 0.950 | 0.950 | 0.950 | 0.50 | 0.50 | 18.3 | 26.7 |

In these cases it is clear that $\pi_z \doteq E(w_z)$, and that the sampling distribution of $\Delta_p$ is quite variable. This is a worse case as regards variability of the distribution of $\Delta_p$ (only two models, and they are nonnested). We can see that if the first case above were reality, we would expect to select the correct model in about 96% of all samples (for $n = 42$). These Monte Carlo results give us added faith in the usefulness of the bootstrap results based on the actual data and add faith in the strength of evidence deduced here from the data using $\Delta_x$ ($= 17$) and from $w_z = 0.9998$.

By looking at the more detailed results (not given here) on average values of estimated parameters, $\hat{\theta}$, and their averaged estimated standard errors, $\widehat{se}(\hat{\theta} \mid M)$, given the selected model we can assess model-selection bias in both point estimates and standard errors. If case 1 above were truth, then in fact the simulations suggested there would be little model-selection bias here when $M_z$ was selected and conditional standard errors applied (to be expected if model $M_z$ is selected 96% of the time). When model $M_x$ was selected, no strong biases in parameter estimators were suggested, but the sample size for this inference was only $m = 314$.

There is another interesting question we can explore with these simulation results. When a model is selected, right ($M_z$) or wrong ($M_x$), how frequently do we then judge the weight of evidence to be strongly in favor of that model? Our interest in such a question is mostly focused on when we make the wrong choice (we will not know this to be the case): Having picked the model that is not the K-L best model, will the data appear strongly to support the selected model as being best, or will the evidence be weak? For case 1 above we make the wrong choice with sampling probability only about 0.03 (314 of 10,000 samples). If we considered $w_x > 0.9$ as strong evidence in favor of the model $M_x$, when it was selected, we find that 101 of the 314 samples produced strong evidence in favor of the wrong model. Thus only 1% of all 10,000 samples would be strongly misleading in this simulated scenario. Conversely, for the 9,688 samples wherein model $M_z$ was selected, 9,223 produced strong evidence ($w_z > 0.9$) in favor of the selected model. Hence we expect that in 92% of all samples (in this particular scenario) we would select the correct model and do so with convincing evidence. Note however, that we cannot tell from the actual data whether it is one if the "1%" strongly misleading samples. We can say, again just for this simulated scenario, that the *estimated* odds are 92 : 1 that we have reached a correct conclusion for these pine wood data.

## 4.4 Summary

Model-selection should not be considered just the search for the best model. Rather, the basic idea ought to be to make more reliable inferences based on the entire set of models that are considered *a priori*. This means that we rank and calibrate the set of models and determine a confidence subset, of the $R$ models, for the K-L best model. Parameter estimation should also make use of all the models when appropriate (e.g., model averaging for prediction) and attempt to use unconditional variances unless the selected best model is strongly supported (say its $w_i > 0.9$).

In general there is a substantial amount of model-selection uncertainty in many practical problems (but see the simulated starling experiment, and real pine wood data, for exceptions). Such uncertainty about what model structure (and associated parameter values) is the K-L best approximating model applies whether one uses hypothesis testing, information-theoretic criteria, dimension-consistent criteria, cross validation; or various Bayesian methods. Often, there is a nonnegligible variance component for estimated parameters (this includes prediction) due to uncertainty about what model to use, and this component should be included in estimates of precision.

Model-selection uncertainty can be quantified in two basic ways: based on the $\Delta_i$ values for the set of models considered, or based on use of the bootstrap method.

The simple computation of the relative likelihood of each model considered, given the data, and rescaling these model likelihoods to be Akaike weights, $w_i$, is often effective and easy to understand and interpret (the $w_i$ provide a basis for a measure of relative support of the data for the models, what we can call model calibration). The bootstrap, while computationally intensive, provides estimates of model-selection probabilities, $\hat{\pi}_i$ (the $w_i$ are a useful approximation to the $\pi_i$).

If there is substantial model-selection uncertainty and if the sampling variance is estimated conditionally on the selected model, the actual precision of estimated parameters will likely be overestimated, and the achieved confidence interval coverage will be below the nominal level (e.g., 80% rather than 95%). Estimates of unconditional standard errors can be made using either Akaike weights or bootstrap selection probabilities. The full set of results for the $R$ models considered over all bootstrap samples may also be useful to compute unconditional standard errors of estimated parameters.

Conditional inference can be relatively poor. There is a need for additional research on general methodology to incorporate model-selection uncertainty as a variance component in the precision of parameter estimators. Model averaging has potential in some applications where interest is concentrated on one parameter that appears (implicitly, at least) in every model in the set. Prediction might often be approached as a problem in (weighted) model averaging where the weights are either Akaike weights ($w_i$) or bootstrap selection probabilities ($\hat{\pi}_i$).

The importance of a *small* number ($R$) of candidate models, defined prior to detailed analysis of the data, cannot be overstated. Small is best conceptualized in contrast to what is not small: If the number of models considered exceeds the sample size, hence $R > n$, then there is *not* a small number of candidate models. This condition often occurs in the commonly used case of considering all possible linear models in variable selection. In that all-models case, when one has $p$ variables, then $R = 2^p$; hence $R = 1024$ for $p = 10$, or $R = 1,048,576$ for $p = 20$. One should have $R$ much smaller than $n$. If the background science is lacking, so that needed *a priori* considerations are deemed impossible, then the analysis should probably be considered to be only exploratory.

Finally, investigators should explain what was actually done in the model-selection (i.e., data analysis). Was it objective model-selection and assessment based on an *a priori* set of models? Alternatively, was the selected model a result of a subjective strategy of seeking a model that fits the data well by introducing new models into consideration as data analysis progresses? In the former case we recommend AIC (or its variants as needed). In the latter case, if the strategy can be implemented as a computer algorithm, then use the bootstrap to assess model-selection uncertainty for such subjective or iterative searches. If the model-selection strategy cannot be represented as an explicitly defined algorithm, one cannot determine model-selection uncertainty.

# 5
# Monte Carlo and Example-Based Insights

## 5.1 Introduction

This chapter gives results from some illustrative exploration of the performance of information-theoretic criteria for model-selection and methods to quantify precision when there is model-selection uncertainty. The methods given in Chapter 4 are illustrated and additional insights are provided based on simulation and real data. Section 5.2 utilizes a chain binomial survival model for some Monte Carlo evaluation of unconditional sampling variance estimation, confidence intervals, and model averaging. For this simulation the generating process is known and can be of relatively high dimension. The generating model and the models used for data analysis in this chain binomial simulation are easy to understand and have no nuisance parameters. We give some comparisons of AIC versus BIC selection and use achieved confidence interval coverage as an integrating metric to judge the success of various approaches to inference.

Section 5.3 focuses on variable selection (equivalent to all-subsets selection) in multiple regression for observational data assuming normally distributed homogeneous errors. A detailed example of $AIC_c$ data analysis and inference is given, and it is shown how to extend the example to a relevant Monte Carlo investigation of methodology. The same Monte Carlo methods are used to generate additional specific illustrative results. We also discuss and illustrate ways to reduce *a priori* the number of variables. A discussion subsection provides details of model-selection bias and other ideas with emphasis on problems of, and better approaches to, all-subsets selection.

Detailed examples of real data analysis are given in Sections 5.4 and 5.5. In both cases we use real data to make a number of general points about K-L–based model-selection. One example uses distance sampling data on kangaroos in Australia; the other uses weather data from South Africa. The reader is encouraged to read these examples, because each is used to convey some general information about data analysis under a model-selection approach. Section 5.6 emphasizes the problem of using AIC in applications where it is imperative to use $AIC_c$ and includes some other thoughts about use of models.

In the following material it is important to distinguish between sampling variances, or standard errors, that are conditional on one particular model, versus those that are unconditional, hence not based on just one specific model. Conditional measures of precision based on a restrictive model $M$ are often denoted by $\text{var}(\hat{\theta}_j \mid \theta_j)$ or $\text{se}(\hat{\theta}_j \mid \theta_j)$; however, they are more properly denoted by $\text{var}(\hat{\theta}_j \mid M)$ or $\text{se}(\hat{\theta}_j \mid M)$. Corresponding unconditional values may be denoted by $\text{var}(\hat{\theta}_j)$ or $\text{se}(\hat{\theta}_j)$, and have (as needed) an added variance component due to model-selection uncertainty. Even such "unconditional" sampling variances do depend on the full set of models considered.

The "as needed" phrase is appropriate because sometimes inference is design-based rather than model-based. This applies, for example to the use of the sample mean from a random sample; or here to the use of just $\hat{S}_i = n_{i+1}/n_i$: There is no restrictive model assumed for this inference. However, model-based inference is usually both necessary and can be very effective for parameter estimation based on complex data.

## 5.2 Survival Models

### 5.2.1 A Chain Binomial Survival Model

We consider here tracking a cohort of animals through the entire survival process of the cohort. Thus at time 1 there are a known number $n_1$ of animals alive. Usually, we are thinking that these are all young of the year, hatched or born at time 1. One year latter there are $n_2$ survivors; in general, at annual anniversary dates $i > 1$ there are $n_i$ survivors. Eventually, we have for some last anniversary year $\ell$, $n_\ell > 0$ but $n_{\ell+1} = 0$. Given the initial cohort size $n_1$, the survival probability in year one (i.e., for the first year of life) is conceptually defined as $S_1 = \text{E}(n_2 \mid n_1)/n_1$. In general, for year $i$, $S_i = \text{E}(n_{i+1} \mid n_i)/n_i$. Thus the obvious general estimator is $\hat{S}_i = n_{i+1}/n_i$; this estimator is not based on any assumed model structure imposed on the set of basic survival parameters. If $n_i$ is large enough, then $\hat{S}_i$ is an acceptable estimator in terms of precision. However, as the cohort dies out, $n_i$ becomes small, and then $\hat{S}_i$ is not reliable. In particular, in the year $\ell$ that the cohort dies out we always get $\hat{S}_\ell = (0/n_\ell) = 0$. This is a terrible estimator of age $\ell$ survival probability. To avoid this latter problem, and to improve age-specific survival estimates for ages $i$ at which $n_i$ is small, we must impose a model structure on the set of parameters $S_1, S_2, \ldots, S_\ell$. First, we consider what to use here for the global model.

## 5.2 Survival Models

The quantity $n_1$ is known at the study initiation, while the subsequent counts $n_2, n_3, n_4, \ldots$, being not then known or predictable, are treated as both random variables and known ancillaries in the eventual data analysis. Given $n_1$ we will generate, in the Monte Carlo study, $n_2$ as binomial$(n_1, S_1)$. Because the survival process is sequential in time, once the survivors (hence $n_2$) at time 2 are known, we can model $n_3$, given $n_2$, as binomial$(n_2, S_2)$. In general, at time $i$ we know $n_i$; hence we can generate $n_{i+1}$ as a binomial$(n_i, S_i)$ random variable. Thus we generate our probability model, hence likelihood, for the data as a chain binomial model with conditional independence of random variable $n_i$ given the temporally preceding $n_1$ to $n_{i-1}$. The likelihood function for the most general possible model is

$$\mathcal{L}(\underline{S} \mid \underline{n}) = \prod_{i=1}^{\ell} \binom{n_i}{n_{i+1}} (S_i)^{n_{i+1}} (1 - S_i)^{n_i - n_{i+1}}, \qquad (5.1)$$

where $\underline{S}$ and $\underline{n}$ are vectors of the survival parameters and data counts, respectively. The underlying parameters explicitly in (5.1) are $S_1, S_2, \ldots, S_\ell$. In this context a (restricted) model for the parameters is some imposed smoothness such as $S_i \equiv S$ for all ages $i \geq 1$ (model $M_1$), or perhaps the $S_i$ are restricted only after age three, hence $S = S_4 = \cdots = S_\ell$, and $S_1$, $S_2$, and $S_3$ are unrestricted (model $M_4$). If no structural restrictions are imposed, we have the most general possible global model, $M_g$, for which the MLEs are

$$\hat{S}_i = \frac{n_{i+1}}{n_i}, \quad i \leq \ell,$$

with conditional sampling variances

$$\text{var}(\hat{S}_i \mid M_g) = \frac{S_i(1 - S_i)}{n_i}.$$

If such data were from moderately long-lived species such as owls, elk, or wolves, one might expect substantial differences in survival for the first few age classes, hence $S_1 < S_2$ and so on for several age classes. There would then be near-equal survival probabilities for adults until a decrease occurs in yearly survival as the surviving animals approach old age (i.e., senescence). Thus, there are some large age-specific effects in survival followed by smaller, tapering effects (this conceptual model is not universally applicable; for example it does not apply to salmon). In real populations, the age-specific survival rates would be confounded with annual environmental effects and also, most likely, with some individual heterogeneity. We will not pursue all of these realities here and instead focus on some points related to conditional and unconditional inference in model selection and some comparison of information-theoretic approaches versus dimension-consistent criteria (e.g., BIC). This examination will illustrate the amount of variability in model-selection and the effects of sample size, $n_1$.

Because of the paucity of data at older ages, some constraint (i.e., model) must be assumed, at least for the older ages, to get reliable results. One simple solution is to pool data beyond some particular age and assume that the pooled ages have a constant survival probability. Thus, for animals of age $r$ and older we assume that

Ritei Shibata, was born in Tokyo, Japan in 1949. He received a B.S. in 1971, an M.S. in 1973, and a Ph.D. in mathematics in 1981 from the Tokyo Institute of Technology. He is currently a professor in data science, Department of Mathematics, at Keio University, in Yokohama, Japan. He has been a visiting professor at the Australian National University, University of Pittsburg, Mathematical Science Research Institute (Berkeley, CA), and Victoria University of Wellington, NZ. He is currently interested in statistical model-selection, time series analysis, S software, and data and description (D&D).

they all have the same survival probability, $S \equiv S_i$, $i \geq r$. This is not to say that truth is no longer age-specific, but rather that a parsimonious *model* of the survival process is not age-specific after some age, $r$. An alternative is to do some modeling (e.g., logistic models) of the survival process and then do model-selection. We will address both approaches in this section.

Our first sequence of models is defined below in terms of restrictions on the age-specific annual survival probabilities:

| Model | K | Parameters | |
|---|---|---|---|
| 1 | 1 | $S \equiv S_i,$ | $i \geq 1$ |
| 2 | 2 | $S_1, S \equiv S_i,$ | $i \geq 2$ |
| 3 | 3 | $S_1, S_2, S \equiv S_i,$ | $i \geq 3$ |
| ⋮ | ⋮ | ⋮ | ⋮ |
| R | R | $S_1, S_2, \ldots, S_{R-1}, S \equiv S_i,$ | $i \geq R$ |

The pattern is obvious; model $M_r$ has $r$ parameters with different survival probabilities for ages 1 to $r-1$, and constant annual survival probability for age $r$ and older. The global model is $M_R$; model $M_r$ has $K = r$ parameters.

This particular set of models is convenient for Monte Carlo simulations because all MLEs exist in closed form. Let the tail sum of numbers of animals alive at and after age $r$ be denoted by $n_{r,+} = n_r + n_{r+1} + \cdots + n_j + \cdots$. Then the MLEs for model $M_r$ ($r \leq \ell$) are

$$\hat{S}_i = \frac{n_{i+1}}{n_i}, \quad i = 1, \ldots, r-1,$$

$$\hat{S} = \frac{n_{r+1,+}}{n_{r,+}} = \frac{n_{r+1} + n_{r+2} + \cdots}{n_r + n_{r+1} + \cdots}.$$

Also, the likelihood for the $r$ parameters given model $M_r$ is (from formula 5.1 constrained by model $M_r$)

$$\mathcal{L}_r(\underline{S} \mid \underline{n}) = \left[\prod_{i=1}^{r-1} \binom{n_i}{n_{i+1}} (S_i)^{n_{i+1}} (1-S_i)^{n_i - n_{i+1}}\right] \left[\binom{n_{r,+}}{n_r} (S)^{n_{r,+}} (1-S)^{n_r}\right]. \tag{5.2}$$

### 5.2.2 An Example

The first example compares the performance of $AIC_c$ and BIC (cf. Section 2.8.2) for sample size $n_1 = 150$. The true survival probabilities $S_1$, $S_2$, and so forth were 0.5, 0.7, 0.75, 0.8, 0.8, and then beyond $S_5$, the age-specific annual survival decreased by 2% per year (e.g., 0.784, 0.768, 0.753, 0.738, and so forth for $S_6$ and beyond). Thus, in the data-generating model the survival parameters vary smoothly, increasing to a maximum at ages 4 and 5, then simulating senescence by a 2% per year decrease in $S_j$. For each Monte Carlo rep the cohort is followed until all $n_1$ animals are dead. For example, the simulated animal counts for one repetition ($n_i$, $i = 1, 2, \ldots, 16 = \ell + 1$) were

150 (fixed), 74, 49, 34, 27, 21, 13, 9, 6, 6, 6, 4, 4, 1, 1, 0.

The set of models considered here is $M_1$ to $M_{10}$; thus $R = 10$. This is a nested set of models within the global model, $M_{10}$. Hence, the global model here allows separate estimates of $S_1, S_2, \ldots, S_9$ but a single "pooled" estimate, $S$, after age 9. The results for 10,000 Monte Carlo reps are given in Table 5.1, where the two selection approaches yield substantial differences in the model selected. On average, $AIC_c$-selected an approximating model with $K = 3.3$ parameters, while BIC selected a model with an average of 2.1 parameters.

The theory underlying BIC is that the true model exists and is in the set of candidate models; this condition is not met here: Truth is not in the set of ten models (the model closest to truth is model $M_{10}$). Asymptotically BIC attempts to estimate the dimension of the true model, a concept that is not even well-defined here because there are nominally an unbounded number of age-specific survival

TABLE 5.1. Comparison of model-selection relative frequencies for $AIC_c$ vs. BIC using models $M_1$ to $M_{10}$ and data from the chain binomial generating model with parameters $S_1$ to $S_{10}$ as 0.5, 0.7, 0.75, 0.8, 0.8, 0.784, 0768, 0753, 0738, and 0.723 (i.e., after $j = 5$, $S_{j+1} = 0.98 \cdot S_j$); results are for sample size $n_1 = 1{,}000$.

| Model | K | $AIC_c$ model selection | | BIC model selection | |
|---|---|---|---|---|---|
| | | percent | cumul. % | percent | cumul. % |
| 1 | 1 | 0 | 0 | 0.2 | 0.2 |
| 2 | 2 | 52.0 | 52.0 | 88.6 | 88.8 |
| 3 | 3 | 22.4 | 74.4 | 10.0 | 98.8 |
| 4 | 4 | 5.6 | 80.0 | 0.5 | 99.3 |
| 5 | 5 | 5.5 | 85.5 | 0.5 | 99.8 |
| 6 | 6 | 5.0 | 90.5 | 0.1 | 99.9 |
| 7 | 7 | 3.4 | 93.9 | 0 | 100.0 |
| 8 | 8 | 2.4 | 96.3 | 0 | 100.0 |
| 9 | 9 | 2.2 | 98.5 | 0 | 100.0 |
| 10 | 10 | 1.5 | 100.0 | 0 | 100.0 |

rates. However, there will not be enough data ever to estimate all those parameters, and increasing $n_1$ does not change this conundrum.

In sharp contrast, $AIC_c$ attempts to select a parsimonious approximating model as a basis for inference about the population sampled. It does not assume that full truth exists as a model, nor does it assume that such a "true model" is in the set of candidates. $AIC_c$ estimates relative expected Kullback–Leibler distance and then selects the approximating model that is closest to unknown truth (i.e., the model with the smallest value of $AIC_c$). Based on the evaluated $E(\Delta_i)$ (good to about two significant digits), it is model $M_2$ that is the $AIC_c$ best model here, on average, but only by a minute winning margin over model $M_3$. Therefore, without loss of generality we can set $E(\Delta_2) = 0$ and thus give the other values of $E(\Delta_i)$ compared to this minimum value. In order, $i = 1$ to 10, we have the $E(\Delta_i)$ as 25.0, 0.0, 0.09, 1.2, 2.0, 2.7, 3.6, 4.6, 5.6, and 6.7.

Further results from the Monte Carlo study are presented in Table 5.2. Specifically, we look at the properties of the estimated age-specific survival rates under $AIC_c$ and BIC model-selection strategies. In Table 5.2, $E(\hat{S}_i)$ is the estimated expected value of $\hat{S}_i$, where $\hat{S}_i$, by rep, is the estimate of age $i$ survival rate based on whatever model was selected as best, for that rep; this estimator is computed for both $AIC_c$ and BIC model-selection methods. For example, if model $M_4$ is selected, then $\hat{S}_i = n_{i+1}/n_i$, $i = 1, 2, 3$, and $\hat{S}_i = n_{5,+}/n_{4,+}$ for $i \geq 4$.

From Table 5.2 the estimators of age-specific survival probabilities were nearly unbiased; however, bias was slightly smaller for the $AIC_c$-selected models than for the BIC-selected models for 8 of the 9 ages where the $E(\hat{S}_i)$ differed. The expected estimated standard errors, conditional (cond.) on the selected model, were generally smaller for BIC selection compared to $AIC_c$ selection, as was expected. The conditional (on selected model) standard errors under both $AIC_c$

TABLE 5.2. Summary of Monte Carlo results for estimated age-specific survival probabilities under $AIC_c$ and BIC model selection (see text for details of data generation); the models used were $M_1$ to $M_{10}$ (see text); these results are based on 10,000 reps; results are for sample size $n_1 = 150$.

| Age $i$ | $S_i$ | $E(\hat{S}_i)$ | | $E(\widehat{se})$, AIC | | $E(\widehat{se})$, BIC | | Coverage, $AIC_c$ | |
|---|---|---|---|---|---|---|---|---|---|
| | | $AIC_c$ | BIC | cond. | unc. | cond. | unc. | cond. | unc. |
| 1 | 0.500 | 0.500 | 0.501 | 0.041 | 0.041 | 0.041 | 0.041 | 0.959 | 0.959 |
| 2 | 0.700 | 0.711 | 0.732 | 0.039 | 0.053 | 0.029 | 0.052 | 0.731 | 0.959 |
| 3 | 0.750 | 0.751 | 0.749 | 0.035 | 0.053 | 0.026 | 0.038 | 0.916 | 0.966 |
| 4 | 0.800 | 0.772 | 0.751 | 0.033 | 0.052 | 0.026 | 0.031 | 0.639 | 0.921 |
| 5 | 0.800 | 0.769 | 0.750 | 0.032 | 0.050 | 0.026 | 0.029 | 0.630 | 0.906 |
| 6 | 0.784 | 0.760 | 0.749 | 0.032 | 0.048 | 0.026 | 0.029 | 0.806 | 0.958 |
| 7 | 0.768 | 0.754 | 0.749 | 0.032 | 0.048 | 0.026 | 0.029 | 0.895 | 0.976 |
| 8 | 0.753 | 0.750 | 0.749 | 0.032 | 0.048 | 0.026 | 0.029 | 0.914 | 0.977 |
| 9 | 0.738 | 0.745 | 0.749 | 0.032 | 0.047 | 0.026 | 0.029 | 0.870 | 0.962 |
| 10 | 0.723 | 0.740 | 0.749 | 0.032 | 0.047 | 0.026 | 0.029 | 0.770 | 0.912 |
| Average | | | | | | | | 0.813 | 0.950 |

and BIC selection are too small, as they do not account for any model-selection uncertainty.

Various comparisons are possible from these simulations, such as achieved empirical standard errors versus expected estimated standard errors based on theory. However, confidence-interval coverage is in integrating measure of how well the methodology is performing. Therefore, we focus on achieved confidence interval coverage for nominal 95% intervals, all computed (in this section) as $\hat{S} \pm 2\,\widehat{se}$. Use of this simple form may have resulted in an additional 1 to 3% failure in coverage of $S_i$ for older ages, but we judge this as irrelevant to the contrast of conditional versus unconditional coverage.

Conditional confidence interval coverage in Table 5.2 of true $S_i$ under $AIC_c$ model-selection is generally below the nominal level, ranging from 0.731 to 0.959 over $S_1$ to $S_{10}$. Adjusting the conditional standard errors of $\hat{S}_i$ to be unconditional (unc.) using formula (4.11) provides much improved coverage (averaging 95.0%), ranging from 0.906 to 0.977 (Table 5.2).

Achieved coverage using conditional standard errors for BIC model-selection averaged 77.8% across the 10 age classes (Table 5.2). Use of formula (4.11) with BIC model-selection (there is no theoretical basis to justify doing this) improved the achieved coverage of true $S_i$ for BIC model-selection to 86.4% (range 60.1 to 98.7%); coverage generally remained below the nominal level (95%). Buckland et al. (1997) present results on a survival model that is similar to the one used here.

This example shows the large amount of uncertainty associated with $AIC_c$ model-selection when sample size is small ($n_1 = 150$) and the generating model has tapering effects. This simple simulation exercise also demonstrates that BIC selection cannot be recommended: It makes simplistic assumptions that are not true (e.g., the true model is in the set of candidates); it requires very large sample sizes to achieve consistency; and typically, BIC results in a selected model that is under-fit (e.g., biased parameter estimates, overestimates of precision and achieved confidence interval coverage below that achieved by $AIC_c$-selected models). Conditional estimates of precision, under either $AIC_c$ or BIC, exclude model-selection uncertainty, and this is often an important omission. Incorporating model-selection uncertainty can bring achieved confidence interval coverage up to approximately the nominal level for model-selection under $AIC_c$ (using formula 4.11) (this applies to AIC also).

The results in Table 5.2 are typical of many similar simulation cases we have examined (with different sample size, various sets of $S_i$, and numbers of models, $R$). To make some more use of this particular example, in terms of the set of true $S_i$, we obtained confidence interval coverage on $S_i$, averaged over all $S_i$, $i = 1$ to $R$, for $AIC_c$ model-selection for some more sample sizes. We also tabulated the 90th, 95th, and 99th percentiles of $\Delta_p$; this information is useful as regards interpretation of the $\Delta_i$. As sample size increases, the $AIC_c$ best model gets larger, thus one needs also to increase the size of the model set, i.e., $R$.

Unconditional confidence interval coverage is much superior to conditional coverage (Table 5.3). When $R$ is large, like 20 for $n_1 = 10,000$, average (over $S_1$ to $S_{20}$) confidence interval coverage suffers as a result of the often extrapolated estimates of, say, $S_{20}$ based on a fitted model averaging around $M_{13}$. The $S_i$ have a shallow peak at $i = 4$ and 5, and this feature of truth requires a large sample size to detect reliably. Thus the theoretically best $AIC_c$ model, though increasing as sample size increases, stalls at model $M_3$ until a threshold sample size is passed, after which that theoretically best $AIC_c$ model is more responsive to increasing sample size. It is this feature that here causes the percentiles of $\Delta_p$ to be so big (e.g., for $n_1 = 500$, a 95th percentile of 10.3 and a 99th percentile of 16.8).

The other notable feature in Table 5.3 is found when $n_1 = 10,000$ and $R = 10$. The theoretically best $AIC_c$ model is then model $M_{10}$, the most general model in the set considered. Model-selection probabilities here include $\pi_{10} = 0.8$ and

TABLE 5.3. Some Monte Carlo results for $AIC_c$ model selection, for the same example set of $S_i$ used in Tables 5.1, 5.2 (see text for details), for varying sample size, $n_1$; $R$ is the number of models considered; achieved confidence interval coverage (nominally 95%) has been averaged over conditional (cond.) and unconditional (unc.) intervals on $S_1$ to $S_R$.

| Sample size $n_1$ | $R$ | Best model $k$ | Confidence interval coverage, % | | percentiles of $\Delta_p$ | | |
|---|---|---|---|---|---|---|---|
| | | | cond. | unc. | 0.90 | 0.95 | 0.99 |
| 100 | 10 | 2 | 84.4 | 95.5 | 4.0 | 5.9 | 10.7 |
| 250 | 10 | 3 | 77.8 | 93.3 | 4.8 | 7.1 | 12.7 |
| 500 | 10 | 3 | 75.5 | 92.2 | 7.6 | 10.3 | 16.8 |
| 1,000 | 10 | 8 | 78.6 | 92.8 | 5.9 | 7.3 | 10.4 |
| 1,000 | 12 | 8 | 76.0 | 90.6 | 6.4 | 7.8 | 12.0 |
| 10,000 | 10 | 10 | 86.2 | 93.0 | 1.0 | 1.9 | 2.0 |
| 10,000 | 20 | 13 | 78.3 | 88.9 | 5.4 | 7.2 | 11.2 |

$\pi_9 = 0.18$; hence, $\Delta_p$ is 0 in 80% of the samples and small in the other 20% of the samples.

Use of BIC is theoretically inappropriate for model-selection in the example of Table 5.3 because the set of models used does not include a true, simple generating model with fixed, small $K$ and all effects being big as sample size increases. However, BIC performs best at large sample size, so one might ask whether BIC performs well here when $n_1 = 10,000$ and $R = 10$ or 20. The achieved coverage (for nominally 95% intervals) for BIC model-selection for these two cases is given below, along with average (over $S_1$ to $S_R$) MSE for both BIC and AIC (AIC coverage is in Table 5.3):

|       |    | coverage, BIC | | average MSE | |
|-------|----|------|------|----------|----------|
| $n_1$ | R  | cond.| unc. | BIC      | AIC      |
| 10,000 | 10 | 77.8 | 83.7 | 0.000245 | 0.000195 |
| 10,000 | 20 | 45.9 | 51.5 | 0.003077 | 0.001484 |

Average mean square error is lower for AIC, and confidence interval coverage is much better. Using BIC, model $M_8$ is selected with probability 0.37 ($R = 10$) or 0.36 ($R = 20$), and the probability of selecting one of models $M_7$, $M_8$, $M_9$, $M_{10}$ is 0.96 ($R = 10$) or 0.93 ($R = 20$).

For some additional results and comparison to Table 5.3, we mimicked Table 5.3 results for a model wherein true $S_i = 0.5 + 0.3/i$. Table 5.4 gives confidence interval coverage results and percentiles of $\Delta_p$ for this example. Because of the monotonicity of the true $S_i$, the percentiles of $\Delta_p$ are much more stable, being about 4 (90th percentile), 5 (95th percentile) and 9 (99th percentile). Again, for the last case ($n_1 = 10,000$ and $R = 20$) the poor coverage is due to averaging coverage over the intervals on all of $S_1$ to $S_{20}$, yet $\hat{S}_{11}$ to $\hat{S}_{20}$ are considerable extrapolations. Generally, $AIC_c$ does not select a model beyond model $M_{10}$. If for this last case we look only at intervals on $S_1$ to $S_{10}$, the coverage levels under $AIC_c$ model-selection are 80.7% (cond.) and 93.3% (unc.).

TABLE 5.4. Some Monte Carlo results for $AIC_c$ model selection under the chain binomial generating model, $S_i = 0.5 + 0.3/i$; the theoretical $AIC_c$ best model is $M_k$; achieved confidence interval coverage (nominally 95%) has been averaged over conditional (cond.) and unconditional (unc.) intervals on $S_1$ to $S_R$.

| Sample size $n_1$ | R | Best model $k$ | Confidence interval coverage, % | | percentiles of $\Delta_p$ | | |
|---|---|---|---|---|---|---|---|
| | | | cond. | unc. | 0.90 | 0.95 | 0.99 |
| 100 | 10 | 3 | 78.0 | 93.4 | 2.7 | 4.3 | 8.2 |
| 250 | 10 | 3 | 74.6 | 92.2 | 3.6 | 5.4 | 9.6 |
| 500 | 10 | 3 | 76.4 | 92.3 | 4.8 | 7.0 | 11.5 |
| 1,000 | 10 | 4 | 76.3 | 92.6 | 3.5 | 5.2 | 9.4 |
| 1,000 | 12 | 4 | 72.0 | 91.5 | 3.6 | 5.3 | 9.6 |
| 10,000 | 10 | 6 | 80.1 | 92.4 | 3.5 | 4.6 | 8.3 |
| 10,000 | 20 | 6 | 59.6 | 85.2 | 3.7 | 5.3 | 9.0 |

## 5.2.3 An Extended Survival Model

A second Monte Carlo example is given here to add additional complexity and realism in the set of approximating models. This example uses the same chain binomial generating model and associated parameters presented in Section 5.2.2; only the set of approximating models is different. Rather than assuming for model $M_r$ the constraint $S \equiv S_i$ for all $i \geq r$, the age-specific survival probabilities at and beyond age $r$ are assumed to follow a logistic model. The 10 models used here are given below:

| | | |
|---|---|---|
| Model $L_1$ | $\text{logit}(S_i) = \alpha + \beta \cdot i,$ | $i \geq 1,$ |
| Model $L_2$ | $S_1, \text{logit}(S_i) = \alpha + \beta \cdot i,$ | $i \geq 2,$ |
| Model $L_3$ | $S_1, S_2, \text{logit}(S_i) = \alpha + \beta \cdot i,$ | $i \geq 3,$ |
| $\vdots$ | $\vdots$ | $\vdots$ |
| Model $L_{10}$ | $S_1, S_2, \ldots, S_9, \text{and logit}(S_i) = \alpha + \beta \cdot i,$ | $i \geq 10.$ |

Each model $L_r$ has an intercept and slope parameter for the logistic regression fitting of age-specific survival rates $S_i$ for ages $i \geq r$. Model $L_r$, for $r \geq 2$, also fits unconstrained age-specific survival rates for ages 1 to $r - 1$. Thus, model $L_r$ has $K = r + 1$ parameters.

The initial population size was $n_1 = 150$ animals, as before, and all animals were followed until the last one died. The true parameters were $S_1 = 0.5$, $S_2 = 0.7$, $S_3 = 0.75$, $S_4 = 0.8$, $S_5 = 0.8$, and for $r > 5$, $S_{r+1} = 0.98 S_r$. The computer generated 10,000 independent repetitions under the true $S_r$, for analysis using the 10 approximating models. In some samples not all 10 models could be fit because all animals died before reaching age 10. As before, let $\ell$ be the last age at which $n_\ell > 0$. If $\ell - 2 \geq 10$, then all ten logistic-based models were fit. Otherwise, only $\ell - 2$ models were fit (if $\ell - 2 < 10$). Thus, the number of models that could be fit to a sample varied somewhat. This did not matter to the overall strategy of either selecting a best model or using model averaging, because for each fitted model it was always possible to compute the derived $\hat{S}_i$ (and its estimated conditional variance) for any value of $i$.

Generally, the logistic approximating models were superior in this situation as compared to the models used in Section 5.2.2. The theoretically best model (under AIC$_c$ selection) was $L_3$ ($K = 4$) with model $L_4$ a close second (E($\Delta_4$) = 0.17 relative to setting E($\Delta_3$) = 0). The model-selection relative frequencies are shown in Table 5.5. In 84.5% of samples either model $L_2$, $L_3$, or $L_4$ was selected. Model-selection uncertainty in this example is similar to that of the example in Section 5.2.2 (compare Tables 5.1 and 5.5), but here the best model to use has 4 parameters, rather than 2, as was the case with the models previously considered (i.e., $M_1$ to $M_{10}$). Use of a better set of models leads to more informative inferences from the data.

For the simulation underlying Table 5.5 results, all 10 models were fitted in 9,491 reps. In 18 reps, only models $L_1$ to $L_7$ could be fit (under the protocol noted above). Only models $L_1$ to $L_8$ were fitted in 128 reps; and only up to model $L_9$ were fitted in 363 reps.

TABLE 5.5. Summary of $AIC_c$ model-selection relative frequencies, based on 10,000 Monte Carlo reps, for the logistic models. The generating model allowed survival probabilities to increase to a maximum at ages 4 and 5 and then decrease slowly with age, exactly as in Section 5.2.2 (see text for details). Model $L_3$ is the theoretically best model; results are for sample size $n_1 = 150$.

| Model | Percent | Cumul. % |
|---|---|---|
| $L_1$ | 00.0 | 00.0 |
| $L_2$ | 35.2 | 35.2 |
| $L_3$ | 29.5 | 64.7 |
| $L_4$ | 19.8 | 84.5 |
| $L_5$ | 6.3 | 90.8 |
| $L_6$ | 3.5 | 94.3 |
| $L_7$ | 2.2 | 96.4 |
| $L_8$ | 1.5 | 98.0 |
| $L_9$ | 1.2 | 99.2 |
| $L_{10}$ | 0.8 | 100.0 |

Table 5.6 summarizes the estimated expected values of $\hat{S}_j$ for $AIC_c$-selected models and under model averaging (MA) using formula (4.6), with $\hat{\pi}_i = w_i$ being the Akaike weights. Also shown in Table 5.6 is the achieved coverage of confidence intervals based on estimated unconditional sampling variances (i.e., using formula 4.11). Model averaging provided a slightly less biased estimator of conditional survival. However, both methods performed relatively well. The achieved confidence interval coverage for these intervals, which includes model-selection uncertainty, is very close to the nominal level (0.95) for both approaches; both approaches use the same estimate of unconditional standard error.

From Table 5.6 the unconditional confidence interval coverage averaged over the intervals on $S_1$ to $S_{10}$, and over all 10,000 reps, was 0.942 under the strategy of inference based on the selected $AIC_c$ best model. Confidence intervals based on the estimated sampling variances conditional on the selected model ("conditional intervals") produced a corresponding overall coverage of 0.860. In the worst case, which was at age 5 (i.e., for $S_5$), conditional coverage was only 0.717, vs. unconditional of 0.917.

In this example, the simulation program also computed the mean squared error (MSE) for estimators of $S_i$ at each age for $AIC_c$ model-selection, for model averaging, and for the simple (almost model-free) estimator of age-dependent survival, $\hat{S}_i = n_{i+1}/n_i$, $i = 1, 2, \ldots, 10$. The MSE for the model-averaged estimator was smaller than that of the $AIC_c$-based estimator (ranging from 11% to 25% smaller, except for $\hat{S}_1$, where the two approaches had the same MSE). In this example, at least, model averaging, using Akaike weights, has important advantages. Both $AIC_c$ model-selection and model averaging are quite superior to the use of the simple estimator of $S_i$. For example, at age nine, the MSEs for $\hat{S}_i$ from $AIC_c$ model-selection, model averaging, and the simple estimator were 0.0041, 0.0035, and 0.0186, respectively. This illustrates the advantage of some appropriate

TABLE 5.6. Summary of Monte Carlo results, based on 10,000 reps, of age-specific survival estimation under $AIC_c$ selection and model averaging (MA). The generating model had parameters $S_i$; 10 logistic models ($L_1, \ldots, L_{10}$) were fit to the simulated age-specific data; results are for sample size $n_1 = 150$.

| Age | $S_i$ | $E(\hat{S})$ | | Coverage | |
| --- | --- | --- | --- | --- | --- |
| $i$ | | $AIC_c$ | MA | $AIC_c$ | MA |
| 1 | 0.500 | 0.499 | 0.499 | 0.957 | 0.957 |
| 2 | 0.700 | 0.708 | 0.707 | 0.907 | 0.925 |
| 3 | 0.750 | 0.758 | 0.757 | 0.920 | 0.934 |
| 4 | 0.800 | 0.789 | 0.790 | 0.926 | 0.946 |
| 5 | 0.800 | 0.785 | 0.786 | 0.917 | 0.943 |
| 6 | 0.784 | 0.772 | 0.773 | 0.951 | 0.964 |
| 7 | 0.768 | 0.759 | 0.760 | 0.964 | 0.974 |
| 8 | 0.753 | 0.744 | 0.745 | 0.962 | 0.974 |
| 9 | 0.738 | 0.727 | 0.728 | 0.962 | 0.970 |
| 10 | 0.723 | 0.709 | 0.709 | 0.957 | 0.964 |
| Average | | | | 0.942 | 0.955 |

*modeling* of the underlying survival probabilities (this is similar to the lesson learned from the starling experiment in Section 3.4).

We computed the pivotal quantities $\Delta_p = AIC_k - \min AIC$ (formula 4.4) for each of the 10,000 repetitions to better understand these values for nested models of this type. Some percentiles of the sampling distribution of $\Delta_p$ for this example are shown below:

| Percentile | $\Delta_p$ |
| --- | --- |
| 50.0 | 1.20 |
| 75.0 | 2.09 |
| 80.0 | 2.29 |
| 85.0 | 3.11 |
| 90.0 | 4.21 |
| 95.0 | 6.19 |
| 97.5 | 8.33 |
| 98.0 | 8.80 |
| 99.0 | 10.63 |

Thus, in approximately 90% of the simulated data sets the difference, on average, between $AIC_c$ for the selected model, for that sample, and the $AIC_c$ for the theoretically best model was $\leq 4.21$. If we adopted the idea of using an approximate 90% confidence set on what is the actual $AIC_c$ theoretically best model, then in this example, fitted models that had $\Delta_i$ values below about 4 should be in that set, and hence might be candidates for some further consideration in making inferences from an individual data set.

To make further use of the generating model and set of approximating models used here, we determined several quantities for differing sample sizes: the theoretically best model, conditional and unconditional coverage under $AIC_c$ model

5.2 Survival Models    171

selection (as opposed to model averaging), and 90th, 95th and 99th percentiles of $\Delta_p$. Regardless of this variation in sample size from 50 to 100,000 the unconditional confidence interval coverage stays between about 93 and 95% (Table 5.7). Also, there is considerable stability of the shown percentiles of the (approximate) pivotal, $\Delta_p$. These same results hold fairly well for other generating models and fitted models that we have examined by Monte Carlo methods. Often the conditional coverage averages less than what occurs in Table 5.7 (more like 70% to 85%).

The true $S_i$ do not exactly fit any of the approximating logistic models. However, model $L_4$ turns out to be an excellent approximation to truth, at least for $i \leq 15$ (we did not plan this to be the case). Here, even if we could know $S_1$ to $S_{15}$ exactly, but we needed a mathematical model to represent these numbers, we would likely fit and use model $L_4$. We would do so because the fitted model then provides a concise, yet almost exact, summary of truth. The lack of fit is both trivial and statistically "significant" given a huge sample size. We do not generally have such huge sample sizes, but if we did, then the usual concerns of small-sample-size statistics would not apply (practical significance then replaces statistical significance), and we might very well select model $L_4$ rather than model $L_8$ as a preferred approximation to truth. This illustrates a philosophical point: Even if we knew truth, we would often prefer to replace said truth by a low-order parsimonious, fitted model because of the advantages this confers about understanding the basic structure of full truth.

### 5.2.4  Model Selection If Sample Size Is Huge, or Truth Known

The results in Table 5.7 provide a motivation for us to mention some philosophical issues about model selection when truth is essentially known, or equivalently in statistical terms, when we have a huge sample size. With a sample size of $n_1 = $

TABLE 5.7. Some Monte Carlo results (10,000 reps per sample size) for $AIC_c$ model selection, for the same generating model and set of models, $L_i$, as used for Table 5.6. Sample size is $n_1$; $R$ is the maximum number of models considered; the theoretical $AIC_c$ best model is $L_k$; achieved confidence interval coverage (nominally 95%) has been averaged over conditional (cond.) and unconditional (unc.) intervals on $S_1$ to $S_R$.

| Sample size $n_1$ | R | Best model k | Confidence interval coverage, % | | percentiles of $\Delta_p$ | | |
|---|---|---|---|---|---|---|---|
| | | | cond. | unc. | 0.90 | 0.95 | 0.99 |
| 50 | 10 | 2 | 84.8 | 92.6 | 3.8 | 5.8 | 10.1 |
| 100 | 10 | 3 | 86.2 | 94.1 | 3.7 | 5.4 | 9.8 |
| 150 | 10 | 3 | 86.0 | 94.2 | 4.2 | 6.2 | 10.6 |
| 200 | 10 | 4 | 86.1 | 94.2 | 3.6 | 4.1 | 7.8 |
| 500 | 10 | 4 | 88.3 | 94.6 | 2.7 | 4.3 | 8.1 |
| 1,000 | 12 | 4 | 90.4 | 95.4 | 2.6 | 4.2 | 7.9 |
| 10,000 | 15 | 4 | 90.3 | 95.5 | 4.9 | 7.1 | 12.3 |
| 100,000 | 15 | 8 | 86.4 | 93.3 | 5.0 | 6.6 | 11.2 |

10,000 model $L_4$ is always theoretically the best model to fit to the survival data. At $n_1 = 100,000$ the optimal model, based on information-theoretic statistical model-selection, is $L_8$. However, at a sample size of 100,000 we find that (1) the parameter estimates under either model $L_4$ or $L_8$ are precise to at least two significant digits (so $\hat{S}_i \doteq E(\hat{S}_i)$ for practical purposes), and (2) the difference (i.e., bias) under either model between $E(\hat{S}_i)$ and the true $S_i$ is also trivial (but not always 0), hence ignorable for $i = 1, \ldots, 15$.

Therefore, at a sample size of 100,000, or if we literally knew the true $S_i$, it is more effective from the standpoint of understanding the basic pattern of variation in the survival probabilities $S_i$ to fit and use model $L_4$ (fit to data if $n_1 = 100,000$, or to truth if truth is in hand). In this situation nothing practical is gained by using model $L_8$ rather than model $L_4$. On the contrary, using the less parsimonious, more complicated model $L_8$ has a cost in terms of ease with which we perceive the basic pattern in the true $S_i$. In essence, we are also saying that even if we knew the true $S_i$, we would be better able to understand the pattern of information therein by fitting (to the $S_i$) and reporting model $L_4$ and using that simple five-parameter model, rather than using either model $L_8$ or the actual $S_i$. (In practice one might find the simple truth here that $S_{i+1} = 0.98 \cdot S_i, i \geq 5$).

The point is that once complex truth, as a set of parameters, is known quite well, such as to two or three leading significant digits (which is quite beyond what we can usually achieve with real sample sizes), we may not need statistical model-selection anymore. Instead, other criteria would be used: Either use truth as is, or use a simple, parsimonious, interpretable model in place of absolute truth when that model explains almost all the variation in truth. The ability to interpret, understand, and communicate one's model is important in all uses of models, and all numerical descriptions of reality should be considered as just a model if for no other reason than we will never know all digits of an empirical parameter.

Another view of this matter is that model $L_8$ is actually no better than model $L_4$ once we can fit either model to truth. Either model fitted to truth would correspond to producing a correlation coefficient $\geq 0.999$ between true $S_i$ and model predicted survival probability. Once you can fit truth that well with a simple model, the gains in understanding and communicating the model, as truth, override that fact that the model is not exactly truth: It is close enough.

As a corollary of this philosophy, model goodness-of-fit based on statistical tests become irrelevant when sample size is huge. Instead, our concern then is to find an interpretable model that explains the information in the data to a suitable level of approximation, as determined by subject-matter considerations. Unfortunately, in life sciences we probably never have sample sizes anywhere near this large, hence statistical considerations of achieving a good fit to the data are important as well as subject matter considerations of having a fitted model that is interpretable.

Data that are deterministic, as may arise in some physical sciences, correspond to having huge sample sizes, hence model-selection therein need not (but could) be based on statistical criteria. There is no theory of how well a deterministic model must fit deterministic truth in order for that model to be useful. Similarly, if truth as a probability distribution is known, but a one seeks an approximating model,

## 5.2 Survival Models

then K-L can be used, but there is no theory about how much information loss is tolerable. That is a subject-matter decision.

### 5.2.5  A Further Chain Binomial Model

Here we consider a generating model with damped oscillations in survival as age increases. It is not intended to be considered as a biological example; it is just a case of a complex truth against which we can examine some aspects of model-selection. We let the initial population size be $n_1 = 1,000$. We generated 10,000 independent samples using the chain binomial data-generating model with true survival probabilities as

$$S_i = 0.7 + (-1)^{i-1}(0.2/i),$$

where $i$ is year of life ("age" for short). The data $(n_1, n_2, \ldots, n_{18}; \ell = 17,$ as $n_{18} = 0$ but $n_{17} > 0)$ for one repetition were

1000 (fixed), 890, 520, 414, 270, 202, 130, 93, 67, 50, 35, 26, 19, 8, 7, 4, 2, 0.

Animals were followed until all were dead. The models used for analysis were the logistic models, $L_i$, of Section 5.2.3. The model set was $L_1, \ldots, L_{15}$ (i.e., $R = 15$). It is because sample size was 1,000 rather than 150 that we increase the size of the model set used. It is a general principle that as sample size increases, one can expect to reliably estimate more parameters, so the size of the set of models considered should depend weakly on sample size.

Results for $AIC_c$ and BIC model-selection are very different (Table 5.8). On average, $AIC_c$-selected a model with 7.6 parameters, while BIC selected a model on average with 5.1 parameters. The theoretically best model to use, in terms of the expected K-L criterion, in this example is $L_6$, which has $K = 7$ parameters. Under BIC selection 95% of the models selected contained 4, 5, or 6 parameters. In contrast, under $AIC_c$ selection approximately 95% of the models selected had between 5 and 13 parameters, inclusive.

That BIC selection produces a more concentrated distribution of selected models certainly seems to be an advantage. However, useful comparisons must focus on bias and precision of parameter estimates, hence on confidence interval coverage (given that we have essentially the shortest intervals possible under the different model-selection strategies). From Table 5.9, $AIC_c$ selection produces average conditional and unconditional coverage of 84.9% and 92.7%, respectively. Under BIC selection we had average conditional and unconditional coverage of 73.9% and 78.5%, respectively. BIC selection often (but not always) did achieve a smaller MSE in these examples. For the example of Table 5.9, MSEs averaged over $S_1$ to $S_{15}$ were 0.00143 under BIC model-selection, whereas for AIC that average MSE was 0.00209. However, part of this cost of smaller MSEs (by BIC) for the $\hat{S}_i$ is poor confidence interval coverage; for $S_1$ to $S_{15}$, unconditional BIC coverage varied from 45% to 95% (see Table 5.9 for AIC coverage).

Comparison of the expected estimated standard errors illustrates the magnitude of the variance component due to model-selection uncertainty under $AIC_c$. For

174   5. Monte Carlo and Example-Based Insights

TABLE 5.8. Comparison of model-selection relative frequencies for $AIC_c$ vs. BIC based on truth as the chain binomial survival model with damped oscillations, $S_i = 0.7 + (-1)^{i-1}(0.2/i)$, and the model set for data analysis as $L_1$ to $L_{15}$; results are for sample size $n_1 = 1,000$.

| Model $i$ | K | $AIC_c$ model selection | | BIC model selection | |
|---|---|---|---|---|---|
| | | percent | cumul. % | percent | cumul. % |
| 1 | 2 | 0 | 0 | 0 | 0 |
| 2 | 3 | 0 | 0 | 0 | 0 |
| 3 | 4 | 0.6 | 0.6 | 23.0 | 23.0 |
| 4 | 5 | 11.5 | 12.1 | 50.3 | 73.3 |
| 5 | 6 | 26.3 | 38.4 | 21.7 | 95.0 |
| 6 | 7 | 20.5 | 58.9 | 4.1 | 99.2 |
| 7 | 8 | 16.0 | 75.0 | 0.8 | 99.9 |
| 8 | 9 | 7.8 | 82.8 | 0.1 | 100.0 |
| 9 | 10 | 6.0 | 88.8 | 0 | 100.0 |
| 10 | 11 | 3.0 | 91.8 | 0 | 100.0 |
| 11 | 12 | 2.5 | 94.4 | 0 | 100.0 |
| 12 | 13 | 1.6 | 96.0 | 0 | 100.0 |
| 13 | 14 | 1.6 | 97.6 | 0 | 100.0 |
| 14 | 15 | 1.1 | 98.7 | 0 | 100.0 |
| 15 | 16 | 1.3 | 100.0 | 0 | 100.0 |

example, for age 10, the expected conditional standard error is 0.028 vs. the expected unconditional standard error of 0.037. Table 5.9 gives these results for ages 1 to 15.

We computed $\Delta_p = AIC_k - \min AIC$ values for each of the 10,000 repetitions of this case where there were damped oscillations in the $S_i$ parameters. Some percentiles of the sampling distribution of $\Delta_p$ are shown below:

| Percentile | $\Delta_p$ |
|---|---|
| 50.0 | 1.7 |
| 75.0 | 3.0 |
| 80.0 | 3.5 |
| 85.0 | 4.2 |
| 90.0 | 5.7 |
| 95.0 | 7.9 |
| 97.5 | 10.1 |
| 98.0 | 10.9 |
| 99.0 | 13.3 |

These results are consistent with percentile values for $\Delta_p$ we have found for other examples.

Table 5.10 gives expected Akaike weights, $E(w_i)$, and model-selection probabilities, $\pi_i$, from the 10,000 Monte Carlo reps for this example. To compare results from this case of a large sample size ($n_1 = 1,000$), Table 5.10 also gives $E(w_i)$ and $\pi_i$ for $n_1 = 50$ under the same data-generating model. For sample size 1,000 we

TABLE 5.9. Some Monte Carlo results, based on 10,000 reps, on estimation of age-specific survival probabilities under $\text{AIC}_c$ model selection; sample size was $n_1 = 1,000$; the generating model allowed damped oscillations in survival as age increased (see text for details). Conditional (cond.) and unconditional (unc.) standard errors and confidence intervals were also evaluated.

| Age $i$ | $S_i$ | $E(\hat{S}_i)$ | $E(\widehat{se}(\hat{S}_i))$ cond. | $E(\widehat{se}(\hat{S}_i))$ unc. | Coverage cond. | Coverage unc. |
|---|---|---|---|---|---|---|
| 1 | 0.900 | 0.900 | 0.009 | 0.009 | 0.950 | 0.950 |
| 2 | 0.600 | 0.600 | 0.016 | 0.016 | 0.955 | 0.955 |
| 3 | 0.767 | 0.767 | 0.018 | 0.018 | 0.951 | 0.951 |
| 4 | 0.650 | 0.651 | 0.023 | 0.024 | 0.904 | 0.932 |
| 5 | 0.740 | 0.734 | 0.023 | 0.027 | 0.821 | 0.882 |
| 6 | 0.667 | 0.678 | 0.025 | 0.031 | 0.701 | 0.860 |
| 7 | 0.729 | 0.716 | 0.024 | 0.032 | 0.671 | 0.880 |
| 8 | 0.675 | 0.690 | 0.024 | 0.034 | 0.711 | 0.899 |
| 9 | 0.722 | 0.706 | 0.025 | 0.035 | 0.767 | 0.926 |
| 10 | 0.680 | 0.693 | 0.028 | 0.037 | 0.838 | 0.937 |
| 11 | 0.718 | 0.699 | 0.030 | 0.039 | 0.871 | 0.944 |
| 12 | 0.683 | 0.691 | 0.035 | 0.043 | 0.887 | 0.941 |
| 13 | 0.715 | 0.692 | 0.040 | 0.048 | 0.900 | 0.949 |
| 14 | 0.686 | 0.685 | 0.046 | 0.054 | 0.902 | 0.942 |
| 15 | 0.713 | 0.681 | 0.052 | 0.061 | 0.908 | 0.955 |
| Average | | | | | 0.849 | 0.927 |

see from Table 5.10 that on average models $L_4$ to at least $L_9$ must be considered in making inference about the population under $\text{AIC}_c$ model-selection. The Akaike weights give some support to models $L_{10}$ and $L_{11}$, while $\pi_{10}$ and $\pi_{11}$ provide somewhat less support to these models.

There is a high degree of model-selection uncertainty for this example, particularly for the larger sample size. The underlying process (damped oscillations in the $S_i$) is complicated, and the set of logistic approximating models were relatively poor. Had some science been brought to bear on this (artificial) problem, hopefully the set of approximating models would have included some models with at least some oscillating features. Thus, model-selection uncertainty would likely have been greatly reduced.

Use of the bootstrap to estimate model-selection probabilities is effective but computationally intensive. Akaike weights can be easily computed and offer a simple and effective alternative. From Table 5.10 the general agreement between the expected Akaike weights and the model-selection probabilities, $\pi_i$, seems relatively good. Note, however, that these are not estimates of the same quantity, so that exact agreement is not expected. Akaike weights are estimates, normalized to add to 1, of the relative likelihood of each fitted model in the set. These weights provide information about the relative support of the data for the various candidate models. Finally, as all the examples in this section (and other

TABLE 5.10. Expected Akaike weights, E($w_i$), compared to AIC$_c$ model-selection relative frequencies, $\pi_i$, based on Monte Carlo simulation (10,000 reps) for the generating model allowing damped oscillations in survival as age increases; models used for analysis were $L_1$ to $L_{15}$; two sample sizes are considered.

| Model $i$ | $n_1 = 50$ E($w_i$) | $\pi_i$ | $n_1 = 1{,}000$ E($w_i$) | $\pi_i$ |
|---|---|---|---|---|
| 1  | 0.050 | 0.038 | 0.000 | 0.000 |
| 2  | 0.293 | 0.431 | 0.000 | 0.000 |
| 3  | 0.283 | 0.325 | 0.011 | 0.006 |
| 4  | 0.160 | 0.096 | 0.091 | 0.115 |
| 5  | 0.104 | 0.064 | 0.177 | 0.263 |
| 6  | 0.056 | 0.026 | 0.176 | 0.205 |
| 7  | 0.033 | 0.014 | 0.153 | 0.160 |
| 8  | 0.015 | 0.004 | 0.109 | 0.078 |
| 9  | 0.006 | 0.002 | 0.084 | 0.060 |
| 10 | 0.000 | 0.000 | 0.058 | 0.030 |
| 11 | 0.000 | 0.000 | 0.044 | 0.025 |
| 12 | 0.000 | 0.000 | 0.032 | 0.016 |
| 13 | 0.000 | 0.000 | 0.026 | 0.016 |
| 14 | 0.000 | 0.000 | 0.020 | 0.011 |
| 15 | 0.000 | 0.000 | 0.019 | 0.013 |

sections) show, the weights $w_i$ are very useful in model averaging (if that is a goal) and computing estimates of unconditional sampling variances, hence obtaining unconditional confidence intervals that do substantially improve coverage after model-selection.

## 5.3 Examples and Ideas Illustrated with Linear Regression

The model-selection literature emphasizes applications in regression and time series, often as selection of variables in regression; this is the same thing as model-selection. There is a book by McQuarrie and Tsai (1998) devoted to, and entitled, *Regression and Time Series Model Selection*. That book was not published by the time the writing of this book was done, so we never saw it during preparation of our book. However, we had already decided not to concentrate on regression, and knowing about the McQuarrie and Tsai book reinforced that decision. In this section we focus on all-subsets model-selection by presenting an extensive example and some Monte Carlo results; also given are other results and thoughts about K-L–based model-selection for all-subsets regression.

## 5.3.1  All-Subsets Selection: A GPA Example

We use an example based on four regressors (Table 5.11) and a sample size of 20 (a larger number of regressors makes it too demanding of space to present full results). The example of Table 5.11 comes from Graybill and Iyer (1994). They use these example data extensively to illustrate model-selection, including all-subsets selection based on several criteria, but not using AIC nor the other ideas we use herein. Also, we note that the Table 5.11 data are "realistic but not real" (H. Iyer, personal communication).

The full model to fit is

$$y = \beta_0 + \beta_1 x_1 + \beta_2 x_2 + \beta_3 x_3 + \beta_4 x_4 + \epsilon.$$

This model, $M_{15}$, is also denoted in tables by {1234} because all four predictor variables are used (see, for example, Table 5.12). As another example of this notation, model $M_6$ uses only the predictors $x_1$ and $x_3$ and is thus also denoted by $\{1 \cdot 3 \cdot\}$:

$$y = \beta_0 + \beta_1 x_1 + \beta_3 x_3 + \epsilon.$$

Including the intercept-only model $M_{16}$, $\{\cdots\cdot\}$, there are 16 models here in the standard all-subsets approach. We use $\text{AIC}_c$ for selection from this set of 16 models, and we also apply the bootstrap (10,000 reps) to these data with full $\text{AIC}_c$ model-selection applied to all 16 models for each bootstrap rep.

The selected best model ($M_{11}$) includes predictors $x_1$, $x_2$, and $x_3$ (Table 5.12). However, support for $M_{11}$ as the only useful model to use here is weak, because its Akaike weight is just 0.454. Model $M_5$, $\{12\cdot\cdot\}$, is also credible, as is, perhaps, model $M_6$, $\{1\cdot3\cdot\}$. The Akaike weights, $w_i$, and bootstrap selection probabilities, $\hat{\pi}_i$, agree to a useful extent, and both demonstrate the model-selection uncertainty in this example. For a confidence set on models here (if we felt compelled in practice to provide one in this sort of application; we do not) we would use a likelihood ratio criterion: the set of all models $M_i$ for which $w_{11}/w_i \leq 8$ (cf. Royall 1997). Thus, we have the set $\{M_{11}, M_5, M_6, M_{15}\}$. This criterion is identical to using as a cutoff, models for which $\Delta_i \leq 4.16$ (4 would suffice).

We can explore for this example the coherence (with the likelihood approach) of such a rule by determining the bootstrap distribution of the pivotal $\Delta_p$ as $\Delta_p^* = \text{AIC}_{c,11}^* - \min \text{AIC}_c^*$. Here, $\min \text{AIC}_c^*$ is the minimum $\text{AIC}_c$ value for the given bootstrap rep, and $\text{AIC}_{c,11}^*$ is the $\text{AIC}_c$ value in that same bootstrap rep for model $M_{11}$. We computed the 10,000 bootstrap values of $\Delta_{p,b}^*$ and give below some percentiles of this random variable:

| percentile | $\Delta_p^*$ |
|---|---|
| 50 | 1.00 |
| 80 | 3.21 |
| 90 | 3.62 |
| 95 | 5.18 |
| 98 | 7.00 |
| 99 | 8.79 |

Thus, we have a good basis to claim that the confidence set $\{M_{11}, M_5, M_6, M_{15}\}$ on the expected K-L best model is about a 95% confidence set. As with any statistical inference, this result depends on sample size, and hence could change if $n$ increased.

Because model selection in regression is often thought of as seeking to identify the importance of each predictor variable, we next computed the evidence for importance of each variable as the sum of the Akaike weights, $w_i$, for each model in which the predictor variable appears (Table 5.13). We also computed the relative frequency of models selected containing each variable (hence, the same sum of the bootstrap $\hat{\pi}_i$). The results (Table 5.13) show that at this sample size, $x_4$ is not important, whereas $x_1$ is very important and $x_2$ and $x_3$ are at least moderately important. Indeed, the selected best model was $M_{11}$, $\{123\cdot\}$.

Standard theory gives us the MLEs for each $\beta_j$ included in a model and the estimated conditional standard error, $\widehat{se}(\hat{\beta}_j \mid M_i)$. We go a step further here and compute the model-averaged estimate of each regression parameter $\hat{\bar{\beta}}_j$, whether or not $\beta_j$ is in the selected best model. However, $\hat{\bar{\beta}}_j$ is computed only over those

TABLE 5.11. First-year college GPA and four predictor variables from standardized tests administered before matriculation. This example comes from Graybill and Iyer (1994: Table 4.4.3).

| First-year GPA | SAT | | High School | |
|---|---|---|---|---|
| | math | verbal | math | English |
| $y$ | $x_1$ | $x_2$ | $x_3$ | $x_4$ |
| 1.97 | 321 | 247 | 2.30 | 2.63 |
| 2.74 | 718 | 436 | 3.80 | 3.57 |
| 2.19 | 358 | 578 | 2.98 | 2.57 |
| 2.60 | 403 | 447 | 3.58 | 2.21 |
| 2.98 | 640 | 563 | 3.38 | 3.48 |
| 1.65 | 237 | 342 | 1.48 | 2.14 |
| 1.89 | 270 | 472 | 1.67 | 2.64 |
| 2.38 | 418 | 356 | 3.73 | 2.52 |
| 2.66 | 443 | 327 | 3.09 | 3.20 |
| 1.96 | 359 | 385 | 1.54 | 3.46 |
| 3.14 | 669 | 664 | 3.21 | 3.37 |
| 1.96 | 409 | 518 | 2.77 | 2.60 |
| 2.20 | 582 | 364 | 1.47 | 2.90 |
| 3.90 | 750 | 632 | 3.14 | 3.49 |
| 2.02 | 451 | 435 | 1.54 | 3.20 |
| 3.61 | 645 | 704 | 3.50 | 3.74 |
| 3.07 | 791 | 341 | 3.20 | 2.93 |
| 2.63 | 521 | 483 | 3.59 | 3.32 |
| 3.11 | 594 | 665 | 3.42 | 2.70 |
| 3.20 | 653 | 606 | 3.69 | 3.52 |

TABLE 5.12. All-subsets selection results based on $AIC_c$ for the example GPA data in Table 5.11. GPA is regressed against all 16 combinations of four predictor variables as indicated by the set notation code such as {1234} for the full model. Here, the estimated model-selection probabilities, $\hat{\pi}_i$, are based on 10,000 bootstrap reps.

| Model $i$ | Predictors used | $AIC_c$ | $\Delta_i$ | $w_i$ | $\hat{\pi}_i$ |
|---|---|---|---|---|---|
| 11 | {123·} | −43.74 | 0.00 | 0.454 | 0.388 |
| 5 | {12··} | −42.69 | 1.05 | 0.268 | 0.265 |
| 6 | {1·3·} | −40.77 | 2.97 | 0.103 | 0.152 |
| 15 | {1234} | −39.89 | 3.85 | 0.066 | 0.022 |
| 12 | {12·4} | −39.07 | 4.67 | 0.044 | 0.017 |
| 13 | {1·34} | −38.15 | 5.58 | 0.028 | 0.064 |
| 1 | {1···} | −38.11 | 5.62 | 0.027 | 0.045 |
| 8 | {1··4} | −35.19 | 8.55 | 0.006 | 0.012 |
| 14 | {·234} | −32.43 | 11.30 | 0.002 | 0.027 |
| 10 | {··34} | −31.59 | 12.15 | 0.001 | 0.004 |
| 7 | {·23·} | −29.52 | 14.21 | 0.000 | 0.001 |
| 9 | {·2·4} | −25.87 | 17.86 | 0.000 | 0.003 |
| 3 | {··3·} | −25.71 | 18.02 | 0.000 | 0.000 |
| 2 | {·2··} | −23.66 | 20.07 | 0.000 | 0.000 |
| 4 | {···4} | −21.70 | 22.03 | 0.000 | 0.000 |
| 16 | {····} | −15.33 | 28.41 | 0.000 | 0.000 |

TABLE 5.13. Evidence for the importance of each regressor variable in the GPA example, based on sums of Akaike weights over models in which the variable occurs and based on relative frequency of occurrence of the variable in the selected model based on 10,000 bootstrap reps.

| Predictor | $\Sigma w_i$ | $\Sigma \hat{\pi}_i$ |
|---|---|---|
| $x_1$ | 0.997 | 0.965 |
| $x_2$ | 0.834 | 0.722 |
| $x_3$ | 0.654 | 0.658 |
| $x_4$ | 0.147 | 0.148 |

models in which variable $x_j$ appears, hence wherein an estimate of $\beta_j$ is computed. Also computed is the corresponding estimated unconditional standard error $\widehat{\text{se}}(\hat{\bar{\beta}}_j)$, which we also use as the estimated unconditional standard error of any estimated $\beta_j$ from the $AIC_c$-selected best model. The needed formulae are given in Section 4.2.6, especially formula (4.10) with $\hat{\pi}_i = w_i$. The numerical inputs to these calculations are given in Table 5.14 for each of the four parameters: note that we can compute a model-averaged estimate of a parameter independently of issues of selecting a single best model. In Table 5.14 we show for each $\beta_j$ its estimate for the eight models it appears in, and the corresponding conditional standard error and Akaike

TABLE 5.14. Parameter estimates, $\hat{\beta}_1, \ldots, \hat{\beta}_4$, by models that include each $x_i$ (hence $\beta_i$), conditional standard errors given the model, and Akaike weights; also shown is the model-averaged $\bar{\hat{\beta}}_i$ (formula 4.6), its unconditional standard error (formula 4.10), and the sum of Akaike weights over the relevant subset of models.

| Model $i$ | Predictors used | Results by model $\hat{\beta}_1$ | $\widehat{\text{se}}(\hat{\beta}_1 \mid M_i)$ | $w_i$ |
|---|---|---|---|---|
| 11 | {123·} | 0.002185 | 0.0004553 | 0.454 |
| 5  | {12··} | 0.002606 | 0.0004432 | 0.268 |
| 6  | {1·3·} | 0.002510 | 0.0004992 | 0.103 |
| 15 | {1234} | 0.002010 | 0.0005844 | 0.066 |
| 12 | {12·4} | 0.002586 | 0.0005631 | 0.044 |
| 13 | {1·34} | 0.002129 | 0.0006533 | 0.028 |
| 1  | {1···} | 0.003178 | 0.0004652 | 0.027 |
| 8  | {1··4} | 0.002987 | 0.0006357 | 0.006 |
| $\bar{\hat{\beta}}_1$ and unc. se: | | 0.002368 | 0.0005350 | 0.997 |

| Model $i$ | Predictors used | Results by model $\hat{\beta}_2$ | $\widehat{\text{se}}(\hat{\beta}_2 \mid M_i)$ | $w_i$ |
|---|---|---|---|---|
| 11 | {123·} | 0.001312 | 0.0005252 | 0.454 |
| 5  | {12··} | 0.001574 | 0.0005555 | 0.268 |
| 14 | {·234} | 0.001423 | 0.0007113 | 0.002 |
| 15 | {1234} | 0.001252 | 0.0005515 | 0.066 |
| 12 | {12·4} | 0.001568 | 0.0005811 | 0.044 |
| 7  | {·23·} | 0.002032 | 0.0007627 | 0.000 |
| 9  | {·2·4} | 0.002273 | 0.0008280 | 0.000 |
| 2  | {·2··} | 0.003063 | 0.0008367 | 0.000 |
| $\bar{\hat{\beta}}_2$ and unc. se: | | 0.001405 | 0.0005558 | 0.834 |

weight. These weights must be renormalized to sum to 1 to apply formulae (4.6) and (4.10). In Table 5.14, the normalizing constant is shown along with the model-averaged estimate and its estimated unconditional standard error. Note that for a given regression coefficient, $\beta_j$, its estimate can be quite different by model. Part of this variation is due to model bias, as discussed in Section 5.3.5. For example, $\hat{\beta}_3$ is 0.18 (se = 0.09) for model $M_{11}$, {123·}; however, for model $M_3$, {··3·}, we have $\hat{\beta}_3 = 0.51$ (se = 0.12). It is because the predictors are correlated here that $\beta_3$ actually varies over the different models (i.e., due to inclusion/exclusion of other predictor variables), and this source of variation contributes to model uncertainty about $\hat{\beta}_3$ under model selection, or model averaging, of the set of $\hat{\beta}_3$ values in Table 5.14.

We also used the 10,000 bootstrap reps to estimate the model-averaged parameter $\bar{\hat{\beta}}_j$ and its unconditional sampling variation se($\bar{\hat{\beta}}_j$), $j = 1, 2, 3, 4$. The results from the analytical approach using model-selection and Akaike weights, and the

## 5.3 Examples and Ideas Illustrated with Linear Regression 181

TABLE 5.14. (*Continued*)

| Model $i$ | Predictors used | $\hat{\beta}_3$ | $\widehat{se}(\hat{\beta}_3 \mid M_i)$ | $w_i$ |
|---|---|---|---|---|
| 11 | {123·} | 0.1799 | 0.0877 | 0.454 |
| 7  | {·23·} | 0.3694 | 0.1186 | 0.000 |
| 6  | {1·3·} | 0.2331 | 0.0973 | 0.103 |
| 15 | {1234} | 0.1894 | 0.0919 | 0.066 |
| 3  | {··3·} | 0.5066 | 0.1236 | 0.000 |
| 13 | {1·34} | 0.2474 | 0.0990 | 0.028 |
| 14 | {·234} | 0.3405 | 0.1045 | 0.002 |
| 10 | {··34} | 0.4171 | 0.1054 | 0.001 |
| $\hat{\bar{\beta}}_3$ and unc. se: | | 0.1930 | 0.0932 | 0.654 |

| Model $i$ | Predictors used | $\hat{\beta}_4$ | $\widehat{se}(\hat{\beta}_4 \mid M_i)$ | $w_i$ |
|---|---|---|---|---|
| 15 | {1234} | 0.08756 | 0.1765 | 0.066 |
| 12 | {12·4} | 0.01115 | 0.1893 | 0.044 |
| 13 | {1·34} | 0.17560 | 0.1932 | 0.028 |
| 8  | {1··4} | 0.09893 | 0.2182 | 0.006 |
| 14 | {·234} | 0.45333 | 0.1824 | 0.002 |
| 10 | {··34} | 0.57902 | 0.1857 | 0.001 |
| 9  | {·2·4} | 0.51947 | 0.2207 | 0.000 |
| 4  | {···4} | 0.77896 | 0.2407 | 0.000 |
| $\hat{\bar{\beta}}_4$ and unc. se: | | 0.09024 | 0.1989 | 0.147 |

bootstrap results, are shown in Table 5.15. The three estimated unconditional standard errors based on Akaike weights and model averaging (i.e., from formula 4.10) are each larger than the corresponding conditional standard error for the selected best model, $M_{11}$, {123·}. Note especially the case for $\beta_1$ where based on analytical methods $\widehat{se}(\hat{\bar{\beta}}_1) = 0.000535$, whereas $\widehat{se}(\hat{\beta}_1 \mid M_{11}) = 0.000455$, and from the bootstrap, $\widehat{se}(\hat{\bar{\beta}}_1) = 0.000652$. This and the other bootstrap results are precise to essentially two decimal places.

The three different point estimates (bootstrap, $AIC_c$-based model-averaged, and best model) for each of $\beta_1$, $\beta_2$, and $\beta_3$ are quite similar, given their standard errors. However, this is not true for the two estimates of $\beta_4$; we do not know whether the bootstrap result (point estimate 0.29, $\widehat{se} = 0.31$) or analytical result (point estimate 0.09, $\widehat{se} = 0.20$) is better in this example.

Note also that here the bootstrap-estimated unconditional standard errors are less than the estimated conditional standard errors for $\hat{\bar{\beta}}_2$ and $\hat{\bar{\beta}}_3$. However, it seems that unconditional standard errors should not be smaller than conditional ones. Unfortunately, this example constitutes only a sample of size 1 as regards

182    5. Monte Carlo and Example-Based Insights

TABLE 5.15. Bootstrap (10,000 reps) and Akaike weight-based results for the GPA example for model-averaged estimated regression coefficients and associated estimated unconditional standard errors, which include model selection uncertainty. Also shown are the estimate from the selected model, $M_{11}$, and its estimated conditional standard error.

|   | Bootstrap results | | $w_j$-model averaged | | $\text{AIC}_c$ best model | |
|---|---|---|---|---|---|---|
| $j$ | $\hat{\hat{\beta}}_j$ | $\widehat{\text{se}}(\hat{\hat{\beta}}_j)$ | $\hat{\hat{\beta}}_j$ | $\widehat{\text{se}}(\hat{\hat{\beta}}_j)$ | $\hat{\beta}_j$ | $\widehat{\text{se}}(\hat{\beta}_j \mid M_{11})$ |
| 1 | 0.00236 | 0.000652 | 0.00237 | 0.000535 | 0.00219 | 0.000455 |
| 2 | 0.00156 | 0.000508 | 0.00141 | 0.000556 | 0.00131 | 0.000525 |
| 3 | 0.2296  | 0.0684   | 0.1930  | 0.0932   | 0.1799  | 0.0877 |
| 4 | 0.2938  | 0.3056   | 0.0902  | 0.1989   |         |        |

comparing bootstrap and information-theoretic analytical methods. It will take a very large Monte Carlo study to make a reliable general comparison of these two approaches to assessing model-selection uncertainty.

### 5.3.2  A Monte Carlo Extension of the GPA Example

Simulation is a very useful way to gain insights into complex model-selection issues. In particular, here we can assume that the five-dimensional vector $(y, x_1, x_2, x_3, x_4)'$ is multivariate normal, $\text{MVN}(\mu, \Sigma)$ (this is now "truth"), generate 10,000 independent simulated sets of data under this generating model, and do full model-selection to learn about selection performance issues. Given the matrix $\Sigma$ we can determine the true regression coefficients (and their approximate true conditional standard errors) under any of the 16 regression models. The $\beta_j$ (and other needed quantities) given a regression model depend only on elements of $\Sigma$, so it suffices to set the general scale to zero: $\mu = \underline{0}$. The needed $5 \times 5$ variance–covariance matrix is taken here as the sample variance–covariance matrix from the GPA data. Thus, our simulation will be under an assumed truth that is close enough to the truth underlying these GPA data to provide useful results and insights about the analysis of this GPA example.

This particular use of Monte Carlo simulation is also called the parametric bootstrap: We use as the generating model that parametric model having its parameters estimated from the actual data. Consequently, we can expect some of the results here to be about the same as those already obtained from the (nonparametric) bootstrap. The advantage of this parametric approach is that we can determine true values of parameters and hence evaluate confidence interval coverage.

Symbolically, the full variance–covariance matrix is partitioned as below for model $M_i$:

$$\Sigma_i = \begin{bmatrix} \sigma_y^2 & \underline{c}' \\ \underline{c} & \Sigma_x \end{bmatrix}. \tag{5.3}$$

The marginal variance of the response variable $y$ is $\sigma_y^2$. For whatever $m$ predictors, $x_j$, are in the regression model, the vector $\underline{c}$ ($m \times 1$, $1 \leq m \leq 4$) gives their

## 5.3 Examples and Ideas Illustrated with Linear Regression

covariances with $y$ (cov($y, x_j$)). The variance–covariance matrix of just the predictors considered (i.e., for any of the 15 models excluding the intercept-only model, $\{\cdots\}$) is given by matrix $\Sigma_x$ ($m \times m$). The vector of true regression parameters, other than the intercept $\beta_0$, is given by

$$\underline{\beta}' = \underline{c}'(\Sigma_x)^{-1} \tag{5.4}$$

(we ignore $\beta_0$). The approximate sampling variance–covariance matrix of $\hat{\underline{\beta}}'$ is given by $\sigma_{y|x}^2 (\Sigma_x)^{-1}$, where

$$\sigma_{y|x}^2 = \sigma_y^2 - \underline{c}'(\Sigma_x)^{-1}\underline{c} \tag{5.5}$$

is the true residual variance in the regression (a good reference for this multivariate theory is Seber 1984).

The actual simulation process generates the rows of the design matrix, $X$, as random, but then we condition on them in the regression model, $\underline{y} = X\underline{\beta} + \underline{\epsilon}$. Conditionally on $X$, $\hat{\underline{\beta}}$ is unbiased; so it is also unconditionally unbiased; hence $E(\hat{\underline{\beta}}') = \underline{c}'(\Sigma_x)^{-1}$. This same argument applies to $\hat{\sigma}_{y|x}^2$. However, conditionally (by rep), the variance–covariance matrix of $\hat{\underline{\beta}}$ is $\sigma_{y|x}^2 (X'X)^{-1}$, and $E(X'X)^{-1} = (\Sigma_x)^{-1}$ holds only asymptotically as sample size gets large. Thus for the simulations, $\sigma_{y|x}^2 (\Sigma_x)^{-1}$ is only an approximation to the true average variance–covariance matrix of $\hat{\underline{\beta}}$.

Rather than show $\Sigma$, which we do use, we show the derived correlation matrix, upper elements only:

|       | $x_1$ | $x_2$ | $x_3$ | $x_4$ |
|-------|-------|-------|-------|-------|
| $y$   | 0.850 | 0.653 | 0.695 | 0.606 |
| $x_1$ |       | 0.456 | 0.559 | 0.663 |
| $x_2$ |       |       | 0.434 | 0.417 |
| $x_3$ |       |       |       | 0.272 |

(the ordered $x_i$ are SATmath, SATverbal, HSmath, HSenglish). No pairwise $x_i, x_j$ correlations are so high that we would need to eliminate any $x_i$. This consideration and a principal components analysis of just the covariance matrix of the predictor variables are always good to do.

The data-generating model used here ($M_{15}$, $\{1234\}$), based on the GPA data, has the pairwise correlations given above and the residual variance $\sigma_{y|x}^2 = 0.05692$ (from formula 5.5). From formula (5.4) we compute the true $\beta$'s for the generating model $M_{15}$, and from $\sigma_{y|x}^2 (\Sigma_x)^{-1}$ we compute the approximate conditional standard errors of the $\hat{\beta}_j$ under model $M_{15}$. We also give the approximate conditional coefficient of variation of each $\hat{\beta}_j$ (Table 5.16). These same quantities are computable for any submodel fitted to the generated data; Table 5.16 shows these theoretical values for models $\{1234\}$, $\{123\cdot\}$, $\{12\cdot\cdot\}$, $\{1\cdot\cdot\cdot\}$, and $\{\cdot\cdot 3\cdot\}$.

The expected results of fitting any model to data reflect both the truth as contained in empirical data and the adequacy of the model. For example, from Table 5.16 if we use model $M_3$, then $E(\hat{\beta}_3) = 0.5066$ (cv = 0.117), whereas when all four

TABLE 5.16. True values of $\sigma^2_{y|x}$ and $\beta$'s, and approximate conditional standard errors and cv's for some models when they are fit to data, for $n = 20$, from the true generating model as $M_{15}$ (as detailed in the text).

Model $M_{15}$, {1234}: $\sigma^2_{y|x} = 0.05692$

| $j$ | $\beta_j$ | se($\hat{\beta}_j \mid M_{15}$) | cv($\hat{\beta}_j \mid M_{15}$) |
|---|---|---|---|
| 1 | 0.002010 | 0.0005061 | 0.252 |
| 2 | 0.001252 | 0.0004776 | 0.381 |
| 3 | 0.1895 | 0.0796 | 0.420 |
| 4 | 0.0875 | 0.1528 | 1.745 |

Model $M_{11}$, {123·}: $\sigma^2_{y|x} = 0.05785$

| $j$ | $\beta_j$ | se($\hat{\beta}_j \mid M_{11}$) | cv($\hat{\beta}_j \mid M_{11}$) |
|---|---|---|---|
| 1 | 0.002185 | 0.0004072 | 0.186 |
| 2 | 0.001312 | 0.0004698 | 0.358 |
| 3 | 0.1799 | 0.0784 | 0.436 |

Model $M_5$, {12··}: $\sigma^2_{y|x} = 0.07307$

| $j$ | $\beta_j$ | se($\hat{\beta}_j \mid M_5$) | cv($\hat{\beta}_j \mid M_5$) |
|---|---|---|---|
| 1 | 0.002606 | 0.0004086 | 0.157 |
| 2 | 0.001574 | 0.0005121 | 0.325 |

Model $M_1$, {1···}: $\sigma^2_{y|x} = 0.10759$

| $j$ | $\beta_j$ | se($\hat{\beta}_j \mid M_1$) | cv($\hat{\beta}_j \mid M_1$) |
|---|---|---|---|
| 1 | 0.003178 | 0.0004413 | 0.139 |

Model $M_3$, {··3·}: $\sigma^2_{y|x} = 0.20001$

| $j$ | $\beta_j$ | se($\hat{\beta}_j \mid M_3$) | cv($\hat{\beta}_j \mid M_3$) |
|---|---|---|---|
| 3 | 0.5066 | 0.1173 | 0.117 |

predictors are included, $\beta_3 = E(\hat{\beta}_3) = 0.1895$ (cv = 0.420). For model $M_{11}$, $E(\hat{\beta}_3) = 0.1799$ (cv = 0.436). Results in Table 5.16 apply when the specified model is always fit to the data; hence no data-based model-selection occurs. When the inference strategy is to first select a model based on the data, then use it for inference, the properties of estimators and other inferences are affected (model-selection biases and uncertainties occur).

Examination of results in Table 5.16 demonstrates that in general, as measured by the conditional coefficient of variation and, for given data, precision of a parameter estimator (for a parameter in all models considered) increase as the number of other parameters in the models decreases. That a given $\beta_j$ varies by model is because the predictors are correlated. This effect (i.e., model variation in $E(\hat{\beta}_j)$) will get more pronounced if correlations get stronger; it does not occur if all predictors are uncorrelated with each other. Leaving an $x_j$ out of fitted model $M_i$ has little effect if that predictor is unimportant as measured by a large cv($\hat{\beta}_j \mid M_i$). For example,

$x_4$ (HSenglish) can be left out of the fitted model {1234}; hence one uses {123·}. Indeed, in this Monte Carlo example the expected K-L best model is $M_{11}$ (based on 10,000 simulation reps).

What we want to illustrate with simulation here (and in Section 5.3.4) are some results under all-subsets $AIC_c$ model-selection in regression. Firstly, we focus on unconditional vs. conditional confidence interval coverage on true $\beta_j$, i.e., the value of $\beta_j$ in the generating model $M_{15}$ in Table 5.16. For confidence intervals we used $\hat\beta \pm 2$ se; hence, we ignored the issue of a t-distribution based multiplier. This affects coverage a little, but the focus is really on the difference between conditional and unconditional coverage.

Secondly, we look at induced model-selection bias in $\hat\sigma^2_{y|x}$. A selected model $M_i$, out of the 16 models fitted (especially at a small sample size, even using $AIC_c$) tends to have a better fit for that data set, hence a smaller residual sum of squares, than would occur on average if model $M_i$ were always fitted. Thus, data-based selection in regression will tend to result in fitting the data a little too well; as a result, we get $E(\hat\sigma^2_{y|x}) < \sigma^2_{y|x}$ (this also has an effect on confidence interval coverage that is not correctable by using unconditional intervals). Confidence intervals depend on $\hat\sigma_{y|x}$; however, the true value of $\sigma_{y|x}$ varies by model, so what we report to assess selection bias is the relative bias,

$$RB = \frac{E(\hat\sigma_{y|x}) - \sigma_{y|x}}{\sigma_{y|x}}. \quad (5.6)$$

Other quantities of interest include the expected value of the model-averaged estimator $\hat{\bar\beta}_j$ and unconditional interval coverage based on this estimator, model-selection variation, and percentiles of $\Delta_p$. From the Monte Carlo results of this example we find the percentiles below:

| percentile | $\Delta_p$ |
|---|---|
| 50 | 0.9 |
| 80 | 3.1 |
| 90 | 3.6 |
| 95 | 5.1 |
| 98 | 7.3 |
| 99 | 9.4 |

When a parameter (hence $\hat\beta_j$) appeared in the selected model we computed several quantities: the model-averaged estimate, $\hat{\bar\beta}_j$, the unconditional standard error, $\widehat{se}(\hat\beta_j)$, and three confidence intervals (nominally 95%). The conditional interval (cond.) is based on $\hat\beta_j$ and its estimated conditional standard error given the selected model. The unconditional interval (unc.) is based on $\hat\beta_j$ and $\widehat{se}(\hat\beta_j)$. The interval based on model averaging (MA) uses $\hat{\bar\beta}_j$ and $\widehat{se}(\hat\beta_j)$. Finally, it needs to be clearly understood that the coverage we refer to is on the true parameter from the actual data-generating model.

In this example, the achieved coverage of the unconditional interval is better than that of the conditional intervals (Table 5.17), especially for $\beta_1$ (89% vs 80%; both coverage percentages increase by about 0.02 if a t-distribution-based interval

TABLE 5.17. Expected values of estimators of $\beta_j$ and confidence interval coverage on true $\beta_j$ under AIC$_c$-based all-subsets model selection from the Monte Carlo generated data (10,000 reps) mimicking the GPA example; occurrence frequency is the number of reps in which the selected model included the indicated $\beta_j$.

| Occur. freq. | $j$ | $\beta_j$ | $E(\hat{\beta}_j)$ | $E(\tilde{\hat{\beta}}_j)$ | Achieved coverage |  |  |
|---|---|---|---|---|---|---|---|
|  |  |  |  |  | cond. | unc. | MA |
| 9544 | 1 | 0.00201 | 0.00229 | 0.00232 | 0.801 | 0.879 | 0.889 |
| 7506 | 2 | 0.00125 | 0.00156 | 0.00157 | 0.906 | 0.927 | 0.930 |
| 6506 | 3 | 0.190 | 0.248 | 0.253 | 0.887 | 0.913 | 0.916 |

is used). A noncorrectable source of lowered coverage (actually, the bootstrap might be used to compute a correction) comes from bias due to model-selection. For example, here model-selection results in the bias $E(\hat{\beta}_1) - \beta_1 = 0.00229 - 0.00201 = 0.00028$. This bias is important only in relation to the unconditional standard error, which is here $se(\hat{\beta}_1) = 0.000514$. Thus, the bias/se ratio is $\delta = 0.54$; this value of $\delta$ will result in a coverage decrease to 92.1% if coverage would be 95% at $\delta = 0$ (see Cochran 1963:Table 1.1).

An unexpected result in this example is that the conditional coverage for $\beta_2$ and $\beta_3$ was as high as (about) 0.9. The unconditional coverage then does improve and without exceeding 95% coverage.

The model-selection bias induced in $\hat{\sigma}_{y|x}$ is negative and depends on the probability that the model will be selected (Table 5.18). Shown in Table 5.18 is the average value of $\hat{\sigma}^2_{y|x}$ when the model is fit to all 10,000 generated data sets (compare to theoretical results in Table 5.16). The relative bias of $\hat{\sigma}_{y|x}$ under model-selection is given by RB from formula (5.6). Good models (under the K-L paradigm) do not correspond to very bad levels of RB. As the model becomes less acceptable (in terms of expected K-L value), it is selected only when the data are an unusually good fit to that model.

While general in their qualitative nature, these numerical results are more extreme for a sample size of 20 than would be the case at a large sample size. In fact, for this generating model the selection bias in $\hat{\sigma}_{y|x}$ is trivial at sample size $n = 50$, and confidence interval coverage is nearly 95% for each type of interval. Even though model-selection can induce biases, under information-theoretic selection and associated unconditional inferences results can be quite good and certainly better (for the sample size) than use of an unnecessarily high-dimension global model that includes all predictors.

### 5.3.3 An Improved Set of GPA Prediction Models

An even better way to improve on the all-predictors global model is first to reduce one's models to a smaller set of *a priori* meaningful models suggested by subject matter or logical considerations. Basically, this means using logical transformations of the predictors (consideration of meaningful model forms is also important)

TABLE 5.18. The relative bias (RB) in $\hat{\sigma}_{y|x}$ (formula 5.6) induced by $AIC_c$ model selection for the Monte Carlo example (10,000 reps) based on the GPA data; $E(\hat{\sigma}^2_{y|x} \mid M_i)$ is the average of the 10,000 values $\hat{\sigma}^2_{y|x}$ when the model is fit to every generated data set (i.e., no selection occurs).

| Model $i$ | Predictors used | $E(\hat{\sigma}^2_{y\mid x} \mid M_i)$ no selection | $\pi_i$ | RB |
|---|---|---|---|---|
| 11 | {1 2 3 ·} | 0.0576 | 0.3786 | −0.0618 |
| 5  | {1 2 · ·} | 0.0728 | 0.2730 | −0.0762 |
| 6  | {1 · 3 ·} | 0.0803 | 0.1451 | −0.1161 |
| 13 | {1 · 3 4} | 0.0762 | 0.0458 | −0.1854 |
| 15 | {1 2 3 4} | 0.0566 | 0.0389 | −0.1878 |
| 1  | {1 · · ·} | 0.1076 | 0.0351 | −0.1724 |
| 12 | {1 2 · 4} | 0.0728 | 0.0301 | −0.1693 |
| 14 | {· 2 3 4} | 0.1017 | 0.0216 | −0.2664 |
| 10 | {· · 3 4} | 0.1272 | 0.0161 | −0.2681 |
| 8  | {1 · · 4} | 0.1063 | 0.0078 | −0.2121 |
| 7  | {· 2 3 ·} | 0.1420 | 0.0064 | −0.2843 |
| 9  | {· 2 · 4} | 0.1695 | 0.0012 | −0.3634 |
| 2  | {· 2 · ·} | 0.2227 | 0.0001 | −0.3837 |
| 3  | {· · 3 ·} | 0.2012 | 0.0001 | −0.5496 |
| 4  | {· · · 4} | 0.2446 | 0.0001 | −0.3913 |

and dropping predictors that are very unlikely to be related to the response variable of interest (investigators often record a variable simply because it is easy to measure). To illustrate this idea we conceived of five different GPA prediction models based on simple, derived predictors that make some sense.

The original four predictors are each just indices to general academic ability, and they are measured with error. That is, a person's test grade would surely vary by circumstances (and luck), such as if they had a cold the day of the exam. Viewing these predictor variables as just semicrude indices, why not just compute a single averaged index? In so doing we average over math and verbal (English) ability, but grades in many courses depend upon both abilities anyway. With a large sample size (say $n > 1,000$) it makes sense to let the regression fit calibrate the relative importance of the four indices. However, with only 20 observations some combining of indices may be advantageous.

The SAT and HS scores are on very different scales. There are several ways to allow for this, such as first to normalize each predictor variable to have a mean of 0 and a standard deviation of 1 and then just average all four adjusted predictors to get a total (*tot*) predictor index. To circumvent that minor nuisance we used geometric means to cope with the scale issue. Thus the five new variables that replace the original four variables are

$$sat = (x_1 \times x_2)^{0.5},$$
$$hs = (x_3 \times x_4)^{0.5},$$
$$math = (x_1 \times x_3)^{0.5},$$

$$engl = (x_2 \times x_4)^{0.5},$$
$$tot = (sat \times hs)^{0.5} = (math \times engl)^{0.5} = (x_1 \times x_2 \times x_3 \times x_4)^{0.25}.$$

These variables are interpretable and seem just as adequate as the original four variables as indices to first-year college GPA.

Next, we would not use, in an *a priori* analysis, any of the original 16 models. The only linear regression models we would (did) consider with these derived predictors are given below, in terms of predictor variables in the model (all models have an intercept and $\sigma^2$). We numbered these as models 17 to 21 in order to compare results to the original 16 models:

| model | K | variables included |
|---|---|---|
| $M_{17}$ | 3 | tot |
| $M_{18}$ | 4 | sat   hs |
| $M_{19}$ | 4 | math   engl |
| $M_{20}$ | 4 | sat |
| $M_{21}$ | 4 | hs |

We conceptualized these models before examining fit of the original 16 models to the GPA data, and no other derived models were considered.

The $AIC_c$ best model of the above is $M_{17}$ (Table 5.19). In fact, model $M_{17}$ is best in the full set of all 21 models (results are not shown for all 21 models in Table 5.19). When adding new models to an existing set, no earlier $AIC_c$ values need to be recomputed: Just reorder the full set from smallest to largest $AIC_c$. The full set of $\Delta_i$ values may need to be recomputed if the best model changes (as here, from $M_{11}$ to $M_{17}$). Given the new set of $\Delta_i$, recompute the Akaike weights, $w_i$.

The results in Table 5.19 illustrate that the best model in a set of models is relative only to that set of models. Kullback–Leibler model-selection does not provide an absolute measure of how good a fitted model is; model $M_{11}$ is only best in the set of 16 all-subsets models. Compared to model $M_{17}$, model $M_{11}$ (and the entire original set of 16 models) can almost be discarded as contenders for expected K-L best model for these data. Correspondingly, we emphasize that any model-based inference is conditional on the set of models considered. The specifics of inferences and computable uncertainties are conditional on the models formally considered.

The models used here are useful only for prediction; they do not relate to any causal process. Hence, we illustrate inclusion of model uncertainty into prediction based on models $M_{17}$–$M_{21}$ (standard aspects of prediction inference given a fitted linear model are assumed here; see, e.g., Graybill and Iyer 1994). As computed in Graybill and Iyer (1994:244), under model $M_{15}$, {1234}, the prediction of expected GPA at $x_1 = 730$, $x_2 = 570$, $x_3 = 3.2$, and $x_4 = 2.7$ is $\hat{E}(y) = 3.185$ with conditional (on model) standard error 0.172. For comparison we note that given model $M_{11}$, {123·}, the corresponding results are $\hat{E}(y) = 3.253$, se $= 0.102$.

The model-averaged predicted expected GPA is $\hat{\overline{E}}(y) = 3.06$ with estimated unconditional standard error of 0.11 (Table 5.20). These results are computed using formulae (4.6) and (4.10). To construct a confidence interval here that allows for the small degrees of freedom of $\hat{\sigma}^2_{y|x}$ we suggest that it suffices in this example to

TABLE 5.19. Results of fitting, to the GPA data, the five new models ($M_{17}$–$M_{21}$) based on transformations of the original test scores ($x_1$–$x_4$); also given are results for some of the 16 models originally considered (see text for details of the new predictors).

| Model $i$ | Predictors used | $AIC_c$ | $\Delta_i$ | $w_i$ |
|---|---|---|---|---|
| 17 | tot | −48.20 | 0.00 | 0.590 |
| 18 | sat hs | −45.97 | 2.23 | 0.193 |
| 20 | sat | −43.96 | 4.24 | 0.071 |
| 11 | {123·} | −43.74 | 4.47 | 0.063 |
| 5 | {12··} | −42.69 | 5.52 | 0.037 |
| 6 | {1·3·} | −40.77 | 7.44 | 0.014 |
| 15 | {1234} | −39.89 | 8.32 | 0.009 |
| 19 | math engl | −39.23 | 8.97 | 0.007 |
| ... | ... | ... | ... | ... |
| 21 | hs | −33.98 | 14.22 | 0.000 |
| ... | ... | ... | ... | ... |

TABLE 5.20. Quantities needed in the computation of model-averaged prediction of expected GPA, $\hat{\bar{E}}(y)$ (= 3.056), and its unconditional standard error (= 0.1076), under models $M_{17}$–$M_{21}$ for the predictors $x_1 = 730$, $x_2 = 570$, $x_3 = 3.2$, and $x_4 = 2.7$.

| Model $i$ | Predictors used | $\Delta_i$ | $w_i$ | $\hat{E}(y)$ | $\widehat{se}(\hat{E}(y)\mid M_i)$ |
|---|---|---|---|---|---|
| 17 | tot | 0.00 | 0.685 | 3.016 | 0.0738 |
| 18 | sat hs | 2.23 | 0.224 | 3.095 | 0.1210 |
| 20 | sat | 4.24 | 0.082 | 3.177 | 0.1993 |
| 19 | math engl | 8.97 | 0.008 | 3.271 | 0.1045 |
| 21 | hs | 14.22 | 0.001 | 2.632 | 0.0832 |
| | weighted results: | | | 3.056 | 0.1076 |

use $3.06 \pm t \times \widehat{se}(\hat{\bar{E}}(y))$, where the multiplier $t = 2.10$ is from the t-distribution on 18 df. Model $M_{17}$ has 18 df for $\hat{\sigma}^2_{y\mid x}$, and the weight on that model is $w_{17} = 0.685$ (more sophisticated procedures will not make a practical difference here). The model-averaged result is distinctly more precise than the prediction based on the fitted global model (standard errors of 0.108 vs. 0.172, model-averaged vs. global model-based). Also, the inclusion of model uncertainty increases the standard error as compared to the result conditional on model $M_{17}$ (se = 0.074).

### 5.3.4 More Monte Carlo Results

The theory for Monte Carlo generation of regression data, with random regressors, was presented in Section 5.3.2. Using that approach we computed a few more simulations. Our motivation was firstly to see whether anything bad occurred in using model averaging and unconditional confidence intervals (it did not), and secondly to see what biases might result from model-selection and what confidence interval coverage could be achieved.

This is far from a full-scale simulation study because we greatly restricted the many factors to consider in the design of an all-subsets model-selection study. For example, we used only $m = 4$ predictors here (but much larger values of $m$ need to be explored). Thus, the global model is model $M_{15}$, {1234}. Given a sample size $n$, one generates a sample from the $(m+1)$-dimensional $MVN(\underline{\mu}, \Sigma)$ generating model. Without loss of generality we can set $\underline{\mu} = \underline{0}$. However, there are still (in general) $(m+1) \times (m+2)/2$ parameters to specify in $\Sigma$. To make this design problem tractable we used the following structure on the generating model: either

$$\underline{c}' = [0\ 0\ 0\ 0] \text{ or } \underline{c}' = [0.8\ 0.6\ 0.4\ 0.2],$$

mostly the latter; and

$$\Sigma_x = \begin{bmatrix} 1 & \rho & \rho & \rho \\ \rho & 1 & \rho & \rho \\ \rho & \rho & 1 & \rho \\ \rho & \rho & \rho & 1 \end{bmatrix}.$$

The regression parameters of the generating model are given by $\underline{\beta}' = \underline{c}'(\Sigma_x)^{-1}$.

Another design factor was taken to be the regression residual variance, $\sigma^2_{y|x} = \sigma^2_y - \underline{c}'(\Sigma_x)^{-1}\underline{c}$. Given values for $\sigma^2_{y|x}$ (we used only 1 and 25) we find the marginal variance of $y$, $\sigma^2_y$. These quantities (i.e., $\underline{c}$, $\Sigma_x$, and $\sigma^2_y$) suffice to compute the full $5 \times 5$ variance–covariance matrix of formula (5.3). The conditional variance–covariance of $\underline{\hat{\beta}}$ is given by

$$\frac{\sigma^2_{y|x}}{n}(\Sigma_x)^{-1}.$$

We see that factors $\sigma^2_{y|x}$ and $n$ are redundant in their effect on the sampling variance. Therefore, for not-small sample sizes it is much more economical to fix $n$ (say $n \geq 28 + K$, for the global model value of $K$) and directly lower $\sigma^2_{y|x}$ to gain precision, rather than to fix $\sigma^2_{y|x}$ and simulate greater precision by increasing sample size. We did not do so here; it is still necessary to have small actual $n$ to explore small-sample-size effects.

Because of the choice of the form of $\Sigma_x$, the estimators $\hat{\beta}_1$, $\hat{\beta}_2$, $\hat{\beta}_3$, and $\hat{\beta}_4$ all have the same conditional variance. In fact, for any $\rho$ it suffices to present the constant diagonal element, $v$, of $(\Sigma_x)^{-1}$, because

$$\text{var}(\hat{\beta}_i \mid M_{15}) = v \times \sigma^2_{y|x} / n.$$

For $\underline{c}' = [0.8, 0.6, 0.4, 0.2]$ we simulated 10,000 reps at each combination of $\rho = 0, 0.2, 0.4, 0.6,$ and $0.8$ crossed with $n = 20, 50,$ and $100$. We focused on $\hat{\beta}_1$, but looked at other parameters in a few cases (based then on another set of 10,000 reps). To these cases 1 to 5 for $\rho$ we added cases 6 and 7, as noted in Table 5.21. In total we looked at 29 simulated "populations."

TABLE 5.21. Values of design factors used in simulation exploration of model selection (see text for details); sample size sets used are $n$-set, a = {20, 50, 100}, b = {20, 50, 100, 200, 500, 1,000}; $\beta_i$ are rounded to two decimal places.

| Case | $\rho$ | $n$-set | $\sigma^2_{y\|x}$ | $v$ | $\beta_1$ | $\beta_2$ | $\beta_3$ | $\beta_4$ |
|---|---|---|---|---|---|---|---|---|
| 1 | 0   | a | 1  | 1.00 | 0.8  | 0.6  | 0.4   | 0.2   |
| 2 | 0.2 | a | 1  | 1.09 | 0.69 | 0.44 | 0.19  | -0.06 |
| 3 | 0.4 | a | 1  | 1.36 | 0.73 | 0.39 | 0.06  | -0.27 |
| 4 | 0.6 | a | 1  | 1.96 | 0.93 | 0.43 | -0.07 | -0.57 |
| 5 | 0.8 | a | 1  | 3.82 | 1.65 | 0.65 | -0.35 | -1.35 |
| 6 | 0   | a | 1  | 1.00 | 0    | 0    | 0     | 0     |
| 7 | 0   | b | 25 | 1.00 | 0.8  | 0.6  | 0.4   | 0.2   |

We tabulated some basic results (Table 5.22) wherein full $AIC_c$ model selection was applied to all 16 possible models (the labeling of models is the same as in Table 5.12). In particular, we tabulated the 90th, 95th, and 99th percentiles of $\Delta_p$. There is one variation here; with no models more general than the generating global model ($M_{15}$) if sample size gets too large, selection converges on model $M_{15}$ and all percentiles of $\Delta_p$ go to 0. This "boundary" effect (i.e., some degree of reduction in percentiles of $\Delta_p$) will not normally occur in real data analysis, so we flagged populations where a boundary effect is occurring. Our recommendations about interpreting $\Delta_i$ are for when no boundary effect occurs. For the 15 populations where no boundary effect occurred, the mean percentiles of $\Delta_p$ in Table 5.22 are 4.7, 6.4, and 10.6 (90th, 95th, and 99th respectively).

Sample size also has an effect on the distribution of $\Delta_p$; it is not a strong effect for $n$ greater than about 20. For case 7 in Table 5.22 the effect of sample size ($20 \leq n \leq 500$) is about 2 units at the 90th and 95th percentiles and about 3 units at the 99th percentile. These are typical of sample size effects we have observed.

There is a lot of model-selection uncertainty in these 29 simulated populations (Table 5.22), for example, as indexed by how low the selection probability, $\pi_k$, is even for the expected $AIC_c$ best model. The other index of model-selection uncertainty used in Table 5.22 is simply a count of the number of models, of the 16, that have selection probabilities $\geq 0.01$ (often $\geq$ one-half of the possible models). For the all too typical application of variable selection with 10 or more variables ($R \geq 1,024$) and not-large sample size, the selection process will be highly unstable (cf. Brieman 1996) as to what model is selected. That is, selection probabilities can be expected to be very low, even for the actual K-L best model, and not exhibit a strong mode. As a result the selected model itself is not at all the basis for a reliable inference about the relative importance of the predictor variables, even if the selected model provides reliable predictions.

A confidence interval for a parameter $\beta_i$ was computed only when that parameter was in the selected model, in which case point estimates computed were the MLE $\hat{\beta}_i$ and the model-averaged $\hat{\hat{\beta}}_i$. Three types of intervals were computed: the classical conditional interval, $\hat{\beta}_i \pm 2\,\widehat{se}(\hat{\beta}_i \mid M_r)$; the corresponding unconditional interval,

192     5. Monte Carlo and Example-Based Insights

TABLE 5.22. Monte Carlo results (10,000 reps used) for simulated populations; column four is the $AIC_c$ best model; column five is the corresponding selection probability, $\pi_k$; column six is the number of models for which $\pi_i \geq 0.01$; also given are percentiles for $\Delta_p$.

| Case | $\rho$ | $n$ | Best $M_k$ | $\pi_k$ best | # $\pi_i$ $\geq 0.01$ | Percentiles of $\Delta_p$ | | | Boundary Effect |
|---|---|---|---|---|---|---|---|---|---|
| | | | | | | 90 | 95 | 99 | |
| 1 | 0   | 20   | 11 | 0.243 | 13 | 5.37 | 6.74 | 10.84 | no |
| 1 | 0   | 50   | 15 | 0.368 | 4  | 2.55 | 2.59 | 4.01  | yes |
| 1 | 0   | 50   | 15 | 0.378 | 4  | 2.56 | 2.59 | 4.25  | yes |
| 1 | 0   | 100  | 15 | 0.668 | 2  | 1.81 | 2.12 | 2.26  | yes |
| 2 | 0.2 | 20   | 5  | 0.342 | 12 | 4.65 | 6.51 | 11.37 | no |
| 2 | 0.2 | 20   | 5  | 0.334 | 12 | 4.58 | 6.41 | 10.77 | no |
| 2 | 0.2 | 50   | 11 | 0.264 | 7  | 2.67 | 3.78 | 7.10  | yes |
| 2 | 0.2 | 100  | 11 | 0.462 | 4  | 2.22 | 3.55 | 7.14  | yes |
| 3 | 0.4 | 20   | 1  | 0.398 | 12 | 7.03 | 9.35 | 14.80 | no |
| 3 | 0.4 | 50   | 12 | 0.353 | 8  | 3.35 | 4.29 | 7.02  | yes |
| 3 | 0.4 | 100  | 12 | 0.611 | 6  | 2.17 | 3.36 | 7.06  | yes |
| 4 | 0.6 | 20   | 8  | 0.227 | 12 | 6.38 | 8.73 | 13.53 | no |
| 4 | 0.6 | 50   | 12 | 0.544 | 8  | 2.47 | 3.65 | 6.41  | yes |
| 4 | 0.6 | 100  | 12 | 0.743 | 4  | 1.74 | 2.89 | 6.60  | yes |
| 5 | 0.8 | 20   | 8  | 0.499 | 10 | 5.43 | 7.68 | 12.58 | no |
| 5 | 0.8 | 50   | 15 | 0.305 | 4  | 3.32 | 4.03 | 4.77  | yes |
| 5 | 0.8 | 50   | 15 | 0.308 | 4  | 3.24 | 4.03 | 4.83  | yes |
| 5 | 0.8 | 100  | 15 | 0.595 | 4  | 2.04 | 2.23 | 2.88  | yes |
| 5 | 0.8 | 100  | 15 | 0.597 | 4  | 2.03 | 2.23 | 2.94  | yes |
| 6 | 0   | 20   | 16 | 0.594 | 11 | 3.43 | 5.11 | 8.92  | no |
| 6 | 0   | 50   | 16 | 0.535 | 11 | 3.35 | 4.96 | 8.71  | no |
| 6 | 0   | 100  | 16 | 0.518 | 11 | 3.41 | 4.90 | 8.65  | no |
| 7 | 0   | 20   | 16 | 0.493 | 11 | 4.55 | 6.47 | 10.55 | no |
| 7 | 0   | 50   | 1  | 0.180 | 13 | 5.62 | 7.48 | 11.42 | no |
| 7 | 0   | 100  | 5  | 0.131 | 16 | 5.11 | 6.90 | 10.86 | no |
| 7 | 0   | 200  | 11 | 0.143 | 15 | 4.08 | 5.20 | 8.42  | no |
| 7 | 0   | 200  | 11 | 0.140 | 15 | 4.10 | 5.20 | 8.43  | no |
| 7 | 0   | 500  | 11 | 0.391 | 8  | 3.42 | 4.93 | 8.92  | no |
| 7 | 0   | 1000 | 15 | 0.380 | 4  | 2.00 | 2.04 | 3.30  | yes |

$\hat{\beta}_i \pm 2\,\widehat{se}(\hat{\bar{\beta}}_i)$; and the interval based on the model-averaged point estimate, $\hat{\bar{\beta}}_i \pm 2\,\widehat{se}(\hat{\bar{\beta}}_i)$. One result was that the coverage for interval types two and three was barely different, but was slightly better for the interval $\hat{\beta}_i \pm 2\,\widehat{se}(\hat{\bar{\beta}}_i)$ (ratio of coverages: 0.995). Hence, we present results only for this latter interval (Table 5.23). We focused on the coverage for $\beta_1$ (i.e., parameter index 1); Table 5.23 gives coverage results for a few instances of a difference parameter ($\beta_3$ or $\beta_4$).

Confidence interval coverage is affected by bias in either the point estimator or its standard error estimator. Therefore, we tabulated information on these biases for the interval based on the model-averaged estimator, $\hat{\bar{\beta}}_i \pm 2\,\widehat{se}(\hat{\bar{\beta}}_i)$. In this context

## 5.3 Examples and Ideas Illustrated with Linear Regression

bias is only important in relation to standard error, which Table 5.23 shows,

$$\delta = \frac{E(\hat{\bar{\beta}}_i) - \beta_i}{E(\widehat{se}(\hat{\bar{\beta}}_i))},$$

which is bias of the model-averaged estimator of $\beta_i$ divided by the expected unconditional standard error of the estimator. The effect on confidence interval coverage is trivial for $|\delta| \leq 0.25$ and ignorable for $|\delta| \leq 0.5$.

The other factor that can affect coverage is bias in the estimator $\widehat{se}(\hat{\bar{\beta}}_i)$; hence we show the ratio

$$\text{se-}r \; \frac{E(\widehat{se}(\hat{\bar{\beta}}_i))}{MC\text{-}se(\hat{\bar{\beta}}_i)}$$

in Table 5.23. Here, MC-se($\hat{\bar{\beta}}_i$) is the actual achieved standard error of $\hat{\bar{\beta}}_i$ over the Monte Carlo reps (out of 10,000) wherein $\hat{\bar{\beta}}_i$ is computed. A value of se-$r$ = 1 is desirable. Coverage would be reduced (other factors being equal) if se-$r$ becomes much less than 1. The effect on coverage is ignorable for $0.9 \leq \text{se-}r \leq 1.1$. For all these results, the relevant sample size is denoted by "Freq." in Table 5.23: the number of reps wherein $\hat{\bar{\beta}}_i$ is computed because the parameter is in the AIC$_c$-selected model.

The biggest surprise was the high achieved coverage of the traditional (conditional) confidence interval (Table 5.23). When that coverage was poor (for example, case 2, $\rho = 0.2, n = 20$, unconditional coverage of 0.755 on $\beta_3$), it was because of severe bias in either the point estimator or its standard error estimator. Moreover, these biases are clearly a form of model-selection bias, and they occurred when the reference parameter was infrequently selected (i.e., infrequent selection of any model containing $\beta_i$), which itself is a result of the predictor variable $x_i$ being unimportant at the given sample size. Confusing the matter, however, not all instances of small frequency of selecting models including $\beta_i$ resulted in deleterious effects on coverage (for example, case 5, $\rho = 0.8, n = 50$, unconditional coverage of 0.925 on $\beta_3$). On the positive side, if a predictor variable was important (as judged by high selection frequency), its unconditional (and conditional) coverage was always good.

In all 29 simulated populations the unconditional interval coverage was greater than or equal to the conditional coverage and provided improved coverage when the conditional coverage was less than 0.95 (for example, case 5, $\rho = 0.8, n = 20$, conditional and unconditional coverage of 0.858 vs. 0.937 on $\beta_1$). For the 23 populations where the parameter was important to the selected model, the average conditional and unconditional confidence interval coverage was 0.930 vs. 0.947 (and $\overline{\delta} = 0.28$, $\overline{\text{se-}r} = 1.1$). The improvement in coverage is not dramatic, but is generally worthwhile.

For the other six populations, the bias in coverage is seen to be caused by strong biases in point estimates or their standard errors, as reflected by $\delta$ and se-$r$($\overline{\text{abs}(\delta)} = 0.64$ and $\overline{\text{se-}r} = 0.63$). These biases are a direct result of model-

194     5. Monte Carlo and Example-Based Insights

TABLE 5.23. Confidence interval coverage from the Monte Carlo simulations; coverage is for $\beta_i$ (hence Parm. index $i$), and is from all selected models containing $\beta_i$: the number of such selected models is denoted by "Freq.;" coverage is for the traditional conditional interval and the interval based on the model-averaged estimator $\hat{\bar{\beta}}_i$; see text for explanation of $\delta$ and se-$r$.

| Case | $\rho$ | $n$ | Parm. index | Freq. | Coverage cond. | Coverage MA | Bias/se $\delta$ | Se-ratio se-$r$ |
|---|---|---|---|---|---|---|---|---|
| 1 | 0 | 20 | 1 | 8864 | 0.932 | 0.948 | 0.22 | 1.07 |
| 1 | 0 | 50 | 1 | 9994 | 0.943 | 0.948 | −0.01 | 0.98 |
| 1 | 0 | 50 | 4 | 4337 | 0.934 | 0.934 | 0.90 | 1.43 |
| 1 | 0 | 100 | 1 | 10000 | 0.948 | 0.950 | 0.01 | 0.98 |
| 2 | 0.2 | 20 | 1 | 8309 | 0.923 | 0.947 | 0.42 | 1.12 |
| 2 | 0.2 | 20 | 3 | 2106 | 0.694 | 0.755 | 1.37 | 0.82 |
| 2 | 0.2 | 50 | 1 | 9970 | 0.945 | 0.955 | 0.07 | 1.02 |
| 2 | 0.2 | 100 | 1 | 10000 | 0.944 | 0.950 | 0.02 | 0.99 |
| 3 | 0.4 | 20 | 1 | 8026 | 0.908 | 0.945 | 0.41 | 1.13 |
| 3 | 0.4 | 50 | 1 | 9912 | 0.918 | 0.947 | 0.03 | 1.00 |
| 3 | 0.4 | 100 | 1 | 9999 | 0.932 | 0.944 | 0.02 | 0.98 |
| 4 | 0.6 | 20 | 1 | 8178 | 0.905 | 0.957 | 0.27 | 1.13 |
| 4 | 0.6 | 50 | 1 | 9953 | 0.905 | 0.942 | 0.00 | 0.97 |
| 4 | 0.6 | 100 | 1 | 10000 | 0.937 | 0.945 | −0.03 | 0.98 |
| 5 | 0.8 | 20 | 1 | 9175 | 0.858 | 0.937 | 0.17 | 1.03 |
| 5 | 0.8 | 50 | 1 | 10000 | 0.918 | 0.945 | −0.02 | 0.99 |
| 5 | 0.8 | 50 | 3 | 3572 | 0.913 | 0.925 | −0.93 | 1.22 |
| 5 | 0.8 | 100 | 1 | 10000 | 0.930 | 0.946 | −0.07 | 0.97 |
| 5 | 0.8 | 100 | 3 | 6055 | 0.957 | 0.960 | −0.61 | 1.41 |
| 6 | 0 | 20 | 1 | 1220 | 0.396 | 0.495 | 0.13 | 0.45 |
| 6 | 0 | 50 | 1 | 1450 | 0.643 | 0.677 | 0.06 | 0.51 |
| 6 | 0 | 100 | 1 | 1501 | 0.658 | 0.676 | −0.06 | 0.51 |
| 7 | 0 | 20 | 1 | 2014 | 0.726 | 0.759 | 1.19 | 0.73 |
| 7 | 0 | 50 | 1 | 3624 | 0.904 | 0.913 | 1.01 | 1.32 |
| 7 | 0 | 100 | 1 | 5594 | 0.952 | 0.955 | 0.73 | 1.47 |
| 7 | 0 | 200 | 1 | 7911 | 0.967 | 0.969 | 0.37 | 1.31 |
| 7 | 0 | 200 | 4 | 2225 | 0.785 | 0.796 | 1.00 | 0.78 |
| 7 | 0 | 500 | 1 | 9828 | 0.969 | 0.969 | 0.06 | 1.06 |
| 7 | 0 | 1000 | 1 | 9998 | 0.956 | 0.956 | 0.02 | 1.01 |

selection, i.e., they are model-selection bias. In those cases where the selected model is not the expected $AIC_c$ best (i.e., expected K-L best) model, rather it includes a variable $x_s$ that is rarely included in the selected model, then inference on $\beta_s$ (an unimportant variable) can be very misleading because of resultant model-selection bias for $\hat{\beta}_s$. Fortunately, this scenario is uncommon, almost by definition, as it is a case of $x_s$ being commonly excluded from the selected model. Also, even then inference on an important parameter was generally sound in these simulations and others we have done even if inference on an unimportant $\beta_s$ was slightly (but not strongly) misleading when its unconditional confidence interval was considered.

## 5.3.5  Linear Regression and Variable Selection

We present here some thoughts on aspects of model selection with specific reference to linear regression and so-called variable selection. This is arguably the most used and misused application area of model-selection. In particular, every conceivable type of model-(variable-) selection method seems to have been tried in the context of having $m$ predictors and using linear multiple regression models (see, for example, Hocking 1976, Draper and Smith 1981, Henderson and Velleman 1981, Breiman and Freedman 1983, Copas 1983, Miller 1990, Hjorth 1994, Breiman 1995, Tibshirani 1996, Raftery et al. 1997). However, almost always the statistical literature approaches the problem as if it is only a matter of "just-the-numbers" data analysis methodology. Firstly, in fact there is always a subject-matter scientific context, and a possible limitation of sample size, that must be brought into the problem, and so doing will make an enormous difference as compared to any naive model-selection approach that does not consider context, prior knowledge, and sample size.

Secondly, there is always a goal of either (1) selecting a best model (this should include ranking competitor models) because one seeks understanding of the relationships (presumably causal) between $\underline{x}$ (independent variables) and $y$, or (2) prediction of $E(y \mid \underline{x})$ at values of $\underline{x}$ (predictors) not in the sample (prediction of $E(y \mid \underline{x})$ for $\underline{x}$ in the sample can be considered just parameter estimation). These goals really are different. That is, if there is a lot of model-selection uncertainty, then selecting the best model under goal (1) and using it for goal (2), prediction, is not optimal.

We recommend that prior to any data analysis full consideration must be given to how the problem (i.e., set of models) should be structured and restricted. This means dropping variables that cannot reasonably be related to $y$ for prediction, or cannot reasonably be causally related at detectable effect levels given the sample size. From the literature and our experience, investigators are far too reluctant to drop clearly irrelevant variables and otherwise apply *a priori* considerations based on logic and theory. This is the "measure everything that is easy to measure and let the computer sort it out" syndrome, and it does not work. Even a good exploratory analysis needs input of investigator insights to reach useful results. In this regard we quote Freedman et al. (1988):

> A major part of the problem in applications is the curse of dimensionality: there is a lot of room in high-dimensional space. That is why investigators need model specifications tightly derived from good theory. We cannot expect statistical modeling to perform at all well in an environment consisting of large, complicated data sets and weak theory. Unfortunately, at present that describes many applications.

An important *a priori* aspect is to consider reducing the number of independent variables by functionally combining them into a smaller set of more useful variables. This may be as simple as computing, by observation, an average of some of the predictor variables (such as in the GPA example of this section), and then that average replaces all the variables that went into it. Use of such derived variables is common (for example, wind chill factor, relative humidity, density as mass per unit volume in physics or animals per resource in ecology, rates of all sorts, and so forth). Consider also any bounds on $y$. For example, often college GPA is bounded on 0 to 4; hence, we do not want our model to be able to make a prediction of 4.2. We could model GPA/4 using a logistic link function and rescale predictions by 4.

As a rule of thumb, and this is liberal, the maximum number of structural parameters to allow in a regression (or other univariate) model should be $n/10$. It is not possible to reliably estimate anything like $n/2$ (or $n/3$) parameters from "noisy" data. Mistakenly, models of such size are often fit to data, and may be selected based on an invalid criterion such as minimum residual variance, or an inappropriate to the situation criterion like adjusted $R^2$ or even AIC (AIC$_c$ correctly adjusts for either small sample size or large $K$).

To illustrate some of this thinking we consider another somewhat classic example of variable selection (Hocking 1976; see also Hendersen and Velleman 1981 for the actual data; we did not read this latter paper before the thinking below): automobile gas mileage ($y$) as MPG (miles per gallon) versus $m = 10$ independent variables (there is clearly causation involved here; when this is not so we use the term predictor variable). Note that $y$ is already a derived variable. The 10 $x_i$ are:

| | |
|---|---|
| 1 | Engine shape (straight or V) |
| 2 | Number of cylinders (4, 6, or 8) |
| 3 | Transmission type (manual or automatic) |
| 4 | Number of transmission speeds |
| 5 | Engine size (cubic inches) |
| 6 | Horsepower |
| 7 | Number of carburetor barrels |
| 8 | Final drive ratio |
| 9 | Weight |
| 10 | Quarter mile time |

The data arose from testing 32 ($= n$) different types of automobiles under standardized conditions. We independently generated *a priori* considerations (it would be better to get an automotive engineer involved). We did not first look at the dependent variable (either as $y$ alone, or as $y$ vs. the $x_i$). In general, given that one has decided that the analysis will only be to look at models for a response variable $y$ based on $x_1, \ldots, x_m$ as predictor variables (upon which the models are all conditional), one next reduces the number of such variables as much as possible by logical and subject-matter considerations. Given the resultant reduced set of predictors, we might then recommend looking at the correlations of the independent variables to be assured that there are no remaining pairs having an extremely high pairwise correlation.

A more comprehensive examination would be a principal components eigenvalue evaluation of the design matrix, $X$. Such results are given in Hocking (1976) for the full set of 10 predictors (but the data are not given there). The eigenvalue analysis suggests that a two-variable model might fit as much as 90% of the variation in the MPG response variable (because the first two eigenvalues add to about 90% of the total of all 10 eigenvalues). This much explained variation is often all we can hope for without over-fitting the data at sample size 32.

The $x_i$ are highly intercorrelated in this observational study because of car design: big cars have bigger engines; are more likely to have 8, not 4 or 6, cylinders; are therefore more likely to have a V-engine design; and so forth. For such observational studies if the issue of interest is causality, there are substantial inference problems (see Draper and Smith 1981, page 295, for some sage cautionary comments on such problems with observational data).

One of us (KPB, who is automotively challenged) proceeded as follows. Because $n = 32$, do not include more than three structural parameters. Gas mileage is strongly dependent on car weight, so always include $x_9$. Given that an intercept will be used here, this leaves room for only one more variable. As a first thought, then, consider the nine models

$$y = \beta_0 + \beta_1 x_9 + \beta_i x_i + \epsilon, \qquad i \neq 9,$$

plus $y = \beta_0 + \beta_1 x_9 + \epsilon$. However, bearing in mind the intercorrelated nature of these variables, consider dropping some on *a priori* grounds. Do not drop any $x_i$ based just on a high correlation unless it is extreme, such as $|r| \geq 0.95$; as then there is a variable redundancy problem (near collinearity). Do eliminate (near) collinearity problems; and do eliminate variables based on knowledge, reasoning and experience.

As a type of thought experiment (because the data did not arise from an experiment) consider whether engine shape ($x_1$) is causally related to MPG. Do we really think that if all car features were held fixed except whether the engine is a straight or V8 that there would be any effect on MPG? Probably either not at all, or at a level we will never care about and could not detect except with an experiment and a huge sample size. Conclusion: Drop variable $x_1$ (we surmise that it was recorded because it is easy to determine—this is not justification for including a variable). Recommendation: Use thought experiments in conjunction with observational studies.

The same reasoning leads KPB to drop variable 2: number of cylinders. Again the thought is that if all else (horsepower, total cylinder displacement, etc.) were fixed, would just number of cylinders (as 4, 6, or 8) alone affect MPG? And again, no; at least not in these data. Variables 2 through 8 would be retained on fundamental grounds. Quarter mile time is in effect a complex derived variable; it might predict MPG well, but it is not causally related. Instead, variable 10 might itself be well predicted based on variables $x_3$–$x_9$. Conclusion: drop $x_{10}$. Thus KPB would consider only models with variable $x_9$ always included and at most one of variables 3–8 (seven models). This is very different from an all-subsets selection over $10^m = 1{,}024$ models, and this sort of thinking can, and should, always be brought to bear on a variable selection problem.

One advantage of doing thinking such as this is that it focuses one's attention on the problem and basic issues. In this case the realization arose (still for KPB) that horsepower itself is a derived variable but one that can be engineered. To some extent horsepower might replace variables 5 and 7, and might be more important than variables 3, 4, 6, and 8. Thus we have the *a priori* hypothesis that the best two-variable model might be based on $x_9$ and $x_6$. If this model was not best, but nearly tied for best with a less-interpretable model, this *a priori* thinking would justify objective selection of these two variables as most important.

Another issue is the suitability of the linear model form. Because MPG is bounded below by zero but weight can be unbounded, a linear model could predict negative MPG. Surely, over a big enough weight range the relationship is curvilinear, such as $E(y \mid x_9) = \beta_0 e^{-\beta_1 x_9}$ or $E(1/y \mid x_9) = \beta_0 + \beta_1 x_9$. We use MPG only by convention, hence *a priori* KPB would fit all models as linear but based on an inverse link function to MPG. One could just fit $1/y$ to several linear models and select a best model. However, $AIC_c$ is then not comparable from models fit to $y$ vs. models fit to $1/y$.

Less time was afforded to this exercise by DRA (who is much more automotively knowledgeable), who independently put forth two *a priori* models. Both include weight ($x_9$);

$$y = \beta_0 + \beta_1(x_9)^2 + \beta_2 z + \epsilon$$

and

$$y = \beta_0 + \beta_1(x_9)^2 + \beta_2 x_{10} + \epsilon.$$

The variable $z$ is a derived variable meant to reflect the combined effect of several variables on MPG:

$$z = \frac{x_2 \times x_5}{x_6}.$$

Similarly, $x_{10}$ is used here as a predictor that summarizes many features of the ability of the car to consume fuel.

Considerations like these based on reasoning and theory must be thought about before data analysis if reliable uncertainty bounds are to be placed on an inference made after model-selection. One can always do data-dependent exploratory analyses after the *a priori* analysis. We just recommend separating the two processes because results of exploratory analyses are not defensible as reliable inferences in the sense that the data cannot both suggest the question (the model) and then reliably affirm the inferential uncertainty of the answer (the same model). There is a saying from the USA Western frontier days: "Shoot first, ask questions later." This strategy often precludes obtaining desired information. Similarly, "compute first, then create models" (or "compute first, think later") is also not a strategy for making reliable inference. The result can be a model that describes the data very well (because it over-fits the data), but is a poor model as an inference to independent data from the same generating process. Over-fitting, if it occurs, cannot be diagnosed with that same data set that has been over-fit.

## 5.3.6 Discussion

A variety of comments and opinions are given here, some because they do not fit well elsewhere in Section 5.3. There is no particular order to the following comments and opinions.

The Monte Carlo simulations of Sections 5.3.2 and 5.3.4 might seem to violate our general philosophy that the actual data-generating model, $M_T$, would (should) in reality be more general than the global model, $M_G$, used as the basis for data analysis (an expanded vector of predictors, $\underline{x}_T$, would apply under the true generating model; the global model does not use all of these predictors). This is only partly true. The part that is not true is thinking that because we generated the data under the global model, no more general model could actually apply. In fact, the residual variation of the global regression model, $\sigma^2_{y|x}$, is a confounding of average (with respect to model $M_T$) model structural variations arising from the differences $E(y \mid x, M_G) - E(y \mid \underline{x}_T, M_T)$, plus the "true" unexplained residual variation, $\sigma^2_\epsilon$ (it might be 0), under model $M_T$. Thus almost all aspects of a conceptually more general data-generating model are swept into $\sigma^2_{y|x}$ of the global model. Hence, it is more economical to simply generate data under an assumed global model.

One way in which this lacks generality is that the numerical values of the components of the true parameter vector, $\beta_T$, that are in the lower dimensional global $\underline{\beta}$ may not exactly equal their counterpart component values in $\underline{\beta}_T$. This would affect confidence interval coverage, which should be relative to the appropriate components of $\beta_T$, not to $\underline{\beta}$. This seems like a small concern in initial Monte Carlo studies intended to explore basic model-selection issues.

The more important lack of generality relates to how we conceive the asymptotic sequence of models as sample size increases. Classical theory holds the model or set of models fixed, independent of sample size. This is not in accord with reality, because as sample size grows we will include more structure in the data and in our models (e.g., in the GPA example, effects of year, high-school type, university attended, major, student age, and so forth). For that reason we should simply have a larger global generating model than the one we used in simulations here, and include more factors so that the size of the selected model can grow without the arbitrary bound of a global model with only four predictors. In a sense the issue becomes one of not strongly "bumping" up against a bound (i.e., large $\pi_i$ for the generating model) as sample size increases, because this feature of data analysis is often unrealistic for observational studies. This problem is solved simple by having a sufficiently general global generating model, and it is then still acceptable (but not required) to have that generating model as also the global model for data analysis.

It is well known that selecting a best model from a set of regression models can lead to important biases in parameter estimates and associated estimated standard errors (see e.g., Miller 1990, Hjorth 1994). The estimated residual variance $\hat{\sigma}^2_{y|x}$ is especially susceptible to being biased low due to the process of selecting a good model, because all selection criteria involve, to some extent, seeking a fitted model with a relatively small residual sum of squares. Use of $AIC_c$ will not entirely protect one from this possible bias, but it helps (see, e.g., Table 5.18; the relative bias in $\hat{\sigma}^2_{y|x}$ is not high for the K-L good models, and this example is for a small sample size). The bias in $\hat{\sigma}^2_{y|x}$ is worse the more infrequent a model is selected because for such poor models, they are selected only when they fit a sample unusually well. In the data analysis phase of a study (hence, sample size is then a given) the best way to avoid serious bias in $\hat{\sigma}^2_{y|x}$ is to keep the candidate set of models small.

When the predictors are intercorrelated and model-selection is used, such selection tends to induce a bias in the estimators of regression coefficients of selected predictors. The less important a predictor $x_i$ is, the less likely it is to be selected, and then when selection occurs, both of the associated estimators $\hat{\beta}_i$ or $\hat{\bar{\beta}}_i$, conditional on the model, tend to be biased away from zero. That is, let $E(\hat{\beta}_i \mid M_r$ always) denote the expected value of $\hat{\beta}_i$ under model $M_r$ when that model is always fit to the data. Let $E(\hat{\beta}_i \mid M_r$ selected) denote the conditional expectation of $\hat{\beta}_i$ when model $M_r$ is selected, as by $AIC_c$. If $E(\hat{\beta}_i \mid M_r$ always) $> 0$, then we usually find that $E(\hat{\beta}_i \mid M_r$ selected) $> E(\hat{\beta}_i \mid M_r$ always); whereas if $E(\hat{\beta}_i \mid M_r$ always) $< 0$; then we find that $E(\hat{\beta}_i \mid M_r$ selected) $< E(\hat{\beta}_i \mid M_r$ always). The strength of the bias depends mostly on the importance of the predictor, as measured by its overall selection probability (and that probability depends on sample size and goes to 1 as $n$ goes to infinity if $|\beta_i| > 0$).

Consider Table 5.17, which gives Monte Carlo results for the simulation mimicking the GPA data. Using that information, and extending it to $\beta_4$, we computed the percent relative bias of $\hat{\beta}_i$, PRBias($\hat{\beta}_i$) below, attributed solely to model-selection. The reference value for computing bias is the true value of $\beta_i$ from the data-generating model, not the parameter value $\beta_{i,r}$ that applies conditionally to model $M_r$ when model $M_r$ is always fit to the data. The relative frequency of occurrence of the given parameter (i.e., predictor) in any selected model is denoted by Pr$\{x_i\}$:

| $i$ | Pr$\{x_i\}$ | PRBias($\hat{\beta}_i$) |
|---|---|---|
| 1 | 0.954 | 14% |
| 2 | 0.751 | 25% |
| 3 | 0.650 | 31% |
| 4 | 0.162 | 267% |

A 30% relative bias in conventional estimators due to model-selection should be of concern (let alone 267%, but $x_4$ is not in the K-L best model).

We have looked at this issue for other models and sample sizes for all-subsets selection, and it is quite clear that this aspect of model-selection bias in estimators is, as above, strongly related to the importance of the predictor: Model-selection bias is less for a predictor always included in the selected model, but it can be very strong for predictors rarely selected. The more predictors one considers, the more likely it is that many are unimportant (especially in the presence of better predictors they correlate with) and the more likely it is that a few of those unimportant predictors will end up in the selected model. When that happens, all

## 5.3 Examples and Ideas Illustrated with Linear Regression

the model-selection biases operate in a direction to make you think that the selected variables are important ("significant" in hypothesis-testing terms). The best way to reduce this risk of misleading results is to have a small list of carefully considered candidate variables. (To guarantee wrongly selecting one or more unimportant variables; just have a large list of poorly conceived variables and a small sample size; see, e.g., Freedman 1983, Rexstad et al. 1988, 1990).

Two undesirable, but mutually exclusive, properties of model-selection strategies particularly relevant to all-subsets regression for observational data are worth noting here: over-fitting the data or over-fitting the model. If your strategy is to always fit and use the global model, you will probably over-fit the model (i.e., include unnecessary variables). This approach to analysis will avoid subjectively tailoring the model to the data, but you probably will greatly inflate standard errors of all the $\hat{\beta}_i$. This loss of precision can be so bad that all the estimates are worthless. Thus, usually one is forced into some sort of model-selection with multivariable observational data (it should be firstly by *a priori* considerations).

If you use a subjective selection procedure of first fitting a model and then examining the results (e.g., residual plots, r-squares, leverages, effect of transformations of variables) in search of a better model based on some vague synthetic criterion of your own choosing, you probably will over-fit the data. Thus, you will include in model structure what are really stochastic aspects of the data, thereby biasing $\hat{\sigma}^2_{y|x}$ quite low and as a result inferring "noise" as real structure. The resultant model may become more of a description of the particular data at hand than a valid inference from those data. All samples have their nearly unique peculiarities as well as their main features that would show up in all, or most, samples you might get for the inference situation at hand. Inference is about correctly identifying the repeatable features of samples. When you over-fit the data, you mistakenly include in model structure uncommon data features that would not be found in most such samples that might arise.

When you have a large number of models for a much smaller number of variables (like $R = 1,024$, $m = 10$, all-subsets) and all those models are fit and considered for selection, you run a high risk that some models will over-fit the data. The use of $AIC_c$ much reduces this risk (because *heuristically*, it looks at model fit penalized by a function of model size, $K$, and sample size, $n$), but does not eliminate it for all-subsets selection: Some degree of selection bias remains. For this reason, and the instability of all-subsets selection, it is critical to properly evaluate model-selection uncertainty under all-subsets regression and base inferences on more than just the selected best model.

For the few cases where we applied the bootstrap to evaluate aspects of uncertainty for all-subsets model selection we found the results problematic. That is, either the bootstrap failed (the cement data, Section 4.3.1), or it produced some peculiar results that the theoretical approach did not produce (GPA example, Section 5.3.1). Two small studies that evaluated the bootstrap method of assessing aspects of model reliability after model-selection in regression expressed pessimism that the bootstrap would always be a reliable method for the task (Freedman et al. 1998, Dijkstra and Veldkamp 1998). The only opinion we can now offer is that for even moderately high-dimensional problems (say $m \geq 7$, hence $R > 100$) one should not blithely think that the bootstrap will be a reliable way to assess model-selection uncertainty for all-subsets selection; the method needs more study.

In fairness, it can also be said that $AIC_c$ and associated methods presented here need more evaluation for their performance under all-subsets selection. However, a more basic issue is whether or not ever to do all-subsets selection (especially when the number of predictors is large) when this means selecting a single best model and ignoring all other models. The practical problems are instability of what model is selected (cf. Brieman 1996) and substantial model-selection biases. Model instability arises when all model-selection probabilities (i.e., the $\pi_i$) are low. For large $R$ (hence if $m$ is at all large, even $m = 7$) even the expected K-L best model might have selection probability less than 0.1, so the set of supposed important regressor variables, as judged by the selected best model, can vary dramatically over samples.

At a fundamental level the question of variable selection ought really to be a question of what is the strength of evidence in the data for the importance of each predictor variable. If the problem is thought of this way, then strict model selection as such is an illegitimate discretization of what ought to be a problem of estimating continuous parameters (the regression coefficients of the global model). The flaw in using model selection is then just like the flaw of using hypothesis testing that makes a problem a reject-or-not dichotomy when it ought to be approached as an evaluation of strength of evidence (the use of P-values rather than strict reject-or-not procedures is also not acceptable; it is a flawed methodology; see, e.g., Harlow et al. 1997). We strongly recommend against doing all-subsets selection only for the purpose of identifying a single "best" model: Promoting this practice is not good science and is a failing of statistical science.

We believe that the only defensible reason for fitting all-subsets of regression models should be to obtain the full set of Akaike weights, and then inferences are based on the full set of models as mediated by their associated Akaike weights. The selected best model constitutes only one subset (of $R$) of the predictor variables. Unless the Akaike weight for that best model is very high (say $w_k \geq 0.9$), we maintain that it is totally misleading to infer that one has found *the* important predictors, and that the predictors not selected are unimportant. As noted above, it is not properly a yes-or-no issue as regards importance of a predictor variable; rather, it is a matter of quantifying predictor importance on a continuous scale, say 0 to 1. Also, this importance value is only with respect to prediction; reliable causal inferences cannot be made from just the data alone when those data are from an observational study.

We are led to believe that the only legitimate application of all-subsets model fitting with purely observational data (and then only after serious reduction of the number of predictors, as discussed in Section 5.3.5) is prediction. For prediction in this context we recommend model averaging. That is, a prediction is made with each model, and the Akaike weights are used to compute a weighted average of these predictions. We do not know who invented model averaging, but we have seen it only in the Bayesian literature, using of course Bayesian-based model weights (see, e.g., Madigan and Raftery 1994, Draper 1995, Hoeting 1996, Raftery et al. 1997).

As noted by Brieman (1996), selection of a best model in all-subsets fitting is inherently unstable in its outcome. The solution proposed by Brieman to produce stabilized inferences is a type of model averaging: Generate many perturbed sets of the data (such as bootstrap samples create), select the best model in each case, and produce some sort of averaged inference. Our solution, for stabilized inference, is a sort of reverse strategy: Keep the one actual data set as is, but find for each fitted model its Akaike weight; then compute inferences as some form of weighted average over all the models.

Interest in regression parameter estimates in conjunction with all-subsets model fitting will no doubt continue. Perhaps there is a legitimate need for this (we are not convinced). Motivated by this need and our recommendation to use model-averaged predictions, we decided to analytically relate such prediction to parameter estimation. This led to some surprising new ideas and issues that we will only outline here. These are issues needing additional research.

The model-averaged prediction (estimate) of $E(y \mid \underline{x})$ is

$$\hat{\bar{E}}(y \mid \underline{x}) = \sum_{r=1}^{R} w_r \hat{E}(y \mid \underline{x}, M_r).$$

We define an indicator function for when a predictor is in a model:

$$I_i(M_r) = \begin{cases} 1 & \text{if predictor } x_i \text{ is in model } M_r, \\ 0 & \text{otherwise.} \end{cases}$$

For model $M_r$ the value of $\beta_i$ is denoted here by $\beta_{i,r}$. A model-averaged parameter estimator is

$$\hat{\bar{\beta}}_i = \frac{\sum_{r=1}^{R} w_r I_i(M_r) \hat{\beta}_{i,r}}{\sum_{r=1}^{R} w_r I_i(M_r)} = \frac{\sum_{r=1}^{R} w_r I_i(M_r) \hat{\beta}_{i,r}}{w_+(i)}, \tag{5.7}$$

$$w_+(i) = \sum_{r=1}^{R} w_r I_i(M_r).$$

An alternative to the above conditional parameter estimator is the full model-averaged estimator over all models wherein if predictor $x_i$ is not in model $M_r$, we simply set $\hat{\beta}_{i,r} = 0$. Thus a new estimator, denoted by $\tilde{\beta}_i$, is

$$\tilde{\beta}_i = w_+(i) \hat{\bar{\beta}}_i.$$

This is just $\hat{\bar{\beta}}_i$ shrunk toward zero by the amount $(1 - w_+(i))\hat{\bar{\beta}}_i$. Moreover, we found, based on empirical results, that we could also consider this shrinkage estimator as

$$\tilde{\beta}_i \equiv \hat{\bar{\beta}}_i - (1 - w_+(i))\hat{\bar{\beta}}_i = \hat{\bar{\beta}}_i - \widehat{\text{model-selection bias}}; \tag{5.8}$$

that is, formula (5.8) is our original model-averaged estimator adjusted for (estimated) model-selection bias. Certainly, the term $(1 - w_+(i))\hat{\bar{\beta}}_i$ is not an unbiased estimator of model-selection bias, but it is a usable estimator of that bias.

Then we realized that $\tilde{\bar{\beta}}_i$ is of fundamental importance because the model-averaged prediction can be expressed as

$$\hat{\bar{E}}(y \mid \underline{x}) = \hat{\bar{\beta}}_0 + \sum_{i=1}^{m} w_+(i)\hat{\bar{\beta}}_i x_i,$$

$$= \hat{\bar{\beta}}_0 + \sum_{i=1}^{m} \tilde{\bar{\beta}}_i x_i.$$

If we accept $\hat{\bar{\beta}}_i$ as the appropriate naive estimate of $\beta_i$ given model-selection, then heuristically, the above suggests that prediction is improved by shrinkage toward zero of each parameter's estimate by a measure of that parameter's unimportance ($= 1 - w_+(i)$). The value of shrinkage is well established in statistics (see, e.g., Copas 1983, Tibshirani 1996), hence this seems like a line of thought worth pursuing.

Thus we have compelling reasons to want to replace, at least in all-subsets regression, the conditional estimator of formula (5.7) by the unconditional estimator of formula (5.8). This would allow us to ignore the issue of what is a best model, and simply make inferences from the full set of models as regards any parameter or prediction. In our limited Monte Carlo evaluation of this idea we have found that $\tilde{\bar{\beta}}_i$ is less biased by model-selection than is $\hat{\bar{\beta}}_i$.

An unresolved matter is the correct sampling variance for $\tilde{\bar{\beta}}_i$ and a confidence interval based on $\tilde{\bar{\beta}}_i$. The expression $\widehat{se}(\tilde{\bar{\beta}}_i) = w_+(i)\,\widehat{se}(\hat{\bar{\beta}}_i)$ is unacceptably poor. We examined use of $\widehat{se}(\tilde{\bar{\beta}}_i) = \widehat{se}(\hat{\bar{\beta}}_i)$ and found the results to be poor. We examined use of formula (4.10) to get $\widehat{se}(\tilde{\bar{\beta}}_i)$ by using $\hat{\beta}_i \equiv 0$ and $\widehat{se}(\hat{\beta}_i \mid M_r) \equiv 0$ when variable $x_i$ is not in model $M_r$; the result was a biased low estimator. However, an ad hoc adjustment showed promise as an estimator of $\widehat{se}(\tilde{\bar{\beta}}_i)$:

$$\sum_{r=1}^{R} w_i \sqrt{\left(\widehat{se}(\hat{\beta}_i \mid M_r)\right)^2 + \left(\hat{\beta}_i - \tilde{\bar{\beta}}_i\right)^2} + |\tilde{\bar{\beta}}_i|(1 - w_+(i)).$$

This formula did not lead to the best confidence interval coverage. For coverage the best results came from using $\widehat{se}(\tilde{\bar{\beta}}_i) = (2 - w_+(i))\,\widehat{se}(\hat{\bar{\beta}}_i)$, but neither this nor the above formula constitutes an ideal approach. So here is an open problem that seems quite promising because it will allow inference about specific regression coefficients based on all models rather than forcing us to tie our inference in any way to the selected best model.

We again consider the Monte Carlo evaluation of the GPA data example. Results in Table 5.17 were extended to $\beta_4$ and $\tilde{\bar{\beta}}_i$ to compare expected values. We need to be clear on what was done in this new Monte Carlo example. If predictor variable

$x_i$ was in the selected $\text{AIC}_c$ best model, then we estimated $\beta_i$ by just $\hat{\beta}_i$ from that selected best model. Otherwise, no inference was made about $\beta_i$; this corresponds to classical model-selection practice. The average of these estimates is $E(\hat{\beta}_i)$; the relevant sample size is $10{,}000 \times \pi_+(i)$ for $\pi_+(i)$ the probability that $x_i$ is in the selected best model. In contrast, $\bar{\bar{\beta}}_i$ was computed for every rep, giving $E(\bar{\bar{\beta}}_i)$. The results are below; $\bar{\bar{\beta}}_i$ has the better performance:

| $i$ | $\pi_+(i)$ | $\beta_i$ | $E(\hat{\beta}_i)$ | $E(\bar{\bar{\beta}}_i)$ |
|---|---|---|---|---|
| 1 | 0.95 | 0.00201 | 0.00229 | 0.00217 |
| 2 | 0.75 | 0.00125 | 0.00156 | 0.00112 |
| 3 | 0.65 | 0.18945 | 0.248 | 0.1577 |
| 4 | 0.16 | 0.08752 | 0.321 | 0.0588 |

## 5.4 Estimation of Density from Line Transect Sampling

### 5.4.1 Density Estimation Background

Animal inventory and monitoring programs often focus on the estimation of population density (i.e., number per unit area). Buckland et al. (1993) provide the theory and application for field sampling and analysis methods using line transects. We will illustrate several aspects of statistical inference in the face of model-selection uncertainty using line transect data collected by Southwell (1994) on the eastern grey kangaroo (*Macropus giganteus*) at Wallaby Creek, in New South Wales, Australia. The program DISTANCE (Laake et al. 1994) was written for the analysis of line transect data, uses AIC in model-selection, and has an option for bootstrapping the sample (see Buckland et al. 1997 for a similar example). Thus, line transect sampling and the program DISTANCE will be used to provide some deeper insights concerning model-selection uncertainty and will serve as another comprehensive example.

In line transect sampling, the estimator of density ($D$) is

$$\hat{D} = \frac{n}{2wL\hat{P}},$$

where $n$ (= 196 in this example) is the number of objects detected on $r$ (= 78) transects of total length $L$ (= 88.8 km) and width $w$ (= 263 m). The unconditional probability of detection, within the strips examined, is $P$. The focus of the estimation of animal density is on the probability of detection, and this is defined as

$$P = \frac{\int_0^w g(x)dx}{w},$$

where $g(x)$ is the detection function (i.e., the probability of detection given that an animal is at perpendicular distance $x$ from the line). The detection function $g(x)$

is a confounding of animal density, environmental factors affecting detection, differences in detectability by individual observer, and differences in detectability by the animals being surveyed. The detection function can be estimated from perpendicular distances taken from the transect line to each object detected. Assumptions required in line transect sampling and other details and theory are given in Buckland et al. (1993).

Substituting the expression for $P$ into the estimator of $D$, and canceling out the $w$ and $1/w$, gives

$$\hat{D} = \frac{n}{2L \int_0^w \hat{g}(x)dx}.$$

Thus, the essence of data analysis here is to find a good approximating model for $g(x)$, the detection function. Buckland et al. (1993) recommend models of the general form

$$g(x) = key(x)[1 + series(x)].$$

The key function alone may be adequate for modeling $g(x)$, especially if sample size is small or the distance data are easily described by a simple model. Often, one or more adjustment terms must be added to achieve an acceptable model for the data. For the purposes of this example we chose to use four specific models of the above form. Each of these models provides a reasonable, but not unique, basis for data analysis in this example.

### 5.4.2  Line Transect Sampling of Kangaroos at Wallaby Creek

Eastern grey kangaroos often occur in family groups; thus an estimate of total animal density is the product of the estimated number of groups and the average group size. In this example, we will focus only on estimating the number of groups of this species of kangaroo. We define the set of candidate models for this example in Table 5.24. The analysis theory of line transect sampling has been the subject of a great deal of work since about 1976; thus the set of candidate models is relatively well based in this example. The program DISTANCE (Laake et al. 1994) was used to compute MLEs of the model parameters in the key functions ($\sigma$, $a$, or $b$, in Table 5.24) and the series expansions (the $a_j$ in Table 5.24), and choose, using AIC, the best model among the four.

### 5.4.3  Analysis of Wallaby Creek Data

The results of the initial analysis of these data are given in Table 5.25. Model 1 was selected using AIC and provides an estimated density of 9.88 groups per km$^2$ (conditional se $= 1.00$ and conditional cv $= 10.12\%$). All four models produce relatively similar point estimates of $D$ for these data; the largest $\Delta_i$ value was 3.86 (model 3). In this example, the estimated log-likelihood values, number of model parameters, and estimated density are similar across models; however, the

## 5.4 Estimation of Density from Line Transect Sampling

TABLE 5.24. Model set for the line transect example on kangaroo data; Hermite polynomials are special polynomials, defined recursively (see Buckland et al. 1993) for use with the half-normal key function.

| Model | $g(x)$ | K | Key function | Series expansion |
|---|---|---|---|---|
| 1 | $\left\{\dfrac{1}{w}\right\}\left\{1+\sum_{j=1}^{2} a_j \cos\left(\dfrac{j\pi x}{w}\right)\right\}$ | 2 | uniform | cosine |
| 2 | $\left\{e^{-x^2/(2\sigma^2)}\right\}\left\{1+\sum_{j=2}^{3} a_j H_{2j}\left(\dfrac{x}{\sigma}\right)\right\}$ | 3 | half-normal | Hermite polynomials |
| 3 | $\left\{1-e^{-(x/a)^{-b}}\right\}\left\{1+\sum_{j=2}^{3} a_j \left(\dfrac{x}{w}\right)^{2j}\right\}$ | 4 | hazard | simple polynomials |
| 4 | $\left\{e^{-x^2/(2\sigma^2)}\right\}\left\{1+\sum_{j=2}^{3} a_j \cos\left(\dfrac{j\pi x}{w}\right)\right\}$ | 3 | half-normal | cosine |

TABLE 5.25. Summary statistics for the line transect data on eastern grey kangaroos at Wallaby Creek, New South Wales, Australia (from Southwell 1994). The AIC-selected model is shown in bold; $\hat{D}_i$ is used to clarify that $\hat{D}$ is based on model $M_i$.

| Model | K | $\log(\mathcal{L}(\hat{\theta}))$ | AIC | $\Delta_i$ | $\exp(-\Delta_i/2)$ | $w_i$ | $\hat{D}_i$ | $\widehat{\text{var}}(\hat{D}_i \mid M_i)$ |
|---|---|---|---|---|---|---|---|---|
| **1** | **2** | **−1,021.725** | **2,047.449** | **0.000** | **1.000** | **0.499** | **9.88** | **0.999** |
| 2 | 3 | −1,021.546 | 2,049.092 | 1.643 | 0.440 | 0.220 | 10.43 | 1.613 |
| 3 | 4 | −1,021.654 | 2,051.307 | 3.858 | 0.145 | 0.072 | 10.83 | 2.761 |
| 4 | 3 | −1,021.600 | 2,049.192 | 1.743 | 0.418 | 0.209 | 10.46 | 1.935 |

estimated conditional sampling variances differ by a factor of almost 3. In this case, all four models contain the same parameter ($D$, or equivalently, $P$); thus, model averaging (formulae 4.6 and 4.10) should be considered. In either analysis, estimates of unconditional variances and associated confidence intervals should be used in making inferences about population density.

Using the Akaike weights, $w_i$, and the conditional sampling variances, $\widehat{\text{var}}(\hat{D}_i \mid M_i)$, for each model we computed the model-averaged estimate of density, $\hat{D}_a = 10.19$, and an estimate of its unconditional sampling variance, $\widehat{\text{var}}(\hat{D}_a) = 1.51$ (formula 4.10). Hence, the (unconditional) standard error of $\hat{D}_a$ is 1.23, and its cv is 12.06%. This unconditional cv is slightly higher than the cv of 10.12% conditional on the AIC-selected model. Inferences would be essentially the same here whether based on the model-averaged results, or based on the density estimate from the AIC-selected model but using for its variance the unconditional estimate of 1.51. In either case the achieved confidence interval coverage can be expected to be better than that based on the conditional standard error, and often very near the nominal level.

## 5.4.4 Bootstrap Analysis

The most obvious advantages of using Akaike weights as the basis to compute estimated unconditional sampling variances (and $\hat{D}_a$) are simplicity and speed. However, the bootstrap method also can be used to make unconditional inferences; the bootstrap is especially useful in complex situations where theory for analytical variances, even given the models, is lacking. Here we do have such analytical theory to compare to the bootstrap results.

We used the program DISTANCE to draw and analyze bootstrap samples (see Section 4.2.7), based on transects as the sampling unit (thus, there were 78 sampling units for the bootstrap), and thereby compute lower and upper confidence limits on $D$ as well as an estimated unconditional sampling variance for $\hat{D}$. We computed 10,000 bootstrap reps; we present first the results from all 10,000 reps. Then we examine the variability inherent here in a "mere" 1,000 bootstrap reps, based on the 10 sets of 1,000 reps each from cases 1–1,000, 1,001–2,000, etc.

The resultant density estimates, by model, and the model-selection frequencies are shown in Table 5.26. The mean of the estimates from the 10,000 bootstrap samples, 10.39, is quite close to the estimate based on the Akaike weights and formulae (4.6) and (4.10) (10.19, se $= 1.23$). Based on all 10,000 values of $\hat{D}^*$, the bootstrap estimate of the unconditional standard error of $\hat{D}$ (and of $\hat{D}_a$, and $\overline{\hat{D}^*}$) is 1.48. The model-selection relative frequencies from the bootstrap procedure are similar to, but do not exactly match, the Akaike weights (this is expected). However, the results are close for the favored model $M_1$: Akaike weight $w_1 = 0.50$ (Table 5.25) and from the bootstrap, $\hat{\pi}_1 = 0.45$ (Table 5.26).

## 5.4.5 Confidence interval on D

There are several options for setting a confidence interval on $D$ based on the estimated density and its estimated unconditional sampling variance. First, there is the usual procedure that assumes that the sampling distribution of the estimator is approximately normal. Hence, an approximate 95% confidence interval is based

TABLE 5.26. Summary of results from 10,000 bootstrap samples of the line transect data for eastern grey kangaroos at Wallaby Creek, New South Wales, Australia (from Southwell 1994): Empirical means of the $\hat{D}^*$ by selected model, and overall; standard error estimates; and selection frequencies.

| Model | $\overline{\hat{D}^*}$ | Standard error estimate | Selection frequency |
|---|---|---|---|
| 1 | 9.97 | 1.10 | 4,529 |
| 2 | 10.63 | 1.41 | 2,992 |
| 3 | 10.92 | 2.34 | 1,239 |
| 4 | 10.75 | 1.38 | 1,240 |
| All | 10.39 | 1.48 | 10,000 |

on

$$\hat{D} \pm 2\,\widehat{\text{se}}(\hat{D}),$$

where $\widehat{\text{se}}(\hat{D})$ ($= 1.23$ from Section 5.4.3) is the estimated (by theory) unconditional standard error. For this example, $9.88 \pm 2 \times 1.23$ gives the interval (7.42, 12.34).

The second method assumes (in this example) that the sampling distribution of $\hat{D}$ is log-normal. This is a plausibly better assumption than a normal sampling distribution for an estimator $\hat{\theta}$ in any context where the parameter $\theta$ is strictly $> 0$, and for fixed sample size, the $\text{cv}(\hat{\theta})$ tends to be independent of the actual value of $\theta$. Then one computes lower and upper bounds as (from Burnham et al. 1987),

$$D_L = \hat{D}/C \quad \text{and} \quad D_U = \hat{D}C,$$

where

$$C = \exp\left[t_{\alpha/2,df}\sqrt{\log[1 + (\text{cv}(\hat{D}))^2]}\,\right].$$

The confidence level is $1 - \alpha$; $t_{\alpha/2,df}$ is the upper $1 - \alpha/2$ percentile point of the $t$-distribution on df degrees of freedom. The degrees of freedom are those of the estimated $\text{var}(\hat{D})$. For an approximate 95% interval, if df are 30 or more, it suffices to use 2 in place of $t_{0.025,df}$.

For this example $\hat{D} = 9.88$ (from the AIC-selected model), with unconditional $\text{cv}(\hat{D}) = 0.124$, and thus $C = 1.28$. Therefore, if we base inference on the AIC-selected model, the approximate 95% confidence interval is 7.72 to 12.65. If we base inference on the model-averaged estimate of density (which increasingly strikes us as the preferred approach), then the results are $\hat{D}_a = 10.19$, again with standard error estimate 1.23, hence $C = 1.272$ and approximate 95% confidence interval 8.01 to 12.96. The bootstrap method would provide a point estimate of $D_a$; hence the corresponding confidence interval is more comparable to the analytical results for model averaging than to the results based on the selected single best model.

A third option is to use the bootstrap to produce a robust confidence interval, for example, based on the percentile method (Efron and Tibshirani 1993, Shao and Tu 1995). Here the 10,000 values of $\hat{D}_b^*$ generated in producing Table 5.26 are sorted in ascending order. Thus we have $\hat{D}_{(1)}^*, \ldots, \hat{D}_{(10000)}^*$. The 2.5 and 97.5 percentiles of the bootstrap sampling distribution are used as the 95% confidence interval endpoints on $D$: $\hat{D}_{(250)}^* \leq D \leq \hat{D}_{(9750)}^*$. The results here were $\hat{D}_{(250)}^* = 7.88 \leq D \leq 13.78 = \hat{D}_{(9750)}^*$.

The interval lower bounds from the three methods are more alike than the upper bounds. Results from the bootstrap in this example estimate more model-selection uncertainty than the results based on use of Akaike weights and formula (4.11); we rectify this matter in Section 5.4.6 below. In general, with either a good analytical approach or the bootstrap, achievement of nominal confidence interval coverage is likely if a good model is selected, if model-selection uncertainty has been incorporated into an estimate of the unconditional standard error, and if nonnormality has been accounted for.

It can be problematic to identify a correct unit of data as the basis for bootstrap resampling. Aside from this fundamental issue, the bootstrap is conceptually simple and can effectively handle model-selection uncertainty if computer software exists or can be written. The program DISTANCE allows bootstrapping in the context of distance sampling (Laake et al. 1994). In contrast, bootstrapping the experimental starling data (Section 3.4) would have been nearly impossible. Specialized software development for just this case would be prohibitive; and the computer time required might be measured in weeks. In these cases, we recommend use of Akaike weights to compute the estimate of an unconditional standard error, and then use of some suitable analytical confidence interval procedure.

### 5.4.6 Bootstrap Reps: 1,000 vs. 10,000

The $B = 10,000$ bootstrap samples were partitioned, in the order they were generated, into 10 sets of 1,000 reps per set, and estimates were computed on a per-set basis. The results are given in Table 5.27. Before discussing these results we need to establish our goals for precision of the bootstrap-based computation (estimate, actually) of quantities such as

$$\widehat{\text{se}}^*(\hat{D} \mid B) = \sqrt{\frac{\sum(\hat{D}_b^* - \overline{\hat{D}^*})^2}{B-1}}.$$

The true bootstrap estimate of the standard error of $\hat{D}$ (given the data) is actually the limit of $\widehat{\text{se}}^*(\hat{D} \mid B)$ as $B$ goes to infinity. We denote that limit simply by $\widehat{\text{se}}(\hat{D})$; however, this bootstrap standard error need not be exactly the same number as the analytically computed standard error of $\hat{D}$ (for which we use the same notation). For any value of $B$ we have $\widehat{\text{se}}^*(\hat{D} \mid B) = \widehat{\text{se}}(\hat{D}) + \epsilon$, where $E(\epsilon)$ goes to 0 quickly as $B$ gets large and $\text{var}(\epsilon) = \phi/B$ ($\phi$ unknown, but estimable). The goal in selecting $B$ should be to ensure that $\sqrt{\phi/B}$ is small relative to the value of $\widehat{\text{se}}(\hat{D})$. Our preference is to achieve a bootstrap uncertainty coefficient of variation of 0.005 or less; hence $\sqrt{\phi/B}/\widehat{\text{se}}(\hat{D}) \leq 0.005$, because this means that we get our result for $\widehat{\text{se}}^*(\hat{D} \mid B)$ (taken as $\widehat{\text{se}}(\hat{D})$) reliable, essentially, to two significant digits. That is, we target a large enough $B$ that the bootstrap result for $\widehat{\text{se}}(\hat{D})$) (or whatever is being computed) is nearly stable in the first two significant digits over all bootstrap samples of size $B$. If the true result should be 100, we want to be assured that generally our bootstrap result will be between about 99 and 101. This does not seem like too much precision to ask for; yet even this precision may require in excess of 10,000 bootstrap samples; it is rarely achieved with $B = 1,000$.

Now consider the variation exhibited in Table 5.27 in bootstrap estimates of $\pi_1$, $D_a$, unconditional standard error of $\hat{D}$, percentile confidence interval endpoints (95%), and the interval width, $\hat{D}_U - \hat{D}_L$. Only $\hat{D} \equiv \hat{D}_a = \overline{\hat{D}^*}$ satisfies our precision criterion for $B = 1,000$. However, we do not do bootstrapping to get $\overline{\hat{D}^*}$: We already have $\hat{D}$, from the best model, and $\hat{D}_a$ from model averaging. It

5.4 Estimation of Density from Line Transect Sampling

is the other quantities in Table 5.27 that we use the bootstrap method to compute. We find (empirically or theoretically) that $\hat{\pi}_1$ for $B = 1{,}000$ falls generally within 0.42 to 0.48; this does not meet our precision criterion. Similarly, none of $\widehat{\text{se}}(\hat{D})$, the confidence interval bounds, or width, in Table 5.27 meet our (modest) precision criterion when $B = 1{,}000$. Based on the variation over the 10 reps in Table 5.27 we estimate that for 10,000 reps the percent coefficients of variation on the bootstrap estimates are as follows: $\text{cv}(\hat{\pi}_1) = 0.005$, $\text{cv}(\widehat{\text{se}}(\hat{D})) = 0.007$, $\text{cv}(\hat{D}_L) = 0.004$, $\text{cv}(\hat{D}_U) = 0.006$, and $\text{cv}(\hat{D}_U - \hat{D}_L) = 0.01$. Thus here $B = 10{,}000$ is not too many reps to produce bootstrap-computed quantities reliable to (almost) 2 significant digits. When using the bootstrap, think in terms of $B = 10{,}000$.

## 5.4.7 Bootstrap vs. Akaike Weights: A lesson on $QAIC_c$

The estimated unconditional standard error of $\hat{D}$ is 1.23 based on analytical formula and use of the Akaike weights. However, based on the computer-intensive bootstrap method we obtained 1.48 for the estimated unconditional standard error of $\hat{D}$. The bootstrap method is telling us that there is more uncertainty in our density estimator than our analytical (i.e., theoretical) formula accounts for. We perceived a need to resolve this issue. Unfortunately, we took the wrong approach: We assumed that the bootstrap result might be wrong, and tried to find out why. It is not wrong, but we mention some of our thinking before giving the correct resolution of this matter.

The correct analytical variance of $\hat{D}$, given a model, is conditioned on total line length, $L$ (88.85 km), and has two parts: $\text{var}(n/L) = \text{var}(n)/L^2$, and $\text{var}(\hat{P})$.

TABLE 5.27. Some bootstrap estimates for 10 independent bootstrap samples of size 1,000 from the line transect data for eastern grey kangaroos at Wallaby Creek, New South Wales, Australia (from Southwell 1994); $\pi_1$ is the selection probability for model $M_1$; standard errors and (percentile) confidence intervals for $D$ are unconditional, and hence include model-selection uncertainty. Results for "All" are based on the full 10,000 reps.

| Set | $\hat{\pi}_1$ | $\hat{D}^*$ | $\widehat{\text{SE}}(\hat{D})$ | 95% Conf. Int. | | Width |
|---|---|---|---|---|---|---|
| 1 | 0.478 | 10.39 | 1.47 | 7.77 | 13.64 | 5.86 |
| 2 | 0.412 | 10.41 | 1.52 | 7.71 | 13.84 | 6.13 |
| 3 | 0.473 | 10.43 | 1.42 | 7.98 | 13.56 | 5.59 |
| 4 | 0.418 | 10.37 | 1.49 | 7.84 | 13.73 | 5.89 |
| 5 | 0.442 | 10.40 | 1.49 | 8.00 | 13.80 | 5.80 |
| 6 | 0.410 | 10.37 | 1.48 | 8.03 | 13.82 | 5.79 |
| 7 | 0.461 | 10.40 | 1.50 | 7.73 | 13.90 | 6.17 |
| 8 | 0.447 | 10.40 | 1.53 | 7.92 | 13.82 | 5.89 |
| 9 | 0.448 | 10.39 | 1.48 | 7.84 | 13.75 | 5.90 |
| 10 | 0.540 | 10.30 | 1.44 | 7.92 | 13.47 | 5.55 |
| All | 0.453 | 10.39 | 1.48 | 7.88 | 13.78 | 5.90 |

The var($\hat{P}$) component is conditional on $n$ (196). Because detections and kangaroo locations may not be independent within line segments, the units used here as the basis for the bootstrapping are the separate line segments (78 of them). The length of these segments varies from 0.5 to 1.6 km. In generating a bootstrap sample the value of $L_b^*$ is not held fixed at 88.85 over bootstrap reps, $b$. Instead, $L_b^*$ varies quite a lot. Also, the value of $n_b^*$ varies a lot over bootstrap reps. Might these aspects of variation incorporated into the bootstrap samples result in an inflated estimate of se($\hat{D}$)? We investigated this issue very intensively for this example and concluded that the bootstrap estimate of se($\hat{D}$) was acceptable here. However, the simple, theoretically computed unconditional se($\hat{D}$) did not account for all uncertainty in $\hat{D}$ (even though it accounts for all model-selection uncertainty).

The resolution of the matter also turned out to be simple: We had forgotten to consider the need for a variance inflation factor, $\hat{c}$ (forgotten because this was only an example, not a full-blown data analysis). The var($\hat{P}$) component above was based on theoretical formulae under ML estimation given the model. However, this variance is underestimated if important assumptions fail: The assumption of independence of detections within a line segment may fail; there may be spatio-temporal variation in true detection probabilities by detection distance $x$; there may be errors in recording detection distances (there usually are). All these problems lead to more variance than theory accounts for. We can adjust the theoretical $\widehat{se}(\hat{D})$ to allow for these sources of variation (in a way analogous to what the bootstrap does). The simplest adjustment is to use $\hat{c} \cdot \widehat{se}(\hat{D})$ as our theory-based unconditional standard error.

When all models considered are subsets of one global model, then $\hat{c}$ for QAIC, and variance inflation, comes from the goodness-of-fit of the global model: $\hat{c} = \chi^2/\text{df}$. However, here we have four models, but there is no global model, so the approach to obtaining $\hat{c}$ must consider the goodness-of-fit of all four models. Below we give the goodness-of-fit chi-square statistic, its degrees of freedom and $\hat{c}$ for these four models, as well as the Akaike weights based on use of AIC:

| Model | $\chi^2$ | df | $\hat{c}$ | $w_i$ |
|---|---|---|---|---|
| 1 | 25.11 | 17 | 1.48 | 0.499 |
| 2 | 23.73 | 16 | 1.48 | 0.220 |
| 3 | 24.66 | 15 | 1.64 | 0.072 |
| 4 | 23.40 | 16 | 1.46 | 0.209 |

The weighted average of $\hat{c}$, weighted by $w_i$, is 1.49. As a general procedure we would use either $\hat{c}$ from the selected model, or this weighted average. It makes no difference here; hopefully, this would be the usual situation. Hence, we use here $\hat{c} = 1.48$, df $= 17$ from model $M_1$.

We should, however, have been using QAIC rather than AIC, because our Akaike weights might then change (along then with other results). From Table 5.25 we obtain $-2\log(\mathcal{L})$ for each model and thus compute QAIC $= (-2\log(\mathcal{L})/\hat{c}) + 2K$, and the associated weights, $w_i$:

| Model | QAIC | $\Delta_i$ | $w_i$ |
|---|---|---|---|
| 1 | 1,384.71 | 0.00 | 0.511 |
| 2 | 1,386.47 | 1.76 | 0.211 |
| 3 | 1,388.61 | 3.90 | 0.073 |
| 4 | 1,386.54 | 1.83 | 0.205 |

The differences between the Akaike weights based on AIC vs. QAIC are here trivial (this is because of a large sample size here). Using the above weights with each $\hat{D}$ from Table 5.25 gives a model-averaged result of 10.18; the original result was 10.19. For an unconditional standard error based on the QAIC derived $w_i$ we get 1.23 (the same as with AIC-based weights). We will stay with the originally computed $\hat{D}_a = 10.19$. In this example, the only effect of using QAIC is to make us realize that we need to use a variance inflation factor with our theoretical standard errors.

The quick way to adjust the theoretical unconditional standard error is to compute $\sqrt{\hat{c}} \cdot \text{se}(\hat{D}) = \sqrt{1.48} \cdot 1.23 = 1.22 \cdot 1.23 = 1.50$; the bootstrap-based result for the unconditional standard error of $\hat{D}$ was 1.48. However, the use of $\sqrt{\hat{c}} \cdot se(\hat{D})$ is not the correct formula (we have used it here for its heuristic epistemological value). Rather, one should adjust each theoretical $\widehat{\text{var}}(\hat{D}_i \mid M_i)$ to be $\hat{c} \cdot \widehat{\text{var}}(\hat{D}_i \mid M_i)$ and then apply formula (4.11), which here becomes

$$\widehat{\text{se}}(\hat{D}) = \sum_{i=1}^{4} w_i \sqrt{\hat{c} \cdot \widehat{\text{var}}(\hat{D}_i \mid M_i) + (\hat{D}_i - \hat{D}_a)^2}. \quad (5.9)$$

The two approaches will give almost same results when the values of $(\hat{D}_i - \hat{D}_a)^2$ are small relative to $\hat{c} \cdot \widehat{\text{var}}(\hat{D}_i \mid M_i)$, as they are here. Applying formula (5.9) using the quantities from Table 5.25 and $\hat{c} = 1.48$, we get as an analytical formula-based result $\widehat{\text{se}}(\hat{D}) = 1.48$. This is exactly the same result as generated by the bootstrap (this may be a coincidence).

The bootstrap method to obtain the unconditional standard error of a parameter estimator will, if done correctly, automatically include all sources of uncertainty in that standard error. Estimation based on theoretical–analytical formulae, for models that do not automatically estimate empirical residual variation, will not automatically include overdispersion variation that exceeds what theory assumes. Thus in these cases we must always consider the need to include an empirical variance inflation factor, $\hat{c}$, in our calculations.

## 5.5 An Extended Binomial Example

### 5.5.1 The Durban Storm Data

Linhart and Zucchini (1986:176–182) apply AIC to storm frequency data from the Botanical Gardens in Durban, South Africa. The detailed data are given in their Table 10.1. By seven day periods in the year ("weeks"), beginning 1 January, they obtained the frequency of weeks with at least one storm event occurring. For example, in 47 consecutive years of data, for January 1-7 there were 6 years with at least one storm event. The data are based on a rigorous definition of a storm: "a rainfall event of at least 30 mm in 24 hours" (Linhart and Zucchini 1986:176). We use here their period I data, $y_i$ ($i$ denotes week, 1 to 52), wherein for the first 22 weeks, the sample size of years is $n_i = 47$; for weeks $23 \leq i \leq 52$, $n_i = 48$. The data are from Jan. 1932 to Dec. 1979. We ignore, as did Linhart and Zucchini (1986), the minor matter of a few weeks needing to have 8 days (such as 26 Feb. to 4 Mar. when a leap year occurs). Listed in order $i = 1$ to 52, the data $y_i$ are

6, 8, 7, 6, 9, 15, 6, 12, 16, 7, 9, 6, 8, 2, 7, 4, 4, 3, 3, 10, 3, 3, 0, 5, 1, 2,
4, 0, 2, 0, 3, 1, 1, 5, 4, 3, 6, 1, 8, 3, 4, 6, 9, 5, 8, 6, 5, 7, 5, 8, 5, 4.

Conceptually, there exists a probability $p_i$ of a storm at the Durban Botanical Gardens in week $i$. Based on these data, what is a "good" estimate of $p_1$ to $p_{52}$? That was the analysis objective of Linhart and Zucchini, and it will be one of our objectives. Our other objective is to reliably assess the uncertainty of our $\hat{p}_i$. A simple estimator is $\hat{p}_i = y_i/n_i$; it is very nonparsimonious, lacks precision, and (most seriously) fails to be a smooth, hence informative, estimator of time trends in the true $p_i$. We expect that anyone considering this problem would strongly believe that the $p_i$ would have a considerable degree of smoothness as a function over the 52 weeks. Therefore, we want to fit some model $p_i(\underline{\theta})$ for a not-large number of parameters represented by $\underline{\theta} = (\theta_1, \ldots, \theta_K)'$.

### 5.5.2 Models Considered

We agree with the general approach taken by Linhart and Zucchini (1986); they construct a likelihood by treating the $y_i$ as a set of independent binomial random variables on sample sizes $n_i$ for parameters $p_i$, and use the structural model as

$$\text{logit}(p_i) = \log(p_i/(1-p_i)) = \sum_{j=1}^{K} \theta_j z_{ji}$$

being some suitable linear model on $\underline{\theta}$, for known "covariates" $z_{ji}$. Essentially, this is a type of logistic regression (we consider theory for AIC model-selection in this situation in Section 6.6.6). Linhart and Zucchini used a finite Fourier series model for the $z_{ji}$ and used TIC for model-selection (which here became essentially the same as AIC). We extend their example by using QAIC and model averaging; also, we compute unconditional confidence intervals on the $\hat{p}_i$.

The structure of the simplest model, model $M_1$, is given by

$$\text{logit}(p_i) = \theta_1, \quad i = 1, \ldots, 52.$$

For model $M_2$:

$$\text{logit}(p_i) = \theta_1 + \theta_2 \cos\left(\frac{2\pi(i-1)}{52}\right) + \theta_3 \sin\left(\frac{2\pi(i-1)}{52}\right), \quad i = 1, \ldots, 52.$$

## 5.5 An Extended Binomial Example

For model $M_3$:

$$\text{logit}(p_i) = \theta_1 + \theta_2 \cos\left(\frac{2\pi(i-1)}{52}\right) + \theta_3 \sin\left(\frac{2\pi(i-1)}{52}\right)$$
$$+ \theta_4 \cos\left(\frac{4\pi(i-1)}{52}\right) + \theta_5 \sin\left(\frac{4\pi(i-1)}{52}\right), \quad i = 1, \ldots, 52.$$

In general, the structure for model $M_r$ (wherein $K = 2r - 1$) is given by

$$\text{logit}(p_i) = \theta_1 + \sum_{j=1}^{r} \left[ \theta_{2j} \cos\left(\frac{2j\pi(i-1)}{52}\right) \right.$$
$$\left. + \theta_{2j+1} \sin\left(\frac{2j\pi(i-1)}{52}\right) \right], \quad i = 1, \ldots, 52.$$

Assuming marginal binomial variation and independence, the form of the likelihood for any model is

$$\mathcal{L}(\underline{\theta}) \propto \Pi_{i=1}^{52} (p_i)^{y_i} (1 - p_i)^{n_i - y_i}.$$

Given the model for $\text{logit}(p_i)$ as a function of $\underline{\theta}$, say $h_i(\underline{\theta}) = \text{logit}(p_i)$, we compute $p_i$ as

$$p_i = \frac{1}{1 + \exp[-h_i(\underline{\theta})]}.$$

The independence assumption may not be true, but it seems likely to be not badly wrong. Similarly, the count $y_i$ may not be the sum of exactly homogeneous Bernoulli events over the $n_i$ years. Truth may correspond more closely to having varying year-to-year weekly probabilities of a storm. A useful way to cope with these types of model inadequacies is to use ideas from quasi-likelihood theory, hence to use a variance inflation factor, $\hat{c} = \chi^2/\text{df}$. This $\hat{c}$ is computed from the global model goodness-of-fit chi-square ($\chi^2$) on degrees of freedom df. Then we use QAIC, rather than AIC; also, conditional sampling variances based on assumed models are multiplied by $\hat{c}$.

Following Linhart and Zucchini (1986) we consider seven models as our set over which model uncertainty and model averaging are computed. For model $M_7$, $K = 13$. We obtained MLEs for these models by using SAS PROC NLIN (SAS Version 6.12); it is easy to adapt PROC NLIN to produce ML estimates (see, e.g., Burnham 1989). For each fitted model we also computed the usual chi-square goodness-of-fit statistic, its significance level (P-value), and $\hat{c}$. For the purpose of a more thorough consideration of model fit we also fit models $M_8$ ($K = 15$), and $M_9$ ($K = 17$). Table 5.28 gives basic results from these fitted models: $K$, $\log(\mathcal{L})$, $\Delta$-AIC, $\chi^2$ goodness-of-fit, and corresponding P-values, $\hat{c}$, and $\Delta$-QAIC. The values of $\Delta$-QAIC are for when model $M_7$ is taken as the global model.

The $\log(\mathcal{L})$ values in Table 5.28 for models 1 through 7 match the values of Linhart and Zucchini (1986) in their Table 10.3 (they did not fit models $M_8$ and $M_9$). The AIC-selected model has 9 parameters, and our MLE $\hat{\underline{\theta}}$ does match the results of Linhart and Zucchini (on their page 182). The parameters, $\underline{\theta}$, in the

TABLE 5.28. Some basic results from models $M_1$ to $M_9$ fitted to the weekly storm incidence data (Linhart and Zucchini 1986); QAIC is based on taking $M_7$ as the global model (hence $\hat{c} = 1.4$); the df of the goodness-of-fit $\chi^2$ are $52 - K$.

| Model | K | log($\mathcal{L}$) | $\Delta$-AIC | $\chi^2$ | P | $\hat{c}$ | $\Delta$-QAIC |
|---|---|---|---|---|---|---|---|
| 1 | 1 | −863.24 | 62.66 | 131.4 | 0.000 | 2.57 | 40.67 |
| 2 | 3 | −833.83 | 7.85 | 76.5 | 0.007 | 1.56 | 2.66 |
| 3 | 5 | −829.17 | 2.53 | 69.3 | 0.019 | 1.47 | 0.00 |
| 4 | 7 | −826.37 | 0.93 | 61.2 | 0.054 | 1.36 | **0.00** |
| 5 | 9 | −823.91 | **0.00** | 55.6 | 0.094 | 1.29 | 0.49 |
| 6 | 11 | −823.89 | 3.95 | 55.6 | 0.064 | 1.36 | 4.45 |
| 7 | 13 | −823.40 | 7.04 | 54.7 | 0.049 | 1.40 | 7.76 |
| 8 | 15 | −822.76 | 9.70 | 54.0 | 0.035 | 1.46 | — |
| 9 | 17 | −822.47 | 13.11 | 53.8 | 0.022 | 1.54 | — |

likelihood of these models do not have intrinsic meaning and are not of direct interest. Therefore, we do not present values of $\hat{\theta}$ from any fitted models, nor their estimated conditional standard errors (standard likelihood theory was applied to obtain the large-sample variance–covariance matrix of the MLE $\hat{\theta}$). Rather, our goal is to estimate well the set of $p_1$ to $p_{52}$, which in effect are parameters in common to all models.

### 5.5.3   Consideration of Model Fit

Before we accept the AIC-selected model, we must consider whether the global model fits. Based on the results in Table 5.28, the global model, $M_7$, is a not a good fit to the data: $P = 0.049$. More importantly, $\hat{c} = 1.4$ on 43 df is sufficiently greater than 1 that we should not accept results of AIC-selection that here require $c \doteq 1$. Even the AIC-selected model has $\hat{c} = 1.29$ (and $P = 0.094$ even though this model is deliberately selected to fit well). To explore this issue further we fit two more models; models $M_8$ and $M_9$ also fit the data poorly. If the problem was an inadequate structural model, we would expect the fit to $M_8$ and $M_9$, compared to model $M_7$, to improve. The results for $\hat{c}$ in Table 5.28 strongly suggest that there is extra binomial variation in these count data. Such a result is common for real count data such as these, as is the value of $\hat{c}$ (i.e., $1 < \hat{c} < \dot\sim 2$).

However, before automatically resorting here to QAIC, there is another issue worth noting. The expected counts from the models fitted here are often small (i.e., the data are sparse in the sense of being small counts). For example, for model $M_5$, $\hat{E}(y_{26})$ to $\hat{E}(y_{32})$ are about 1.5; these are the smallest estimated expected count values here; the largest estimated expected values are about 10. Perhaps even if the global model is structurally true, the plethora of small count values will invalidate the usual central chi-square null distribution of the goodness-of-fit statistic.

We explored this matter by Monte Carlo methods (also called the parametric bootstrap method). We generated data based on truth being $\hat{p}_i$ from the AIC-selected model, $M_5$. That is, independent $y_i^*$ were generated as binomial($n_i$, $\hat{p}_i$) based on fitted model $M_5$. For each such data set we then fitted model $M_5$ and computed the chi-square goodness-of-fit statistic to see whether its distribution was noticeably different from that of a central chi-square on 43 df. We used only a sample of size 100 such generated data sets because we were looking for a big effect: For this situation is $c = 1$ or 1.29?

The answer was clear: If the model truly fits, then on average we will get $\hat{c} = 1$; i.e., the usual null distribution holds well here despite small counts. The average of the 100 $\chi^2$ goodness-of-fit values was 41.9 (theoretically it is 43). The largest and smallest of the 100 values were 71.6 ($P = 0.004$) and 21.8 ($P = 0.997$); these are not unusual for a sample of 100 such test statistics. When each test statistic was converted to a P-value, the set of 100 P-values fit a uniform (0, 1) distribution. Finally, the average of the set of 100 values of $\hat{c}$ was 0.98 ($\widehat{se} = 0.21$). While the possibility remains that for these data $c = 1$ is appropriate (and we just happened to get an unusual realized sample), this is quite unlikely based on the Monte Carlo evidence. Moreover, experience-based general statistical wisdom for real data supports the belief that we should accept that extra binomial variation often exists in count data. We therefore will use QAIC, not AIC, with $\hat{c} = 1.4$ as our basis for model-selection.

When sufficient precision is used in the calculations, we find that model $M_4$ is the QAIC best model, although for practical purposes models $M_3$ and $M_4$ are tied for best (and model $M_5$ is almost as good, based on QAIC). Figure 5.1 gives a plot of the fitted $\hat{p}_i$ for both models $M_3$ and $M_4$. Also shown are the approximate 95% confidence bands on $p_i$ based on $\hat{p}_{L,i}$ and $\hat{p}_{U,i}$ for each week $i$. We next explain the calculation of these confidence intervals.

### 5.5.4 Confidence Intervals on Predicted Storm Probability

Basically, $\hat{p}_{L,i}$ and $\hat{p}_{U,i}$ arise as back-transformed lower and upper confidence limits on logit($p_i$). However, we used SAS PROC NLIN to directly gives us the estimated MLE-based theoretical $\widehat{se}_t(\hat{p}_i \mid M)$ that is computed assuming $c = 1$. The first step is then to form the correct (inflated) estimated standard error: $\sqrt{\hat{c}} \cdot \widehat{se}_t(\hat{p}_i \mid M) = 1.183 \, \widehat{se}_t(\hat{p}_i \mid M) = \widehat{se}(\hat{p}_i \mid M)$. The interval $\hat{p}_i \pm 2 \, \widehat{se}(\hat{p}_i \mid M)$ could be used. However, it is better to use here what is basically an appropriate back-transformed logit-based interval (Burnham et al. 1987:214):

$$\hat{p}_{L,i} = \frac{\hat{p}_i}{\hat{p}_i + (1 - \hat{p}_i)C},$$

$$\hat{p}_{U,i} = \frac{\hat{p}_i}{\hat{p}_i + (1 - \hat{p}_i)/C},$$

where

$$C = \exp\left[\frac{t_{\alpha/2, df} \, \widehat{se}(\hat{p}_i \mid M)}{\hat{p}_i(1 - \hat{p}_i)}\right]$$

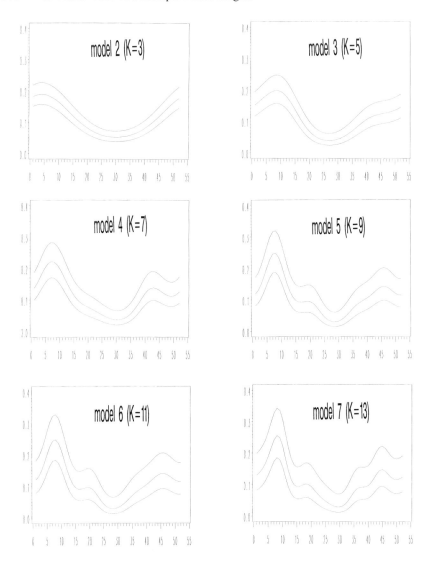

FIGURE 5.1. Plots of the predicted probability of one or more storms per week, $\hat{p}_i$ ($y$-axis), from models $M_2$ to $M_7$ fitted to the Durban storm data from Linhart and Zucchini (1986) (see text for details). Also shown are approximate 95% confidence bands on $p_i$; these bands are conditional on the model (see text for details).

(acceptable as long as $\hat{p}_i$ does not get too close to 0 or 1). The confidence bands in Fig. 5.1 were computed in this manner and are thus conditional on the model. We used $t_{\alpha/2}$, df = 43, because the df that apply here are those of $\hat{c}$; thus here df = 43.

Estimates of unconditional standard errors require the Akaike weights (or the bootstrap), in this case based on $\Delta_i$ from QAIC (Table 5.28). Formula (4.2) applies, and we find that $w_1, \ldots, w_7$ are

$$0.0000, 0.0833, 0.3149, 0.3149, 0.2465, 0.0340, 0.0064.$$

Then for each week we find $\hat{p}_i$ under models $M_1$ to $M_7$ and apply formula (4.6) (using $\hat{\pi}_i = w_i$) to find the model-averaged $\hat{p}_{a,i}$. Next we apply formula (4.10) to obtain the unconditional $\widehat{\text{se}}(\hat{p}_{a,i})$; this unconditional standard error also applies to $\hat{p}_i$ from the selected model as well as to $\hat{p}_{a,i}$. In cases like this where the $\underline{\theta}$ are not of direct interest we recommend that the $\hat{p}_i$ to use are the model-averaged values, $\hat{p}_{a,i}$. Moreover, here we use the unconditional $\widehat{\text{se}}(\hat{p}_{a,i})$ and we base confidence bands on the above formula for $\hat{p}_{L,i}$ and $\hat{p}_{U,i}$, however, for C based on $\widehat{\text{se}}(\hat{p}_{a,i})$. The resulting $\hat{p}_{a,i}$, $\hat{p}_{L,i}$, and $\hat{p}_{U,i}$ are shown in Fig. 5.2.

### 5.5.5 Precision Comparisons of Estimators

We now give some considerations about (estimated) standard errors for different versions of the $\hat{p}_i$. Common practice would be to select a model and use the standard errors conditional on the model. In this case that would mean using model $M_4$ (by the slimmest of margins). We computed and examined the ratios

$$r_i(\text{se}) = \frac{\widehat{\text{se}}(\hat{p}_i \mid M_4)}{\widehat{\text{se}}(\hat{p}_{a,i})}, \qquad i = 1, \ldots, 52$$

and

$$r(\overline{\text{se}}) = \frac{\sum \widehat{\text{se}}(\hat{p}_i \mid M_4)}{\sum \widehat{\text{se}}(\hat{p}_{a,i})}.$$

These ratios are less than 1 if the unconditional is larger than the conditional standard error (the notation used for these ratios has no special meaning; we just need to represent them somehow).

We obtained $0.78 \leq r_i(\text{se}) \leq 1.02$ and $r(\overline{\text{se}}) = 0.90$. Thus the proper unconditional standard errors are on average 1.11 times the standard errors that are conditional on the model, and hence ignore model uncertainty. Also, we note that the average of the 52 values of $\widehat{\text{se}}(\hat{p}_{a,i})$ was 0.0214 ($0.012 \leq \widehat{\text{se}}(\hat{p}_{a,i}) \leq 0.035$). This is good absolute precision; the actual average width of the 52 confidence intervals was 0.084.

An alternative that avoids model-selection is to use $\hat{p}_i = y_i/n_i$. This is not a very useful parameter-saturated model. Estimated standard errors under this model are given by $\widehat{\text{se}}_s(\hat{p}_i) = \sqrt{\hat{p}_i(1-\hat{p}_i)/n_i}$ (no adjustment by any $\hat{c}$ is used here, as there is no basis on which to compute a variance inflation factor given this model). We computed and examined the ratio $\left[\sum \widehat{\text{se}}_s(\hat{p}_i)\right] / \left[\sum \widehat{\text{se}}(\hat{p}_{a,i})\right]$ and considered

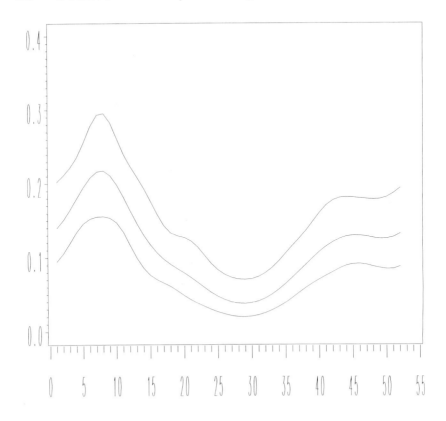

Probability of a storm, by week

FIGURE 5.2. Plot of the model-averaged (from models $M_2$ to $M_7$) predicted probability of one or more storms per week, $\hat{p}_{a,i}$, from the Durban storm data from Linhart and Zucchini (1986) (see text for details). Also shown are approximate 95% confidence bands on $p_i$; these bands include model-selection uncertainty.

the separate $\widehat{\text{se}}_s(\hat{p}_i)$ and $\widehat{\text{se}}(\hat{p}_{a,i})$. We obtained

$$\frac{\sum \widehat{\text{se}}_s(\hat{p}_i)}{\sum \widehat{\text{se}}(\hat{p}_{a,i})} = 2.31;$$

so on average the unconditional standard errors of the model-averaged $\hat{p}_{a,i}$ were more precise by a multiplicative factor $0.433 = 1/2.31$ compared to the much less useful parameter-saturated model estimates. Also, we observed that $0 \leq \widehat{\text{se}}_s(\hat{p}_i) \leq 0.082$ (and a variance estimate of 0 is quite wrong), whereas $0.012 \leq \widehat{\text{se}}(\hat{p}_{a,i}) \leq 0.035$; thus the standard errors for the model-averaged $\hat{p}_{a,i}$ are much more stable than is the case for the parameter-saturated model.

Linhart and Zucchini (1986) used TIC for model-selection, not AIC. The only difference between the two methods is the use of $\text{tr}(JI^{-1}) = K$ rather than estimating this term, which Linhart and Zucchini denote by tr $\Omega_n^{-1}\Sigma_n$. In fact, in Section 6.6.6 we demonstrate by theory and example that tr($JI^{-1}$) is very near $K$ unless the structural model is truly terrible. In fact, we think that it is better for count data (and generally) simply to use tr($JI^{-1}$) = $K$ rather than estimate this quantity. Linhart and Zucchini (1986:181) give tr($JI^{-1}$) in their Table 10.3. For models $M_1$ to $M_7$ the ratios of the estimated trace term to $K$ are 0.95, 0.97, 0.97, 0.98, 0.98, 0.98, 0.98.

## 5.6 Lessons from the Literature and Other Matters

### 5.6.1 Use $AIC_c$, Not AIC, with Small Sample Sizes

It is far too common that papers examining AIC, by itself or compared to BIC, fail to use $AIC_c$ when the latter must be used because the number of parameters, at least for some models considered, is not small relative to sample size. In a recent paper Chatfield (1996) considered model-selection issues and used a time series example with $n = 132$ and $R = 12$ *a priori* designated models wherein $K$ ranged from 6 to 61. Overall we commend the paper; however, in this particular example the conclusion that AIC performed poorly is misleading. AIC did do poorly; but it is well known, documented, and commented on in the literature on K-L–based model selection that in such an example it is imperative to use $AIC_c$, not AIC (e.g., Sakamota et al. 1986, Bozdogan 1987, Hurvich and Tsai 1989, Hurvich et al. 1990).

Recall that

$$AIC_c = AIC + 2\frac{K(K+1)}{n-K-1} = -2\log(\mathcal{L}) + 2K + 2\frac{K(K+1)}{n-K-1}.$$

Consider for $n = 132$ the effect of the bias-correction term for $K = 6$ and 61 for these two models with likelihoods denoted by $\mathcal{L}_6$ and $\mathcal{L}_{61}$:

| $K$ | AIC | $AIC_c$ |
|---|---|---|
| 6 | $-2\log(\mathcal{L}_6) + 12$ | $-2\log(\mathcal{L}_6) + 12.672$ |
| 61 | $-2\log(\mathcal{L}_{61}) + 122$ | $-2\log(\mathcal{L}_{61}) + 230.057$ |

The difference here between AIC and $AIC_c$ is huge, and this will greatly affect which model is selected.

We present in Table 5.29 the results that Chatfield (1996) should have presented as regards K-L–based model selection vs. BIC. In so doing we also use $\Delta_i$ values, not absolute values of these model-selection criteria. We do show for comparison the $\Delta$-AIC values implicitly used by Chatfield. The results in Table 5.29 are based on the results in Table 1 of Chatfield (1996). The nature of the models need not concern us, so we label them just 1 to 12, but keep them in the same order as used in Table 1 of Chatfield. The $\Delta_i$ values in Table 5.29 for the $AIC_c$ criterion do have here the interpretations and uses described in Section 4.2. Those interpretations

222   5. Monte Carlo and Example-Based Insights

TABLE 5.29. The $\Delta$-$AIC_c$ that must be used for K-L model selection on the 12 models considered in Table 1 of Chatfield (1996), and corresponding $\Delta$-AIC and $\Delta$-BIC values; also, the Akaike weights based on $AIC_c$.

| Model | K | $\Delta$-$AIC_c$ | $\Delta$-AIC | $\Delta$-BIC | $w_i$-$AIC_c$ |
|---|---|---|---|---|---|
| 1  | 6  | 4.5   | 68.0  | **0.0** | 0.048 |
| 2  | 11 | **0.0** | 62.0  | 13.0  | 0.459 |
| 3  | 21 | 2.5   | 58.3  | 47.0  | 0.132 |
| 4  | 9  | 86.1  | 148.8 | 92.2  | 0.000 |
| 5  | 17 | 98.6  | 157.4 | 130.9 | 0.000 |
| 6  | 41 | 155.5 | 181.4 | 246.0 | 0.000 |
| 7  | 11 | 83.7  | 145.7 | 96.7  | 0.000 |
| 8  | 21 | 94.7  | 150.5 | 139.3 | 0.000 |
| 9  | 13 | 0.5   | 61.6  | 20.0  | 0.358 |
| 10 | 25 | 10.1  | 62.0  | 65.7  | 0.003 |
| 11 | 31 | 15.9  | 60.3  | 86.8  | 0.000 |
| 12 | 61 | 43.9  | **0.0** | 139.8 | 0.000 |

are not true in this example for the $\Delta_i$ derived here from AIC, because for large $K$ relative to $n$, AIC is too biased of an estimator of the expected K-L distance.

As noted by Chatfield, in this example AIC and BIC lead to very different selected models. However, $AIC_c$ (which must be used here) gives (seemingly) acceptable results. In fact, the Akaike weights here (see Table 5.29) show that only 4 fitted models have any plausibility in this set of 12 fitted models. Those models are (in order of their likelihood) $M_2$ ($K = 11$), $M_9$ ($K = 13$), $M_3$ ($K = 21$), and $M_1$ ($K = 6$). Thus the evidence, as interpreted with $AIC_c$, definitely eliminates 8 of the 12 models, because relative to other models in the set, they are extremely implausible. Also, most of the weight of evidence is put on models with low $K$. Based on comparing AIC to BIC, Chatfield (1996) concluded that BIC was a better criterion than AIC. That conclusion is not justified in that here one must compare $AIC_c$ to BIC for a proper comparison of K-L information-theoretic model selection vs. BIC.

## 5.6.2   Use $AIC_c$, Not AIC, When K Is Large

Leirs et al. (1997) report the analysis of an extensive set of capture–recapture data from Tanzania on the rat *Mastomys natalensis*. The objective of their data analysis was to examine factors potentially important in the population survival dynamics of the species. The data were collected between October 1986 and February 1989 by live trapping on a 1 ha grid of 100 live trapping positions (several traps per position). There were three consecutive nights of trapping each month (hence 29 primary trapping periods). There were a total of 6,728 captures of 2,481 individual animals. We take the relevant sample size to be the latter, i.e., $n = 2,481$. Leirs et al. (1997) carefully formulate six *a priori* models to represent how environmental (rainfall) and population density factors might affect survival probabilities ($S$) and

## 5.6 Lessons from the Literature and Other Matters

the probability of subadults maturing to adults ($\psi$). Capture probabilities ($p$) are another subset of parameters in these models. Data analysis was by ML methods for multistate capture–recapture models (see, e.g., Brownie et al. 1993, Nichols and Kendall 1995) with incorporation of covariates for rainfall and population density. The goodness-of-fit of the global model was quite acceptable ($P > 0.9$).

The global model used by Leirs et al. (1997) allows full (unexplained) temporal variation in all model parameters (hence $S$ and $p$ vary by time and age, and $\psi$ varies by time). In their Table 1, this is model $M_1$ with $K = 113$ parameters. Their model $M_2$ is the most restricted model: no temporal variation in the parameters ($S$ and $p$ vary by age only, subadult vs. adult, and there is only one maturation probability parameter, $\psi$). Model 3 allows capture probabilities to vary by time, but $S$ and $\psi$ are not time-varying. Models 4, 5, and 6 allow structured time variation in the three classes of parameters. These latter, quite complex, models are based on population dynamics models melded with the general capture–recapture model. Model 4 has temporal parameter variation as functions only of population density (internal factors only for population regulation); Model 5 has functions only of rainfall (i.e., external factors for population regulation); Model 6 has temporal parameter variation as functions of both population density and rainfall. Leirs et al. (1997) used AIC (not $AIC_c$) for model-selection. In Table 5.30 we present the $\Delta$-AIC values from their analyses, as well as results for $AIC_c$, which we computed, and Akaike weights.

Even though the sample size is large here (2,481), the fact of having a model with 113 parameters means that $AIC_c$ should be used (and it then must be used for all models). The term added to AIC, to get $AIC_c$, for model $M_1$ is $(2 \times 113 \times 114)/2367 = 10.9$ is not trivial. Clearly, using $AIC_c$ here results in a different interpretation of the relative evidence for model $M_1$ vs. $M_6$. Despite using AIC, Leirs et al. (1997) opted to select model $M_6$ ($K = 64$) as a useful model and therefore to infer that there were population dynamics occurring that could be substantially explained only by both external (rainfall) and internal (population density) factors. They worried some about this selection (J. D. Nichols, pers. comm.); they did not need to. Using $AIC_c$, as is correct here, model $M_6$ is a tenable model.

TABLE 5.30. Summary model-selection results from, or based on, Table 1 of Leirs et al. (1997); $AIC_c$ should be used here for model selection, not AIC.

| Model | K | AIC results | | $AIC_c$ results | |
|---|---|---|---|---|---|
| | | $\Delta_i$ | $w_i$ | $\Delta_i$ | $w_i$ |
| 1 Global | 113 | 0.0 | 0.99 | 0.0 | 0.76 |
| 2 No effects | 5 | 540.7 | 0.00 | 529.8 | 0.00 |
| 3 No dynamics | 49 | 207.4 | 0.00 | 198.5 | 0.00 |
| 4 Density effects | 52 | 205.8 | 0.00 | 197.2 | 0.00 |
| 5 Rainfall effects | 55 | 25.9 | 0.00 | 17.6 | 0.00 |
| 6 Rainfall and Density | 64 | 9.7 | 0.01 | 2.3 | 0.24 |

The bias reduction term (as regards estimation of relative expected K-L) added to AIC to get $AIC_c$ is derived under an assumption of linear structure, normal errors models. We need other such generalized bias reduction versions of AIC, especially for discrete date (such as capture–recapture). However, when sample size is large, the normality assumption will hold well (for MLEs), and we therefore think that use of $AIC_c$ is quite suitable for discrete data such as here.

### 5.6.3 Inference from a Less Than Best Model

We continue with some ideas, exemplified by the above example of Leirs et al. (1997), about inference from other than the K-L best model. In some circumstances this is justified, especially if (1) the model, say $M_{(2)}$ (as generic notation for the second-best AIC model), used for inference is nested within the best model, $M_{(1)}$; and (2) the unexplained "effects" in the data represented by the additional parameters added to model $M_{(2)}$ to generate model $M_{(1)}$ are small relative to the explained effects represented by model $M_{(2)}$. Conceptually, this assumes a parametrization of the models as $g(\underline{x} \mid \underline{\theta}_2, \underline{\theta}_1)$ for model $M_1$ with model $M_2$ arising under the imposed constraint $\underline{\theta}_1 = 0$.

We elaborate these ideas further using the Leirs et al. (1997) example. Their $AIC_c$ best model ($M_1$) was not interpretable in its entirety, but their second-best model ($M_6$) was interpretable, and because $\Delta_6 = 2.3$, that model is a plausible model for the data. Moreover, that second-best model is nested within the best model here. In principle here, the best model, $M_{(1)}$, could be parametrized as being the model $M_{(2)}$ structure plus an additional 49 parameters structurally additive to (and preferably orthogonal to) the 64 parameters of that second-best fitted model. Therefore, their results provide overwhelming support (in the set of models used) for the joint importance of rainfall and population density as at least good predictors of the observed population variation in survival and capture probabilities (if not outright support for a causal link to those variables).

In choosing to make inferences based on model $M_{(2)}$ (their model 6) and ignoring model $M_{(1)}$ (their model 1), Leirs et al. (1997) are in effect saying that they cannot interpret the meaning of the additional 49 parameters that constitute the difference between their best and second-best models. This does not in any way invalidate inference from the second-best model in this situation where $M_{(2)}$ is nested within $M_{(1)}$. This sort of argument holds in general if the models are nested.

The only pressing concern here, in ignoring the best model, i.e., ignoring the 49 "effects" defining the difference here between the best and second-best models, is the issue of the relative magnitude of the two sets of effects. In analysis of variance terms this issue is about the partition of the total variation of effects represented by the difference in their fitted model $M_1$ vs. $M_2$ into a sum of squares for effects of $M_6$ vs. $M_2$ plus a sum of squares for effects of $M_1$ vs. $M_6$. Analogous to ANOVA, we can use here analysis of deviance (ANODEV) (see, e.g., McCullagh and Nelder 1989, Skalski et al. 1993, and Smith et al. 1994) to accomplish a useful partition.

In this example, ANODEV proceeds as follows to measure the relative importance of the ignored effects left unexplained in model $M_1$ beyond the explained effects in model $M_6$. First, some baseline "no effects" model is needed; here that baseline is model $M_2$ of Leirs et al. (1997). Note the nesting, $M_2 \subset M_6 \subset M_1$, and corresponding values of $K$: 5, 64, and 113. The ANODEV proceeds by obtaining the log-likelihood values and computing the partition of total deviance of model $M_2$ vs. $M_1$ as

$$\left[2\log(\mathcal{L}(\hat{\underline{\theta}} \mid M_1) - 2\log(\mathcal{L}(\hat{\underline{\theta}} \mid M_2)\right] = \left[2\log(\mathcal{L}(\hat{\underline{\theta}} \mid M_6) - 2\log(\mathcal{L}(\hat{\underline{\theta}} \mid M_2)\right]$$
$$+ \left[2\log(\mathcal{L}(\hat{\underline{\theta}} \mid M_1) - 2\log(\mathcal{L}(\hat{\underline{\theta}} \mid M_6)\right].$$

The result here is $756.8 = 649.0 + 107.8$. The above three bracketed differences are also interpretable as likelihood ratio test statistics on 108, 59, and 49 df.

The above partitions a measure of the magnitude of the total effects (756.8, on 108 df) represented by fitted model $M_1$ into a measure of the effects explained by model $M_6$ alone (649.0, on 59 df), plus the additional measure of effects (107.8, on 49 df) explained by the added 49 parameters that "create" model $M_1$ from model $M_6$. Based on this partition we can define a type of multiple coefficient of determination, $R^2$, as (here)

$$R^2 = \frac{2\log(\mathcal{L}(\hat{\underline{\theta}} \mid M_6)) - 2\log(\mathcal{L}(\hat{\underline{\theta}} \mid M_2))}{2\log(\mathcal{L}(\hat{\underline{\theta}} \mid M_1)) - 2\log(\mathcal{L}(\hat{\underline{\theta}} \mid M_2))} = \frac{649.0}{756.8} = 0.858.$$

The interpretation is that 86% of the total structural information about parameter variation in model $M_1$ is contained in model $M_6$. Thus, in some sense 14% of potentially interpretable effects has been lost by making inferences based only on model $M_6$ (i.e., the second-best $AIC_c$ model), rather than based on model $M_1$. However, that other 14% of information was left as not interpretable. It was judged to be real information, as evidenced by $AIC_c$ selection of model $M_1$ as the best model, but ignoring it does not invalidate the inferences made from model $M_6$.

Clearly, the addition to model $M_6$ of all the structure represented by the additional 49 parameters (to get model $M_1$) does, for the data at hand, lead to the K-L best-fitted model. However, in principle there is some intermediate model, between models $M_6$ and $M_1$ that adds far fewer than 49 parameters and would produce an even smaller $AIC_c$ than model $M_1$. Such an additional model would extract additional useful information from the data; it might be some form of random effects model, or some interaction effect of rainfall and population density. The situation faced here was, essentially, considered in Sections 2.7.2, 3.5.5, and 3.5.6, where we pointed out that if there are two models, one nested in the other and differing by a large number of parameters (say 10 or more), then anomalies can arise in data analysis based on K-L model-selection.

In general, there are situations where choosing to make inferences based on other than the $AIC_c$ best model can be justified. However, this situation is not satisfied if the $AIC_c$ best model has many additional parameters compared to the model one uses for the basis of inference. If we find ourselves in this situation, it suggests that we did not think hard enough *a priori* about our set of models, because we probably left out at least one good model. Now some a posteriori (to

### 5.6.4 Are Parameters Real?

Consideration of what is a parameter seems important, inasmuch as we are focused entirely on parametric models. With only one class of exceptions we regard a parameter as a hypothetical construct. Hence, a parameter is usually the embodiment of a concept and does not have the reality of a directly recordable variable. As such, a parameter in a statistical setting is (usually) just a useful, virtually essential, conceptual abstraction based on the fundamental concept of the expected value of a measurable variable that is not fully predictable. There also needs to be a large number of actual occurrences possible for this measurable variable, or at least a well-defined conceptually possible large number of occurrences. Then the concept of an average of observed values converging to some stable number is at the heart of the concept of a statistical parameter. As such, a statistical parameter cannot be determined exactly by one, or a few, simple measurements. There is no instrument, or simple protocol, to record the exact value of a parameter used in a statistical model. (The exception occurs in measurement error models where the quantity measured is real but becomes the parameter of interest because each recorded measurement is recognized to be imprecise at a nonignorable level of imprecision).

We go a step further and recognize two classes of parameters in statistical models: (1) parameters that appear in the log-likelihood; these may or may not have any associated physical or biological reality; and (2) parameters as noted above that are directly related to expectations of measurable, hence predictable, variables. The second class of parameters are tied to measurable reality, but need not appear in the likelihood (they often do appear).

As an example, consider the analysis of cohort survival data, such as represented by examples in Section 5.2. The age-specific survival probability parameters $S_r$ cannot be directly measured (such as the weight of an animal can be). However, the concept represented by $S_r$ has clear and obvious ties to a measurable event: survival of an animal over a defined time interval. The event can be repeated based on a sample of animals (from a large, if not conceptually infinite, population of animals). These survival probability parameters are in the second class of parameters above. To provide both a useful representation of a set of age-specific survival probabilities, $\{S_r\}$, and provide the basis for parsimonious estimation of this set of parameters from limited data, statistical science adopts smooth, deterministic, parametric mathematical functions ("models" for short) such as

$$S_r = \frac{1}{1 + \exp[-(\theta_1 + \theta_2 \cdot r + \theta_3 \cdot r^2)]}$$

(as emphasized in this book, we should not pretend that exact equality really holds). The parameters $\theta_1, \theta_2$, and $\theta_3$ appear in the likelihood function, $\mathcal{L}(\underline{\theta})$. These parameters are in our first class of parameters above, and they need not have any direct physical or biological reality. In this context the $\theta_i$ are very useful in making parsimonious predictions of the $S_r$, which now become derived parameters based on the interpretable and parsimonious parametric model. Often, interpretability is as important as parsimony, and it is fortuitous that the two criteria of model usefulness are complementary, rather than in conflict. (Interpretability is a subject-matter criterion, not a statistical one, so we have not focused on it here).

The relationship of a parameter to prediction and expectation (which are themselves concepts) is straightforward in a simple linear model like

$$E(y \mid x) = \beta_o + \beta_1 x.$$

If we can measure the values of $y$ when separately $x$ and $x + 1$ occur (we may be able to control $x$), then

$$\beta_1 = E(y \mid x + 1) - E(y \mid x).$$

Hence, measurements directly relatable to the parameter $\beta_1$ can be made. However, $\beta_1$ remains as the embodiment of a concept, whereas specific instances of $y$ can be discovered by direct measurement. As Mayr (1997) notes, concepts are often the driving force in science, much more so than specific discoveries. The concept of parametric models in statistical science is, and remains, a powerful force.

## 5.7 Summary

Modeling is very important, as illustrated in general by examples in this book, and as demonstrated by the much improved estimation results (better precision, less bias) for the chain binomial survival data examples of Section 5.2 and the storm data of Section 5.5. For example, rather than try to separately estimate survival rate for every age, or storm probability for each week, one should produce smoothed estimates of these parameters by using suitable parametric models. For such observational data (this applies to the other examples here—GPA data, Kangaroo data) we would rarely, if ever, know *a priori* the single best model to use for the analysis.

However, in all such cases the investigator can and should postulate *a priori* a small set of suitable candidate models for data analysis. Then $AIC_c$- or $QAIC_c$-based model selection can be very effective at providing a ranking of the models based on Akaike weights. If it makes sense to select a best model (if the models mean something as alternative scientific or mechanistic explanations), one can use the expected K-L best model to draw inferences (bearing in mind that the selection of that model as best is itself an inference). Sampling standard errors of estimated parameters can and should include model-selection uncertainty.

If the models are only a means to the end of "smoothing" the data, as is the case for prediction, then we recommend computing model-averaged parameter (prediction) estimators and their unconditional sampling standard errors based on the Akaike weights. Monte Carlo methods showed that this procedure worked well for the chain binomial models; unconditional confidence interval coverage is generally close to the nominal 95%, while traditional intervals conditioned on the selected best model may achieve only 70 to 80% coverage. Monte Carlo studies in this chapter also show that there is substantial model uncertainty but that the Akaike weights are effective at measuring this uncertainty. The sampling distribution of $\Delta_p$ was examined for many situations, and we found that generally a value $\geq 10$ corresponds to at least the 95th percentile and more often at least the 99th percentile. This supports our contention that an observed $\Delta_i \geq 10$ is strong evidence against model $M_i$.

For reliable results from simulation we recommend at least 10,000 Monte Carlo reps at each set of conditions used to generate data. This holds true for the bootstrap also: For the results to be stable to two significant digits one must often use at least 10,000 bootstrap reps. Too many applications of these simulation methods do not use enough replications.

We do not recommend the dimension-consistent criteria (e.g., BIC, HQ) for model-selection in the biological sciences or medicine. Such criteria are not estimates of K-L information, are based on poor assumptions, and perform poorly even when sample size is quite large. We do not recommend using any form of hypothesis testing for model-selection

The choice of models to examine is important. The chain binomial examples demonstrated that a class of logistic models produced better results than a model class that assumed constant survival rate after a given age. The GPA example demonstrates that in variable-selection problems, thoughtful considerations can lead to much better models than unthoughtful all-subsets selection.

In Section 5.3 we note that model-selection bias occurs in variables selection: Regression coefficient estimators, $\hat{\beta}_i$, are biased away from 0 because the variable $x_i$ is included in the model only when that variable seems to be important (i.e., when $\hat{\beta}_i$ is sufficiently different from 0). The less important a variable, the more biasing effect model-selection has on $\hat{\beta}_i$. Estimated error mean square, $\hat{\sigma}^2_{y|x}$, is biased low by model-selection. The use of $AIC_c$, more so than other methods, provides some protection against both model-selection biases. The best way to minimize model-selection bias is to reduce the number of models fit to the data by thoughtful *a priori* model formulation.

Usually, selection of a best model is needed if scientific understanding is the goal. However, often it is better to think in terms of making an inference based on the full set of models, rather than selecting just one model and basing inferences on that single model. This is especially true in all-subsets variable selection as practiced in regression, because the selected best model is highly variable. Model averaging is then particularly useful, as is computing the importance of a variable as the sum of the Akaike weights over all models in which that variable appears.

Erroneous results have stemmed from the frequent misuse of Monte Carlo simulation in judging various model-selection approaches. In many cases, the generating model has had a few parameters (very often $< 8$ and often $< 5$) with no or few tapering effects, and the objective has been to see which selection method most often chooses the generating model. This conceptualization is in the context of BIC and is a sterile exercise as regards real world applications; hence results do not often apply to real biological problems.

Sometimes authors and analysts have AIC values such as 5,000, 5010, and 5020 for three models under consideration and conclude that the models are a short distance apart and "one model is nearly as good as the other two." This is a poor interpretation and probably influenced by the large sample size that contributes to the fact that AIC values in this case are in the 5,000 range. The focus of attention must always be on the differences in AIC values, the $\Delta_i$ and the associated Akaike weights, $w_i$, and the ranking and calibration of the models based on the $\Delta_i$.

Robust confidence intervals can be established using formula (4.10) with either Akaike weights ($w_i$) or bootstrap estimated selection probabilities ($\hat{\pi}_i$). In all the examples we have examined, such intervals have excellent achieved coverage. There are surely cases where this simple approach does not perform well, but we have not found any during our investigations.

The two extended examples (kangaroo data, storm data) show the importance of QAIC (or $QAIC_c$) for real data analysis. In both cases a variance inflation factor, $\hat{c}$, is needed in the analysis and model selection; results are misleading if $c$ is just assumed to be 1. Another way the literature is misleading is that often AIC is used with small sample size (or with large $K$) when in fact it is imperative to use $AIC_c$. This mistake has led to many erroneous conclusions about the value of K-L–based model selection in general and in comparison to BIC, in particular: BIC has often been preferred over K-L–based methods for invalid reasons.

# 6
# Statistical Theory

This chapter contains theory and derivations relevant to Kullback–Leibler information-theory-based model selection. We have tried to make the other chapters of this book readable by a general audience, especially graduate students in various fields. Hence, we have reserved this chapter for the theoretical material we feel it is important to make available to statisticians and quantitative biologists. For many, it will suffice to know that this theory exists. However, we encourage persons, especially if they have some mathematical–statistical training, to read and try to understand the theory given here, because that understanding provides a much deeper knowledge of many facets of K-L–based model selection in particular, and of some general model selection issues also.

The material given here is a combination of our distillation and interpretation of the existing literature and what we feel are clarifications and extensions of the existing theory. In the former case we have not drawn heavily or directly from any one source; hence there is no particular reference we could cite for these derivations. We have not indicated what results might be truly new to the literature about the estimation of expected K-L information, partly because this is sometimes not clear even to us.

## 6.1 Useful Preliminaries

The sole purpose of this section is to provide a summary of the basic notation, concepts, and mathematical background needed to produce and understand the derivation of AIC and related issues that follow after this section. Even persons

6.1 Useful Preliminaries     231

who totally understand the mathematics involved will benefit from this section in that it establishes much of the notation and conventions to be used in Section 6.2 and beyond in this chapter.

As a model selection criterion, it is clear what AIC is: $-2\log(\mathcal{L}(\hat{\theta})) + 2K$ for a model with $K$ estimated parameters, $\hat{\theta}$ being the MLE of those parameters, computed from the data $\underline{x}$, under an assumed model (i.e., pdf) $g(\underline{x} \mid \underline{\theta})$. However, we need more detailed notation than just $\mathcal{L}(\hat{\theta})$, and in the derivations we need to alternate between the likelihood and the pdf interpretations of the model. Therefore, without loss of generality we take the likelihood of $\underline{\theta}$ as $\mathcal{L}(\underline{\theta} \mid \underline{x}) = g(\underline{x} \mid \underline{\theta})$ by simply then interpreting $g$ as a function of $\underline{\theta}$ given $\underline{x}$. If instead of using this convention we had constantly switched notation between $g(\underline{x} \mid \underline{\theta})$ and $\mathcal{L}(\underline{\theta} \mid \underline{x})$, that would be more confusing than simply staying with the single notation $g(\underline{x} \mid \underline{\theta})$. This dual usage of the notation $g(\underline{x} \mid \underline{\theta})$ is thus noted; the reader must follow the mathematics with an eye to which usage is being made at any point.

A second dual usage of notation for the random variable $\underline{x}$ arises: Sometimes $\underline{x}$ denotes the data (as a random variable), and sometimes $\underline{x}$ denotes the variable of integration, always wrt to $f(\underline{x})$, under an integral sign (over an $n$-dimensional space). Because we are dealing with random variables, integration is usually denoted in terms of the statistical expectation operator, but that operator is just an integral. At times we must have both an integral (hence $\underline{x}$) and, separately, data, say $\underline{y}$. But the notation for data versus integrand variable is arbitrary and sometimes must be switched back and forth in the derivations. It becomes impossible always to use $\underline{x}$ for a variable of integration and $\underline{y}$ for data; hence, we do not try to do so, and instead we often use $\underline{x}$ to denote data even though at other times $\underline{x}$ is an integrand variable and $\underline{y}$ are the data. *Always*, however, the data, no matter how denoted ($\underline{x}$ or $\underline{y}$, or otherwise), actually arise from truth, $f(\cdot)$, not from $g(\cdot \mid \underline{\theta})$ (when $f \neq g$); this is a critically important point.

AIC has been motivated, justified, and derived in a variety of ways (see, for example, Akaike 1973, Sawa 1978, Sugiura 1978, Chow 1981, Stone 1982, Shibata 1989, Bozdogan 1987), but these derivations are often cryptic and thus difficult to follow. Here we give a general derivation in some detail, but without being rigorous about all required conditions (they are not very restrictive). We do note where approximations are made. The data have some sample size $n$, and the general result is justified for "large" $n$. That is, the result is justified asymptotically as $n \to \infty$. Also, the integrals and expectations shown are over an $n$-dimensional sample space, although that fact is not fully denoted by the notation used.

The most general approach to deriving AIC uses the Taylor series expansion to second order. An elementary introduction to the Taylor series is given in Peterson (1960) (or any introductory calculus book); a more rigorous treatment, including results for real-valued multivariable functions, is given by Apostol (1957) (or any rigorous book on real analysis). If $h(\underline{\theta})$ is a real-valued function on $K$ dimensions, then the Taylor series expansion about some value $\underline{\theta}_o$ near to $\underline{\theta}$ is given below:

$$h(\underline{\theta}) = h(\underline{\theta}_o) + \left[\frac{\partial h(\underline{\theta}_o)}{\partial \underline{\theta}}\right]'[\underline{\theta} - \underline{\theta}_o] + \frac{1}{2}[\underline{\theta} - \underline{\theta}_o]'\left[\frac{\partial^2 h(\underline{\theta}_o)}{\partial \underline{\theta}^2}\right][\underline{\theta} - \underline{\theta}_o] + Re \quad (6.1)$$

($\underline{\theta}$ and $\underline{\theta}_o$ are just two different points in the space over which $h(\cdot)$ is defined). Here, $Re$ represents the exact remainder term for the quadratic Taylor series expansion; the exact nature of $Re$ is known (see Apostol 1957). Various approximations for the error that results from ignoring $Re$ can be given. For its heuristic value only, we can claim an approximation to this error, $Re$, is of order

$$O(\|\underline{\theta} - \underline{\theta}_o\|^3).$$

Here, for any vector argument $\underline{z} - \underline{w}$,

$$\|\underline{z} - \underline{w}\| = \sqrt{\sum_{i=1}^{K}(z_i - w_i)^2}$$

denotes the Euclidian distance between the two points in the $K$-dimensional space. Thus, the order of the approximation error is the cube of the Euclidian distance between $\underline{\theta}$ and $\underline{\theta}_o$. This is quite a simplification of what $Re$ is, but it makes the point that the error of approximation is quite small if this distance is small.

The notation $O(x)$ denotes an unspecified (but possibly complicated) function of the scalar argument $x$ that satisfies the condition that $O(x)$ is approximately equal to $cx$ for small $x$, where $c$ is a constant. Hence, $O(x)$ goes to 0 at least at a linear rate in $x$ as $x$ gets near 0. In the case of (6.1) the quadratic approximation to $h(\underline{\theta})$ "near" $\underline{\theta}_o$ is arbitrarily good, as $\underline{\theta}$ is nearer to $\underline{\theta}_o$ for $h(\cdot)$ a suitably smooth and bounded function.

In (6.1) the notation

$$\left[\frac{\partial h(\underline{\theta}_o)}{\partial \underline{\theta}}\right]$$

denotes a $K \times 1$ column vector of the first partial derivatives of $h(\underline{\theta})$ with respect to $\theta_1, \ldots, \theta_K$, evaluated at $\underline{\theta} = \underline{\theta}_o$; hence,

$$\left[\frac{\partial h(\underline{\theta}_o)}{\partial \underline{\theta}}\right] = \begin{bmatrix} \frac{\partial h(\underline{\theta})}{\partial \theta_1} \\ \vdots \\ \frac{\partial h(\underline{\theta})}{\partial \theta_K} \end{bmatrix}_{|\underline{\theta} = \underline{\theta}_o}$$

The notation

$$\left[\frac{\partial^2 h(\underline{\theta}_o)}{\partial \underline{\theta}^2}\right] = \left\{\frac{\partial^2 h(\underline{\theta})}{\partial \theta_i \partial \theta_j}\right\}_{|\underline{\theta} = \underline{\theta}_o}, \quad i = 1, \ldots, K, \quad j = 1, \ldots, K, \quad (6.2)$$

denotes the $K \times K$ matrix of second mixed partial derivatives of $h(\underline{\theta})$ with respect to $\theta_1, \ldots, \theta_K$, evaluated at $\underline{\theta} = \underline{\theta}_o$. This matrix is often called the Hessian of $h(\underline{\theta})$.

The expansion in (6.1) when terminated at the quadratic term is only an approximation to $h(\underline{\theta})$. In this deterministic case, as indicated above, the error of approximation is related roughly to the cube of the Euclidean distance between $\underline{\theta}$ and $\underline{\theta}_o$. For a sufficiently small distance, this is a good order of approximation. For the cases of interest, $h(\cdot)$ will be a log-likelihood function based on a probability distribution. One special value of $\underline{\theta}$, denoted by $\underline{\theta}_o$, needed in this expansions is the large sample (hence approximate) expected value of the MLE, $\hat{\underline{\theta}}$; that is, $E(\hat{\underline{\theta}}) \doteq \underline{\theta}_o$ for large $n$ (the exact nature of $\underline{\theta}_o$ in relation to K-L information will be given below). The approximation here is often of order $1/n$, denoted by $O(1/n)$. This notation means that the error of approximation in $E(\hat{\underline{\theta}}) \doteq \underline{\theta}_o$ is less than or equal to a constant divided by the sample size for large sample sizes (the constant might even be 0).

Stronger statements about large-sample limits are possible. In particular, as sample size $n \to \infty$, $\hat{\underline{\theta}} \to \underline{\theta}_o$ with probability 1, and the Taylor series approximation given by (6.1) is quite good. In this case (6.1) becomes (as one usage)

$$h(\hat{\underline{\theta}}) = h(\underline{\theta}_o) + \left[\frac{\partial h(\underline{\theta}_o)}{\partial \underline{\theta}}\right]' [\hat{\underline{\theta}} - \underline{\theta}_o] + \frac{1}{2}[\hat{\underline{\theta}} - \underline{\theta}_o]' \left[\frac{\partial^2 h(\underline{\theta}_o)}{\partial \underline{\theta}^2}\right]' [\hat{\underline{\theta}} - \underline{\theta}_o] + O_p(1/n). \tag{6.3}$$

Now the error of approximation in (6.3) is stochastic, but its expectation is generally on the order of $1/n$ with probability going to 1 as $n \to \infty$, hence the added "$p$" notation of the form $O_p(\cdot)$. The exact size of the expected error of approximation in expansions like (6.3) is not known (in general), but it is negligible for large sample sizes, subject to mild regularity conditions of the same type needed to ensure that the MLE is well behaved (see, for example, Lehmann 1983).

In the context of parametric MLE the standard approach is to assume that the data are generated by one specific member of a family of models. That family of models, denoted here by $g(\underline{x} \mid \underline{\theta})$, is a set of probability distributions indexed by an unknown parameter that may be estimated as any value in the parameter space $\Theta$. By assumption, truth corresponds to one specific (but unknown) value of $\underline{\theta}$, which we could for clarity denote by $\underline{\theta}_o$. One would not ask where $\underline{\theta}_o$ comes from; it simply exists as (unknown) truth. Thus even when we assume that the known model structure of $g$ is true, there is still a fundamental concept of an underlying unknown truth to the problem of inference from data (and we cannot know, metaphysically, where this truth, $\underline{\theta}_o$, comes from).

When we acknowledge that $g$ is just a model of truth, hence must be misspecified, the issue arises as to what unique parameter in $\Theta$, hence what unique distribution in the class $g$, we are estimating. In fact, there is a unique parameter in $\Theta$ that the MLE, $\hat{\underline{\theta}}$, is estimating given the parametric class of models and given the concept of a fixed underlying unknown truth, as a pdf $f(\underline{x})$. As part of this essential conceptualization of the inference problem we must assume that the data arose from some deep truth, denoted without loss of generality (wlg) by $f$. Now, one cannot usefully ask where truth, $f$, comes from, in the same metaphysical sense that one cannot ask where $\underline{\theta}_o$ comes from under the assumption that $g(\underline{x} \mid \underline{\theta}_o)$ is truth, but we just do not happen to know true $\underline{\theta}_o$.

Given this essential framework of $f$ as truth (rather than any model structure as truth) we can, and must, ask whether there is a unique model $g(\cdot \mid \underline{\theta}_o)$ in the class of models $g(\cdot \mid \underline{\theta})$ that best describes the data? Hence, given the set of models $g(\cdot \mid \underline{\theta})$, $\underline{\theta} \in \Theta$, is there a unique $\underline{\theta}_o$ that the MLE is estimating, and is this $g(\cdot \mid \underline{\theta}_o)$ a best model in some sense? In fact, the MLE is (for large samples) estimating a unique parameter value that we will denote by $\underline{\theta}_o$; it is this parameter value that indexes our target model under likelihood inference (we will say more on this below).

Approached theoretically, ignoring issues of data and estimation, the best approximating model $g$ in the class of models considered, under the (compelling) K-L information measure, is simply the model that produces the minimum K-L value over $\Theta$. Hence we look for a unique value of $\underline{\theta} \in \Theta$, which we will denote by $\underline{\theta}_o$, that provides the K-L best approximating model. Therefore, $\underline{\theta}_o$ is the solution to the optimization problem

$$\min_{\underline{\theta} \in \Theta} [I(f, g)] = \int f(\underline{x}) \log \left( \frac{f(\underline{x})}{g(\underline{x} \mid \underline{\theta}_o)} \right) d\underline{x}.$$

Clearly, $g(\underline{x} \mid \underline{\theta}_o)$ is the best model here, and this serves, in fact, to define truth as a target $\underline{\theta}_o$ given $f$ and given the class of models, $g$. As we will discuss below, the MLE of $\underline{\theta}$ under model $g$ is estimating $\underline{\theta}_o$.

Given the assumed regularity conditions on the model, $\underline{\theta}_o$ satisfies the vector equations

$$\frac{\partial}{\partial \underline{\theta}} \int f(\underline{x}) \log \left( \frac{f(\underline{x})}{g(\underline{x} \mid \underline{\theta}_o)} \right) d\underline{x} = \underline{0}. \tag{6.4}$$

Rewriting (6.4) using that $\log(a/b) = \log(a) - \log(b)$, we have

$$\frac{\partial}{\partial \underline{\theta}} \int f(\underline{x}) \log(f(\underline{x})) - \frac{\partial}{\partial \underline{\theta}} \int f(\underline{x}) \log(g(\underline{x} \mid \underline{\theta})) d\underline{x} = \underline{0}.$$

Because $\underline{\theta}$ is not involved in $f(\cdot)$, the first term of the above is $\underline{0}$. The second term (ignoring the minus sign) can be written as

$$\int f(\underline{x}) \left[ \frac{\partial}{\partial \underline{\theta}} \log(g(\underline{x} \mid \underline{\theta})) \right]_{|\underline{\theta} = \underline{\theta}_o} d\underline{x} = E_f \left[ \left[ \frac{\partial}{\partial \underline{\theta}} \log(g(\underline{x} \mid \underline{\theta})) \right]_{|\underline{\theta} = \underline{\theta}_o} \right] = \underline{0}.$$

In a more compact way to denote this result we can write

$$E_f \left[ \frac{\partial}{\partial \underline{\theta}} \log(g(\underline{x} \mid \underline{\theta}_o)) \right] = \underline{0}. \tag{6.5}$$

The well-known asymptotic consistency property of MLEs and strong convergence of means of *iid* random variables allow, in conjunction with (6.5), another interpretation of $\underline{\theta}_o$. If $\underline{x}$ represents an *iid* sample of size $n$ from pdf $f(\underline{x}) \equiv \prod_{i=1}^{n} f(x_i)$ and we consider the MLE under model $g(\underline{x} \mid \underline{\theta}) \equiv \prod_{i=1}^{n} g(x_i \mid \underline{\theta}_o)$, then for every $n$ we have the $K$ likelihood equations (expressed as a mean, wlg)

$$\frac{1}{n} \left[ \sum_{i=1}^{n} \frac{\partial}{\partial \underline{\theta}} \log(g(x_i \mid \underline{\hat{\theta}})) \right] = \underline{0}.$$

As $n \to \infty$ two limits are approached with probability one (almost sure convergence). The sequence of MLEs, $\underline{\hat{\theta}}(n)$ (adding notation to denote the MLE as a function of sample size), converges to something. In fact, $\underline{\hat{\theta}}(n)$ has to converge to

$\underline{\theta}_o$ because the means on the left-handed sides of the above likelihood equations converge (as $n$ gets large) to their expected values, which must all equal 0. Under suitable regularity conditions, (6.5) is satisfied only for the unique value of $\underline{\theta} = \underline{\theta}_o$. Hence, the sequence $\hat{\underline{\theta}}(n)$ must converge almost surely to $\underline{\theta}_o$, which is the K-L minimizer (see, e.g., White 1994).

Some deep ideas and philosophy are involved in the above results. In particular, we have the distinction that unknown truth, $f(\underline{x})$, implicitly incorporates the numerical values on the (often only) conceptual, but unknown, parameter, $\underline{\theta}$, of interest to us. Yet $f(\underline{x})$ is not a mathematical function of $\underline{\theta}$. Only our model $g(\underline{x} \mid \underline{\theta})$ is a mathematical function of $\underline{\theta}$, because therein $\underline{\theta}$ is unknown, but interpretable, hence useful to consider, and varies over a defined parameter space. Even if we think that $g(\cdot)$ represents truth, this is only the case at the single point $\underline{\theta}_o$ in the parameter space (in a frequentist philosophy of statistics). Hence, in this case we would be saying that $f(\underline{x}) \equiv g(\underline{x} \mid \underline{\theta}_o)$, where $\underline{\theta}_o$ is a single fixed point; thus even in this context $\underline{\theta}$ is not a variable in $f(\underline{x})$. Therefore, in this or any case, $f(\underline{x})$ is not a function of $\underline{\theta}$, and therefore

$$\frac{\partial}{\partial \underline{\theta}} \int f(\underline{x}) \log(f(\underline{x})) d\underline{x} = \underline{0}.$$

The derivation of AIC occurs in the context of probability distributions and expectations of functions of random variables. Such expectations are just types of integrals, but the notation and "machinery" of statistical expectations are more convenient to use here than the explicit notation of integration. One particular aspect of this matter that needs to be noted is the validity of interchanging the order of taking two expectations of the form $E_{\underline{x}} E_{\underline{y}}[h(\underline{x}, \underline{y})]$ when $\underline{x}$ and $\underline{y}$ denote random variables. The function $h(\cdot, \cdot)$ is arbitrary. From basic calculus of integrals as linear operators, $E_{\underline{x}} E_{\underline{y}}[h(\underline{x}, \underline{y})] = E_{\underline{y}} E_{\underline{x}}[h(\underline{x}, \underline{y})]$. This interchange equivalence is true for the case of $\underline{x}$ and $\underline{y}$ having the same or different probability distributions and whether or not $\underline{x}$ and $\underline{y}$ are independent.

Another aspect of preliminaries concerns (6.2). If $h(\cdot)$ is the log-likelihood, $\log(g(\underline{x} \mid \underline{\theta}))$, then (6.2) is

$$\left\{ \frac{\partial^2 \log(g(\underline{x} \mid \underline{\theta}))}{\partial \theta_i \partial \theta_j} \right\}_{\mid \underline{\theta} = \underline{\theta}_o},$$

which is related to the Fisher information matrix

$$\mathcal{I}(\underline{\theta}_o) = E_g \left\{ -\frac{\partial^2 \log(g(\underline{x} \mid \underline{\theta}))}{\partial \theta_i \partial \theta_j} \right\}_{\mid \underline{\theta} = \underline{\theta}_o} \tag{6.6}$$

(expectation here is wrt $g(\cdot)$). If $g(\cdot)$ is the true model form (which it is if $f$ is a special case of $g$, or if $g = f$), then the sampling variance–covariance matrix, $\Sigma$, of the MLE is (for large samples) $\Sigma = [\mathcal{I}(\underline{\theta}_o)]^{-1}$. That is, $\Sigma = E(\hat{\underline{\theta}} - \underline{\theta}_o)(\hat{\underline{\theta}} - \underline{\theta}_o)'$ is $[\mathcal{I}(\underline{\theta}_o)]^{-1}$. If $g$ is not the true model for $\underline{x}$ (it may be less general than the true model, or otherwise different from $f$), then in general we must expect that $\Sigma \neq [\mathcal{I}(\underline{\theta}_o)]^{-1}$. In fact, in deriving AIC, we take expectations with respect to $f$, not $g$. Hence, we define

$$I(\underline{\theta}_o) = E_f \left\{ -\frac{\partial^2 \log(g(\underline{x} \mid \underline{\theta}))}{\partial \theta_i \partial \theta_j} \right\}_{|\underline{\theta}=\underline{\theta}_o}. \tag{6.7}$$

In the case that $f = g$ or $f$ is a special case of $g$, then and only then do we have $\mathcal{I}(\underline{\theta}_o) = I(\underline{\theta}_o)$. We will not generally make this distinction in our notation as to whether the situation allows $\mathcal{I}(\underline{\theta}_o) = I(\underline{\theta}_o)$ or not. It is an important matter, however, to be always cognizant of whether the expectation defining any given $I(\underline{\theta}_o)$ is wrt $f$ or $g$.

Additional notation useful here is the empirical, but unknown, matrix

$$\hat{I}(\underline{\theta}_o) = \left\{ -\frac{\partial^2 \log(g(\underline{x} \mid \underline{\theta}))}{\partial \theta_i \partial \theta_j} \right\}_{|\underline{\theta}=\underline{\theta}_o}.$$

For simpler notation we will use

$$I(\underline{\theta}_o) = E_f \left[ -\frac{\partial^2 \log(g(\underline{x} \mid \underline{\theta}_o))}{\partial \underline{\theta}^2} \right],$$

which means exactly the same as (6.7), and hence simpler notation for the $\hat{I}(\underline{\theta}_o)$ is

$$\hat{I}(\underline{\theta}_o) = -\frac{\partial^2 \log(g(\underline{x} \mid \underline{\theta}_o))}{\partial \underline{\theta}^2}.$$

It is obvious that $E_f[\hat{I}(\underline{\theta}_o)] = I(\underline{\theta}_o)$. When $\underline{x}$ is a random sample from $f(\cdot)$, $\hat{I}(\underline{\theta}_o)$ converges to $I(\underline{\theta}_o)$ as $n \to \infty$. We can express this alternatively as

$$\hat{I}(\underline{\theta}_o) = I(\underline{\theta}_o) + Re, \text{ and usually } Re \text{ is } O(1/n).$$

An actual estimator of $I(\underline{\theta}_o)$ is $\hat{I}(\hat{\underline{\theta}})$ (the negative Hessian of the log-likelihood equations):

$$\hat{I}(\hat{\underline{\theta}}) = -\frac{\partial^2 \log(g(\underline{x} \mid \hat{\underline{\theta}}))}{\partial \underline{\theta}^2}. \tag{6.8}$$

Because $\hat{\underline{\theta}}$ is the MLE under the model $g(\underline{x} \mid \underline{\theta})$, $\hat{\underline{\theta}}$ converges to $\underline{\theta}_o$ as $n \to \infty$, and hence $\hat{I}(\hat{\underline{\theta}})$ converges to $I(\underline{\theta}_o)$. Thus, $\hat{I}(\hat{\underline{\theta}}) \doteq I(\underline{\theta}_o)$; the order of this approximation is at worst $O(1/\sqrt{n})$, and in most common applications it will be $O(1/n)$. If we could determine the analytical form (under actual analysis of data) of $I(\underline{\theta}_o)$, an alternative estimator would be $I(\hat{\underline{\theta}})$, i.e., (6.7) evaluated at the MLE; $I(\hat{\underline{\theta}})$ is often not the same as $\hat{I}(\hat{\underline{\theta}})$. Note also that the commonly used estimator $\mathcal{I}(\hat{\underline{\theta}})$ (i.e., formula 6.6 evaluated at the MLE) is not always the same as either $I(\hat{\underline{\theta}})$ or $\hat{I}(\hat{\underline{\theta}})$ and may not converge to $I(\underline{\theta}_o)$.

There are two ways to compute the Fisher information matrix, $\mathcal{I}(\underline{\theta})$, of (6.6) when $f = g$. This additional material, and more, is needed below, hence is given here. Because the model is a probability distribution,

$$\int g(\underline{x} \mid \underline{\theta}) d\underline{x} = 1,$$

## 6.1 Useful Preliminaries

and therefore (under the same mild regularity conditions already assumed)

$$\int \frac{\partial g(\underline{x} \mid \underline{\theta})}{\partial \underline{\theta}} d\underline{x} = \underline{0}.$$

Next, we use in the above the result

$$\frac{\partial \log(g(\underline{x} \mid \underline{\theta}))}{\partial \underline{\theta}} = \frac{1}{g(\underline{x} \mid \underline{\theta})} \left[ \frac{\partial g(\underline{x} \mid \underline{\theta})}{\partial \underline{\theta}} \right],$$

and hence we get

$$\int g(\underline{x} \mid \underline{\theta}) \left[ \frac{\partial}{\partial \underline{\theta}} \log(g(\underline{x} \mid \underline{\theta})) \right] d\underline{x} = \underline{0}. \tag{6.9}$$

Now take the partial derivative vector of (6.9) with respect to $\underline{\theta}$ to get (6.10); this derivation uses the chain rule of differentiation and some of the above algebraic results:

$$\int g(\underline{x} \mid \underline{\theta}) \left[ \frac{\partial}{\partial \underline{\theta}} \log(g(\underline{x} \mid \underline{\theta})) \right] \left[ \frac{\partial}{\partial \underline{\theta}} \log(g(\underline{x} \mid \underline{\theta})) \right]' d\underline{x}$$
$$+ \int g(\underline{x} \mid \underline{\theta}) \frac{\partial^2 \log(g(\underline{x} \mid \underline{\theta}))}{\partial \underline{\theta}^2} d\underline{x} = O \tag{6.10}$$

($O$ is a $K \times K$ matrix of zero elements). We can rewrite (6.10) as

$$E_g \left[ \left[ \frac{\partial}{\partial \underline{\theta}} \log(g(\underline{x} \mid \underline{\theta})) \right] \left[ \frac{\partial}{\partial \underline{\theta}} \log(g(\underline{x} \mid \underline{\theta})) \right]' \right] = E_g \left[ -\frac{\partial^2 \log(g(\underline{x} \mid \underline{\theta}))}{\partial \underline{\theta}^2} \right],$$

or

$$E_g \left[ \left[ \frac{\partial}{\partial \underline{\theta}} \log(g(\underline{x} \mid \underline{\theta})) \right] \left[ \frac{\partial}{\partial \underline{\theta}} \log(g(\underline{x} \mid \underline{\theta})) \right]' \right] = \mathcal{I}(\underline{\theta}).$$

We will denote the left-hand side of the above as $\mathcal{J}(\underline{\theta})$; hence define

$$\mathcal{J}(\underline{\theta}) = E_g \left[ \left[ \frac{\partial}{\partial \underline{\theta}} \log(g(\underline{x} \mid \underline{\theta})) \right] \left[ \frac{\partial}{\partial \underline{\theta}} \log(g(\underline{x} \mid \underline{\theta})) \right]' \right]. \tag{6.11}$$

Thus, $\mathcal{I}(\underline{\theta}) = \mathcal{J}(\underline{\theta})$, but the expectations underlying this result are taken wrt $g(\underline{x} \mid \underline{\theta})$, not wrt to $f(\underline{x})$. One implication of this is that the inverse Fisher information matrix may not be the theoretically correct conditional variance–covariance matrix of the MLE if the model is misspecified.

What we need more than (6.11) is

$$J(\underline{\theta}) = E_f \left[ \left[ \frac{\partial}{\partial \underline{\theta}} \log(g(\underline{x} \mid \underline{\theta})) \right] \left[ \frac{\partial}{\partial \underline{\theta}} \log(g(\underline{x} \mid \underline{\theta})) \right]' \right]. \tag{6.12}$$

We can expect $J(\underline{\theta}) = \mathcal{J}(\underline{\theta})$ only when $f = g$, or $f$ is a special case of $g$. Although $\mathcal{I}(\underline{\theta}) = \mathcal{J}(\underline{\theta})$, there is no such general equality between $I(\underline{\theta})$ and $J(\underline{\theta})$ when $g$ is only an approximation to $f$, hence when the K-L discrepancy between $f$ and $g$, $I(f, g)$, is $> 0$. Heuristically, however, we can expect near equalities of the sort $I(\underline{\theta}_o) \doteq J(\underline{\theta}_o), \mathcal{I}(\underline{\theta}_o) \doteq I(\underline{\theta}_o)$, and $\mathcal{J}(\underline{\theta}_o) \doteq J(\underline{\theta}_o)$ when $I(f, g) \doteq 0$, hence when a good approximating model is used.

There is a large-sample relationship among $I(\underline{\theta}_o)$, $J(\underline{\theta}_o)$, and $\Sigma$ that is worth knowing, and perhaps should be used more:

$$I(\underline{\theta}_o)\Sigma = J(\underline{\theta}_o)[I(\underline{\theta}_o)]^{-1}, \tag{6.13}$$

and hence

$$\Sigma = [I(\underline{\theta}_o)]^{-1} J(\underline{\theta}_o)[I(\underline{\theta}_o)]^{-1}, \tag{6.14}$$

where $\Sigma$ is the true large-sample variance–covariance matrix of the MLE of $\underline{\theta}$ derived from model $g$ when $f$ is truth. It suffices to derive (6.14), although it is (6.13) that we will use more directly in deriving AIC.

Expanding the likelihood equations evaluated at $\underline{\theta}_o$ as a first-order Taylor series about the MLE, we have

$$\frac{\partial}{\partial \underline{\theta}} \log(g(\underline{x} \mid \underline{\theta}_o)) \doteq \frac{\partial}{\partial \underline{\theta}} \log(g(\underline{x} \mid \hat{\underline{\theta}})) + \left[\frac{\partial^2 \log(g(\underline{x} \mid \hat{\underline{\theta}}))}{\partial \underline{\theta}^2}\right](\underline{\theta}_o - \hat{\underline{\theta}}).$$

The MLE satisfies

$$\frac{\partial}{\partial \underline{\theta}} \log(g(\underline{x} \mid \hat{\underline{\theta}})) = \underline{0};$$

hence we have

$$\frac{\partial}{\partial \underline{\theta}} \log(g(\underline{x} \mid \underline{\theta}_o)) \doteq \left[-\frac{\partial^2 \log(g(\underline{x} \mid \hat{\underline{\theta}}))}{\partial \underline{\theta}^2}\right](\hat{\underline{\theta}} - \underline{\theta}_o)$$

$$= \hat{I}(\hat{\underline{\theta}})(\hat{\underline{\theta}} - \underline{\theta}_o) \doteq I(\underline{\theta}_o)(\hat{\underline{\theta}} - \underline{\theta}_o).$$

From the above we get

$$[I(\underline{\theta}_o)]^{-1}\left[\frac{\partial}{\partial \underline{\theta}} \log(g(\underline{x} \mid \underline{\theta}_o))\right] \doteq (\hat{\underline{\theta}} - \underline{\theta}_o). \tag{6.15}$$

Transpose (6.15) and use that transposed result along with, again, (6.15) to derive

$$[I(\underline{\theta}_o)]^{-1}\left[\frac{\partial}{\partial \underline{\theta}} \log(g(\underline{x} \mid \underline{\theta}_o))\right]\left[\frac{\partial}{\partial \underline{\theta}} \log(g(\underline{x} \mid \underline{\theta}_o))\right]'[I(\underline{\theta}_o)]^{-1} \doteq (\hat{\underline{\theta}} - \underline{\theta}_o)(\hat{\underline{\theta}} - \underline{\theta}_o)'.$$

Now take the expectation of the above with respect to $f(\underline{x})$ to get (see 6.13)

$$[I(\underline{\theta}_o)]^{-1} J(\underline{\theta}_o)[I(\underline{\theta}_o)]^{-1} \doteq E_f(\hat{\underline{\theta}} - \underline{\theta}_o)(\hat{\underline{\theta}} - \underline{\theta}_o)' = \Sigma;$$

hence, we have (6.14) as a large-sample result.

The above likelihood-based results under either a true data-generating model or under model misspecification (i.e., truth is $f$, the model used is $g$) are all in the statistical literature. For very rigorous derivations see White (1994).

To take expectations of the quadratic forms that are in expansions like (6.3) we will need to use an equivalent expression of that form:

$$\underline{z}' A \underline{z} = \text{tr}\left[A \underline{z}\underline{z}'\right].$$

Here "tr" stands for the matrix trace function: The sum of the diagonal elements of a square matrix. The trace function is a linear operator; therefore, when the quadratic is a stochastic variable in $\underline{z}$, its expectation can be written as

$$\mathrm{E}_{\underline{z}}\left[\underline{z}'A\underline{z}\right] = \mathrm{tr}\left[\mathrm{E}_{\underline{z}}\left[A\underline{z}\underline{z}'\right]\right].$$

If $A$ is fixed (or stochastic but independent of $\underline{z}$), then $\mathrm{E}_{\underline{z}}\left[A\underline{z}\underline{z}'\right] = A\mathrm{E}_{\underline{z}}\left[\underline{z}\underline{z}'\right]$. If $\underline{z}$ has mean $\underline{0}$ (such as $\underline{z} = \hat{\underline{\theta}} - \mathrm{E}(\hat{\underline{\theta}}))$, then $\mathrm{E}_{\underline{z}}[\underline{z}\underline{z}'] = \Sigma$ is the variance–covariance matrix of $\underline{z}$; hence then

$$\mathrm{E}_{\underline{z}}\left[\underline{z}'A\underline{z}\right] = \mathrm{tr}\left[A\Sigma\right].$$

If $A$ is stochastic but independent of $\underline{z}$, then we can use

$$\mathrm{E}_A\mathrm{E}_{\underline{z}}\left[\underline{z}'A\underline{z}\right] = \mathrm{tr}\left[\mathrm{E}_A\mathrm{E}_{\underline{z}}\left[A\underline{z}\underline{z}'\right]\right] = \mathrm{tr}\left[\mathrm{E}_A(A)\mathrm{E}_{\underline{z}}(\underline{z}\underline{z}')\right].$$

A final aspect of notation, and of concepts, reemphasizes some ideas at the start of this section: The notation for a random variable (i.e., data point) is arbitrary in taking expectations over the sample space. What is not arbitrary for such an expectation (i.e., integration) is the model used, which refers to its form, its assumptions, its parameters, and the distribution of unexplained residuals (i.e., "errors"). Moreover, it is just a convenience to switch thinking modes between integrals and expectations in this probabilistic modeling and data-analysis framework. Because an expectation is a type of integral over a defined space, the result of the integration is not dependent on the notation used in the integrand. Thus

$$\mathrm{E}_f\left[\log(g(\underline{x}\mid\hat{\underline{\theta}}(\underline{x})))\right] = \int f(\underline{x})\log(g(\underline{x}\mid\hat{\underline{\theta}}(\underline{x})))d\underline{x}$$
$$\equiv \int f(\underline{y})\log(g(\underline{y}\mid\hat{\underline{\theta}}(\underline{y})))d\underline{y}$$
$$= \mathrm{E}_f\left[\log(g(\underline{y}\mid\hat{\underline{\theta}}(\underline{y})))\right].$$

Changing notation for the integrand (i.e., $\underline{x}$ to $\underline{y}$) has no effect on the result; this type of useful notation change is required in derivations below, because in places, at a conceptual level, we recognize two independent samples, hence have two notations, $\underline{x}$ and $\underline{y}$. In fact, these derivations are about average frequentist properties of data-analysis methods, but there is no real data literally being used in these derivations. Rather, in these theoretical derivations the possible "data" are just points in an $n$-dimensional sample space that arise in accordance with some true probability distribution $f(\cdot)$.

## 6.2 A General Derivation of AIC

We now give a general conceptual and then mathematical derivation of AIC starting from K-L information for the best approximating model in the class of models

$g(\underline{x} \mid \underline{\theta})$:

$$I(f, g(\cdot \mid \underline{\theta}_o)) = \int f(\underline{x}) \log \left( \frac{f(\underline{x})}{g(\underline{x} \mid \underline{\theta}_o)} \right) d\underline{x}. \tag{6.16}$$

Note that while for the model we do not know $\underline{\theta}$, the target K-L information value for the class of models is appropriately taken as $I(f, g)$ evaluated at $\underline{\theta}_o$ (i.e., formula 6.16), because the parameter value we will be estimating is $\underline{\theta}_o$. Also, note the expanded notation in (6.16), so we can represent $I(f, g)$ as dependent, in general, on the unknown parameter value, given the model form. However, $I(f, g)$ does not involve any data, nor any value of $\underline{x}$, as $\underline{x}$ has been integrated out.

Given that we have data $\underline{y}$ as a sample from $f(\cdot)$, the logical step would be to find the MLE $\hat{\underline{\theta}} = \hat{\underline{\theta}}(\underline{y})$ and compute an estimate of $I(f, g(\cdot \mid \underline{\theta}_o))$ as

$$I(f, g(\cdot \mid \hat{\underline{\theta}}(\underline{y}))) = \int f(\underline{x}) \log \left( \frac{f(\underline{x})}{g(\underline{x} \mid \hat{\underline{\theta}}(\underline{y}))} \right) d\underline{x}.$$

This $I(f, g(\cdot \mid \hat{\underline{\theta}}(\underline{y})))$ remains conceptual, as we do not know $f$. Still, it is useful to push ahead, and we shall do so with two different conceptual approaches, both will lead to the same basis for AIC (there is no unique path from K-L to AIC).

If we could find the $\underline{\theta}_o$ that minimizes K-L (for a given $g$), we would know that our target for a perfect model would be $I(f, g) = 0$. We could then judge how good any model is relative to this absolute value of zero. But matters change when we have only an estimate of $\underline{\theta}$. Even if our model structure was (miraculously) truth, hence $g(\underline{x} \mid \underline{\theta}_o) = f(\underline{x})$, our estimator $\hat{\underline{\theta}}(\underline{y})$ would not equal $\underline{\theta}_o$ almost surely for continuous parameters and distributions, and at best for some discrete distributions the equality would be with probability $\ll 1$. Any value of $\hat{\underline{\theta}}(\underline{y})$ other than $\underline{\theta}_o$ results in $I(f, g(\cdot \mid \hat{\underline{\theta}}(\underline{y}))) > I(f, g(\cdot \mid \underline{\theta}_o))$. Thus, even if we had the correct model structure, because we must estimate $\underline{\theta}$ we should think in terms of the (essentially estimated) K-L as taking, on average, a value $> 0$. This motivates us to revise our idea of what our target must be as a measure of perfect agreement of fitted model with truth, $f$.

In the context of repeated sampling properties as a guide to inference we would expect our estimated K-L to have on average the positive value $E_{\underline{y}}\left[I(f, g(\cdot \mid \hat{\underline{\theta}}(\underline{y})))\right]$. We should therefore readjust our idea of perfection of the model to be not the minimizing of $I(f, g(\cdot \mid \underline{\theta}_o))$ (given $g$), but the slightly larger value, on average, given by

$$E_{\underline{y}}\left[I(f, g(\cdot \mid \hat{\underline{\theta}}(\underline{y})))\right] > I(f, g(\cdot \mid \underline{\theta}_o))$$

(and repeating ourselves because it is an important point: All expectations here are wrt $f$ regardless of the notation for random variables involved, such as $\underline{x}$, $\underline{y}$ or $\hat{\underline{\theta}}$). Thus, given the reality that we must *estimate* $\underline{\theta}$, we must adopt the criterion

$$\text{"select the model } g \text{ to minimize } E_{\underline{y}}\left[I(f, g(\cdot \mid \hat{\underline{\theta}}(\underline{y})))\right]." \tag{6.17}$$

Hence our goal must be to minimize the expected value of this (conceptually) estimated K-L information value. (If we could compute the value of $\underline{\theta}_o$ for each model, we could stay with the goal of minimizing K-L itself. For the curious we note here that the large sample difference is $E_{\underline{y}}\left[I(f, g(\cdot\mid\hat{\underline{\theta}}(\underline{y})))\right] - I(f, g(\cdot\mid\underline{\theta}_o)) = \frac{1}{2} \operatorname{tr}\left[J(\underline{\theta}_o)I(\underline{\theta}_o)^{-1}\right]$, which does not depend on sample size, $n$.)

Rewriting the basis of this new target to be minimized, (6.17), we have

$$E_{\underline{y}}\left[I(f, g(\cdot\mid\hat{\underline{\theta}}(\underline{y})))\right] = \int f(\underline{x})\log(f(\underline{x}))d\underline{x} - E_{\underline{y}}\left[\int f(\underline{x})\log[g(\underline{x}\mid\hat{\underline{\theta}}(\underline{y}))]d\underline{x}\right];$$

hence

$$E_{\underline{y}}\left[I(f, g(\cdot\mid\hat{\underline{\theta}}(\underline{y})))\right] = \text{Constant} - E_{\underline{y}}E_{\underline{x}}\left[\log[g(\underline{x}\mid\hat{\underline{\theta}}(\underline{y}))]\right]. \quad (6.18)$$

It turns out that we can estimate $E_{\underline{y}}E_{\underline{x}}\left[\log[g(\underline{x}\mid\hat{\underline{\theta}}(\underline{y}))]\right]$, and therefore we can select a model to minimize the expected estimated relative K-L information value given by (6.18). In most of our writing here about this matter we find it much simpler just to say that we are selecting an estimated relative K-L best model by use of AIC.

There is a second, less compelling, approach that we can take in going from K-L to AIC: Start with

$$I(f, g(\cdot\mid\underline{\theta}_o)) = \text{Constant} - E_{\underline{x}}\left[\log(g(\underline{x}\mid\underline{\theta}_o))\right]$$

and see whether we can compute (or estimate) $E_{\underline{x}}\left[\log(g(\underline{x}\mid\hat{\underline{\theta}}(\underline{y})))\right]$ based on Taylor series expansions. As will be made evident below, we can derive the result

$$E_{\underline{x}}\left[\log(g(\underline{x}\mid\hat{\underline{\theta}}(\underline{y})))\right] \doteq E_{\underline{x}}\left[\log(g(\underline{x}\mid\hat{\underline{\theta}}(\underline{x})))\right] - \frac{1}{2}\operatorname{tr}\left[J(\underline{\theta}_o)I(\underline{\theta}_o)^{-1}\right]$$
$$- \frac{1}{2}(\hat{\underline{\theta}}(\underline{y}) - \underline{\theta}_o)'I(\underline{\theta}_o)(\hat{\underline{\theta}}(\underline{y}) - \underline{\theta}_o).$$

On the rhs above, the only component that absolutely cannot be estimated or computed (in any useful way) is the quadratic term involving $(\hat{\underline{\theta}}(\underline{y}) - \underline{\theta}_o)$ (and it is pointless therein to use $\hat{\underline{\theta}}_o = \hat{\underline{\theta}}(\underline{y})$). But if we take the expectation of both sides above wrt $\underline{y}$, we get a quantity we can estimate:

$$E_{\underline{y}}E_{\underline{x}}\left[\log(g(\underline{x}\mid\hat{\underline{\theta}}(\underline{y})))\right] \doteq E_{\underline{x}}\left[\log(g(\underline{x}\mid\hat{\underline{\theta}}(\underline{x})))\right] - \operatorname{tr}\left[J(\underline{\theta}_o)I(\underline{\theta}_o)^{-1}\right].$$

Thus, either line of derivation demonstrates that we have to change our objective from model selection based on minimum K-L with known $\underline{\theta}_o$ given $g$, to selecting the model with estimated $\underline{\theta}$ based on minimizing an expected K-L information measure. It is still the case that only a relative minimum can be found based on $E_{\underline{y}}E_{\underline{x}}\left[\log(g(\underline{x}\mid\hat{\underline{\theta}}(\underline{y})))\right]$ as the target objective function to be maximized; the constant $E_{\underline{x}}[f(\underline{x})\log(f(\underline{x}))]$ cannot be computed or estimated.

Only some of the literature is clear that AIC model selection is based on the concept of minimizing the expected K-L criterion, $E_{\underline{y}}\left[I(f, g(\cdot\mid\hat{\underline{\theta}}(\underline{y})))\right]$ (see, e.g.,

Sawa 1978, Sugiura 1978, Bozdogan 1987 (page 351), Bonneu and Milhaud 1994). It is the relative value of this criterion that is estimated over the set of models. That is, we want to estimate without bias, as our model selection criterion (denoted below by $T$ for target) for each approximating model, the value of

$$T = \int f(\underline{y}) \left[ \int f(\underline{x}) \log(g(\underline{x} \mid \hat{\underline{\theta}}(\underline{y}))) d\underline{x} \right] d\underline{y}. \tag{6.19}$$

The change from conceptual model selection based on minimum K-L to actual model selection based on maximizing an estimate of $T$ in (6.19) is forced on us because we must estimate the parameters in $g$ based on a finite amount of data.

Sometimes the criterion given by (6.19), hence AIC, is motivated by the concept of Akaike's predicative likelihood $E_p[\log(\mathcal{L}(\hat{\theta}))] = E_y E_x[\log(\mathcal{L}(\hat{\theta}(y) \mid x))] \equiv T$, which has a heuristic interpretation in terms of cross validation and independent random variables $\underline{x}$ and $\underline{y}$. However, the quantity $T$, and selection by maximizing $\hat{T}$ (or minimizing $-2\hat{T}$), does arise from a pure K-L approach to the problem of model selection without ever invoking the idea of cross validation.

In a slightly simplified, but obvious, notation, the K-L–based model-selection problem is now to find a useful expression for, and estimator of, the target

$$T = E_{\hat{\underline{\theta}}} E_{\underline{x}} \left[ \log(g(\underline{x} \mid \hat{\underline{\theta}})) \right], \tag{6.20}$$

where it is understood that the MLE $\hat{\underline{\theta}}$ is based on sample $\underline{y}$, and the two expectations are for $\underline{x}$ and $\underline{y}$ (hence $\hat{\underline{\theta}}$) both with respect to truth, $f$. It is because $T$ is also a double expectation based, conceptually, on two independent samples, that AIC-based model selection is asymptotically equivalent to cross validation (see, e.g., Stone 1977); cross validation is a well-accepted basis of model selection.

Step 1 is an expansion of the form (6.3) applied to $\log(g(\underline{x} \mid \hat{\underline{\theta}}))$ around $\underline{\theta}_o$ for any given $\underline{x}$:

$$\log(g(\underline{x} \mid \hat{\underline{\theta}})) \doteq \log(g(\underline{x} \mid \underline{\theta}_o)) + \left[ \frac{\partial \log(g(\underline{x} \mid \underline{\theta}_o))}{\partial \underline{\theta}} \right]' [\hat{\underline{\theta}} - \underline{\theta}_o]$$

$$+ \frac{1}{2} [\hat{\underline{\theta}} - \underline{\theta}_o]' \left[ \frac{\partial^2 \log(g(\underline{x} \mid \underline{\theta}_o))}{\partial \underline{\theta}^2} \right] [\hat{\underline{\theta}} - \underline{\theta}_o]. \tag{6.21}$$

Truncation at the quadratic term entails an unknown degree of approximation (but it is an error of approximation that goes to zero as $n \to \infty$). To relate (6.21) to (6.20) we first take the expected value of (6.21) with respect to $\underline{x}$:

$$E_{\underline{x}} \left[ \log(g(\underline{x} \mid \hat{\underline{\theta}})) \right] \doteq E_{\underline{x}} \left[ \log(g(\underline{x} \mid \underline{\theta}_o)) \right] + E_{\underline{x}} \left[ \frac{\partial \log(g(\underline{x} \mid \underline{\theta}_o))}{\partial \underline{\theta}} \right]' [\hat{\underline{\theta}} - \underline{\theta}_o]$$

$$+ \frac{1}{2} [\hat{\underline{\theta}} - \underline{\theta}_o]' \left[ E_{\underline{x}} \frac{\partial^2 \log(g(\underline{x} \mid \underline{\theta}_o))}{\partial \underline{\theta}^2} \right] [\hat{\underline{\theta}} - \underline{\theta}_o]. \tag{6.22}$$

The vector multiplier of $[\hat{\underline{\theta}} - \underline{\theta}_o]$ in that linear term above is exactly the same as (6.5). It is just that for clarification $E_{\underline{x}}$ is used to mean $E_f$ over the function of the random variable $\underline{x}$ (and keep remembering that $\hat{\underline{\theta}} \equiv \hat{\underline{\theta}}(\underline{y})$ is independent

## 6.2 A General Derivation of AIC

of $\underline{x}$). Therefore, upon taking this expectation, the linear term vanishes; that is, (6.5) applies:

$$E_{\underline{x}}\left[\frac{\partial \log(g(\underline{x}\mid\underline{\theta}_o))}{\partial\underline{\theta}}\right] = \underline{0}.$$

Also, for the quadratic term in (6.22), definition (6.7) applies; hence we can write

$$E_{\underline{x}}\left[\log(g(\underline{x}\mid\underline{\hat{\theta}}))\right] \doteq E_{\underline{x}}\left[\log(g(\underline{x}\mid\underline{\theta}_o))\right] - \frac{1}{2}[\underline{\hat{\theta}}-\underline{\theta}_o]'I(\underline{\theta}_o)[\underline{\hat{\theta}}-\underline{\theta}_o]. \quad (6.23)$$

Now we can take the expectation of (6.23) with respect to $\underline{\hat{\theta}}$ (i.e., $\underline{y}$). Here is where the trace function is used, yielding

$$E_{\underline{\hat{\theta}}}E_{\underline{x}}\left[\log(g(\underline{x}\mid\underline{\hat{\theta}}))\right] \doteq E_{\underline{x}}\left[\log(g(\underline{x}\mid\underline{\theta}_o))\right] - \frac{1}{2}\operatorname{tr}\left[I(\underline{\theta}_o)E_{\underline{\hat{\theta}}}\left[[\underline{\hat{\theta}}-\underline{\theta}_o][\underline{\hat{\theta}}-\underline{\theta}_o]'\right]\right].$$

The left-hand side above is $T$ from (6.20), and $E_{\underline{\hat{\theta}}}\left[[\underline{\hat{\theta}}-\underline{\theta}_o][\underline{\hat{\theta}}-\underline{\theta}_o]'\right] = \Sigma$ is the correct large-sample theoretical sampling variance of the MLE, because the expectation herein is taken with respect to truth, $f$, not with respect to $g$. Thus we have

$$T \doteq E_{\underline{x}}\left[\log(g(\underline{x}\mid\underline{\theta}_o))\right] - \frac{1}{2}\operatorname{tr}\left[I(\underline{\theta}_o)\Sigma\right]. \quad (6.24)$$

Step 2 starts with the realization that we have not yet derived what we need: a relationship between $T$ and $E_{\underline{x}}\left[\log[g(\underline{x}\mid\underline{\hat{\theta}}(x))]\right]$, which is the expectation of the actual log-likelihood at the MLE. We now do a second expansion, this time of $\log(g(\underline{x}\mid\underline{\theta}_o))$ about $\underline{\hat{\theta}}(x)$, treating $\underline{x}$ as the sample data, hence getting the MLE of $\underline{\theta}$ for this $\underline{x}$. This procedure is acceptable, because all we are after is an expected value, which means taking an integral over all possible points in the sample space. Therefore, it does not matter what notation we use for these sample points: $\underline{x}$ or $\underline{y}$. Applying the Taylor series approximation (6.3) (but with the roles of $\underline{\hat{\theta}}$ and $\underline{\theta}_o$ switched; also note well that here, $\underline{\hat{\theta}} \equiv \underline{\hat{\theta}}(x)$):

$$\log(g(\underline{x}\mid\underline{\theta}_o)) \doteq \log(g(\underline{x}\mid\underline{\hat{\theta}})) + \left[\frac{\partial \log(g(\underline{x}\mid\underline{\hat{\theta}}))}{\partial\underline{\theta}}\right]'[\underline{\theta}_o-\underline{\hat{\theta}}]$$

$$+ \frac{1}{2}[\underline{\theta}_o-\underline{\hat{\theta}}]'\left[\frac{\partial^2 \log(g(\underline{x}\mid\underline{\hat{\theta}}))}{\partial\underline{\theta}^2}\right][\underline{\theta}_o-\underline{\hat{\theta}}]. \quad (6.25)$$

The MLE $\underline{\hat{\theta}}$ is the solution of, hence satisfies, the equations

$$\frac{\partial \log(g(\underline{x}\mid\underline{\hat{\theta}}))}{\partial\underline{\theta}} = \underline{0}.$$

Therefore, the linear term in (6.25) vanishes. Taking the needed expectation we can write

$$E_{\underline{x}}\left[\log(g(\underline{x}\mid\underline{\theta}_o))\right] \doteq E_{\underline{x}}\left[\log(g(\underline{x}\mid\underline{\hat{\theta}}))\right] - \frac{1}{2}\operatorname{tr}\left[E_{\underline{x}}\left[\hat{I}(\underline{\hat{\theta}})\right][\underline{\theta}_o-\underline{\hat{\theta}}][\underline{\theta}_o-\underline{\hat{\theta}}]'\right].$$

$$(6.26)$$

See (6.8) for $\hat{I}(\hat{\theta})$, the Hessian of the log-likelihood evaluated at the MLE.

To make analytical progress with (6.26) we use the approximation $\hat{I}(\hat{\theta}) \doteq I(\underline{\theta}_o)$; hence we obtain

$$\mathrm{E}_{\underline{x}}\left[\hat{I}(\hat{\theta})\right][\underline{\theta}_o - \hat{\underline{\theta}}][\underline{\theta}_o - \hat{\underline{\theta}}]' \doteq \left[I(\underline{\theta}_o)\right]\left[\mathrm{E}_{\underline{x}}[\underline{\theta}_o - \hat{\underline{\theta}}][\underline{\theta}_o - \hat{\underline{\theta}}]'\right]$$
$$= \left[I(\underline{\theta}_o)\right]\left[\mathrm{E}_{\underline{x}}[\hat{\underline{\theta}} - \underline{\theta}_o][\hat{\underline{\theta}} - \underline{\theta}_o]'\right]$$
$$= \left[I(\underline{\theta}_o)\right]\Sigma. \quad (6.27)$$

The approximation made in (6.27) is often good to $O(1/n)$, hence is justified. However, there are circumstances where the approximation may not be this good, and the overall approximation in (6.27) is equivalent to using $\hat{I}(\hat{\theta}) \doteq I(\underline{\theta}_o)$ after first writing the approximation

$$\mathrm{E}_{\underline{x}}\left[\hat{I}(\hat{\theta})\right][\underline{\theta}_o - \hat{\underline{\theta}}][\underline{\theta}_o - \hat{\underline{\theta}}]' \doteq \left[\mathrm{E}_{\underline{x}}\left[\hat{I}(\hat{\theta})\right]\right]\left[\mathrm{E}_{\underline{x}}[\underline{\theta}_o - \hat{\underline{\theta}}][\underline{\theta}_o - \hat{\underline{\theta}}]'\right]$$
$$= \left[[I(\underline{\theta}_o)]\right]\Sigma \quad (6.28)$$

to arrive at the same result as (6.27). In any case, (6.28) does improve with sample size, but the overall error involved in this approximation to $\mathrm{E}_{\underline{x}}\left[\hat{I}(\hat{\theta})\right][\underline{\theta}_o - \hat{\underline{\theta}}][\underline{\theta}_o - \hat{\underline{\theta}}]'$ is hard to assess, in general. (The matter is revisited below for the exponential family of distributions, and (6.28) is found to be there a good approximation to $O(1/n)$.)

Using either (6.27), or (6.28), along with (6.26), we have

$$\mathrm{E}_{\underline{x}}\left[\log(g(\underline{x}\mid\underline{\theta}_o))\right] \doteq \mathrm{E}_{\underline{x}}\left[\log(g(\underline{x}\mid\hat{\underline{\theta}}(\underline{x})))\right] - \frac{1}{2}\mathrm{tr}\left[I(\underline{\theta}_o)\Sigma\right]. \quad (6.29)$$

Recall what (6.24) was:

$$T \doteq \mathrm{E}_{\underline{x}}\left[\log(g(\underline{x}\mid\underline{\theta}_o))\right] - \frac{1}{2}\mathrm{tr}\left[I(\underline{\theta}_o)\Sigma\right].$$

Substituting (6.29) into (6.24) we have a key result that is known in the literature:

$$T \doteq \mathrm{E}_{\underline{x}}\left[\log(g(\underline{x}\mid\hat{\underline{\theta}}(\underline{x})))\right] - \mathrm{tr}\left[I(\underline{\theta}_o)\Sigma\right]. \quad (6.30)$$

The literature usually presents not (6.30), but rather an alternate equivalent form based on (6.13):

$$T \doteq \mathrm{E}_{\underline{x}}\left[\log(g(\underline{x}\mid\hat{\underline{\theta}}(\underline{x})))\right] - \mathrm{tr}\left[J(\underline{\theta}_o)[I(\underline{\theta}_o)]^{-1}\right]. \quad (6.31)$$

The notation $\hat{\underline{\theta}}(\underline{x})$ rather than just $\hat{\underline{\theta}}$ is used above only to emphasize that on the right-hand side of (6.31) only one random variable $\underline{x}$ appears, and it can be taken to refer to the actual data. From (6.30) or (6.31), we can infer that a criterion for model selection (i.e., a nearly unbiased estimator of $T$) is structurally of the form

$$\hat{T} \doteq \log(g(\underline{x}\mid\hat{\underline{\theta}})) - \widehat{\mathrm{tr}}\left[I(\underline{\theta}_o)\Sigma\right], \quad (6.32)$$

or

$$\hat{T} \doteq \log(g(\underline{x}\mid\hat{\underline{\theta}})) - \widehat{\mathrm{tr}}\left[J(\underline{\theta}_o)[I(\underline{\theta}_o)]^{-1}\right]. \quad (6.33)$$

## 6.2 A General Derivation of AIC

Simple, direct estimation of $\Sigma$ from one sample is not possible, because there is only one $\hat{\theta}$ available (a bootstrap estimator of $\Sigma$ is possible), whereas both $J(\theta_o)$ and $I(\theta_o)$ are directly estimable from the single sample. We note that (6.33), but not (6.32), requires a parametrization wherein $I(\theta_o)$ is of full rank, whence its inverse exists. There is no loss in generality if we assume that all the probability distribution models have fully identifiable parameters, and hence are of full rank.

The maximized log-likelihood, $\log(g(\underline{x} \mid \hat{\theta}))$, in (6.31) is an unbiased estimator of its own expectation, $E_x[\log(g(\underline{x} \mid \hat{\theta}))]$ (but is biased as an estimator of $T$). Hence, the only problem left is to get a reliable (low, or no, bias) estimator of the trace term, or at least an estimator with small mean square error. Then the best model to use is the one with the largest value of $\hat{T}$, because this would produce a model with the smallest estimated expected K-L distance. As a matter of convention the criterion is often stated as that of minimizing

$$-2\log(g(\underline{x} \mid \hat{\theta})) + 2\widehat{\mathrm{tr}}\left[J(\theta_o)[I(\theta_o)]^{-1}\right]. \tag{6.34}$$

If $f$ is a subset of $g$ (i.e., if $g \equiv f$ or $f$ is contained within $g$ in the sense of nested models), then $I(\theta_o) \equiv \mathcal{I}(\theta_o) = \mathcal{J}(\theta_o) = J(\theta_o) = \Sigma^{-1}$, and hence $\mathrm{tr}\left[I(\theta_o)\Sigma\right] = K$. Even if $g$ is just a good model (i.e., a good approximation) for $f$, the literature supports the idea that our best estimator is probably to use $\widehat{\mathrm{tr}}\left[I(\theta_o)\Sigma\right] = K$ (Shibata 1989, in prep.).

When the model is too restrictive to be good, the term "$-2\log(g(\underline{x} \mid \hat{\theta}))$" will be much inflated (compared to this same term for a "good" model), and we will not select that model. In this case having a good estimate of the trace term should not matter. The practical key to making AIC (wherein we have assumed $\widehat{\mathrm{tr}}\left[I(\theta_o)\Sigma\right] = K$) work is then to have some good models in the set considered, but not too many good, but over parametrized, models. By a "good" model we mean one that is close to $f$ in the sense of having a small K-L value, in which case such "closeness" also means that the use of $\widehat{\mathrm{tr}}\left[I(\theta_o)\Sigma\right] = K$ is itself a parsimonious estimator. This matter of estimation of the trace term and closeness of $g$ to $f$ is explored further in Section 6.6. It is those Section 6.6 derivations, and the above ideas in this paragraph, that to us justifies AIC, which is seen as a special case of (6.34):

$$\mathrm{AIC} = -2\log(g(\underline{x} \mid \hat{\theta})) + 2K.$$

The generalization given by (6.34) leads to Takeuchi's (1976) information criterion (TIC) for model selection (Shibata 1989). The result (6.32) suggests that we might use the bootstrap to compute $\widehat{\mathrm{tr}}\left[I(\theta_o)\Sigma\right]$ and hence implement the TIC criterion via

$$-2\log(g(\underline{x} \mid \hat{\theta})) + 2\widehat{\mathrm{tr}}\left[I(\theta_o)\Sigma\right] \tag{6.35}$$

or even use more exact forms for the trace term. These ideas are pursued a bit in the next section.

First, however, there is one more crucial point on which the reader must be clear: it is not required that truth, $f$, be in the set of models to which we apply AIC model selection. Many derivations of AIC are quite misleading by making the assumption (often implicitly, hence without realizing it) that $f \equiv g$ (or $f \subset g$).

Kei Takeuchi was born in 1933 in Tokyo, Japan, and graduated in 1956 from the University of Tokyo. He received a Ph.D. in economics in 1966 (Keizaigaku Hakushi) and his research interests include mathematical statistics, econometrics, global environmental problems, history of civilization, and Japanese economy. He is the author of many books on mathematics, statistics, and the impacts of science and technology on society. He is currently a professor on the Faculty of International Studies at Meiji Gakuin University and Professor Emeritus, University of Tokyo (recent photograph).

Such derivations lead directly to AIC, hence bypass the completely general result of (6.33), which does not require $f \subset g$. Once one has (6.33), then it is possible to see how a proper philosophy of having a set of good approximating models to complex truth in conjunction with the parsimonious choice of $\widehat{\text{tr}}\left[I(\underline{\theta}_o)\Sigma\right] = K$ justifies use of AIC.

There are a few odds and ends worth considering at this point. First, we stated the result

$$E_{\underline{y}}\left[I(f, g(\cdot \mid \hat{\underline{\theta}}(\underline{y})))\right] - I(f, g(\cdot \mid \underline{\theta}_o)) = \frac{1}{2}\text{tr}\left[J(\underline{\theta}_o)I(\underline{\theta}_o)^{-1}\right].$$

The proof is simple, because the left-hand side of the above reduces to

$$E_{\underline{y}}\left[E_{\underline{x}}[\log(g(\underline{x} \mid \underline{\theta}_o))] - E_{\underline{x}}[\log(g(\underline{x} \mid \hat{\underline{\theta}}(\underline{y})))]\right].$$

Now substitute (6.23) for $E_{\underline{x}}[\log(g(\underline{x} \mid \hat{\underline{\theta}}(\underline{y})))]$ in the above to get the result

$$\frac{1}{2}E_{\underline{y}}\left[\hat{\underline{\theta}}(\underline{y}) - \underline{\theta}_o]'I(\underline{\theta}_o)[\hat{\underline{\theta}}(\underline{y}) - \underline{\theta}_o\right],$$

which becomes $\frac{1}{2}\,\mathrm{tr}\left[I(\underline{\theta}_o)\Sigma\right] = \frac{1}{2}\,\mathrm{tr}\left[J(\underline{\theta}_o)I(\underline{\theta}_o)^{-1}\right]$.

It should be almost obvious (and it is true) that this trace term, $\mathrm{tr}\left[J(\underline{\theta}_o)I(\underline{\theta}_o)^{-1}\right]$, does not depend upon sample size. Rather, for good models it is about equal to $K$ (these matters are explored in other Chapter 6 sections below). In stark contrast, quantities such as the log-likelihood, expected log-likelihood, and both of K-L, $I(f, g(\cdot \mid \underline{\theta}_o))$, and the expected K-L, $\mathrm{E}_{\underline{y}}\left[I(f, g(\cdot \mid \hat{\underline{\theta}}(\underline{y})))\right]$, increase linearly in sample size, $n$. As a result, for large sample sizes, and $K/n$ small, the ratio

$$\frac{\mathrm{E}_{\underline{y}}\left[I(f, g(\cdot \mid \hat{\underline{\theta}}(\underline{y})))\right]}{I(f, g(\cdot \mid \underline{\theta}_o))}$$

is essentially 1 even though the difference between expected and actual K-L is > 0. Thus, on an absolute scale TIC and AIC (when thoughtfully applied) model selection are producing the model estimated to provide the minimum K-L model from the set of models considered if sample size is large and $K/n$ is small.

The reason that the criterion for practical model selection gets changed from minimum K-L to minimum expected K-L, as $\mathrm{E}_{\underline{y}}\left[I(f, g(\cdot \mid \hat{\underline{\theta}}(\underline{y})))\right]$, is because we must estimate $\underline{\theta}$ by the model-based MLE. This seemingly innocent little fact has deep ramifications. It is why the K-L–based conceptual motivation (at the start of this section) virtually forces us to adopt $\mathrm{E}_{\underline{y}}\left[I(f, g(\cdot \mid \hat{\underline{\theta}}(\underline{y})))\right]$ to be minimized, hence $T$, i.e., (6.20), to be maximized.

In this regard there is a nominally puzzling result people might find. If we just start with K-L as

$$I(f, g(\cdot \mid \underline{\theta}_o)) = \mathrm{Constant} - \mathrm{E}_{\underline{x}}\left[\log(g(\underline{x} \mid \underline{\theta}_o))\right],$$

and no actual data in hand, hence no estimate of $\underline{\theta}$, we might notice a direct Taylor series expansion of $\log(g(\underline{x} \mid \underline{\theta}_o))$ about what would be the MLE given any value of the variable of integration, $\underline{x}$ (which is *not* data). After taking the expectation over the sample space of the random variable $\underline{x}$, the result is

$$\mathrm{E}_{\underline{x}}\left[\log(g(\underline{x} \mid \underline{\theta}_o))\right] = \mathrm{E}_{\underline{x}}\left[\log(g(\underline{x} \mid \hat{\underline{\theta}}(\underline{x})))\right] - \frac{1}{2}\,\mathrm{tr}\left[J(\underline{\theta}_o)I(\underline{\theta}_o)^{-1}\right].$$

The above would suggest K-L model selection could be based on maximizing $\log(g(\underline{x} \mid \hat{\underline{\theta}}(\underline{x}))) - \frac{1}{2}\,\mathrm{tr}\left[\hat{J}(\underline{\theta}_o)\hat{I}(\underline{\theta}_o)^{-1}\right]$; it cannot be so based.

This conclusion is not valid, because there are no data lurking anywhere. There never was a valid MLE of $\hat{\underline{\theta}}$ injected into the process. The K-L criterion has already been integrated over the sample space, and properly there is no $\underline{x}$ and no data involved in K-L. Data cannot be manufactured by a Taylor series expansion on a random variable. Thus, the intriguing result is mathematically correct, by conceptually wrong for what we are trying to do, and hence misleading.

## 6.3 General K-L–Based Model Selection: TIC

### 6.3.1 Analytical Computation of TIC

There are other alternatives to estimation of relative K-L (not much used) that try to provide a data-based estimator of the trace term. These methods are computationally much more intense, and the resultant estimator of the trace term can be so variable, and may have its own biases, that it is questionable whether such approaches are worth applying (unless perhaps $n$ is huge). Takeuchi (1976) proposed TIC (see also Shibata 1989, and Konishi and Kitagawa 1996): Select the model that minimizes (6.34) for specific estimators of $J(\theta_o)$ and $I(\theta_o)$, hence getting an estimator of tr $\left[J(\theta_o)[I(\theta_o)]^{-1}\right]$. The estimator of $I(\theta_o)$ is (6.8), the empirical Hessian:

$$\hat{I}(\theta_o) = \hat{I}(\hat{\theta}) = -\frac{\partial^2 \log(g(\underline{x} \mid \hat{\theta}))}{\partial \underline{\theta}^2}. \tag{6.36}$$

General estimation of $J(\theta_o)$ relies on recognizing the sample as structured on $n$ independent units of information. In the simplest case we would have $\underline{x}$ as an iid sample, $x_1, \ldots, x_n$. It is required only that the sample be recognized as having $n$ conditionally independent components so that the log-likelihood can be computed as the sum of $n$ terms; hence we have

$$\log(g(\underline{x} \mid \hat{\theta})) = \sum_{i=1}^{k} \log(g_i(x_i \mid \hat{\theta})).$$

For the *iid* sample case, $g_i(x_i \mid \hat{\theta}) \equiv g(x_i \mid \hat{\theta})$. Using here $g(\cdot \mid \theta)$ for both the basic sample-size one pdf and for the probability distribution function of the full sample of size $n$ is a minor abuse of notation. However, we think that the reader will understand the meaning of the formulae and that it is better to minimize notation to facilitate comprehension of concepts.

A general estimator of $J(\theta_o)$, for TIC, can be derived from (6.12):

$$J(\theta_o) = \mathrm{E}_f\left[\left[\frac{\partial}{\partial \underline{\theta}} \log(g(\underline{x} \mid \theta_o))\right]\left[\frac{\partial}{\partial \underline{\theta}} \log(g(\underline{x} \mid \theta_o))\right]'\right].$$

For the case of a random sample,

$$J(\theta_o) = \mathrm{E}_f\left[\left[\sum_{i=1}^{k}\frac{\partial}{\partial \underline{\theta}} \log(g(x_i \mid \theta_o))\right]\left[\sum_{i=1}^{k}\frac{\partial}{\partial \underline{\theta}} \log(g(x_i \mid \theta_o))\right]'\right]$$

$$= \sum_{i=1}^{k}\mathrm{E}_f\left[\frac{\partial}{\partial \underline{\theta}} \log(g(x_i \mid \theta_o))\right]\left[\frac{\partial}{\partial \underline{\theta}} \log(g(x_i \mid \theta_o))\right]'.$$

Therefore, we are led to use

$$\hat{J}(\theta_o) = \sum_{i=1}^{k}\left[\frac{\partial}{\partial \underline{\theta}} \log(g(x_i \mid \hat{\theta}))\right]\left[\frac{\partial}{\partial \underline{\theta}} \log(g(x_i \mid \hat{\theta}))\right]'. \tag{6.37}$$

## 6.3 General K-L–Based Model Selection: TIC

A general version of TIC can be defined based on (6.36) and (6.37) (see, e.g., Shibata 1989:222):

$$\text{TIC} = -2\log(g(\underline{x} \mid \hat{\underline{\theta}})) + 2\,\text{tr}\left[\hat{J}(\underline{\theta}_o)[\hat{I}(\underline{\theta}_o)]^{-1}\right]. \tag{6.38}$$

One selects the model that produces the smallest TIC. Because $-\text{TIC}/2 = \hat{T}$ is for each model an asymptotically unbiased estimator of $\text{E}_{\underline{y}}\left[I(f, g(\cdot \mid \hat{\underline{\theta}}(\underline{y})))\right] -$ Constant, the underlying optimization criterion is that we select the model that on average (over the set of models) minimizes this expected K-L information loss. For large $n$ this expected criterion is almost the same as minimizing the criterion $I(f, g) -$ Constant; thus using (6.38), we are essentially targeting selecting the K-L best model of the set of models, and this is regardless of whether or not $f$ is in the model set.

The estimator $\hat{J}(\underline{\theta}_o)$ converges to $J(\underline{\theta}_o)$, and $\hat{I}(\underline{\theta}_o)$ converges to $I(\underline{\theta}_o)$, so TIC is asymptotically unbiased (i.e., consistent) as a selection criterion for the minimum expected K-L model. In practice this estimator of the trace term is so variable (and is not unbiased), even for large $n$, that it seems better to just use the parsimonious "estimator" $\hat{\text{tr}}\left[J(\underline{\theta}_o)[I(\underline{\theta}_o)]^{-1}\right] = K$ (cf. Shibata 1989) (we will consider the matter further in later sections). This seems especially appropriate if we have done a good job of specifying our set of models from which to select a best-fitting model.

### 6.3.2 Bootstrap Estimation of TIC

The primary value of the bootstrap method herein is to assess model-selection uncertainty based on applying an analytical model-selection criterion (e.g., AIC, $\text{AIC}_c$, $\text{QAIC}_c$, or TIC based on formulae 6.33, 6.34, and 6.35). However, a second and quite different use of the bootstrap can be made: Use some bootstrap method to estimate directly the quantity $T = \text{E}_{\underline{x}}\text{E}_{\hat{\underline{\theta}}}\left[\log(g(\underline{x} \mid \hat{\underline{\theta}}))\right]$; the K-L best model is the one that maximizes $\hat{T}$. Variations on this theme involve more direct bootstrap estimation of the key quantity $\text{tr}\left[I(\underline{\theta}_o)\Sigma\right]$ (or equivalently, $\text{tr}\left[J(\underline{\theta}_o)[I(\underline{\theta}_o)]^{-1}\right]$). We will describe a method designed to minimize the impact of approximations made in deriving (6.35).

From (6.24) and (6.26) (wherein $\hat{\underline{\theta}}$ denotes $\hat{\underline{\theta}}(\underline{x})$) we derive

$$T \doteq \text{E}_{\underline{x}}\left[\log(g(\underline{x} \mid \hat{\underline{\theta}}))\right] - \frac{1}{2}\text{tr}\left[I(\underline{\theta}_o)\Sigma\right] - \frac{1}{2}\text{tr}\left[\text{E}_{\underline{x}}\left[\hat{I}(\hat{\underline{\theta}})\right][\underline{\theta}_o - \hat{\underline{\theta}}][\underline{\theta}_o - \hat{\underline{\theta}}]'\right].$$

Hence, a model-selection criterion can be based on

$$\hat{T} = \log(g(\underline{x} \mid \hat{\underline{\theta}})) - \frac{1}{2}\text{tr}\left[\hat{I}(\underline{\theta}_o)\hat{\Sigma}\right] - \frac{1}{2}\text{tr}\left[\hat{\text{E}}_{\underline{x}}\left[\hat{I}(\hat{\underline{\theta}})\right][\underline{\theta}_o - \hat{\underline{\theta}}][\underline{\theta}_o - \hat{\underline{\theta}}]'\right]. \tag{6.39}$$

Additional approximations applied to (6.39), or to the basic derivations, lead to

$$\hat{T} = \log(g(\underline{x} \mid \hat{\underline{\theta}})) - \text{tr}\left[\hat{I}(\underline{\theta}_o)\hat{\Sigma}\right],$$

which could also be the basis for a bootstrap estimator (as could 6.35).

We assume that the sample structure allows a meaningful bootstrap sampling procedure (easily done in the *iid* sample case). Let a bootstrap sample be denoted by $\underline{x}^*$ with corresponding bootstrap MLE $\hat{\underline{\theta}}^*$. The needed likelihood second partial derivatives will have to be determined either analytically or numerically. To avoid more notation, we do not index the bootstrap samples, but rather just note that needed summations are over $B$ bootstrap reps.

In the bootstrap estimators, the MLE $\hat{\underline{\theta}}$ plays the role of $\underline{\theta}_o$. Hence bootstrap estimators of $I(\underline{\theta}_o)$, $\Sigma$, and $\mathrm{E}_{\underline{x}}\left[\hat{I}(\hat{\underline{\theta}})\right][\underline{\theta}_o - \hat{\underline{\theta}}][\underline{\theta}_o - \hat{\underline{\theta}}]'$ are

$$\hat{I}(\underline{\theta}_o) = -\frac{1}{B}\left[\sum_B \frac{\partial^2 \log(g(\underline{x}^* \mid \hat{\underline{\theta}}))}{\partial \underline{\theta}^2}\right], \quad (6.40)$$

$$\hat{\Sigma} = \frac{1}{B}\left[\sum_B [\hat{\underline{\theta}}^* - \hat{\underline{\theta}}][\hat{\underline{\theta}}^* - \hat{\underline{\theta}}]'\right], \quad (6.41)$$

$$\hat{\mathrm{E}}_{\underline{x}}\left[\left[\hat{I}(\hat{\underline{\theta}})\right][\underline{\theta}_o - \hat{\underline{\theta}}][\underline{\theta}_o - \hat{\underline{\theta}}]'\right] \quad (6.42)$$

$$= \frac{1}{B}\left[\sum_B \left[-\frac{\partial^2 \log(g(\underline{x}^* \mid \hat{\underline{\theta}}^*))}{\partial \underline{\theta}^2}\right][\hat{\underline{\theta}}^* - \hat{\underline{\theta}}][\hat{\underline{\theta}}^* - \hat{\underline{\theta}}]'\right].$$

These estimators mimic the expectation over $f$, because the sample arises from $f$, the bootstrap resamples the sample, and under any model our best estimator of $\underline{\theta}_o$ is the MLE $\hat{\underline{\theta}}$ (note that $\underline{\theta}_o$ varies by model, $g$). One should use the same $B$ bootstrap samples with every model in the set of models over which selection is made.

The above suffices to compute TIC as

$$\mathrm{TIC} = -2\log(g(\underline{x} \mid \hat{\underline{\theta}})) + 2\,\mathrm{tr}\left[\hat{I}(\underline{\theta}_o)\hat{\Sigma}\right], \quad (6.43)$$

using (6.40) and (6.41) (this is estimating the same quantity as TIC, so we call it TIC here, because $J(\underline{\theta}_o)[I(\underline{\theta}_o)]^{-1} = I(\underline{\theta}_o)\Sigma$). To use (6.39) for bootstrap-based model selection, base the estimation of its second and third components on (6.40), (6.41), and (6.42); or in a form analogous to AIC and TIC, the model-selection criterion to minimize is

$$-2\log(g(\underline{x} \mid \hat{\underline{\theta}})) + \mathrm{tr}\left[\hat{I}(\underline{\theta}_o)\hat{\Sigma}\right] + \mathrm{tr}\,\hat{\mathrm{E}}_{\underline{x}}\left[\left[\hat{I}(\hat{\underline{\theta}})\right][\underline{\theta}_o - \hat{\underline{\theta}}][\underline{\theta}_o - \hat{\underline{\theta}}]'\right]. \quad (6.44)$$

It may well be that (6.43), i.e., TIC, would suffice and (6.44) is not a better estimator of $-2T$.

Recent work on this usage of the bootstrap to find $\hat{T}$ for K-L–based model selection is found in Ishiguro and Sakamoto (1991), Ishiguro, et al. (1991), Cavanaugh and Shumway (1994), Shao (1996) and Chung et al. (1996). Shibata (1997a) has considered, in a general context, theoretical properties of many alternative implementations of the bootstrap to estimate the needed model-selection criterion, $T$. He notes that there is no unique way to do this bootstrapping to estimate the relative K-L model-selection criterion, but that all reasonable bootstrap implementations are asymptotically equivalent to TIC. This use of the bootstrap has the advantage of bypassing concerns about all approximations used to get TIC or AIC. Despite this apparent advantage, Shibata (1997a, page 393) concludes that there is no reason to use the bootstrap this way to compute $\hat{T}$. It probably suffices to use a simple nonbootstrap computation of $\hat{T}$ (in particular, $AIC_c$).

It should thus be clear that there are two very different ways to use the bootstrap in model selection. Not much used is the case of getting a single estimate of $T$ for each model based on the full set of bootstrap samples. The more common (and more useful) use of the bootstrap in model selection is first to accept some easily computable model-selection criterion, such as AIC, and then to apply that criterion to all models considered for all the bootstrap samples created (and tabulate results like frequency of selection of each model). This use of the bootstrap leads to information about inference uncertainties after model selection. [There is also a large literature on use of the bootstrap under non–K-L–based model selection; see, e.g., Breiman 1992; Efron 1983, 1986; Hjorth 1994; Linhart and Zucchini 1986; and Shao 1996].

## 6.4 $AIC_c$: A Second-Order Improvement

### 6.4.1 Derivation of $AIC_c$

The results above are completely general, and as such do not lead to some of the more specific results in the literature. In particular, if we assume a univariate linear structural model with homogeneous, normally distributed errors, conditional on any regressor variables, we can get the results of Hurvich and Tsai (1989, 1995) (see also Sugiura 1978). We let the model structure be

$$\mu_i = E(x_i \mid \underline{z}) = \sum_{j=1}^{K-1} z_{ij}\beta_j, \qquad i = 1, \ldots, n.$$

More specifically (but without explicitly denoting the conditioning on "regressors" $\underline{z}_i$),

$$x_i = \sum_{j=1}^{K-1} z_{ij}\beta_j + \epsilon_i, \qquad i = 1, \ldots, n,$$

where the $\epsilon_i$ are *iid* normal$(0, \sigma^2)$. There are thus $K$ parameters making up $\underline{\theta}$ ($\sigma^2$ is the $K$th one), and $g(\underline{x} \mid \underline{\theta})$ is given by the multivariate–normal$(\underline{\mu}, \sigma^2 I)$ distribution (MVN); $I$ is the $n$ by $n$ identity matrix. If we let $f \equiv g$ or $f \subset g$, then we can derive the $AIC_c$ results of Hurvich and Tsai. This last notation means that either $g$ is the true data-generating "model," or $f$ is actually the same distribution and structural form as *model g* but with one or more elements of $\underline{\theta}$ set to 0 (hence there are superfluous parameters). The superfluous parameters serve only to increase $K$, hence the simplest way to get $AIC_c$ is to assume this regression model, $g$, and assume that $f \equiv g$. The derivation is given below in some detail because of the importance of $AIC_c$.

252  6. Statistical Theory

Matrix notation is simpler to use, and hence $\underline{X} = Z\underline{\beta} + \underline{\epsilon}$, and $E(\underline{X}) = \underline{\mu}$. Without loss of generality we assume that $Z$ ($n$ by $K - 1$) is of full rank. The likelihood is

$$g(\underline{x} \mid \underline{\theta}) = \left[\frac{1}{\sqrt{2\pi}}\right]^n \left[\frac{1}{\sigma^2}\right]^{n/2} \exp\left[-\frac{1}{2}\frac{(\underline{X} - Z\underline{\beta})'(\underline{X} - Z\underline{\beta})}{\sigma^2}\right],$$

and we are here taking $f \equiv g$. Ignoring additive constants, and simplifying, the log-likelihood can be taken as

$$\log(g(\underline{x} \mid \underline{\theta})) = -\frac{n}{2}\log(\sigma^2) - \frac{1}{2}\frac{(\underline{X} - Z\underline{\beta})'(\underline{X} - Z\underline{\beta})}{\sigma^2}.$$

The MLEs are well known here:

$$\hat{\underline{\beta}} = (Z'Z)^{-1}Z'\underline{X},$$

$$\hat{\sigma}^2 = \frac{(\underline{X} - Z\hat{\underline{\beta}})'(\underline{X} - Z\hat{\underline{\beta}})}{n}.$$

Therefore,

$$\log(g(\underline{x} \mid \hat{\underline{\theta}}(\underline{x}))) = -\frac{n}{2}\log(\hat{\sigma}^2) - \frac{1}{2}\frac{(\underline{X} - Z\hat{\underline{\beta}})'(\underline{X} - Z\hat{\underline{\beta}})}{\hat{\sigma}^2};$$

hence, the maximized log-likelihood is

$$\log(g(\underline{x} \mid \hat{\underline{\theta}}(\underline{x}))) = -\frac{n}{2}\log(\hat{\sigma}^2) - \frac{n}{2}$$

(the constant $-n/2$ can be dropped in practice).

We want to determine the bias if we use $\log(g(\underline{x} \mid \hat{\underline{\theta}}(\underline{x})))$ as an estimator of our target,

$$T = E_{\underline{x}} E_{\hat{\underline{\theta}}(\underline{y})}\left[\log(g(\underline{x} \mid \hat{\underline{\theta}}(\underline{y})))\right],$$

where $\underline{x}$ and $\underline{y}$ are two independent random samples of size $n$. To make the evaluation here we actually use the specified form of the model (and of course take expectations with respect to $f \equiv g$). Hence, we want (a simplified notation is used here)

$$T = E_{\underline{x}} E_{\hat{\underline{\theta}}(\underline{y})}\left[\log(g(\underline{x} \mid \hat{\underline{\theta}}(\underline{y})))\right]$$

$$= E_{\hat{\underline{\theta}}(\underline{y})} E_{\underline{x}}\left[-\frac{n}{2}\log(\hat{\sigma}_y^2) - \frac{1}{2}\frac{(\underline{X} - Z\hat{\underline{\beta}}_y)'(\underline{X} - Z\hat{\underline{\beta}}_y)}{\hat{\sigma}_y^2}\right].$$

The order of integration was reversed for the right-hand side above. Thus our first task is to evaluate

$$E_{\underline{x}}\left[(\underline{X} - Z\hat{\underline{\beta}}_y)'(\underline{X} - Z\hat{\underline{\beta}}_y)\right]$$

$$= E_{\underline{x}}\left[((\underline{X} - Z\underline{\beta}) + (Z\underline{\beta} - Z\hat{\underline{\beta}}_y))'((\underline{X} - Z\underline{\beta}) + (Z\underline{\beta} - Z\hat{\underline{\beta}}_y))\right]$$

## 6.4 AIC$_c$: A Second-Order Improvement

$$= E_{\underline{x}}\Big[(\underline{X} - Z\underline{\beta})'(\underline{X} - Z\underline{\beta})\Big] + E_{\underline{x}}\Big[2(Z\underline{\beta} - Z\hat{\underline{\beta}}_y)'(\underline{X} - Z\underline{\beta})\Big]$$
$$+ E_{\underline{x}}\Big[(Z\underline{\beta} - Z\hat{\underline{\beta}}_y)'(Z\underline{\beta} - Z\hat{\underline{\beta}}_y)\Big]$$
$$= E_{\underline{x}}\Big[(\underline{X} - Z\underline{\beta})'(\underline{X} - Z\underline{\beta})\Big] + \Big[2(Z\underline{\beta} - Z\hat{\underline{\beta}}_y)'(E_{\underline{x}}(\underline{X}) - Z\underline{\beta})\Big]$$
$$+ \Big[(Z\underline{\beta} - Z\hat{\underline{\beta}}_y)'(Z\underline{\beta} - Z\hat{\underline{\beta}}_y)\Big].$$

The middle term above vanishes because $E_{\underline{x}}(\underline{X}) = Z\underline{\beta}$. Also, the first of the three terms above is identical to $E_{\underline{x}}(\underline{\epsilon}'\underline{\epsilon}) = n\sigma^2$. So we have the result

$$E_{\underline{x}}\Big[(\underline{X} - Z\hat{\underline{\beta}}_y)'(\underline{X} - Z\hat{\underline{\beta}}_y)\Big] = n\sigma^2 + \Big[(Z\underline{\beta} - Z\hat{\underline{\beta}}_y)'(Z\underline{\beta} - Z\hat{\underline{\beta}}_y)\Big].$$

Using this partial result we have

$$T = E_{\hat{\theta}(y)}\left[-\frac{n}{2}\log(\hat{\sigma}_y^2)\right] - \frac{1}{2}E_{\hat{\theta}(y)}\left[\frac{n\sigma^2 + \Big[(Z\underline{\beta} - Z\hat{\underline{\beta}}_y)'(Z\underline{\beta} - Z\hat{\underline{\beta}}_y)\Big]}{\hat{\sigma}_y^2}\right].$$

The first term above does not need to be evaluated further because it is also the leading term in the expected log-likelihood. Also, at this point, we can drop the designation of $\theta$ as being based on sample $\underline{y}$. The designations $\underline{x}$ or $\underline{y}$ are really just dummy arguments in integrals. Consequently, in the above, the notation could be in terms of $\underline{y}$ or $\underline{x}$, or this notation can just be dropped. Thus we have

$$T = E_{\hat{\theta}}\left[-\frac{n}{2}\log(\hat{\sigma}^2)\right] - \frac{1}{2}E_{\hat{\theta}}\left[\frac{n\sigma^2 + \Big[(Z\underline{\beta} - Z\hat{\underline{\beta}})'(Z\underline{\beta} - Z\hat{\underline{\beta}})\Big]}{\hat{\sigma}^2}\right]. \quad (6.45)$$

Now we make use of another well-known result in theoretical statistics: Under a linear model structure with errors as *iid* normal$(0, \sigma^2)$, the MLE's $\hat{\beta}$ and $\hat{\sigma}^2$ are independent random variables. Therefore, the second expectation term in (6.45) partitions into two multiplicative parts, as below:

$$T = E\left[-\frac{n}{2}\log(\hat{\sigma}^2)\right] - \frac{1}{2}E_{\hat{\beta}}\Big[n\sigma^2 + \Big[(Z\underline{\beta} - Z\hat{\underline{\beta}})'(Z\underline{\beta} - Z\hat{\underline{\beta}})\Big]\Big]E_{\hat{\sigma}^2}\left[\frac{1}{\hat{\sigma}^2}\right].$$

As a next step, rewrite the needed expectation of the quadratic form in the above as

$$E\Big[(Z\underline{\beta} - Z\hat{\underline{\beta}})'(Z\underline{\beta} - Z\hat{\underline{\beta}})\Big] = \text{tr}\Big[(Z'Z)E\Big[(\hat{\underline{\beta}} - \underline{\beta})(\hat{\underline{\beta}} - \underline{\beta})'\Big]\Big].$$

The expectation on the right-hand side above, i.e., $E[(\hat{\underline{\beta}} - \underline{\beta})(\hat{\underline{\beta}} - \underline{\beta})']$, is the sampling variance–covariance matrix of $\hat{\underline{\beta}}$, which is known to be $\sigma^2(Z'Z)^{-1}$. Thus, for the $K - 1$ square identity matrix $I$,

$$E\Big[(Z\underline{\beta} - Z\hat{\underline{\beta}})'(Z\underline{\beta} - Z\hat{\underline{\beta}})\Big] = \text{tr}[\sigma^2 I] = \sigma^2(K - 1).$$

254   6. Statistical Theory

Putting it all together to this point in the derivation, we have

$$T = \mathrm{E}\left[-\frac{n}{2}\log(\hat{\sigma}^2)\right] - \frac{1}{2}\left[(n+K-1)\sigma^2\right] \mathrm{E}_{\hat{\sigma}^2}\left[\frac{1}{\hat{\sigma}^2}\right]. \qquad (6.46)$$

To finish the process we relate $\hat{\sigma}^2$ to a central chi-squared random variable, namely $\chi^2_{df}$ on $n - (K-1)$ degrees of freedom, df. These results also are well known in statistical theory:

$$\frac{n\hat{\sigma}^2}{\sigma^2} \sim \chi^2_{n-k+1}.$$

So we now rearrange (6.46) to be

$$T = \mathrm{E}\left[-\frac{n}{2}\log(\hat{\sigma}^2)\right] - \frac{1}{2}[(n+K-1)n]\,\mathrm{E}\left[\frac{1}{n\hat{\sigma}^2/\sigma^2}\right],$$

$$T = \mathrm{E}\left[-\frac{n}{2}\log(\hat{\sigma}^2)\right] - \frac{n}{2}(n+K-1)\mathrm{E}\left[\frac{1}{\chi^2_{n-K+1}}\right].$$

Yet another known exact result is

$$\mathrm{E}\left[\frac{1}{\chi^2_{df}}\right] = \frac{1}{df-2}$$

(assuming df > 2).

Using the last result above we have reduced (6.46) to

$$T = \mathrm{E}\left[-\frac{n}{2}\log(\hat{\sigma}^2)\right] - \frac{n}{2}(n+K-1)\left[\frac{1}{n-K-1}\right]. \qquad (6.47)$$

This result is exact. No approximations were made in its derivation; however, it only applies to the particular context of its derivation, which includes the constraint that $f \subseteq g$. Some more simplification of (6.47):

$$T = \mathrm{E}\left[-\frac{n}{2}\log(\hat{\sigma}^2)\right] - \frac{n}{2}\left[\frac{n+K-1}{n-K-1}\right]$$

$$= \mathrm{E}\left[-\frac{n}{2}\log(\hat{\sigma}^2)\right] - \frac{n}{2}\left[1 + \frac{2K}{n-K-1}\right]$$

$$= \mathrm{E}\left[-\frac{n}{2}\log(\hat{\sigma}^2)\right] - \frac{n}{2} - \frac{nK}{n-K-1}$$

$$= \mathrm{E}\left[-\frac{n}{2}\log(\hat{\sigma}^2) - \frac{n}{2}\right] - \frac{nK}{n-K-1}.$$

The term above within the expectation operator is the maximized log-likelihood, thus we have

$$T = \mathrm{E}\left[\log(g(\underline{x}\,|\,\underline{\hat{\theta}}(\underline{x})))\right] - \frac{nK}{n-K-1}$$

$$= \mathrm{E}\left[\log(g(\underline{x}\,|\,\underline{\hat{\theta}}(\underline{x})))\right] - \frac{(n-K-1+K+1)K}{n-K-1}$$

$$= \mathrm{E}\left[\log(g(\underline{x} \mid \hat{\underline{\theta}}(\underline{x})))\right] - K - \frac{K(K+1)}{n-K-1}.$$

If we convert this to be an AIC result, we have, as an exact result in this context,

$$-2T = -2\mathrm{E}\left[\log(g(\underline{x} \mid \hat{\underline{\theta}}(\underline{x})))\right] + 2K + \frac{2K(K+1)}{n-K-1}$$

$$= \mathrm{E(AIC)} + \frac{2K(K+1)}{n-K-1} = \mathrm{E(AIC}_c). \tag{6.48}$$

This result thus motivates use of the term $2K(K+1)/(n-K-1)$ as a small-sample size bias correction term added to AIC. The result assumes a fixed-effects linear model with normal errors and constant residual variances. Under different sorts of models, a different small sample correction to AIC arises (the matter is explored some in the next subsection). However, the result given by (6.48) seems useful in other contexts, especially if $n$ is large but $K$ is also large relative to $n$. Without exception, if sample size $n$ is small, some sort of "AIC$_c$" is required for good model-selection results, and we recommend (6.48) unless a more exact small-sample correction to AIC is known.

### 6.4.2 Lack of Uniqueness of AIC$_c$

The result in (6.48) is not universal in that other assumed univariate (and more so for multivariate) models, with $g = f$ to facilitate a derivation, or ways of deriving a small-sample adjustment to AIC, will lead to different adjustment terms. This section is just a brief elaboration of this idea.

The simplest case to present arises for a situation analogous to one-way ANOVA, but we let the within-subgroup variance differ for each subgroup. This can be generalized to having $m$ subsets of data, each of sample size $n_i$, and the full model is as used in Section 6.4.1 above, but with parameter set $\{\underline{\theta}\} = \{\beta_i, \sigma_i^2, i = 1, \ldots, m\}$ (this might be a global model, in some cases). Let each parameter subset be of size $K_i$; hence $K = K_1 + \cdots + K_m$. It should be almost obvious, after some thought, that for this situation the small-sample correction to AIC is

$$-2T = -2\mathrm{E}\left[\log(g(\underline{x} \mid \hat{\underline{\theta}}(\underline{x})))\right] + 2\sum_{i=1}^{m}\left[K_i + \frac{2K_i(K_i+1)}{n_i-K_i-1}\right];$$

hence,

$$\mathrm{AIC}_c = -2\log(g(\underline{x} \mid \hat{\underline{\theta}}(\underline{x}))) + 2K + \sum_{i=1}^{m}\left[\frac{2K_i(K_i+1)}{n_i-K_i-1}\right].$$

The reason that there are $m$ "correction" terms is that we had to estimate $m$ different variance parameters. One can thus envision many other models where the form of AIC$_c$ must differ from that of the simple normal-model case with only one estimated $\sigma^2$.

Another informative exact calculation of the bias term, $T - \mathrm{E}_{\underline{x}}\left[\log(g(\underline{x} \mid \hat{\underline{\theta}}))\right]$, is obtained for the case of the model and truth being the one-parameter negative exponential distribution (hence $K = 1$):

$$g(x \mid \lambda) = \frac{1}{\lambda} e^{-x/\lambda}.$$

For an iid sample from $g(x \mid \lambda)$, let $S = x_1 + \cdots + x_n$. Then

$$\log(g(\underline{x} \mid \lambda)) = -n\log(\lambda) - S/\lambda.$$

The MLE is $\hat{\lambda} = S/n$, so

$$\log(g(\underline{x} \mid \hat{\lambda})) = -n\log(\hat{\lambda}) - n.$$

The target to be unbiasedly estimated is

$$T = \mathrm{E}_{\underline{x}} \mathrm{E}_{\underline{y}} \left[ -n\log(\hat{\lambda}) - S_{\underline{y}}/\hat{\lambda} \right],$$

where the sum $S_{\underline{y}}$ is based on an independent sample of size $n$, while $\hat{\lambda}$ is based on sample $\underline{x}$. It is easy to evaluate the above $T$ to be

$$T = \mathrm{E}\left[\log(g(\underline{x} \mid \hat{\lambda}))\right] + n - n^2 \lambda \mathrm{E}\left[\frac{1}{S}\right]. \tag{6.49}$$

The sum $S$ is a random variable with a simple gamma distribution; hence the expectation of $1/S$ in (6.49) is known to be exactly $1/(\lambda(n-1))$. It is thus easy to derive the exact result, expressed in "AIC" form (i.e., as $-2T$):

$$-2T = -2\mathrm{E}(\log(g(\underline{x} \mid \hat{\lambda}))) + 2 + \frac{2}{n-1}.$$

Recall that here $K = 1$, so the corresponding total bias correction term under the AIC$_c$ form would be $2 + 4/(n-2)$. The point is that the exact form of AIC would be $2K$ plus a small-sample correction term that would vary accordingly to the model assumed. It is reasonable to think that this small-sample correction term should be $O(1/n)$.

Theoretically, when $f \subset g$ the error in using $K$ as the bias correction to $\hat{T} = \log(g(\underline{x} \mid \hat{\theta}))$ is always $O(1/n)$, and Hurvich and Tsai's form seems like a good general choice. There is, however, a lot of scope for research on improved bias terms for AIC$_c$-type criteria. In this regard, an area offering research opportunities is that of when the random variable is discrete (see, e.g., Sugiura 1978, Shibata 1997b), such as Poisson, binomial, or Bernoulli (hence also logistic regression), because then we can get parameter MLEs taking on the value 0. This creates a problem in evaluating the theoretical target model-selection criterion because we encounter the need to compute $y \cdot \log(0)$, which is not defined (see, e.g., Burnham et al. 1994). AIC is still defined, but its small-sample properties are now more problematic, as is the small-sample bias correction term needed to define an AIC$_c$. Operating characteristics of AIC-based model selection for count-type data need more study for small sample sizes.

## 6.5 Derivation of AIC for the Exponential Family of Distributions

A generalization of normality-based models is found in the exponential family of distributions. The realizations that (1) many common applications of statistical analyses are based on exponential family models, and (2) normality-based regression is in the exponential family and leads to exact K-L model selection results (i.e., $AIC_c$) motivated us to show the derivation of AIC theory under this restricted but very useful case. The canonical representation of an exponential family pdf involves sums of functions of the sample values. It is convenient to denote these sums by $S_j$.

A suitable canonical representation for the exponential family of probability distributions is

$$g(\underline{x} \mid \underline{\theta}) = \exp\left[\left[\sum_{j=1}^{K} S_j \theta_j\right] + H(\underline{\theta}) + G(\underline{S})\right]$$
$$= \exp\left[\underline{S}'\underline{\theta} + H(\underline{\theta}) + G(\underline{S})\right]. \quad (6.50)$$

Each $S_j$ is a function (hence $S_j(\underline{x})$) of the full sample $\underline{x} = \underline{\mu}(\underline{\theta}) + \underline{\epsilon}$ and any covariates, $Z$ involved in representing $\underline{\mu}(\underline{\theta})$, on which we condition. The $K$-element vector of sufficient statistics is $\underline{S} = (S_1, \ldots, S_K)'$.

In the canonical representation of (6.50) the parameter $\underline{\theta}$ is generally some 1-to-1 transformation of another $K$-dimensional parameter of direct interest. There is no loss of generality in allowing any such 1-to-1 transformation. We will revisit this matter and show why it is so at the end of this section.

Our goal is to evaluate

$$T = E_{\underline{x}} E_{\underline{y}} \left[\log(g(\underline{x} \mid \hat{\underline{\theta}}_y))\right] = E_{\underline{x}} E_{\underline{y}} \left[\underline{S}'_x \hat{\underline{\theta}}_y + H(\hat{\underline{\theta}}_y) + G(\underline{S}_x)\right]. \quad (6.51)$$

Here, $\underline{S}_x$ and $\hat{\underline{\theta}}_y$ are thought of as based on independent samples $\underline{x}$ and $\underline{y}$. We also simplified the notation, now using $\hat{\underline{\theta}}_y$ rather than $\hat{\underline{\theta}}(y)$.

Formula (6.51) above can be rewritten as

$$T = E_{\underline{x}} E_{\underline{y}} \left[(\underline{S}_x - \underline{S}_y + \underline{S}_y)'\hat{\underline{\theta}}_y + H(\hat{\underline{\theta}}_y) + G(\underline{S}_x)\right]$$
$$= E_{\underline{x}} E_{\underline{y}} \left[(\underline{S}_x - \underline{S}_y)'\hat{\underline{\theta}}_y + \underline{S}'_y\hat{\underline{\theta}}_y + H(\hat{\underline{\theta}}_y) + G(\underline{S}_x)\right]$$
$$= E_{\underline{x}} E_{\underline{y}} \left[\underline{S}'_y\hat{\underline{\theta}}_y + H(\hat{\underline{\theta}}_y) + G(\underline{S}_x)\right] + E_{\underline{x}} E_{\underline{y}} \left[(\underline{S}_x - \underline{S}_y)'\hat{\underline{\theta}}_y\right]$$
$$= \left[E_{\underline{y}}(\underline{S}'_y\hat{\underline{\theta}}_y + H(\hat{\underline{\theta}}_y)) + E_{\underline{x}}(G(\underline{S}_x))\right] + E_{\underline{y}}\left[(E_{\underline{x}}(\underline{S}_x) - \underline{S}_y)'\hat{\underline{\theta}}_y\right].$$

The interchangeability of integration arguments now is used. This is correct to do because both expectations are with respect to $f$; hence $E_{\underline{x}}(G(\underline{S}_x)) = E_{\underline{y}}(G(\underline{S}_y))$. Also, for simplicity we will use $E_{\underline{x}}(\underline{S}_x) = E(\underline{S})$, and now we get

$$T = E_{\underline{y}}\left[\underline{S}'_y\hat{\underline{\theta}}_y + H(\hat{\underline{\theta}}_y)) + G(\underline{S}_y)\right] + E_{\underline{y}}\left[(E(\underline{S}) - \underline{S}_y)'\hat{\underline{\theta}}_y\right].$$

Changing the argument from $y$ to $x$ in the first part above, just to emphasize the result for this exponential family case, we have

$$T = E_x(\log(g(\underline{x}\,|\,\hat{\underline{\theta}}))) + E_y\left[(E(\underline{S}) - \underline{S}_y)'\hat{\underline{\theta}}_y\right].$$
$$\equiv E_x(\log(g(\underline{x}\,|\,\hat{\underline{\theta}}))) - E_y\left[(\underline{S}_y - E(\underline{S}))'\hat{\underline{\theta}}_y\right]. \quad (6.52)$$

Formula (6.52) is an exact result and clearly shows the bias to be subtracted from $E_x(\log(g(\underline{x}\,|\,\hat{\underline{\theta}})))$ to get $T$:

$$\text{Bias} = E_y\left[(\underline{S}_y - E(\underline{S}))'(\hat{\underline{\theta}}_y - \underline{\theta}_*)\right].$$

The notation used here is $E(\hat{\underline{\theta}}_y) = \underline{\theta}_*$ to denote the exact expectation of the MLE for the given sample size $n$ and model $g$; $\underline{\theta}_o \doteq \underline{\theta}_*$ with asymptotic equality.

To further simplify notation, now that only one "sample" is involved, we use

$$\text{Bias} = E\left[(\underline{S} - E(\underline{S}))'(\hat{\underline{\theta}} - \underline{\theta}_*)\right], \quad (6.53)$$
$$\text{Bias} = \text{tr}\, E\left[(\hat{\underline{\theta}} - \underline{\theta}_*)(\underline{S} - E(\underline{S}))'\right] = \text{tr}\left[\text{COV}(\hat{\underline{\theta}}, \underline{S})\right].$$

Hence for the exponential family *an exact result* is

$$T = E_x(\log(g(\underline{x}\,|\,\hat{\underline{\theta}}))) - \text{tr}\left[\text{COV}(\hat{\underline{\theta}}, \underline{S})\right] \quad (6.54)$$

(something similar appears in Bonneu and Milhaud 1994). The $K \times K$ matrix of covariance elements, $\text{COV}(\hat{\underline{\theta}}, \underline{S})$, can be approximated by Taylor series methods. If the exact covariance matrix can be found, then we have an exact result for the needed bias term above (Hurvich and Tsai 1989, in effect, did such an exact evaluation for the normal distribution case). The result (6.54) may seem not very useful because it seems to apply only to the canonical form of the exponential family. This is not true; the matter of generality of the canonical result will be addressed below.

Before further evaluation of the bias term, we consider the MLEs and the Hessian. First,

$$\log(g(\underline{x}\,|\,\underline{\theta})) = \underline{S}'\underline{\theta} + H(\underline{\theta}) + G(\underline{S}),$$

so

$$\frac{\partial \log(g(\underline{x}\,|\,\underline{\theta}))}{\partial \underline{\theta}} = \underline{S} + \frac{\partial H(\underline{\theta})}{\partial \underline{\theta}},$$
$$\frac{\partial^2 \log(g(\underline{x}\,|\,\underline{\theta}))}{\partial \underline{\theta}^2} = \frac{\partial^2 H(\underline{\theta})}{\partial \underline{\theta}^2},$$

and thus

$$I(\underline{\theta}_o) = E_f\left[-\frac{\partial^2 \log(g(\underline{x}\,|\,\underline{\theta}_o))}{\partial \underline{\theta}^2}\right] = -\frac{\partial^2 H(\underline{\theta}_o)}{\partial \underline{\theta}^2}. \quad (6.55)$$

## 6.5 Derivation of AIC for the Exponential Family of Distributions

It follows that the MLE satisfies

$$\underline{S} = -\frac{\partial H(\hat{\underline{\theta}})}{\partial \underline{\theta}}.$$

It is worth noting here that $\underline{\theta}_o$ satisfies

$$E_f(\underline{S}) = -\frac{\partial H(\underline{\theta}_o)}{\partial \underline{\theta}}.$$

This is an exact result, whereas $E(\hat{\underline{\theta}}) \doteq \underline{\theta}_o$ is (in general) only $O(1/\sqrt{n})$.

The formula for $J(\underline{\theta}_o)$, based on (6.12), becomes

$$J(\underline{\theta}_o) = E_f\left[\underline{S} - E_f(\underline{S})\right]\left[\underline{S} - E_f(\underline{S})\right]', \quad (6.56)$$

which is the true variance–covariance matrix of $\underline{S}$. The $n$ iid observations produce a set of statistics $\underline{s}_i$ that sum to $\underline{S}$; hence an estimator of $J(\underline{\theta}_o)$ as given by (6.56) is

$$\hat{J}(\underline{\theta}_o) = \frac{n}{n-1}\left[\sum_{i=1}^{n}[\underline{s}_i - \bar{\underline{s}}][\underline{s}_i - \bar{\underline{s}}]'\right]. \quad (6.57)$$

Returning now to the evaluation of the bias term, a first-order Taylor series expansion gives us

$$-\frac{\partial H(\hat{\underline{\theta}})}{\partial \underline{\theta}} \doteq -\frac{\partial H(\underline{\theta}_o)}{\partial \underline{\theta}} - \frac{\partial^2 H(\underline{\theta}_o)}{\partial \underline{\theta}^2}(\hat{\underline{\theta}} - \underline{\theta}_o);$$

hence

$$\underline{S} \doteq E(\underline{S}) + I(\underline{\theta}_o)(\hat{\underline{\theta}} - \underline{\theta}_o), \quad O_p(1/\sqrt{n}). \quad (6.58)$$

Inserting (6.58) into the exact result (6.53) as well as also using $\underline{\theta}_o$ to approximate $\underline{\theta}_*$ (inasmuch as we are now replacing an exact result with an approximate result anyway), we have

$$\begin{aligned}
\text{Bias} &\doteq E\left[\left[I(\underline{\theta}_o)(\hat{\underline{\theta}} - \underline{\theta}_o)\right]'(\hat{\underline{\theta}} - \underline{\theta}_o)\right] \\
&= E\left[(\hat{\underline{\theta}} - \underline{\theta}_o)'I(\underline{\theta}_o)(\hat{\underline{\theta}} - \underline{\theta}_o)\right] \\
&= E\,\text{tr}\left[I(\underline{\theta}_o)(\hat{\underline{\theta}} - \underline{\theta}_o)(\hat{\underline{\theta}} - \underline{\theta}_o)'\right] \\
&= \text{tr}\left[I(\underline{\theta}_o)E\left[(\hat{\underline{\theta}} - \underline{\theta}_o)(\hat{\underline{\theta}} - \underline{\theta}_o)'\right]\right] = \text{tr}\left[I(\underline{\theta}_o)\Sigma\right].
\end{aligned}$$

Thus we have shown that in this common case of an exponential family model,

$$T \doteq E_{\underline{x}}(\log(g(\underline{x}\,|\,\hat{\underline{\theta}}))) - \text{tr}\left[I(\underline{\theta}_o)\Sigma\right] \quad (6.59)$$

(the approximation is to $O(1/n)$). Note that this derivation did not encounter any problems like those in approximation (6.28) in the general derivation of AIC in Section 6.2.

These results can be extended to any parametrized form of an exponential family model, because then we just have a 1-to-1 transformation from $\underline{\theta}$ to (say) $\underline{\beta}$ via some set of $K$ functions, denoted here by $\underline{W}(\underline{\theta}) = \underline{\beta}$. Now $\underline{W}(\underline{\theta}_o) = \underline{\beta}_o$, and let $\Sigma_\theta$ and $\Sigma_\beta$ be the variance–covariance matrices for the MLEs under the two parametrizations. An expected matrix of mixed second partial derivatives, as per (6.55), exist for the $\beta$ parametrization; denote it by $I(\underline{\beta})$. Let the $K \times K$ Jacobian of $\underline{W}$, evaluated at $\underline{\theta}_o$, be

$$J_w = \left\{ \frac{\partial W_i(\underline{\theta}_o)}{\partial \theta_j} \right\}.$$

Then

$$J_w \Sigma_\theta J'_w = \Sigma_\beta,$$

and

$$(J'_w)^{-1} I(\underline{\theta}_o)(J_w)^{-1} = I(\underline{\beta}_o).$$

Both the K-L–based target and the expected log-likelihood are invariant to 1-to-1 parameter transformations, so this must also be true for the theoretical bias correction. That is, any likelihood and MLE-based model-selection criterion ought to be invariant to 1-to-1 reparametrizations of the models used. This is the case here:

$$\begin{aligned}
\operatorname{tr}(I(\underline{\beta}_o)\Sigma_\beta) &= \operatorname{tr}\left[(J'_w)^{-1} I(\underline{\theta}_o)(J_w)^{-1} J_w \Sigma_\theta J'_w\right] \\
&= \operatorname{tr}\left[(J'_w)^{-1} I(\underline{\theta}_o)\Sigma_\theta J'_w\right] \\
&= \operatorname{tr}\left[I(\underline{\theta}_o)\Sigma_\theta J'_w (J'_w)^{-1}\right] = \operatorname{tr}\left[I(\underline{\theta}_o)\Sigma_\theta\right].
\end{aligned}$$

Note, however, that if we were to estimate this trace term, the estimator might perform better under some parametrizations than under others.

One last point here: It is certainly still true that

$$\operatorname{tr}\left[I(\underline{\theta}_o)\Sigma\right] = \operatorname{tr}\left[J(\underline{\theta}_o)[I(\underline{\theta}_o)]^{-1}\right].$$

So an alternative to (6.59) is

$$T \doteq \mathrm{E}_{\underline{x}}(\log(g(\underline{x} \mid \hat{\underline{\theta}}))) - \operatorname{tr}\left[J(\underline{\theta}_o)[I(\underline{\theta}_o)]^{-1}\right].$$

This could be directly proven here, based on the simple result

$$\frac{\partial \log(g(\underline{x} \mid \underline{\theta}))}{\partial \underline{\theta}} = \underline{S} - \mathrm{E}(\underline{S})$$

and (6.58) to derive $[I(\underline{\theta}_o)]^{-1} J(\underline{\theta}_o)[I(\underline{\theta}_o)]^{-1} = \Sigma$.

For TIC we can use $\hat{J}(\underline{\theta}_o)$ from (6.57) and from (6.55),

$$\hat{I}(\underline{\theta}_o) = -\frac{\partial^2 H(\hat{\underline{\theta}})}{\partial \underline{\theta}^2},$$

getting an estimator of $\operatorname{tr}\left[J(\underline{\theta}_o)[I(\underline{\theta}_o)]^{-1}\right]$ that can be used (because of invariance) even if the parametrization of interest (and used for MLEs) is some $\underline{\beta}_o = \underline{W}(\underline{\theta}_o)$, not $\underline{\theta}$.

Working with exponential family cases facilitates some informative evaluation of both tr$[J(\underline{\theta}_o)[I(\underline{\theta}_o)]^{-1}]$, relative to the value $K$, and the variability of the estimator tr$[\hat{J}(\underline{\theta}_o)[\hat{I}(\underline{\theta}_o)]^{-1}]$. These topics, and others, are explored in the next section.

## 6.6 Evaluation of tr$(J(\underline{\theta}_o)[I(\underline{\theta}_o)]^{-1})$ and Its Estimator

The general derivation of a K-L–based model-selection criterion results in formula (6.31) and hence (6.33). By "a general derivation," we mean a derivation in which there is no assumption that the true data-generating distribution, $f$, is one of the models, $g$, in the set of models considered. Hence, the general result for K-L–based model selection does not appear to be AIC. Rather, the large-sample bias correction term subtracted from the expected maximized log-likelihood to then get $T$ is tr$(J(\underline{\theta}_o)[I(\underline{\theta}_o)]^{-1})$ (Takeuchi 1976). In deriving this result there is not, and need not be, any assumption that any of the candidate models represent truth. However, in general we know with certainty that tr$(J(\underline{\theta}_o)[I(\underline{\theta}_o)]^{-1}) = K$ only when $J(\underline{\theta}_o) = I(\underline{\theta}_o)$ (this is sufficient but not necessary), and the latter equality is only certain, in general, when $f$ is a special case of $g$, hence, when model $g$ equals or is a generalization of "truth." This condition is unrealistic to expect, so how good is the approximation tr$(J(\underline{\theta}_o)[I(\underline{\theta}_o)]^{-1}) = K$ when the truth is more general than the model, but the model is a good approximation to truth? We make here some limited, but useful, progress on this issue. Extensive theory, simulation studies, and experience (e.g., Linhart and Zucchini 1986:176–182, especially results such as in their Table 10.3) are needed to give us full confidence in when we can expect reliable results from AIC, versus when we might have to use TIC. Below, we establish theory and results for some models within the exponential family of distributions.

### 6.6.1 Comparison of AIC vs. TIC in a Very Simple Setting

We consider two simple, one-parameter distributions: negative-exponential and half-normal. For each distribution we can determine the trace, tr$(J(\underline{\theta}_o)[I(\underline{\theta}_o)]^{-1})$, assuming either that the distribution is truth, hence $f = g$, or that the other distribution is truth, $f$, and the given distribution is a model, $g$ (so $f \neq g$). We will also examine the estimators of the traces that can be used in TIC model selection and contrast TIC selection with AIC selection for these two distributions as models. This is a convenient situation to explore, partly because both distributions are in the exponential family.

For the negative-exponential distribution let $S = x_1 + \cdots + x_n = n\bar{x}$; then

$$g(\underline{x} \mid \lambda) = \frac{1}{\lambda^n} e^{-S/\lambda},$$
$$\log(g(\underline{x} \mid \lambda)) = -n\log(\lambda) - S/\lambda,$$

$$\frac{\partial}{\partial \lambda} \log(g(\underline{x} \mid \lambda)) = -n/\lambda + S/\lambda^2,$$

so $\hat{\lambda} = S/n = \bar{x}$; also here $E(x) = \lambda$.

Direct verification yields the results below:

$$I(\lambda) = \frac{n}{\lambda^2},$$

$$\hat{I}(\hat{\lambda}) = \frac{n}{\hat{\lambda}^2},$$

i.e., the empirical Hessian here is the same as $I(\hat{\lambda})$, and

$$\hat{J}(\lambda) = \sum_{i=1}^{n} \left[ \frac{-1}{\lambda} + \frac{x_i}{\lambda^2} \right]^2 = \left[ \sum_{i=1}^{n} \frac{(x_i)^2}{\lambda^4} \right] - \frac{n}{\lambda^2}.$$

The true $J(\lambda) = \hat{E}_f(\hat{J}(\lambda))$, and $\hat{J}(\hat{\lambda})$ is the empirical estimator of $J(\lambda)$. If we assume that the negative-exponential model is truth, then from the above (because $E(x^2) = 2\lambda^2$),

$$J(\lambda) = \frac{n}{\lambda^2}.$$

Clearly, if $f = g$, then here $\text{tr}[J(\lambda)[I(\lambda)]^{-1}] = 1$. The direct empirical estimator of this trace is

$$\text{tr}[\hat{J}(\hat{\lambda})[\hat{I}(\hat{\lambda})]^{-1}] = \frac{1}{n\bar{x}^2} \left[ \sum_{i=1}^{n} (x_i)^2 \right] - 1,$$

or

$$\text{tr}[\hat{J}(\hat{\lambda})[\hat{I}(\hat{\lambda})]^{-1}] = \frac{\sum_{i=1}^{n}(x_i - \bar{x})^2}{n\bar{x}^2}. \tag{6.60}$$

This estimator of the trace is the same as $\frac{n-1}{n}(\widehat{cv})^2$ (when we use $n - 1$ in the denominator of the sample $s^2$ as per convention), which we can expect to be quite variable. The trace estimator of (6.60) does converge to 1 as $n \to \infty$ when $f = g$. This trace estimator is scale-invariant, so we can calculate its distributional properties by Monte Carlo methods with a single run of 100,000 reps at each $n$, and for each case of truth being either the negative-exponential (any value of $\lambda$ can be used) or half-normal distribution (any $\sigma$ can be used). For the case where the negative-exponential model is truth, the simulation motivated a revised, nearly unbiased version of (6.60):

$$\widehat{\text{tr}}[J(\lambda)[I(\lambda)]^{-1}] = \frac{n}{n-1}(\widehat{cv})^2. \tag{6.61}$$

## 6.6 Evaluation of $\mathrm{tr}(J(\underline{\theta}_o)[I(\underline{\theta}_o)]^{-1})$ and Its Estimator

The estimated mean and standard deviation of the estimated trace function based on (6.60) and (6.61) are given below, based on 100,000 reps:

|  | formula (6.60) | | formula (6.61) | |
| --- | --- | --- | --- | --- |
| $n$ | mean | st.dev. | mean | st.dev. |
| 20 | 0.90 | 0.37 | 1.00 | 0.41 |
| 50 | 0.96 | 0.26 | 1.00 | 0.27 |
| 100 | 0.98 | 0.19 | 1.00 | 0.20 |
| 500 | 1.00 | 0.09 | 1.00 | 0.09 |

Notice the substantial standard deviation of either trace estimator; it is this sort of large variability in the trace estimator that has given theoretical pause to routine use of TIC. We also see that the direct estimator has bias for modest sample sizes; analytical reduction of that bias would not be practical in nontrivial applications, and yet such bias in any trace estimator might be important.

Consider now the half-normal distribution:

$$f(x \mid \sigma^2) = \sqrt{\frac{2}{\pi \sigma^2}} \exp\left[-\frac{1}{2}\left(\frac{x}{\sigma}\right)^2\right].$$

Under this distribution $E(x^2) = \sigma^2$ and direct integration gives

$$E(x) = \sigma\sqrt{\frac{2}{\pi}}.$$

If the half-normal distribution is truth and the negative-exponential is a model, then $\lambda_o$ as a function of $\sigma^2$ is found from

$$E_f(x) = \lambda_o = \sigma\sqrt{\frac{2}{\pi}}.$$

For example, with $\sigma^2 = 1$, $\lambda_o = 0.79788$ is the K-L best choice of $\lambda$.

Note the usage and concepts here: We denote the K-L best value of $\lambda$ by $\lambda_o$ to distinguish that the corresponding negative-exponential distribution (i.e., that based on $\lambda_o$) is the K-L best negative-exponential distribution to use as the model for the underlying truth. By denoting this value of $\lambda$ by $\lambda_o$ we are emphasizing that all we have is the K-L best negative-exponential model, but it may be a poor model; it certainly may not be truth.

The above expectations producing $J(\lambda)$ and $I(\lambda)$ were wrt the negative-exponential as $g$ (ignoring what $f$ might be), but now we want to take those expectations wrt $f$ as half-normal. Direct verification yields $I(\lambda_o) = n/\lambda_o^2$, whereas

$$J(\lambda_o) = nE_f\left[\frac{x}{\lambda_o^2} - \frac{1}{\lambda_o}\right]^2$$

$$= I(\lambda_o)\left[\frac{\sigma^2}{\lambda_o^2} - 1\right]$$

$$= I(\lambda_o)\left[\frac{\pi}{2} - 1\right].$$

264     6. Statistical Theory

Hence, when $f$ is half-normal and $g$ is negative-exponential,

$$\text{tr}\left[J(\lambda_o)[I(\lambda_o)]^{-1}\right] = \frac{\pi}{2} - 1 = 0.5708.$$

This trace term is not very close to 1, the number AIC would assume. This big relative difference (i.e., 0.5708 vs. 1) results because the negative-exponential model is a very poor approximation to the half-normal distribution. Note that this trace term is $< 1$ ($K$ is 1 here).

Next we determine the trace estimator under the half-normal as the model. We therein have

$$\hat{\sigma}^2 = \frac{\sum_{i=1}^{n}(x_i)^2}{n},$$

$$I(\sigma^2) = \frac{n}{2\sigma^4},$$

$$\hat{J}(\sigma^2) = \sum_{i=1}^{n}\left[\frac{-1}{2\sigma^2} + \frac{(x_i)^2}{2\sigma^4}\right]^2$$

$$= I(\sigma^2)\left[\frac{1}{2n}\right]\sum_{i=1}^{n}\left[\frac{(x_i)^2}{\sigma^2} - 1\right]^2.$$

The key part of the empirical estimator $\hat{J}(\hat{\sigma}^2)$ is again a squared coefficient of variation, but here it is for the variable $x^2$. Denote this $\widehat{cv}$ by $\widehat{cv}(x^2)$, and we can use the notation

$$\hat{J}(\hat{\sigma}^2) = \hat{I}(\hat{\sigma}^2)\left[\frac{n-1}{2n}\right][\widehat{cv}(x^2)]^2.$$

Thus, for the half-normal distribution being the model, the TIC estimator of the trace is

$$\text{tr}[\hat{J}(\hat{\sigma}^2)[\hat{I}(\hat{\sigma}^2)]^{-1}] = \left[\frac{n-1}{2n}\right][\widehat{cv}(x^2)]^2. \quad (6.62)$$

If truth is the half-normal model, this quantity will converge to 1; and again, (6.62) is scale-invariant. For the half-normal model as truth we computed the mean and standard deviation of (6.62) by Monte Carlo methods with 100,000 reps at each $n$. This led to a nearly unbiased version of (6.62),

$$\widehat{\text{tr}}[J(\sigma^2)[I(\sigma^2)]^{-1}] = \frac{1}{2}\left[\frac{n}{n-1}\right]^2[\widehat{cv}(x^2)]^2, \quad (6.63)$$

that we then also used in the simulations. The results are below:

| | formula (6.62) | | formula (6.63) | |
|---|---|---|---|---|
| $n$ | mean | st.dev. | mean | st.dev. |
| 20 | 0.87 | 0.39 | 1.01 | 0.45 |
| 50 | 0.94 | 0.30 | 1.00 | 0.32 |
| 100 | 0.97 | 0.23 | 1.00 | 0.23 |
| 500 | 0.99 | 0.11 | 1.00 | 0.11 |

### 6.6 Evaluation of tr$(J(\underline{\theta}_o)[I(\underline{\theta}_o)]^{-1})$ and Its Estimator

The main point from the above is how variable the trace estimator is.

To complete a set of analytical results (useful for validating Monte Carlo results) we computed the value of tr$[J(\sigma_o^2)[I(\sigma_o^2)]^{-1}]$ for the half-normal model when truth is the negative-exponential distribution. First we need $\sigma_o^2 = E(x^2) = 2\lambda^2$, because the expectation of $x$ must be taken wrt the negative-exponential distribution. We find that for the half-normal model and the negative-exponential as truth,

$$J(\sigma_o^2) = I(\sigma_o^2)\left[\frac{1}{2}\right] E\left[\frac{x^2}{\sigma_o^2} - 1\right]^2 = I(\sigma_o^2)2.5.$$

Thus when truth is the negative-exponential, and the model is half-normal (a terrible model in this case), tr$[J(\sigma_o^2)[I(\sigma_o^2)]^{-1}] = 2.5$ (not 1 as AIC assumes). Done the other way around we had the trace $< 1$. It turns out that this trace function under model misspecification can be either above or below $K$; it varies by situation, and in some situations the trace function can equal $K$ even with a misspecified model.

We can now compare AIC versus TIC model selection when the choices and truth are negative-exponential or half-normal.

For the negative-exponential model:

$$\text{AIC} = 2n[\log(\bar{x}) + 1] + 2,$$

$$\text{TIC} = 2n[\log(\bar{x}) + 1] + 2\frac{n-1}{n}(\widehat{\text{cv}}(x))^2,$$

$$\text{TICu} = 2n[\log(\bar{x}) + 1] + 2\frac{n}{n-1}(\widehat{\text{cv}}(x))^2.$$

For the half-normal model (using $\hat{\sigma}^2 = $ mean of the $x_i^2$):

$$\text{AIC} = n\left[\log(\hat{\sigma}^2) + 1 - \log(2/\pi)\right] + 2,$$

$$\text{TIC} = n\left[\log(\hat{\sigma}^2) + 1 - \log(2/\pi)\right] + \left[\frac{n-1}{n}\right]^2 \left[\widehat{\text{cv}}(x^2)\right]^2,$$

$$\text{TICu} = n\left[\log(\hat{\sigma}^2) + 1 - \log(2/\pi)\right] + \left[\frac{n}{n-1}\right]^2 \left[\widehat{\text{cv}}(x^2)\right]^2.$$

In both cases here TICu just means that the estimator of tr$[J(\theta_o))[I(\theta_o)]^{-1}]$ is almost unbiased, as opposed to the direct, biased, plug-in estimators of the needed coefficients of variation.

Table 6.1 shows some results. The point of this brief comparison was to learn something about AIC vs. TIC in a simple setting, especially whether or not they would give greatly different results. The context here is so simple that only two models are compared. Moreover, one or the other model was used as the data-generating distribution (i.e., truth). We did not consider prediction here, so the only possible criterion to use to compare performance of AIC vs. TIC is rate of selection of the true model. We did not wish to do, in this book, any serious evaluation of AIC vs. TIC under full-blown realistic conditions of complex truth, and a set of approximating models, wherein the correct basis of evaluation is how

TABLE 6.1. Percentage of correct selection when one of the models (negative-exponential or half-normal) is truth, based on 100,000 reps; see text for AIC, TIC, and TICu formulae; average percent correct is based on equal weighting of the two cases.

| sample size, $n$ | selection criterion | truth negative expon. | truth half-normal | average percent correct |
|---|---|---|---|---|
| 20 | AIC | 64 | 85 | 75 |
|    | TIC | 73 | 77 | 75 |
|    | TICu | 75 | 75 | 75 |
| 50 | AIC | 82 | 92 | 87 |
|    | TIC | 87 | 87 | 87 |
|    | TICu | 87 | 87 | 87 |
| 100 | AIC | 93 | 97 | 95 |
|     | TIC | 95 | 95 | 95 |
|     | TICu | 95 | 95 | 95 |
| 500 | AIC | 100 | 100 | 100 |
|     | TIC | 100 | 100 | 100 |
|     | TICu | 100 | 100 | 100 |

well a selection procedure does at selecting the K-L best model (technically, we would be selecting the expected K-L best model).

Several inferences supported by Table 6.1, and by all other sample sizes examined for this situation, surprised us. For the case that the negative-exponential model is true, the selection results based on TIC were uniformly as good or better than those under AIC. The improvement is not large except at small sample sizes, wherein an "$AIC_c$" should be used anyway. Conversely, for the case that the half-normal model is true, the selection results based on AIC were uniformly as good or better than those under TIC. In either case, bias-correction of the trace estimator makes no real difference. A priori we would not know which (if either) model was true. If we compute an average percent-correct selection based on the idea that we have no information to justify any weighting other than a 50:50 weighting of these results, we get, on average, no advantage at all for TIC over AIC. Clearly, we do not know the extent to which these results would generalize.

## 6.6.2 Evaluation Under Logistic Regression

Logistic regression is used often; therefore we illustrate that it is a case of an exponential family model and we explore the above trace question for this model. Let $x_i$ be a Bernoulli random variable with true probability $\mu_t$ of being 1 (and probability $1 - \mu_t$ of being 0). For a sample of $n$ independent $x_i$ we base analysis on some assumed model for the $\mu_i$. In order to distinguish truth from model we

## 6.6 Evaluation of $\text{tr}(J(\underline{\theta}_o)[I(\underline{\theta}_o)]^{-1})$ and Its Estimator

adopt the notation for the model as $p_i \equiv p_i(\underline{\theta})$, for some structure imposed on these $p_i$, as a function of a $K$-dimensional parameter vector, $\underline{\theta}$. The relevant pdf, or likelihood (the same notation continues to serve this dual role), for the model is

$$g(\underline{x}\,|\,\underline{\theta}) = \prod_{i=1}^{n}(p_i)^{x_i}(q_i)^{1-x_i}.$$

We assume known covariates $\underline{z}_i$, as $K \times 1$ column vectors, are associated with each observation, $x_i$, and an explanatory structural model is

$$p_i = \frac{1}{1+e^{-\underline{z}_i'\underline{\theta}}}, \text{ or } q_i = \frac{e^{-\underline{z}_i'\underline{\theta}}}{1+e^{-\underline{z}_i'\underline{\theta}}},$$

which is equivalent to

$$\log[p_i/(1-p_i)] = \underline{z}_i'\underline{\theta}.$$

A modest amount of algebra gives the result

$$g(\underline{x}\,|\,\underline{\theta}) = \exp\left[\left[\sum_{i=1}^{n}(x_i \underline{z}_i)\right]'\underline{\theta} + \sum_{i=1}^{n}\left(-\log\left(1+e^{\underline{z}_i'\underline{\theta}}\right)\right)\right],$$

which is in the canonical form of the exponential family for

$$H(\underline{\theta}) = \sum_{i=1}^{n}\left(-\log\left(1+e^{\underline{z}_i'\underline{\theta}}\right)\right) \tag{6.64}$$

and

$$\underline{S} = \sum_{i=1}^{n}(x_i \underline{z}_i) = \sum_{i=1}^{n}\underline{S}_i$$

($\underline{G}(\underline{\theta}) = \underline{0}$). We will need the true expectation of $\underline{S}$:

$$E_f(\underline{S}) = \left[\sum_{i=1}^{n}(\mu_i \underline{z}_i)\right].$$

Also, from (6.64), $H(\underline{\theta}) = \sum_{i=1}^{n}\log(1-p_i)$ is an equivalent form for $H(\underline{\theta})$. Two key quantities we need are

$$I(\underline{\theta}) = -\frac{\partial^2 H(\underline{\theta})}{\partial \underline{\theta}^2}$$

(see formula 6.55) and $-\partial H(\underline{\theta})/\partial\underline{\theta} = E_f(\underline{S})$. Some straightforward mathematics leads to the results

$$-\frac{\partial H(\underline{\theta})}{\partial \underline{\theta}} = \sum_{i=1}^{n} p_i \underline{z}_i$$

and

$$I(\underline{\theta}) = -\frac{\partial^2 H(\underline{\theta})}{\partial \underline{\theta}^2} = \sum_{i=1}^{n} p_i q_i \underline{z}_i \underline{z}_i'.$$

These formulae can be put in matrix notation. To do so we define an $n \times 1$ column vector $\underline{P}(\theta)$ with $i$th element $p_i(\theta)$, and an $n \times n$ diagonal matrix $V_p$ with $i$th diagonal element $p_i q_i$, and an $n \times K$ matrix $Z$ where the $i$th row is $\underline{z}'_i$. Then

$$E_f(\underline{S}) = Z'\underline{\mu},$$

$$-\frac{\partial H(\theta)}{\partial \theta} = Z'\underline{P}(\theta), \qquad (6.65)$$

and

$$I(\theta) = Z'V_p Z. \qquad (6.66)$$

The MLE $\hat{\theta}$ is found by setting (6.65) to $\underline{S}$ and solving the resultant $K$ nonlinear equations for $\theta$, hence solving $\underline{S} = Z'\underline{P}(\hat{\theta})$. The true parameter value $\theta_o$ that applies here, given truth $\underline{\mu}$ and the model, is found by solving the same $K$ equations but with $\underline{S}$ replaced by its true expectation, hence solving

$$Z'\underline{\mu} = Z'\underline{P}(\theta_o),$$

or

$$Z'(\underline{\mu} - \underline{P}(\theta_o)) = \underline{0}.$$

In partly nonmatrix notation, we solve

$$\sum_{i=1}^{n}(\mu_i - p_i(\theta_o))\underline{z}_i = \underline{0}.$$

If truth, $\underline{\mu}$, is not given exactly by the assumed model evaluated at $\theta_o$, then $\underline{\mu} = \underline{P}(\theta_o)$ will not hold even though the above equations will have a unique solution in $\theta_o$, just as the MLE equations will have a unique solution as $\hat{\theta}$.

To proceed we also need to know the general formula for $J(\theta_o)$. From (6.12) we have

$$J(\theta_o) = E_f\left[[\underline{S} - Z'\underline{P}(\theta_o)][\underline{S} - Z'\underline{P}(\theta_o)]'\right]$$

In partly nonmatrix form this formula is

$$J(\theta_o) = E_f\left[\sum_{i=1}^{n}(\underline{s}_i - p_i\underline{z}_i)\right]\left[\sum_{i=1}^{n}(\underline{s}_i - p_i\underline{z}_i)\right]'$$

$$= \sum_{i=1}^{n}\sum_{j=1}^{n}E_f(\underline{s}_i - p_i\underline{z}_i)(\underline{s}_j - p_j\underline{z}_j)'.$$

Here, using $\underline{s}_i = x_i\underline{z}_i$ and $E_f(x_i) = \mu_i$ the above becomes

$$J(\theta_o) = \sum_{i=1}^{n}\sum_{j=1}^{n}E_f(x_i - p_i)(x_j - p_j)[\underline{z}_i\underline{z}'_j]$$

$$= \sum_{i=1}^{n}E_f(x_i - p_i)^2[\underline{z}_i\underline{z}'_i] + \sum_{i \neq j}^{n}\sum E_f(x_i - p_i)(x_j - p_j)[\underline{z}_i\underline{z}'_j]$$

$$= \sum_{i=1}^{n} \left[(\mu_i(1-\mu_i)) + (\mu_i - p_i)^2\right][\underline{z}_i \underline{z}_i']$$

$$+ \sum_{i \neq j}^{n} \sum^{n} (\mu_i - p_i)(\mu_j - p_j)[\underline{z}_i \underline{z}_j'].$$

Completing the square in the trailing term above, we get

$$J(\underline{\theta}_o) = \sum_{i=1}^{n} \left[(\mu_i(1-\mu_i)) + (\mu_i - p_i)^2\right][\underline{z}_i \underline{z}_i']$$

$$+ \left[\sum_{i=1}^{n}(\mu_i - p_i(\underline{\theta}_o))\underline{z}_i\right]\left[\sum_{i=1}^{n}(\mu_i - p_i(\underline{\theta}_o))\underline{z}_i\right]'$$

$$- \sum_{i=1}^{n}\left[(\mu_i - p_i)^2\right][\underline{z}_i \underline{z}_i'].$$

The middle term of the above is zero because of the equation defining $\underline{\theta}_o$, and the third term cancels with part of the first term, so we have

$$J(\underline{\theta}_o) = \sum_{i=1}^{n} \mu_i(1-\mu_i)[\underline{z}_i \underline{z}_i'],$$

or in pure matrix terms,

$$J(\underline{\theta}_o) = Z'V_\mu Z. \tag{6.67}$$

Here, $V_\mu$ is an $n \times n$ diagonal matrix with $i$th diagonal element $\mu_i(1-\mu_i)$. Contrast (6.67) to $I(\underline{\theta}_o) = Z'V_p Z$.

It is easy, but not very informative, to now write

$$\operatorname{tr}\left[J(\underline{\theta}_o)[I(\underline{\theta}_o)]^{-1}\right] = \operatorname{tr}\left[(Z'V_\mu Z)(Z'V_p Z)^{-1}\right] \tag{6.68}$$

$$= K + \operatorname{tr}\left[(Z'(V_\mu - V_p)Z)(Z'V_p Z)^{-1}\right].$$

The above makes it easier to realize that the trace term is exactly $K$ if for all $i$ $\mu_i(1-\mu_i) = p_i(\underline{\theta}_o)(1 - p_i(\underline{\theta}_o))$. However, these equalities may fail to hold, yet we can still get $\operatorname{tr}\left[J(\underline{\theta}_o)[I(\underline{\theta}_o)]^{-1}\right] = K$; hence this latter equality can hold with a model that does not match truth, i.e., where $g \subset f$ with strict inequality.

The above results are totally general, so they apply to the case where, say, $w$ replicate observations are taken at each of $j = 1, \ldots, r$ covariate values. The total sample size is then $n = r * w$, but we will have only $r$ different values of $\mu_j$ to specify for truth and only $r$ values of $p_j(\underline{\theta})$ to consider under any model. Hence, to gain some insights here we used the simple model $\log\left[p_j/(1-p_j)\right] = a + bj$ for $j = 1, \ldots, r$, with $w$ replicate observations at each $j$. Thus $K = 2$, $\underline{\theta} = (a, b)'$, and $\underline{z}_j = (1, j)'$. In fact, for numerical or analytical results we do not have to specify what $w$ is (just $r$ is needed), and we can actually proceed as if $n = r$ with just one rep at each value of $j$. However, the results so derived apply reasonably well only to cases where $n = r * w$ would be "large," say 100 or more (given $K = 2$). Thus to explore the trace term under this use of a simple logistic regression model we need only specify a set of $\mu_1, \ldots, \mu_r$, solve

$$\sum_{i=1}^{r}(\mu_i - (a_o + b_o i))\underline{z}_i = \underline{0}$$

for $\underline{\theta}_o = (a_o, b_o)'$, and compute $J(\underline{\theta}_o)$ (6.67), $I(\underline{\theta}_o)$ (6.66), and then $\text{tr}\left[J(\underline{\theta}_o)[I(\underline{\theta}_o)]^{-1}\right]$ (also denoted as "bias"). In doing this we focused on sets of $\mu$ that were near to fitting the logistic structural model, either by generating a $p(\underline{\theta})$ vector that fit the model, then perturbing some (or all) of the $p_j$, or by starting with $\mu_j = j/(r+1)$, which is not a logistic regression structural model but is not too far from fitting such a structural model.

With at most modest deviation of truth from any actual simple logistic regression model structure we found that the trace term value stayed near $K = 2$ (between about 1.8 and 2, sometimes going a little above 2, say to 2.1). For the case of truth being the simple linear model ($\mu_j = j/(r+1)$), the trace term varied monotonically from 1.98 at $r = 5$, to 1.91 at $r = 50$. Table 6.2 gives some results for $r = 10$, based on truth being perturbed values from the logistic model $\text{logit}(p_j) = 3.0 - 0.5j$. The first line of Table 6.2 gives the true $\mu_j$ computed from this model (scaled by 1,000).

TABLE 6.2. Some values of trace $= \text{tr}\left[J(\underline{\theta}_o)[I(\underline{\theta}_o)]^{-1}\right]$ for the simple logistic model $\text{logit}(p_j) = a + bj, (j = 1, \ldots, 10)$ fit to $\mu_j$ as perturbed values of $p_j$ from $\text{logit}(p_j) = 3 - 0.5j$; values of $\mu_j$ are shown, scaled by 1,000; case one (i.e., the first line) exactly fits the logistic model, but none of the other cases are a perfect fit to the assumed model form. The results are reasonably applicable if $w$ is at least 10 or 20.

| $\mu_1$ | $\mu_2$ | $\mu_3$ | $\mu_4$ | $\mu_5$ | $\mu_6$ | $\mu_7$ | $\mu_8$ | $\mu_9$ | $\mu_{10}$ | trace |
|---|---|---|---|---|---|---|---|---|---|---|
| 924 | 881 | 818 | 731 | 622 | 500 | 378 | 269 | 182 | 119 | 2 |
| 900 | 900 | 900 | 900 | 622 | 500 | 378 | 269 | 182 | 119 | 2.063 |
| 924 | 881 | 970 | 757 | 725 | 530 | 410 | 231 | 186 | 134 | 2.047 |
| 900 | 944 | 898 | 773 | 666 | 522 | 349 | 264 | 176 | 127 | 2.046 |
| 864 | 993 | 971 | 821 | 564 | 514 | 366 | 236 | 150 | 135 | 2.023 |
| 894 | 990 | 870 | 826 | 583 | 457 | 372 | 252 | 216 | 127 | 2.00011 |
| 924 | 720 | 818 | 796 | 749 | 632 | 492 | 188 | 260 | 164 | 1.993 |
| 924 | 881 | 768 | 831 | 622 | 400 | 378 | 269 | 282 | 119 | 1.980 |
| 924 | 952 | 650 | 838 | 638 | 448 | 278 | 189 | 204 | 132 | 1.967 |
| 874 | 874 | 874 | 558 | 558 | 558 | 558 | 190 | 190 | 190 | 1.925 |
| 924 | 881 | 818 | 731 | 400 | 600 | 378 | 269 | 182 | 119 | 1.925 |
| 924 | 881 | 818 | 731 | 300 | 700 | 378 | 269 | 182 | 119 | 1.857 |
| 924 | 881 | 818 | 731 | 622 | 500 | 378 | 269 | 182 | 119 | 1.828 |
| 924 | 881 | 568 | 831 | 622 | 100 | 378 | 269 | 432 | 119 | 1.774 |
| 924 | 881 | 818 | 731 | 200 | 800 | 378 | 269 | 182 | 119 | 1,768 |
| 674 | 981 | 818 | 731 | 622 | 500 | 128 | 669 | 182 | 119 | 1.765 |
| 924 | 881 | 818 | 731 | 622 | 500 | 900 | 900 | 900 | 900 | 1.494 |

## 6.6 Evaluation of tr($J(\theta_o)[I(\theta_o)]^{-1}$) and Its Estimator

What one can see in Table 6.2 (and other computations we did corroborate this) is that the true $\mu_j$ have to be here a very poor approximation to an exact simple logistic model before the trace term deviates much from $K = 2$. Thus, if the data seem at all well fit by a logistic model, then the use of trace $= K$ (as opposed to any attempted estimation of the trace) seems quite suitable. This is especially important here because $V_\mu$, hence $J(\theta_o)$, cannot be estimated at all unless there is replication at each $\underline{z}_i$, and there would need to be substantial such replication; this condition rarely occurs with logistic regression.

Formula (6.68) was corroborated by direct Monte Carlo evaluation of the target bias (trace term) for a few cases in Table 6.2. The completely general, and hence most direct, way to do this is to evaluate using simulation the value of

$$\text{bias} = E_{\underline{x}} E_{\hat{\theta}(\underline{y})} \left[\log(g(\underline{x} \mid \hat{\theta}(\underline{y})))\right] - E_{\underline{x}} \left[\log(g(\underline{x} \mid \hat{\theta}(\underline{x})))\right]. \tag{6.69}$$

Hence for one Monte Carlo replicate (generating *iid* $\underline{x}$ and $\underline{y}$) we get

$$\widehat{\text{bias}} = \log(g(\underline{x} \mid \hat{\theta}(\underline{y}))) - \log(g(\underline{x} \mid \hat{\theta}(\underline{x}))).$$

Averaged over many reps ($m$), if large sample size $n$ is used, the average $\widehat{\text{bias}}$ will equal tr$\left[J(\theta_o)[I(\theta_o)]^{-1}\right]$.

For many models the first term on the right-hand side of (6.69) will be linear in $\underline{x}$, so we can analytically take the expectation wrt $\underline{x}$. For this logistic example we thus get, expressed in basic form,

$$\widehat{\text{bias}} = \sum_{i=1}^{n} E_{\underline{y}} \left[n_i \mu_i \log(p_i(\hat{\theta})) + n_i(1 - \mu_i) \log(q_i(\hat{\theta}))\right]$$
$$- \sum_{i=1}^{n} E_{\underline{y}} \left[y_i \log(p_i(\hat{\theta})) + (n_i - y_i) \log(q_i(\hat{\theta}))\right]. \tag{6.70}$$

Here the MLE $\hat{\theta}$ is based on data $\underline{y}$.

For direct Monte Carlo evaluation we used replicate covariate values, as noted above with the same number of replicates, $w$, for each $j = 1, \ldots, r$. Then we generated a large number, $m$, of independent samples (i.e., reps) from the true generating model, fit the model-based MLE to each sample, computed $\widehat{\text{bias}}$ using (6.70), by rep, and got its average and empirical standard error.

All this is quite obvious; where we are going here is that this direct Monte Carlo evaluation is poor in the sense of needing a huge number of reps. The problem is that as $w$ (i.e., $n$) increases, the number of reps needed to get a small standard error (like 0.005) on the estimate bias increases because the variance of $\widehat{\text{bias}}$, for one rep, increases with increasing sample size, and that variance can be quite large. For a large sample size $n$ (which is required for the trace approximation to hold very well) it can take one million Monte Carlo reps to get even a moderately small standard error on estimated bias. For example, for the case in Table 6.2 where the trace is computed to be 1.774, for $w = 100$ (hence sample size $n = 1,000$ Bernoulli trials) for one representative set of 10,000 ($= m$) Monte Carlo reps we got average $\widehat{\text{bias}} = 1.528$ with an estimated standard error of 0.135. Other runs verified that it takes about one million Monte Carlo reps to get a standard error (on estimated bias) of about 0.014 with $w = 100$ in this example. But we might need a bigger $w$ for the trace formula to apply exactly; for $w = 1,000$ and 10,000 Monte Carlo reps we got the average $\widehat{\text{bias}} = 1.982$ with $\widehat{\text{se}} = 0.424$. This phenomenon is the reverse of what we expect; i.e., we expect to get increasing precision (for the same number, $m$, of Monte Carlo reps) as sample size $n$ increases. The reverse phenomenon occurs here because the expected difference in likelihoods involved in direct computation of $\widehat{\text{bias}}$ (i.e., formula 6.70) is constant independent of sample size $n$, but the variance of each of those two likelihood sums in (6.70) is proportional to $n$ and the two terms are not highly correlated. Thus as sample size $n$ increases, the precision of the estimated bias, given a fixed number of Monte Carlo samples ($m$), actually decreases. So to evaluate well, with this brute-force approach, the adequacy of the trace term approximation at large sample sizes, it takes a huge number of Monte Carlo reps.

With models that are in the exponential family there is an alternative way to do exact Monte Carlo evaluation of the bias that must be subtracted from the maximized log-likelihood for exact K-L based model selection. Formula (6.54) is an exact result for any sample size:

$$\text{bias} = \text{tr}\left[\text{COV}(\hat{\underline{\theta}}, \underline{S})\right].$$

While $\underline{\theta}$ and $\underline{S}$ are only for the canonical form of the model, the result will apply for any parametrization of the assumed model because of the invariance of the result to 1-to-1 transformations of $\underline{\theta}$ (see end of Section 6.5). Thus the alternative Monte Carlo evaluation is simply to take for each rep the already computed MLE and minimal sufficient statistic and, from this set of records of size $m$, estimate the covariances $\text{cov}(\hat{\theta}_i, S_i)$, $i = 1, \ldots, K$, then sum these $K$ estimates. The result is $\widehat{\text{bias}}$, and this approach is much more efficient. For the same case in Table 6.2 (i.e., trace $= 1.774$), using $w = 100$ ($r = 10$, hence $n = 1,000$) and 10,000 Monte Carlo reps we got $\widehat{\text{bias}} = 1.804$ and its $\widehat{\text{se}} = 0.019$ using the covariance approach. Based on this and other runs there was a clear suggestion that $w = 100$ was not quite big enough for the trace formula (6.68) to apply reliably to three digits (it was then reliable to two digits). Using $w = 1,000$ and 10,000 Monte Carlo reps we got $\widehat{\text{bias}} = 1.771$, $\widehat{\text{se}} = 0.017$. This result held up on more study: (6.68) seemed to be excellent for $w = 1,000$ (which here meant $n = 10,000$).

As another example consider the last case in Table 6.2, where trace $= 1.494$ ($=$ bias). Using $w = 100$, for one run of 10,000 Monte Carlo reps (the only such run made) we got the direct result based on (6.70) as $\widehat{\text{bias}} = 1.959$, $\widehat{\text{se}} = 0.171$. In contrast, for that same simulated data set the covariance approach yielded $\widehat{\text{bias}} = 1.516$, $\widehat{\text{se}} = 0.016$. Clearly, in working with models in the exponential family, Monte Carlo or bootstrap evaluation of the needed K-L trace term should be based on formula (6.54).

It is worth noting a basis for the estimated standard error of $\widehat{\text{bias}} = \widehat{\text{tr}}\left[\text{COV}(\hat{\underline{\theta}}, \underline{S})\right]$. For the point estimate, use all the simulation reps to compute means; then for component $i$,

$$\widehat{\mathrm{cov}}(\hat{\theta}_i, S_i) = \frac{\sum_{j=1}^{m}(\hat{\theta}_{i,j} - \overline{\hat{\theta}}_i)(S_{i,j} - \overline{S}_i)}{m-1},$$

and

$$\widehat{\mathrm{bias}} = \sum_{i=1}^{K} \widehat{\mathrm{cov}}(\hat{\theta}_i, S_i).$$

However, to estimate the standard error we must partition the set of $m$ reps, say into 25 equal-sized subsets (for $m = 10{,}000$ then each subset has size 400). Compute by the above formulae $\widehat{\mathrm{bias}}_s$ for each subset $s$, then estimate the standard error of $\widehat{\mathrm{bias}}$ from these 25 independent estimates (whose mean will almost equal $\widehat{\mathrm{bias}}$, but will not be equal due to nonlinearities).

The standard error of $\widehat{\mathrm{bias}}$ from this covariance approach is stable as a function of data sample size $n$ because of how the product involved behaves. It suffices to consider the product $(\hat{\theta}_i - \theta_{o,i})(S_i - E_f(S_i))$ (for any component, $i$). This product has variance virtually independent of $n$ because the first term converges (in $n$) at rate proportional to $1/\sqrt{n}$, while the second term converges at rate proportional to $\sqrt{n}$. As a result, the standard error of this covariance-based bias (hence trace) estimator is almost independent of sample size. This is much better behavior (as a function of $n$) than the standard error of the estimator of bias based directly on the likelihood function. The latter method (i.e., formula 6.69) also requires more calculations beyond first getting $\hat{\theta}$ and $\underline{S}$.

### 6.6.3 Evaluation Under Multinomially Distributed Count Data

We here assume that we have count data $n_1, \ldots, n_r$ that sum to the sample size $n$. Truth is the multinomial distribution $\mathrm{mult}(n, \mu_1, \ldots, \mu_r)$ with cell probabilities $\mu_i$ summing to 1, and $0 < \mu_i < 1$. To know truth in this context we only need to know the true $\mu_i$ (assuming that the counts are multinomially distributed; they could have overdispersion, which violates this assumption). We might totally fail to know how these true probabilities arise in general in relation to any explanatory variables, or what would happen if the cells were defined in some other way. Thus, deeper truth may exist regarding the situation, but it is irrelevant to model-selection purposes once we restrict ourselves to a particular multinomial setting.

For a constrained model we assume that cell probabilities $p_i(\underline{\theta})$ are known functions of a $K$-dimensional parameter, $\underline{\theta}$, with $1 \leq K < r - 1$ ($K = r - 1$ is not to be considered, as then the fitted model matches truth in the sense of being a perfect match to the data). The theory in Sections 6.1 and 6.2 is now used; note that here

$$\log(g(\underline{n} \mid \underline{\theta})) = \sum_{i=1}^{r} n_i \log(p_i(\underline{\theta})).$$

## 6. Statistical Theory

First, $\underline{\theta}_o$ is determined as the solution to (6.5), which here becomes

$$\sum_{i=1}^{r} \frac{\mu_i}{p_i(\underline{\theta}_o)} \frac{\partial p_i(\underline{\theta}_o)}{\partial \underline{\theta}} = \underline{0}. \tag{6.71}$$

In (6.71) if we replace $\mu_i$ by $n_i$, we have the likelihood equations. Thus one can treat the $\mu_i$ as data and find $\underline{\theta}_o$ by MLE methods. Equivalently, $\underline{\theta}_o$ is the MLE when the data are replaced by their true expected values, $E_f(n_i) = n\mu_i$.

Second, applying (6.7) we directly get

$$I(\underline{\theta}_o) = n \left[ \sum_{i=1}^{r} \frac{\mu_i}{[p_i(\underline{\theta}_o)]^2} \left( \frac{\partial p_i(\underline{\theta}_o)}{\partial \underline{\theta}} \right) \left( \frac{\partial p_i(\underline{\theta}_o)}{\partial \underline{\theta}} \right)' \right]$$

$$- n \left[ \sum_{i=1}^{r} \frac{\mu_i}{p_i(\underline{\theta}_o)} \left( \frac{\partial^2 p_i(\underline{\theta}_o)}{\partial \underline{\theta}^2} \right) \right].$$

Finally, applying the definition in (6.12), we have

$$J(\underline{\theta}_o) = E_f \left[ \sum_{i=1}^{r} \frac{n_i}{p_i(\underline{\theta}_o)} \left( \frac{\partial p_i(\underline{\theta}_o)}{\partial \underline{\theta}} \right) \right] \left[ \sum_{i=1}^{r} \frac{n_i}{p_i(\underline{\theta}_o)} \left( \frac{\partial p_i(\underline{\theta}_o)}{\partial \underline{\theta}} \right) \right]'.$$

The evaluation of $J(\underline{\theta}_o)$ does take some algebra and knowledge of the multinomial distribution, but it is mostly a straightforward exercise, so we just give the result:

$$J(\underline{\theta}_o) = n \left[ \sum_{i=1}^{r} \frac{\mu_i}{[p_i(\underline{\theta}_o)]^2} \left( \frac{\partial p_i(\underline{\theta}_o)}{\partial \underline{\theta}} \right) \left( \frac{\partial p_i(\underline{\theta}_o)}{\partial \underline{\theta}} \right)' \right]. \tag{6.72}$$

Define the matrix $A$ as

$$A = n \left[ \sum_{i=1}^{r} \frac{\mu_i}{p_i(\underline{\theta}_o)} \left( \frac{\partial^2 p_i(\underline{\theta}_o)}{\partial \underline{\theta}^2} \right) \right],$$

and we have $I(\underline{\theta}_o) = J(\underline{\theta}_o) - A$. Furthermore, if the model is truth, then $\mu_i = p_i(\underline{\theta}_o)$, and $A$ reduces to the null matrix; hence then $I(\underline{\theta}_o) = J(\underline{\theta}_o)$.

Using these results we can write

$$\text{tr}\left[J(\underline{\theta}_o)[I(\underline{\theta}_o)]^{-1}\right] = \text{tr}\left[(J(\underline{\theta}_o) - A + A)[I(\underline{\theta}_o)]^{-1}\right],$$

whence,

$$\text{tr}\left[J(\underline{\theta}_o)[I(\underline{\theta}_o)]^{-1}\right] = K + \text{tr}\left[A[I(\underline{\theta}_o)]^{-1}\right].$$

For a long time a nagging question for us was whether the trace term would always be either $> K$ or $< K$ when the model did not exactly match truth yet the model is logically known to be simpler than truth (i.e., $g \subset f$ in some general sense). Stated differently, if the Kullback–Leibler discrepancy is positive, i.e., K-L $= I(f, g) > 0$, then must $\text{tr}\left[J(\underline{\theta}_o)[I(\underline{\theta}_o)]^{-1}\right] > K$ (or maybe $< K$) always occur when the model is some form of constrained truth (hence the model can be said to approximate, but not equal, truth). The answer is no, as was indicated by the logistic regression examples in Section 6.6.2); however, a more convincing

### 6.6 Evaluation of tr$(J(\underline{\theta}_o)[I(\underline{\theta}_o)]^{-1})$ and Its Estimator

answer is given here: The trace can be either $> K$ or $< K$ and there need be no consistency as to which will occur. A related question also explored below is, If tr$[J(\underline{\theta}_o)[I(\underline{\theta}_o)]^{-1}] = K$ must $I(f, g) = 0$? That answer is also no.

Because the cell probabilities sum to 1, the sum of the matrices of second partials is the null matrix, $O$, of all zeros (the vector of first partials also sums to a null vector). Therefore, an equivalent expression for matrix $A$ is

$$A = n \left[ \sum_{i=1}^{r} \frac{\mu_i - p_i(\underline{\theta}_o)}{p_i(\underline{\theta}_o)} \left( \frac{\partial^2 p_i(\underline{\theta}_o)}{\partial \underline{\theta}^2} \right) \right].$$

The weights in this linear combination of second partial derivative matrices must be either identically zero (hence K-L is 0), or some are negative and some positive. This would suggest that $A$ might not always have the same sign, unless the second partials are very strangely related to the model and truth. But a more detailed case is need to get an example, and it seems best to use $K = 1$ for an example, such as by using a binomial model.

Let us further assume that the data arise from $n$ independent samples of an integer random variable, $y$, taking values 0 to $r - 1$. The data are then just the frequency counts $n_i$ of times $y = i - 1$. A very simple model for the cell probabilities, $\mu_i$, is thus to assume that this underlying random variable is a binomial random variable. This corresponds to imposing an ordering on the multinomial cells, wlg, and thus the model for the cell $i$ probability is

$$p_i(\theta) = \binom{r-1}{i-1} \theta^{i-1}(1-\theta)^{r-i}, \qquad i = 1, \ldots, r.$$

Thus we have, as our model, an assumed underlying binomial random variable $y \sim \text{bin}(r - 1, \theta)$ and a random sample of size $n$ of this random variable. In fact, $y \,(= 0, 1, \ldots, r - 1)$ has the distribution given by the $\mu_{y+1}$ as its true distribution. We will need the functions below, involving first and second partial derivatives:

$$P1_i = \frac{1}{p_i(\theta)} \left( \frac{\partial p_i(\theta)}{\partial \theta} \right) = \frac{(i-1) - \theta(r-1)}{\theta(1-\theta)},$$

$$P2_i = \frac{1}{p_i(\theta)} \left( \frac{\partial^2 p_i(\theta)}{\partial \theta^2} \right) = \left[ \frac{(i-1) - \theta(r-1)}{\theta(1-\theta)} \right]^2$$
$$- \left[ \frac{(i-1)(1 - 2\theta) + \theta^2(r-1)}{(\theta(1-\theta))^2} \right].$$

We solve (6.71), which is $\sum \mu_i P1_i = 0$, to find $\theta_o$; this is exactly the same process as finding an MLE (again, the only tricky aspect is the indexing assumed here):

$$\theta_o = \frac{\sum_{i=1}^{r} \mu_i(i-1)}{r - 1}.$$

276   6. Statistical Theory

This $\theta_o$ is the true expected value of $y/(r-1)$ regardless of any assumed model. We compute (6.72) as $n \sum \mu_i (P1_i)^2$:

$$J(\theta_o) = n \left[ \sum_{i=1}^{r} \mu_i \left( \frac{(i-1) - \theta_o(r-1)}{\theta_o(1-\theta_o)} \right)^2 \right] \equiv n E_f \left( \frac{y - E_f(y)}{\theta_o(1-\theta_o)} \right)^2.$$

We find matrix $A$ as $n \sum \mu_i P2_i$:

$$A = J(\theta_o) - \frac{n(r-1)}{\theta_o(1-\theta_o)};$$

hence

$$I(\theta_o) = \frac{n(r-1)}{\theta_o(1-\theta_o)}.$$

It is now easy to find the trace:

$$\text{tr}\left[J(\theta_o)[I(\theta_o)]^{-1}\right] = \sum_{i=1}^{r} \mu_i \frac{[(i-1) - \theta_o(r-1)]^2}{(r-1)\theta_o(1-\theta_o)}. \quad (6.73)$$

For the case of $\mu_i = p_i(\theta_o)$, then (6.73) is 1 (this can be directly verified); hence using $\mu_i \equiv p_i(\theta_o) + (\mu_i - p_i(\theta_o))$ in (6.73) we obtain

$$\text{tr}\left[J(\theta_o)[I(\theta_o)]^{-1}\right] = 1 + \sum_{i=1}^{r} (\mu_i - p_i(\theta_o)) \frac{[(i-1) - \theta_o(r-1)]^2}{(r-1)\theta_o(1-\theta_o)},$$

whereupon it should be essentially obvious that the term added to 1 ($= K$) can be either positive or negative. However, we will give numerical examples, mostly for $r = 3$ because this is the smallest $r$ we can use for our purposes here, and small $r$ is desirable when we need to display truth.

Our model is thus bin$(2, \theta)$; hence $p_1 = (1-\theta)^2$, $p_2 = 2\theta(1-\theta)$, and $p_3 = \theta^2$. The approach is to specify the $\mu_i$ and compute $\theta_o = (\mu_2/2) + \mu_3$, and from (6.73), for $r = 3$,

$$\text{tr}\left[J(\theta_o)[I(\theta_o)]^{-1}\right] = \frac{4\mu_1(\theta_o)^2 + \mu_2(1-2\theta_o)^2 + 4\mu_3(1-\theta_o)^2}{2\theta_o(1-\theta_o)}. \quad (6.74)$$

We will also consider the values of $I(f, g)$, so we note that this K-L discrepancy is here $\sum(\mu_i) \log[\mu_i/p_i(\theta_o)]$. Numerical values are given below for three cases of truth (the $\mu_i$) in relationship to the K-L best approximating binomial model. Case one exactly fits a binomial model. Cases two and three are fit terribly by even the K-L best approximating binomial model. In all three cases, $\theta_o = 0.5$. In what is below, "Trace" means the value computed from formula (6.74) for $\theta_o = 0.5$, and K-L is the Kullback–Leibler information discrepancy between truth and the best approximating binomial model ("Bias-MC" is explained below):

| $\mu_1$ | $\mu_2$ | $\mu_3$ | Trace | K-L | Bias-MC |
|---|---|---|---|---|---|
| 0.25 | 0.50 | 0.25 | 1.000 | 0.000 | 1.003 |
| 0.05 | 0.90 | 0.05 | 0.200 | 0.368 | 0.194 |
| 0.45 | 0.10 | 0.45 | 1.800 | 0.368 | 1.816 |

## 6.6 Evaluation of $\text{tr}(J(\underline{\theta}_o)[I(\underline{\theta}_o)]^{-1})$ and Its Estimator

(we note that in this situation the trace term (6.73) seems to be bounded above by 2). Clearly, this bias-correction trace term can be either less than or greater than 1 when the model does not match truth. This is because the theoretical variance of $y$ can be either larger or smaller than the theoretical binomial variance for $y$ implied by the K-L best-fitting binomial model.

We build on this example by doing an exact Monte Carlo evaluation of the expected log-likelihood and the K-L–based target model-selection criterion to verify the asymptotic derivation of the bias as being the trace term. In the above, "Bias-MC" denotes the results (accurate to two decimal places), for sample size $n = 200$, from one million Monte Carlo reps to evaluate, without mathematical approximations, the bias that the trace term measures based on asymptotic theory.

For a truth that cannot be well approximated here by a binomial model it is clear that the trace (equation 6.74) can be far from 1. Rather than explore this example for models that are arbitrarily poor (like cases two and three, above) we should consider models that are closer to truth, because with AIC (or TIC) the term $-2 \log(\mathcal{L})$ will prevent the selection of really poor models (hence for those models a choice between the use of $K$ or $\widehat{\text{trace}}$ is irrelevant) if the set of models has some good candidates.

So we looked at one set of cases where a binomial model was not terrible wrong to use. We chose a $\theta$, generated $p_1 = (1 - \theta)^2$, $p_2 = 2\theta(1 - \theta)$, and $p_3 = \theta^2$, then perturbed these cell probabilities to get a truth that was close to a binomial model by setting $\mu_i = p_i + \epsilon_i$, where $\epsilon_i = \delta_i - \bar{\delta}$ for $\delta_i \sim$ iid uniform$(-h, h)$. Inadmissible sets of $\mu_i$ were not generated. Given a set of $\mu_1$, $\mu_2$, and $\mu_3$, formula (6.74) was evaluated; thus, this is not a Monte Carlo study. Rather, we use Monte Carlo methods only as a convenience in generating sets of true $\mu_i$ that are close to a binomial model.

For $\theta = 0.5$ and $h = 0.1$ (and 1,000 generated sets of truth) we got the following results for the trace given by (6.74): min $= 0.747$, max $= 1.234$, and mean $= 0.996$. These results support practical use of $K$ rather than $\widehat{\text{tr}}\left[J(\theta_o)[I(\theta_o)]^{-1}\right]$. However, it is fair to ask about estimating this trace term here (hence using TIC), as can be done by plugging $\hat{\theta}_o$ and $\hat{\mu}_i = n_i/n$ into (6.74); after simplification,

$$\widehat{\text{tr}}\left[J(\theta_o)[I(\theta_o)]^{-1}\right] = \frac{4\hat{\mu}_1(\hat{\theta}_o)^2 + \hat{\mu}_2(1 - 2\hat{\theta}_o)^2 + 4\hat{\mu}_3(1 - \hat{\theta}_o)^2}{2\hat{\theta}_o(1 - \hat{\theta}_o)}. \quad (6.75)$$

A small Monte Carlo evaluation of this estimator was done to see whether it was badly biased or highly variable. Variables in this study are the three $\mu_i$ and sample size $n$. Results, given in Table 6.3 based on one million reps, are the theoretical trace value, and the mean and standard deviation of (6.75) evaluated by simulation, accurate to two decimal places. In Table 6.3 if the trace is 1, then the binomial distribution is truth; otherwise it is not truth.

From Table 6.3 it appears that the trace estimator has good properties, so it is reasonable to consider using TIC rather than AIC; at least the comparison of the two seems worth doing here. For the sets of true $\mu_i$ considered above we compared AIC to TIC for the binomial model ($K = 1$; hence a reduced model, R) and the

TABLE 6.3. Some Monte Carlo results evaluating the formula (6.75) estimator of tr $\left[J(\underline{\theta}_o)[I(\underline{\theta}_o)]^{-1}\right]$ for the case of an assumed binomial(2, $\theta$) model when truth $(\mu_1, \mu_2, \mu_3)$ may be more general; the true trace value is known for these cases; the mean and standard deviation of (6.75) are given based on one million reps.

| $\mu_1$ | $\mu_2$ | $\mu_3$ | $n$ | Trace | Mean | St. dev. |
|---|---|---|---|---|---|---|
| 0.25 | 0.50 | 0.25 | 50  | 1.000 | 0.990 | 0.142 |
| 0.25 | 0.50 | 0.25 | 100 | 1.000 | 0.995 | 0.101 |
| 0.25 | 0.50 | 0.25 | 200 | 1.000 | 0.997 | 0.071 |
| 0.04 | 0.32 | 0.64 | 200 | 1.000 | 0.997 | 0.072 |
| 0.20 | 0.55 | 0.25 | 50  | 0.897 | 0.887 | 0.141 |
| 0.20 | 0.60 | 0.20 | 50  | 0.800 | 0.790 | 0.139 |
| 0.30 | 0.40 | 0.30 | 50  | 1.200 | 1.190 | 0.140 |
| 0.30 | 0.45 | 0.25 | 50  | 1.098 | 1.088 | 0.140 |
| 0.05 | 0.90 | 0.05 | 200 | 0.200 | 0.199 | 0.043 |
| 0.45 | 0.10 | 0.45 | 200 | 1.800 | 1.799 | 0.043 |

parameter-saturated general model ($K = 2$; G). Let the corresponding maximized likelihoods be $\mathcal{L}_R$ and $\mathcal{L}_G$. Hence,

$$\text{AIC}_R = -2\log(\mathcal{L}_R) + 2,$$
$$\text{AIC}_G = -2\log(\mathcal{L}_G) + 4,$$
$$\text{TIC}_R = -2\log(\mathcal{L}_R) + 2\,\widehat{\text{tr}}\left[J(\theta_o)[I(\theta_o)]^{-1}\right],$$
$$\text{TIC}_G = -2\log(\mathcal{L}_G) + 4,$$

where $\widehat{\text{tr}}\left[J(\theta_o)[I(\theta_o)]^{-1}\right]$ is given by (6.75). Because there are only two models here, and because we want to keep matters simple, we just compared selection methods based on how often they selected the same model and how often they selected the correct data-generating model. Results are based on 10,000 Monte Carlo reps, which suffices here to get standard errors $\leq 0.005$ for estimated proportions. "Truth" denotes the correct generating model. The "AIC" and "TIC" columns denote the proportion of reps for which these methods selected the correct model. "Match" denotes the proportion of reps in which both methods selected the same model, regardless of which model it was.

We looked at a lot more results than are given above to compare AIC and TIC in this limited context; there was then no change from the above in the obvious conclusion: no meaningful difference in performance here of AIC vs. TIC. More study is surely warranted; this limited look was done in the spirit that maybe something dramatic would result. It did not; as a tentative conclusion (based on all the considerations we have done on the matter, not just those of this section), it seems that simplicity strongly favors use of AIC over TIC.

We return to an interesting theoretical question posed above. It is known that if $f = g$, then tr$\left[J(\theta_o)[I(\theta_o)]^{-1}\right] = K$. However, if tr$\left[J(\theta_o)[I(\theta_o)]^{-1}\right] = K$, does this mean $f = g$?

6.6 Evaluation of tr$(J(\underline{\theta}_o)[I(\underline{\theta}_o)]^{-1})$ and Its Estimator    279

TABLE 6.4. Some Monte Carlo results evaluating AIC versus TIC model selection for a binomial(2, $\theta$) model ($K = 1$) versus a saturated multinomial model ($K = 2$); the true generating model varied (R for the binomial, G for the multinomial); each case is based on 10,000 reps; column AIC (or TIC) denotes the proportion of cases where AIC (or TIC) selected the correct data-generating model; column "Match" means that both criteria selected the same model whether or not it was the data-generating model (see text for more details).

| $\mu_1$ | $\mu_2$ | $\mu_3$ | $n$ | Truth | AIC | TIC | Match |
|---|---|---|---|---|---|---|---|
| 0.25 | 0.50 | 0.25 | 50 | R | 0.83 | 0.84 | 0.93 |
| 0.25 | 0.50 | 0.25 | 100 | R | 0.85 | 0.84 | 0.96 |
| 0.25 | 0.50 | 0.25 | 200 | R | 0.84 | 0.84 | 0.97 |
| 0.04 | 0.32 | 0.64 | 200 | R | 0.84 | 0.83 | 0.99 |
| 0.20 | 0.55 | 0.25 | 50 | G | 0.69 | 0.76 | 0.91 |
| 0.20 | 0.60 | 0.20 | 50 | G | 0.57 | 0.53 | 0.89 |
| 0.30 | 0.40 | 0.30 | 50 | G | 0.46 | 0.53 | 0.92 |
| 0.30 | 0.45 | 0.25 | 50 | G | 0.24 | 0.28 | 0.93 |
| 0.05 | 0.90 | 0.05 | 200 | G | 1.00 | 1.00 | 1.00 |
| 0.45 | 0.10 | 0.45 | 200 | G | 1.00 | 1.00 | 1.00 |
| | | | | means | 0.73 | 0.75 | 0.95 |

A counterexample shows that the assertion is false; hence there are situations wherein truth is more complex than the models used for analysis and yet AIC is appropriate to use (as opposed to TIC, which would then unnecessarily be just estimating $K$).

For this multinomial context we have shown that $I(\underline{\theta}_o) = J(\underline{\theta}_o) - A$ (considerations here are for any value of $r$). So if $A = O$ (i.e., is all zeros), then $I(\underline{\theta}_o) = J(\underline{\theta}_o)$ and tr$[J(\underline{\theta}_o)[I(\underline{\theta}_o)]^{-1}] = K$ regardless of whether or not the model is truth (which requires $\mu_i = p_i(\underline{\theta}_o)$ for all $i$). Recall that

$$A = n \left[ \sum_{i=1}^{r} \frac{\mu_i}{p_i(\underline{\theta}_o)} \left( \frac{\partial^2 p_i(\underline{\theta}_o)}{\partial \underline{\theta}^2} \right) \right];$$

therefore, if all second partial derivatives of the model cell probabilities are zero, we do get $A = O$. This will occur for any linear model of the cell probabilities; that is, $p_i(\underline{\theta}) = \underline{x}_i'\underline{\theta}$ for a set of known vectors, $\underline{x}_i$. Of course, such models are discouraged because they can generate fitted cell estimates out of range.

As an example we revert to the case of $r = 3$ and use the model structure $\mu_1 = \mu_3 = \theta/2$ and $\mu_2 = 1 - \theta$; so $\log(g(\underline{n} \mid \underline{\theta})) = (n_1 + n_3) \log(\theta/2) + n_3 \log(1 - \theta)$. Here, $\theta_o = 1 - \mu_2$. Upon computing $I(\underline{\theta}_o)$ and $J(\underline{\theta}_o)$ from their basic definitions, we do in fact get $I(\underline{\theta}_o) = J(\underline{\theta}_o) = n/[\theta_o(1 - \theta_o)]$ irrespective of the values of the $\mu_i$. This means that here is a situation and a model where AIC rather than TIC is the correct selection procedure even though the model does differ from truth (i.e., $f \subseteq g$ is not true, yet this condition is sometimes cited as always required for the theoretical validity of AIC).

For the case of general $r$ and the binomial model we can use formula (6.73) to investigate this trace term and AIC vs. TIC. But even (6.73) is too complex to derive any insights from it directly, and numerical methods are needed. So all we really need are usable computational formulae to compute TIC, i.e., estimate $\text{tr}[J(\theta_o)[I(\theta_o)]^{-1}]$. We can get the needed formulae for any postulated model for multinomial data. First, we can find $\hat{\theta}_o$ by solving

$$\sum_{i=1}^{r} \frac{n_i/n}{p_i(\underline{\theta}_o)} \frac{\partial p_i(\underline{\theta}_o)}{\partial \underline{\theta}} = \underline{0},$$

which we do anyway, as this is just our MLE of $\underline{\theta}$ under the assumed model. We do have to compute the set of first and second partial derivatives of the model cell structures evaluated at the MLE, but even that can be done numerically. Thus we can get, hence use and explore, TIC:

$$\hat{I}(\underline{\theta}_o) = n \left[ \sum_{i=1}^{r} \frac{n_i/n}{[p_i(\underline{\hat{\theta}}_o)]^2} \left( \frac{\partial p_i(\underline{\hat{\theta}}_o)}{\partial \underline{\theta}} \right) \left( \frac{\partial p_i(\underline{\hat{\theta}}_o)}{\partial \underline{\theta}} \right)' \right]$$

$$- n \left[ \sum_{i=1}^{r} \frac{n_i/n}{p_i(\underline{\hat{\theta}}_o)} \left( \frac{\partial^2 p_i(\underline{\hat{\theta}}_o)}{\partial \underline{\theta}^2} \right) \right],$$

$$\hat{J}(\underline{\theta}_o) = n \left[ \sum_{i=1}^{r} \frac{n_i/n}{[p_i(\underline{\hat{\theta}}_o)]^2} \left( \frac{\partial p_i(\underline{\hat{\theta}}_o)}{\partial \underline{\theta}} \right) \left( \frac{\partial p_i(\underline{\hat{\theta}}_o)}{\partial \underline{\theta}} \right)' \right].$$

Clearly, we can also compute theoretical values of these quantities for any postulated truth and model. Such studies would be informative, but are beyond the intention of this book.

In a paper on model selection for multinomial distributions De Leeuw (1988) assumes a certain general model selection criterion and pursues it. He does so under the philosophy we espouse here: models used for data analysis are not truth; full truth is very complex; one's analytic goal should be to find a best approximating fitted model. De Leeuw concludes that the most reasonable (essentially, compelling) explicit model-selection criterion to use is AIC. In particular, he says (De Leeuw 1988:132), "This gives a justification for using the AIC, even if the model is not true." On an important related issue it also seems worth quoting De Leeuw (1988:136–137): "The independence assumption, for example, which is at the basis of most work in statistics, cannot really be falsified. As we have seen, the independence assumption merely corresponds with a particular framework of replication, for which we have to decide whether it is relevant or not."

### 6.6.4 Evaluation Under Poisson-Distributed Data

The purpose of this subsection is to see whether a result for multinomial count data extends to the Poisson-distributional case. We assume a sample of size $r$ of observed Poisson counts with unknown means, $\mu_i$ (= truth, *assuming* that the data

## 6.6 Evaluation of tr$(J(\underline{\theta}_o)[I(\underline{\theta}_o)]^{-1})$ and Its Estimator

are Poisson distributed). The model for these means is $\lambda_i(\underline{\theta}_o)$, $i = 1, \ldots, r$. Some results:

$$\log(g(\underline{n} \mid \underline{\theta})) = \sum_{i=1}^{r} \left[ -\lambda_i(\underline{\theta}) + n_i \log(\lambda_i(\underline{\theta})) \right];$$

$\underline{\theta}_o$ is determined as the solution to

$$\sum_{i=1}^{r} \left[ \frac{\mu_i}{\lambda_i(\underline{\theta}_o)} - 1 \right] \frac{\partial \lambda_i(\underline{\theta}_o)}{\partial \underline{\theta}} = \underline{0};$$

$$I(\underline{\theta}_o) = \left[ \sum_{i=1}^{r} \frac{\mu_i}{[\lambda_i(\underline{\theta}_o)]^2} \left( \frac{\partial \lambda_i(\underline{\theta}_o)}{\partial \underline{\theta}} \right) \left( \frac{\partial \lambda_i(\underline{\theta}_o)}{\partial \underline{\theta}} \right)' \right]$$
$$- \left[ \sum_{i=1}^{r} \left( \frac{\mu_i}{\lambda_i(\underline{\theta}_o)} - 1 \right) \left( \frac{\partial^2 \lambda_i(\underline{\theta}_o)}{\partial \underline{\theta}^2} \right) \right],$$

$$J(\underline{\theta}_o) = \left[ \sum_{i=1}^{r} \frac{\mu_i}{[\lambda_i(\underline{\theta}_o)]^2} \left( \frac{\partial \lambda_i(\underline{\theta}_o)}{\partial \underline{\theta}} \right) \left( \frac{\partial \lambda_i(\underline{\theta}_o)}{\partial \underline{\theta}} \right)' \right].$$

We define the $K \times K$ matrix $B$ as

$$B = \sum_{i=1}^{r} \left( \frac{\mu_i}{\lambda_i(\underline{\theta}_o)} - 1 \right) \left( \frac{\partial^2 \lambda_i(\underline{\theta}_o)}{\partial \underline{\theta}^2} \right);$$

then $I(\underline{\theta}_o) = J(\underline{\theta}_o) - B$ and tr$\left[ J(\underline{\theta}_o)[I(\underline{\theta}_o)]^{-1} \right] = K +$ tr$\left[ B[I(\underline{\theta}_o)]^{-1} \right]$. If matrix $B$ is zero, then regardless of how much $\mu_i$ and $\lambda_i(\underline{\theta}_o)$ differ for the $r$ pairs of these values (hence $I(f, g) > 0$ occurs), we still have tr$\left[ J(\underline{\theta}_o)[I(\underline{\theta}_o)]^{-1} \right] = K$, so AIC is justified. This same result, which also holds for multinomial models, is achieved here by any linear model of the form $\lambda_i(\underline{\theta}) = \underline{x}_i'\underline{\theta}$.

### 6.6.5 Evaluation for Fixed-Effects Normality-Based Linear Models

The fixed-effects linear model based on $n$ iid normally distributed residuals is so common that it seems almost mandatory that we consider tr$\left[ J(\underline{\theta}_o)[I(\underline{\theta}_o)]^{-1} \right]$ under this model for some tractable "truth." The model is $\underline{Y} = X\underline{\beta} + \underline{\epsilon}, \underline{\epsilon} \sim$ multivariate-normal$(\underline{0}, \sigma^2 I)$, and the $n \times (K-1)$ matrix $X$ is assumed of full rank wlg. Truth has a structural component, $E(\underline{Y}) = \underline{\mu}$ (which can be estimated by $\underline{Y}$), and a stochastic component for $\underline{\epsilon} = \underline{Y} - \underline{\mu}$, distributed in some unknown way, the properties of which cannot be estimated without strong assumptions. If the model is truth, then both structural (i.e., $\underline{\mu} = X\underline{\beta}$) and distributional assumptions of the model are true.

In reality, the $\epsilon_i$ may not be independent, may not be identically distributed, and may not be normally distributed. In fact, they may not exist, in the sense that some or all $\epsilon_i$ are zero with probability 1. In this latter case truth is deterministic; that is, there is some sufficiently complex computing algorithm (perhaps a formula, with many covariates) such that if we knew that algorithm, we could predict $\underline{Y}$ with certainty (measurement error would become problematic before this level of model accuracy was reached). Hence, for unknown truth we cannot, for cases of real data, evaluate K-L–based model selection for models of continuous random variables.

We can, however, derive informative results under general models for truth that are better approximations to reality (by assumption) than the model to be used for data analysis. Therefore, we assume here that truth is $\underline{Y} = \underline{\mu} + \underline{\epsilon}$, $\underline{\epsilon} \sim$ multivariate-normal($\underline{0}, \tau^2 I$), where $\tau^2$ may be zero. If fact, we can even drop the full distributional assumption, as we will demonstrate below, because the relevant evaluations require only the first four moments of the true distribution. More generally, results could be gotten under the assumption of $\underline{\epsilon} \sim$ multivariate-normal($\underline{0}, \Sigma$) for given $\Sigma$, but the more restricted framework will suffice.

We first need basic notation and results: $\underline{\theta}$ denotes the $K \times 1$ vector $(\beta', \sigma^2)'$, for $\underline{\beta}$ a $(K-1) \times 1$ vector of the structural parameters. We take $\sigma^2$, not $\sigma$, as the parameter to estimate. The model pdf for the data is

$$g(\underline{y} \mid \underline{\theta}) = \frac{1}{\sqrt{2\pi\sigma^2}} \exp\left[-\frac{1}{2\sigma^2}(\underline{Y} - X\underline{\beta})'(\underline{Y} - X\underline{\beta})\right];$$

and we take, wlg,

$$\log(g(\underline{y} \mid \underline{\theta})) = -\frac{n}{2}\log(\sigma^2) - \frac{1}{2\sigma^2}(\underline{Y} - X\underline{\beta})'(\underline{Y} - X\underline{\beta}),$$

$$\frac{\partial \log(g(\underline{y} \mid \underline{\theta}))}{\partial \underline{\beta}} = \frac{1}{\sigma^2}X'(\underline{Y} - X\underline{\beta}), \qquad (6.76)$$

$$\frac{\partial \log(g(\underline{y} \mid \underline{\theta}))}{\partial \sigma^2} = -\frac{n}{2\sigma^2} + \frac{1}{2(\sigma^2)^2}(\underline{Y} - X\underline{\beta})'(\underline{Y} - X\underline{\beta}). \qquad (6.77)$$

As per theory, we take the expectations of (6.76) and (6.77) wrt to what is here just assumed truth, $f$ (evaluation under absolute truth now being impossible) to get the equation

$$\frac{1}{\sigma_o^2}X'(\underline{\mu} - X\underline{\beta}_o) = \underline{0}$$

from (6.76), and then from (6.77) we derive

$$-\frac{n}{2\sigma_o^2} + \frac{1}{2(\sigma_o^2)^2}E_f(\underline{Y} - \underline{\mu} + \underline{\mu} - X\underline{\beta}_o)'(\underline{Y} - \underline{\mu} + \underline{\mu} - X\underline{\beta}_o)$$

$$= -\frac{n}{2\sigma_o^2} + \frac{1}{2(\sigma_o^2)^2}\left[E_f(\underline{\epsilon}'\underline{\epsilon}) + (\underline{\mu} - X\underline{\beta}_o)'(\underline{\mu} - X\underline{\beta}_o)\right]$$

$$= -\frac{n}{2\sigma_o^2} + \frac{1}{2(\sigma_o^2)^2}\left[n\tau^2 + \|\underline{\mu} - X\underline{\beta}_o\|^2\right] = 0.$$

It is now a simple matter to find

$$\underline{\beta}_o = (X'X)^{-1}X'\underline{\mu},$$

$$\sigma_o^2 = \tau^2 + \frac{\|\underline{\mu} - X\underline{\beta}_o\|^2}{n}. \tag{6.78}$$

These parameter values define the vector $\underline{\theta}_o = (\underline{\beta}_o', \sigma_o^2)'$. Formula (6.78) shows that lack-of-fit variation, from the assumed structural model, ends up as part of residual (unexplained) variation.

To find $I(\underline{\theta}_o)$ we need the expected second mixed partials, from (6.76) and (6.77), as below:

$$E_f\left[-\frac{\partial^2 \log(g(\underline{y}\mid\underline{\theta}_o))}{\partial\underline{\beta}^2}\right] = \frac{1}{\sigma_o^2}X'X,$$

$$E_f\left[-\frac{\partial^2 \log(g(\underline{y}\mid\underline{\theta}_o))}{\partial\sigma^2\partial\sigma^2}\right] = -\frac{n}{2(\sigma_o^2)^2} + \frac{1}{(\sigma_o^2)^3}n\sigma_o^2 = \frac{n}{2(\sigma_o^2)^2},$$

and

$$E_f\left[-\frac{\partial^2 \log(g(\underline{y}\mid\underline{\theta}_o))}{\partial\underline{\beta}\partial\sigma^2}\right] = \frac{1}{\sigma_o^2}X'(\underline{\mu} - X\underline{\beta}_o) = \underline{0}.$$

The last vector above is zero because of the defining equation for $\underline{\beta}_o$. Thus we have

$$I(\underline{\theta}_o) = \begin{bmatrix} \frac{1}{\sigma_o^2}X'X & \underline{0} \\ \underline{0}' & \frac{n}{2(\sigma_o^2)^2} \end{bmatrix}.$$

The evaluation of $J(\underline{\theta}_o)$ is harder, and more dependent upon assumed $f$. Evaluation of $I(\underline{\theta}_o)$ required only the second moment of $f$, whereas evaluation of $J(\underline{\theta}_o)$ also requires third and fourth moments. Both derivations rely critically on the independence of the $\epsilon_i$. The upper left $(K-1)\times(K-1)$ submatrix of $J(\underline{\theta}_o)$ is

$$E_f\left[\frac{1}{(\sigma_o^2)^2}X'(\underline{Y} - X\underline{\beta}_o)(\underline{Y} - X\underline{\beta}_o)'X\right] = \frac{\tau^2}{(\sigma_o^2)^2}X'X.$$

The last $(K-1)\times 1$ column vector (of the first $K-1$ rows) is

$$E_f\left[\frac{1}{\sigma_o^2}X'(\underline{Y} - X\underline{\beta}_o)\left[-\frac{n}{2\sigma_o^2} + \frac{1}{2(\sigma_o^2)^2}(\underline{Y} - X\underline{\beta}_o)'(\underline{Y} - X\underline{\beta}_o)\right]\right].$$

Making use of $X'(\underline{\mu} - X\underline{\beta}_o) = \underline{0}$, and some algebra, we can reduce the above to

$$E_f\left[\frac{1}{2(\sigma_o^2)^3}X'(\underline{Y} - \underline{\mu})\left[(\underline{Y} - \underline{\mu})'(\underline{Y} - \underline{\mu}) + 2(\underline{Y} - \underline{\mu})'(\underline{\mu} - X\underline{\beta}_o)\right]\right],$$

and then to

$$E_f\left[\frac{1}{2(\sigma_o^2)^3}\left[X'(\underline{Y}-\underline{\mu})[(\underline{Y}-\underline{\mu})'(\underline{Y}-\underline{\mu})] + 2X'(\tau^2 I)(\underline{\mu}-X\underline{\beta}_o)\right]\right]$$

$$= E_f\left[\frac{1}{2(\sigma_o^2)^3}\left[X'(\underline{Y}-\underline{\mu})[(\underline{Y}-\underline{\mu})'(\underline{Y}-\underline{\mu})]\right]\right].$$

Now write the needed expectation in terms of the hypothetical residuals, which are iid $N(0, \tau^2)$; hence

$$E_f\left[(\underline{Y}-\underline{\mu})[(\underline{Y}-\underline{\mu})'(\underline{Y}-\underline{\mu})]\right] = E_f\left[\underline{\epsilon}'\left[\sum_{i=1}^n (\epsilon_i)^2\right]\right].$$

The $j$th element of this vector is $E_f((\epsilon_j)^3 + \sum_{i \neq j} \epsilon_j(\epsilon_i)^2)$, which by virtue of the mutual independence is $E_f(\epsilon_j)^3$. Because the $\epsilon_j$ are assumed to be normally distributed, their third central moment is 0. Hence, we have

$$E_f\left[\underline{\epsilon}'\left[\sum_{i=1}^n (\epsilon_i)^2\right]\right] = \underline{0},$$

and the desired part of $J(\underline{\theta}_o)$ is $\underline{0}$.

The final needed element is

$$J_{KK}(\underline{\theta}_o) = E_f\left[-\frac{n}{2\sigma_o^2} + \frac{1}{2(\sigma_o^2)^2}(\underline{Y}-X\underline{\beta}_o)'(\underline{Y}-X\underline{\beta}_o)\right]^2.$$

Several straightforward steps reduce the above to

$$J_{KK}(\underline{\theta}_o) = \frac{1}{(2\sigma_o^2)^2}\left[-n^2 + \frac{1}{(\sigma_o^2)^2}E_f\left[(\underline{Y}-X\underline{\beta}_o)'(\underline{Y}-X\underline{\beta}_o)\right]^2\right].$$

Define the $i$th row vector of $X$ as $\underline{x}_i'$. Then

$$E_f\left[(\underline{Y}-X\underline{\beta}_o)'(\underline{Y}-X\underline{\beta}_o)\right]^2$$

$$= E_f\left[\sum_{i=1}^n (y_i - \underline{x}_i'\underline{\beta}_o)^2\right]^2$$

$$= E_f\left[\sum_{i=1}^n \sum_{j=1}^n (y_i - \underline{x}_i'\underline{\beta}_o)^2(y_j - \underline{x}_j'\underline{\beta}_o)^2\right]$$

$$= E_f\left[\sum_{i=1}^n (y_i - \underline{x}_i'\underline{\beta}_o)^4 + \sum_{i \neq j}(y_i - \underline{x}_i'\underline{\beta}_o)^2(y_j - \underline{x}_j'\underline{\beta}_o)^2\right].$$

By virtue of mutual independence, the expectation of the second summation above is easily found, giving

$$E_f\left[(\underline{Y}-X\underline{\beta}_o)'(\underline{Y}-X\underline{\beta}_o)\right]^2$$

## 6.6 Evaluation of $\mathrm{tr}(J(\underline{\theta}_o)[I(\underline{\theta}_o)]^{-1})$ and Its Estimator

$$= E_f\left[\sum_{i=1}^n (y_i - \underline{x}_i'\underline{\beta}_o)^4\right] + (n\sigma_o^2)^2 - \sum_{i=1}^n \left[\tau^2 + (\mu_i - \underline{x}_i'\underline{\beta}_o)^2\right]^2.$$

For the case $\tau^2 = 0$ note that $\underline{Y} = \underline{\mu}$, and so the above directly gives $E_f\left[(\underline{Y} - X\underline{\beta}_o)'(\underline{Y} - X\underline{\beta}_o)\right]^2 = (n\sigma_o^2)^2$; hence

$$J_{KK}(\underline{\theta}_o) = \frac{1}{(2\sigma_o^2)^2}\left[-n^2 + \frac{1}{(\sigma_o^2)^2}(n\sigma_o^2)^2\right] = 0.$$

It is thus clear that if $\tau^2 = 0$, then $J(\underline{\theta}_o) = O$.
The next steps are valid only if $\tau^2 > 0$. Let

$$\sqrt{\lambda_i} = \frac{\mu_i - \underline{x}_i'\underline{\beta}_o}{\tau},$$

and

$$z_i = \frac{y_i - \mu_i}{\tau}.$$

The $z_i$ are *iid* normal$(0, 1)$, and we have

$$E_f\left[\sum_{i=1}^n (y_i - \underline{x}_i'\underline{\beta}_o)^4\right] = \tau^4\left[\sum_{i=1}^n E_f\left[z_i + \sqrt{\lambda_i}\right]^4\right].$$

The needed expectation is now easily found, because it is just the fourth moment of a normal random variable with a nonzero mean; or it can be expressed as a function of the first four moments of a standard normal random variable. We find it easier to note that the needed expectation is that of the square of a noncentral chi-square random variable on 1 df and noncentrality parameter $\lambda_i$. The result is

$$E_f\left[\sum_{i=1}^n (y_i - \underline{x}_i'\underline{\beta}_o)^4\right] = \tau^4\left[\sum_{i=1}^n [3 + 6\lambda_i + \lambda_i^2]\right].$$

Now, by carefully constructing the full result from all the above pieces and simplifying it, we get

$$J_{KK}(\underline{\theta}_o) = \frac{n}{2(\sigma_o^2)^2}\left[\frac{2\tau^2\sigma_o^2 - \tau^4}{(\sigma_o^2)^2}\right];$$

While derived only for $\tau > 0$, the above result can also be validly used for the case of $\tau^2 = 0$.

Finally,

$$J(\underline{\theta}_o) = \begin{bmatrix} \frac{\tau^2}{(\sigma_o^2)^2}X'X & 0 \\ \underline{0}' & \frac{n}{2(\sigma_o^2)^2}\left[\frac{2\tau^2\sigma_o^2 - \tau^4}{(\sigma_o^2)^2}\right] \end{bmatrix}.$$

The result we sought can now be found:

$$\text{tr}\left[J(\underline{\theta}_o)[I(\underline{\theta}_o)]^{-1}\right] = \frac{\tau^2}{\sigma_o^2}\left[K + 1 - \frac{\tau^2}{\sigma_o^2}\right]. \tag{6.79}$$

If model equals truth, we have $\underline{\mu} = X\underline{\beta}$, so that $\sigma_o^2 = \tau^2$ (otherwise $\sigma_o^2 > \tau^2$), and the trace term equals $K$. By continuity in $\tau^2$ we must also define this trace term as $K$ when $\tau^2 = 0$ if the model is true. However, there are deep philosophical issues and problems associated with a truth in which $\tau^2 = 0$, so we will consider only the situation wherein even for truth there is substantial unexplainable uncertainty. In particular, if true replication is used in an experiment (or study), we suggest that it is most useful to consider that $\tau^2$ is then the variance within true replicates (assuming variance homogeneity). This is a definition of convenience, as even truth can be at different levels, and we are mostly interested in structural truth of our models in the face of nontrivial, irreducible uncertainty inherent in data for finite sample sizes.

Surprisingly enough, we see from (6.79) that for this limited evaluation and context, $\text{tr}\left[J(\underline{\theta}_o)[I(\underline{\theta}_o)]^{-1}\right] < K$ under a misspecified model (the key limitation here was assuming truth as a normal distribution). If a good model could achieve, say, $\sigma_o^2 \leq 1.2\tau^2$, then the trace would be well within 80% of $K$; hence use of AIC (rather than TIC) seems acceptable, and should err, on average, on the side of parsimony. Moreover, estimation of this trace term (hence, TIC) seems very problematic, as $\tau^2$ cannot be estimated in a study lacking true replication (cf. Linhart and Zucchini 1986:78). Even what we call true replication in an experiment provides only an estimate of $\tau^2$, by definition, if we restrict our concept of truth to what we can predict under the design structure and independent variables used in the given experiment. Philosophically, we might be able to predict some (hence a smaller $\tau^2$) or all (hence $\tau^2 = 0$) of the observed differences among replicate responses if we knew ultimate truth.

This example can be easily generalized; that is, we retain the assumed model and generalize truth somewhat. Whereas we assumed truth as $\epsilon \sim \text{normal}(0, \tau^2)$, the only way this entered the derivations was via the first four central moments of $\epsilon$. If we retain the *iid* assumption, we can derive generalized results; note that we retain $E(\epsilon) = 0$ wlg. We could allow an asymmetric distribution for $\epsilon$; we will not do so: We assume $E(\epsilon^3) = 0$. Thus all we need is the fourth moment of $\epsilon$, which we will express in standardized form as

$$\gamma = E(\epsilon^4)/[E(\epsilon^2)]^2.$$

For assumed normal truth, $\gamma = 3$. For $f$ as a logistic distribution, $\gamma = 4.2$; for a Laplace distribution, $\gamma = 6$; and for a uniform$(-h, h)$ distribution, $\gamma = 1.8$. The last two are extreme cases; one might think that $\gamma$ lies approximately in the range 2 to 4.

Redoing the derivations for this more general way of representing truth is straightforward; the results are the same for $I(\underline{\theta}_o)$; but for $J(\underline{\theta}_o)$,

$$J(\underline{\theta}_o) = \begin{bmatrix} \frac{\tau^2}{(\sigma_o^2)^2} X'X & \underline{0} \\ \underline{0}' & \frac{n}{2(\sigma_o^2)^2} \left[ \frac{2\tau^2\sigma_o^2 + \tau^4 \left(\frac{\gamma-1}{2} - 2\right)}{(\sigma_o^2)^2} \right] \end{bmatrix}.$$

For the trace function we get

$$\operatorname{tr}\left[J(\underline{\theta}_o)[I(\underline{\theta}_o)]^{-1}\right] = \frac{\tau^2}{\sigma_o^2}\left[K + 1 + \frac{\tau^2}{\sigma_o^2}\left[\frac{\gamma-1}{2} - 2\right]\right]. \tag{6.80}$$

Hence, for these fixed-effects linear models assuming normality, the effect of structural misspecification appears in $\sigma_o^2$ as manifest via the ratio $\tau^2/\sigma_o^2$. However, the effect of error distribution misspecification is only via the fourth moment, $\gamma$ (assuming symmetric errors). We have stressed a focus on complex models wherein $K$ will not be trivially small; from (6.80) we see that as $K$ gets large the effect of error distribution misspecification upon this trace function becomes trivial. In contrast, the effect of structural misspecification, in the sense of then having the ratio $\tau^2/\sigma_o^2 < 1$, remains equally important at any $K$. Once one achieves a good structural fit, then the effect of minor to modest misspecification of the error distribution becomes trivial for large $K$ as regards use of the approximation $\operatorname{tr}\left[J(\underline{\theta}_o)[I(\underline{\theta}_o)]^{-1}\right] = K$, hence further justifying use of AIC rather than TIC. These musings seem likely to apply also to general linear models.

One last point: How good is use of $[I(\underline{\theta}_o)]^{-1}$ for the variance–covariance matrix of $\hat{\underline{\theta}}$ under model misspecification here (ignoring selection uncertainty, that is). As was shown (and is known in general) in Section 6.1, the correct asymptotic variance–covariance matrix is $V(\hat{\underline{\theta}}) = [I(\underline{\theta}_o)]^{-1} J(\underline{\theta}_o)[I(\underline{\theta}_o)]^{-1}$. Thus, here, $V(\hat{\underline{\beta}}) = \tau^2(X'X)^{-1}$, and

$$V(\hat{\sigma}^2) = \frac{2(\sigma_o^2)^2}{n}\left[\frac{2\tau^2\sigma_o^2 + \tau^4\left(\frac{\gamma-1}{2} - 2\right)}{(\sigma_o^2)^2}\right].$$

If a good structural fit has been achieved (so $\tau^2 \doteq \sigma_o^2$), then we have

$$V(\hat{\sigma}^2) = \frac{2(\sigma_o^2)^2}{n}\left[\frac{\gamma-1}{2}\right].$$

The drastic bias induced by a $\gamma$ not near 3 (but assumed as 3) might motivate one to use an estimator of $\gamma$, and with the same estimator to then use TIC, not AIC. This approach can be recommended only weakly, at best, because the estimator of the fourth moment is so highly variable.

288   6. Statistical Theory

## 6.7   Additional Results and Considerations

### 6.7.1   Selection Simulation for Nested Models

The detailed stochastic characteristics of the model-selection process have to be studied mostly by Monte Carlo simulation methods. Currently, it seems as if the only completely general approach is to specify a data generation process and a set of models to be fit to each generated sample, and then generate samples and do all the calculations associated with model fitting and selection. This is very useful, but is not a study of properties of model-selection strategies in the abstract. Rather, each application has some underlying specific models and type of truth (as generated data), and may require extensive computations that are peripheral to the heart of the model-selection process.

An exception arises if we restrict ourselves to a single chain of nested models and to selection methods based on log-likelihood differences between fitted models. This scenario includes simulation of AIC, $AIC_c$, BIC, and likelihood ratio testing-based methods (and can be easily adapted to simulate QAIC and $QAIC_c$ model selection). All we need to generate are the independent, noncentrally distributed log-likelihood chi-square random variables between adjacent models. It is the selection process itself we then study without reference (at least for large $n$) to any specific models. Maximum likelihood estimation is assumed, but no parameters need actually be postulated or estimated (a disadvantage is that we cannot simultaneously study properties of parameter estimators). The important restriction here is that we can correctly generate the needed random variables only for what would be a single chain of nested models. The advantage of the method is speed and generality; this allows quick insights into properties of some model-selection procedures.

At the heart of this procedure we have (conceptual) pairs of models $M_i$ and $M_{i+j}$; model $M_i$ is nested in model $M_{i+j}$, and the difference in number of parameters is $j$. The method can be developed in general, but we will only give it, and use it, with $j = 1$. That is, our conceptual set of models satisfies $M_1 \subset M_2 \subset \cdots \subset M_R$, and each incremented model (i.e., $M_i$ vs $M_{i+1}$) has only 1 added parameter. We assume that large sample theory and ML parameter estimation conceptually underlie such Monte Carlo simulations. Hence, here the usual likelihood ratio test statistic has, in general, a noncentral chi-square distribution on 1 df; denote that random variable by $\chi_1^2(\lambda_i)$. The noncentrality parameter for model $M_i$ vs. $M_{i+1}$ is $\lambda_i$.

For our set of $R$ models we have $R - 1$ noncentrality parameters, $\lambda_1, \ldots, \lambda_{R-1}$. These $\lambda_i$ would be functionally related to the true data-generating model, the model structures assumed, $g_i(\underline{x} \mid \cdot)$, and the specific parameter values, $\underline{\theta}_{o,i}$, that specify the actual best approximating model in each family of models. However, we will be able to bypass all of those specifications in the simulation method below. We do need to be able to interpret sets of the $\lambda_i$. A $\lambda_i > 0$ would reflect the failure of at least model $M_i$ to perfectly match truth. If we had $\lambda_{i-1} > 0$ and $\lambda_i = \cdots = \lambda_{R-1} = 0$, we would interpret this as model $M_i$ being the true data generating model (we will ignore pathologies that might invalidate this

## 6.7 Additional Results and Considerations

interpretation). Also, it is possible to have real situations where, for example, $\lambda_1 = 0$ but $\lambda_2 > 0$. Then both models $M_1$ and $M_2$ seem to be equally bad approximations to truth, because model $M_3$ improves as an approximation to truth compared to model $M_2$, but $M_2$ does not improve over $M_1$. Often, for a real situation we would have the set of $\lambda_i$ monotonically decreasing, and the issue is that of which model provides the AIC best model, i.e., the expected K-L best model, when parameter estimation occurs. We continue now considering how to do simulation in this context.

Let the fitted model log-likelihoods be $\log(\mathcal{L}_i)$. From basic theory,

$$2\log(\mathcal{L}_{i+1}) - 2\log(\mathcal{L}_i) \sim \chi_1^2(\lambda_i).$$

Let the number of parameters in the simplest model be $K_1$. Then we can write the above as

$$-(\text{AIC}_{i+1} - 2K_1 - 2i) + (\text{AIC}_i - 2K_1 - 2i + 2) \sim \chi_1^2(\lambda_i),$$

or

$$\text{AIC}_i - \text{AIC}_{i+1} \sim \chi_1^2(\lambda_i) - 2. \tag{6.81}$$

Alternatively, from result (6.81) we can write a symbolic equation relating random variables:

$$\text{AIC}_i = \text{AIC}_{i+1} + \chi_1^2(\lambda_i) - 2. \tag{6.82}$$

Thus for purposes of a simulation study, if we know $\lambda_i$ and $\text{AIC}_{i+1}$, we can generate $\text{AIC}_i$. We just need to be able to generate a noncentral 1 df chi-square random variable (there are routines for this, such as CINV in SAS).

Based on formula (6.82), we can do a backwards recursive generation of $\text{AIC}_{R-1}$ to $\text{AIC}_1$ starting with $i = R - 1$, given a value for $\text{AIC}_R$. This idea reduces to the formula

$$\text{AIC}_i = \text{AIC}_R + \sum_{j=i}^{R-1}(\chi_1^2(\lambda_j) - 2); \tag{6.83}$$

we just need a value for $\text{AIC}_R$. Because everything we care about under AIC model selection depends only on the relative differences, such as $\text{AIC}_i - \text{AIC}_j$ or $\Delta_i$ or $\Delta_p$, it suffices to set $\text{AIC}_R = R$ (any constant would suffice, this one has advantages) for every sample of AICs generated. A sample now corresponds to a realization of a set of independent $\chi_1^2(\lambda_i)$, $i = 1, \ldots, R-1$.

We have used this approach to do simulation studies of model selection under AIC and other likelihood-based methods. To evaluate BIC we use $\text{BIC}_i = \text{AIC}_i - 2i + i\log(n)$. Because $n$ should now vary, one must also define the noncentrality parameters on a per-unit sample size basis, hence be able to compute $\lambda_{i,n} = n\lambda_{i,1}$. The $\lambda_{i,1}$ should be very small, but otherwise their scale is arbitrary. To mimic $\text{AIC}_c$ selection, use

$$\text{AIC}_{c,i} = \text{AIC}_i + 2\frac{K_i(K_i+1)}{n - K_i - 1}, \tag{6.84}$$

where $K_i = K_1 + i$. Now one must specify $K_1$, the number of parameters envisioned in model $M_1$, as well as use $\lambda_{i,n} = n\lambda_{i,1}$.

To mimic QAIC model selection is a little more involved. Specify a true value of $c$ (variance inflation factor; $c \geq 1$) and its df (in reality, the df may vary over samples). Generate $\hat{c} = \chi^2_{\text{df}}(\text{df}(c-1))/\text{df}$ for each sample (i.e., df$(c-1)$ is the noncentrality parameter for this chi-square random variable). Then use

$$\text{QAIC}_i = \text{QAIC}_R + \sum_{j=i}^{R-1}\left[\frac{\chi_1^2(c-1+c\cdot\lambda_j)}{\hat{c}} - 2\right], \quad i = 1,\ldots,R-1,$$

and $\text{QAIC}_R = R$. Also,

$$\text{QAIC}_{c,i} = \text{QAIC}_i + 2\frac{K_i(K_i+1)}{n-K_i-1}.$$

A few results are given below using this simulation approach to gain insights into model selection. For the most part, however, it is not practical to publish extensive tables of simulation results. We encourage interested persons to do their own extensive simulations, based on (6.83) and (6.84), and learn from them.

From this setup that allows simulating model selection for nested models we can also compute theoretical expected AIC *differences*, hence determine the expected AIC best model exactly. From formula (6.83) we get, for models incrementing by just one parameter,

$$E(\text{AIC}_i) = E(\text{AIC}_R) + \sum_{j=i}^{R-1}(E[\chi_1^2(\lambda_j)] - 2),$$

$$E(\text{AIC}_i) = E(\text{AIC}_R) + \sum_{j=i}^{R-1}(\lambda_j - 1).$$

Let $\lambda_{i+} = \lambda_i + \cdots + \lambda_{R-1}$, $i = 1,\ldots,R-1$, and $\lambda_{R+} = 0$, and we get

$$E(\text{AIC}_i) = (E(\text{AIC}_R) - R) + (\lambda_{i+} + i).$$

Then compute the set of values, $V_i$, given by

$$V_i = \lambda_{i+} + i, \quad i = 1,\ldots,R, \tag{6.85}$$

find their minimum, $V_{\min}$, and then compute

$$E(\Delta_i) = V_i - V_{\min}. \tag{6.86}$$

As an example, if $R = 10$ and (in order) we have $\lambda_i$ as 2, 6, 10, 6, 3, 1.5, 0.8, 0.4, and 0.2, then the $E(\Delta_i)$ are, in order, 22.5, 21.5, 16.5, 7.5, 2.5, 0.5, **0**, 0.2, 0.8, and 1.6. Thus the best expected AIC selected model is $M_7$. Some theoretical variances can also be computed, but nothing directly useful to random minima like $\Delta_p$.

### 6.7.2 Simulation of the Distribution of $\Delta_p$

The random variable $\Delta_p = \text{AIC}_k - \min \text{AIC}$ was introduced in Chapter 4 (Section 4.2.3). For a set of models indexed $i = 1,\ldots,R$, a given sample size $n$, and a

## 6.7 Additional Results and Considerations

conceptually well-defined repeated sampling framework (hence, a sample space), we let model $M_k$ represent the best model, on average, to fit under the AIC selection criterion. Monte Carlo simulation can be used to determine this actual best model (sometimes theory suffices). In applications we are not saying that model $M_k$ is truth; it is just that one of the $R$ models must be the best model, on average, to use for all possible samples, and that is the truth that model $M_k$ represents.

For each simulation-generated sample we can compute $\Delta_p = \text{AIC}_k - \min \text{AIC}$. This min AIC and the value of $\text{AIC}_k$ vary by sample. However, the value of $k$ is fixed for all samples; for example, model $M_4$ might be the actual best model to always use (hence $k = 4$). If model $M_k$ is selected as best in the sample, then $\Delta_p = 0$; otherwise, $\Delta_p > 0$. We can compute the probability distribution, hence percentiles, of this pivotal under the simulation scenario of Section 6.7.1. However, there are too many variables to make extensive tabulations of general results feasible (at a minimum we must specify values for $R$ and $\lambda_1, \ldots, \lambda_{R-1}$; if we use $\text{AIC}_c$, we also need $K_1$ and $n$).

Results about $\Delta_p$ under AIC model selection, for a few values of $R$ with all $\lambda_i = 0$, are feasible to show. In this case model $M_1$ is the true data-generating model ($k = 1$ in $\Delta_p$). This scenario is clearly at odds with what we believe applies to real data analysis (all models for data analysis are just approximations to truth). However, it can be used as a benchmark for percentiles of $\Delta_p$. To the extent that this situation is too simple, it may serve only as a lower bound on the percentiles of the cumulative distribution function of $\Delta_p$, at least for nested models, or for real problems where there is substantial nesting of many of the models considered.

For the case of $R = 2$ and large sample size, the $q$th percentile of $\Delta_p$ ($0 < q < 1$), $\Delta_{p.q}$, is

$$\Delta_{p.q} = \max\{0, (\chi^2_{1.q} - 2)\}$$

(easily derivable from formula 6.81). Here, $\chi^2_{1.q}$ is the $q$th percentile of a central chi-square random variable on 1 df. For example, $\chi^2_{1,0.95} = 3.84$; hence $\Delta_{p,0.95} = 1.84$. We used Monte Carlo simulation to determine some percentiles of $\Delta_p$ for values of $R > 2$. One million reps were used for each value of $R$ (as four independent runs of 250,000 reps, so we can estimate precision). Results below for $R > 2$ have a cv of about 0.5%:

| | percentiles of $\Delta_p$ | | | |
|---|---|---|---|---|
| R | 80% | 90% | 95% | 99% |
| 2 | 0.00 | 0.71 | 1.84 | 4.63 |
| 3 | 0.11 | 1.37 | 2.67 | 5.77 |
| 4 | 0.35 | 1.71 | 3.33 | 6.40 |
| 5 | 0.49 | 1.93 | 3.40 | 6.86 |
| 10 | 0.75 | 2.34 | 3.97 | 7.61 |
| 20 | 0.82 | 2.47 | 4.15 | 8.05 |

We have done many of these simulations to find the distribution of $\Delta_p$ for sets of noncentrality parameters wherein $\lambda_i > 0$; the percentiles are then somewhat larger as compared to the case where all $\lambda_i = 0$. For example, let $R = 10$ and

$\lambda_1, \ldots, \lambda_9$ be 2, 6, 10, 6, 3, 1.5, 0.8, 0.4, 0.2. Now truth, $f$, is not in the set of models, $g$ (actually, $M_R$ could be truth; we cannot rule that out). Based on 20,000 Monte Carlo reps (two sets of 10,000), we find that model $M_7$ is the expected K-L (i.e., AIC) best model. The averages of the sample $\Delta_i$ values (rescaled so their minimum is 0), in order, are 22.6, 21.5, 16.5, 7.5, 2.5, 0.5, 0, 0.2, 0.8, 1.6 (reliable to $\pm 0.1 = 2 * \text{se}$); compare these values to their theoretical expectations from the end of Section 6.7.1: 22.5, 21.5, 16.5, 7.5, 2.5, 0.5, 0, 0.2, 0.8, and 1.6.

Based on $\Delta_p = \text{AIC}_7 - \min \text{AIC}$ over these 20,000 reps, some percentiles of $\Delta_p$ are 3.3 (80%), 4.6 (90%), 6.4 (95%), 10.6 (99%) (cv's are about 1%). From these sorts of simulations, and others with explicit models (especially linear regression or capture–recapture models), we have risked saying that in real applications with at least several models ($R \geq 5$) and some nested sequences, a model $M_i$ for which $\Delta_i \doteq 4$ is implausible as the actual K-L best model structure, and $\Delta_i \doteq 7$ is strong evidence against model structure $M_i$ as being the K-L best model (and $\Delta_i \geq 10$ is very strong evidence against model $M_i$). Additional study might lead to refined interpretations of the $\Delta_i$.

### 6.7.3 Does AIC Over-fit?

The conceptual framework underlying valid use of AIC is one where truth has infinitely many parameters. Over-fitting is often defined in a framework where there is a simple true model, with a finite number of parameters, and that true model is in the set of models considered. Then if the true model structure is nested within the selected model structure, the selected model is said to over-fit: One has estimated more parameters than are in the true model.

This simplistic concept of over-fitting does not apply in the K-L model-selection framework. However, there is a best expected K-L model, which is the model we should use as our basis for data analysis. If that target model is nested within the selected model, might we claim that AIC has selected an over-fit model? We decline to use this definition because there is natural variability in the model selected. If we miss the target model by a few parameters (or, what is the same, that the structure of the selected model is not quite the same as that of the target model), we should not say that we failed. This issue is philosophically the same as being concerned that a parameter estimator, $\hat{\theta}$, may sometimes give a point estimate far away from the true parameter. We are upset only by cases where $\hat{\theta}$ is quite far from $\theta$. But if this happens only with suitably small probability, we consider $\hat{\theta}$ as an acceptable estimator. The argument becomes somewhat circular at this point because we have mentally accustomed ourselves to being satisfied if $\hat{\theta}$ is within about $\pm 2 \, \text{se}(\hat{\theta})$ of $\theta$; hence $\hat{\theta}$ is only unacceptable with a probability of about 0.05. Similarly, practitioners of null hypothesis testing typically are willing to accept (at least de facto) a 0.05 probability of type I error (and probably a higher type II error probability in most applications).

Something similar should apply to possible over-fitting for AIC model selection (under-fitting is at best a minor concern with AIC model selection). We need some idea of how far from the actual K-L best model a fitted model can be before it

## 6.7 Additional Results and Considerations

is regarded as an over-fit model. We have to allow that often we would do well to select a model within, say, 1 or 2 parameters of the actual K-L best model. In contrast, if the selected model has 10 or 20 parameters more than the target model, we think that most people will agree that the model is over-fit. We can use simulation (in some cases theory exists, see Shibata 1984, Speed and Yu 1993) to find the probability distribution of the selected model index, say $\hat{k}$, for a nested sequence of models $M_1$ to $M_R$ and actual K-L best model as model $M_k$. We want to know about the tail of the probability distribution of $\hat{k}$, hence about selection probabilities when $\hat{k} - k$ gets large.

Here is a typical example of what can happen. For a sequence of nested models with $R = 10$ and the $\lambda_i$ as 2, 6, 10, 6, 3, 1.5, 0.8, 0.4, 0.2, the K-L best model was found to be $M_7$. We extend this to $R$ being 20 and then 30 with all additional $\lambda_i = 0$ (and assume that sample size will be quite large). This means that the true data-generating model is model $M_{10}$, so models $M_{11}$ to $M_{30}$ are over-fit, if selected, in the sense that they do contain truly superfluous structure. The K-L best model remains model $M_7$ even for $R > 10$. Below we give the model-selection frequencies, in model order $M_1$ to $M_R$, based on 10,000 Monte Carlo reps:

```
R                                   M_7
10: 11 5 50 746 2282 2635 1924 1161 735 451
20: 16 3 20 405 1557 2399 1970 1373 816 542 274 192 125 93 67 51 39 21 18 19
30: 15 3 23 407 1689 2351 1954 1360 800 522 237 155 121 90 74 37 32 27 19 18
                                                         11 8 10 11 9 5 2 4 2 4
```

The long-tailed nature of the distribution of selected models is typical of AIC when there are a lot of "big" models (models with too many unneeded parameters) containing the K-L best model structure. For the case of $R = 30$ we have a probability of about 0.01 of selecting a model with 19 or more parameters (models $M_{19}$ to $M_{30}$), hence having estimated 12 or more unneeded parameters. For both $R = 20$ and 30 there is about a 0.06 probability of selecting a model with five or more unneeded parameters (i.e. models $M_{12}$ or higher). In general, if we say that we can accept a procedure that has about 5% of its cases a bit misleading, then we should not be upset that AIC can over-fit the K-L best model by about 5 or more parameters with probability of roughly 0.06 if lots of such too-general models are in the set of models considered.

Here is a worst-case scenario for one linear sequence of nested models: All $\lambda_i = 0$, so model $M_1$ is the K-L best model. Based on one million Monte Carlo reps, the estimated model-selection probabilities (good to a standard error $\leq 0.05\%$) are below:

```
R           π̂_i × 100%, in order i = 1, 2, ...
3    78.7  13.3  8.0
4    76.0  12.5  6.7  4.8
5    74.4  12.0  6.4  4.1  3.1
10   71.8  11.4  5.8  3.5  2.4  1.7  1.2  .9  .7  .6
20   71.2  11.2  5.7  3.4  2.3  1.6  1.1  .8  .6  .5  .4  .3  .2  .2  .1  .1  .1  .1  .1  .1
```

(For this case of a single nested sequence of models, these selection probabilities are known theoretically for large $n$ and all $\lambda_i = 0$; see, e.g., Shibata 1981, 1989). The long tail is disturbing, yet with probability about 0.94 the selected model will, in this worst-case scenario, be within four parameters of the K-L best model. Based on a lot of simulations that attempted to mimic key features of realistic AIC model selection, we claim that this is a typical result for AIC (about 0.06 probability of over-fitting by five or more superfluous parameters if such general, overparametrized models are in the set considered). Given models with a lot of unneeded structure (parameters), AIC can select over-fitted models, but the probability of a seriously over-fit model is arguably less than the total error probabilities (type I plus type II) in traditional hypothesis testing.

What will reduce the probability of getting a badly over-fit model? Use $AIC_c$, which helps a lot when sample size, relative to $K$, is not large. Otherwise, the only recommendation we have to avoid the uncommon event of a much over-fit AIC-selected model is to be very thoughtful about the *a priori* set of models considered. In particular, do not casually include models with a great many parameters more than you think are really needed. In regression variable selection this would mean do not simply consider every imaginable regressor variable and include it for possible selection. If you do this, you risk having large numbers of variables that have no explanatory value (they have a $\lambda \doteq 0$), and that leaves you with a small but real probability of selecting a model with lots of such worthless variables.

### 6.7.4 Can Selection Be Improved Based on All the $\Delta_i$?

Given the potentially long-tailed nature of K-L–based model selection (it depends on the set of models), it seems natural to ask whether there might be information in the full set of $\Delta_i$ values that would allow us to identify those cases where we have selected a very over-fit model. If so, can we change our selection to a better model based on information in the entire set of $\Delta_i$ values? We have explored this matter for a single series of nested models (as considered in the above three subsections). The idea was that perhaps the pattern in the $\Delta_i$ would be like that below in the event of selecting a badly over-fit model (line one is model number $i$; line two is $\Delta_i$):

```
 1   2  3   4   5  6   7  8  9   10 11  12  13 14 15  16 17 18 19
30  10  5  0.1  1 2.5  2 1.5 2.7  3 0.8 2.3  1  0 2.4  4  5  6  7
```

Here, AIC has selected model $M_{14}$, but we might suspect that the better model to use is $M_4$ ($\Delta_4 = 0.1$ and none of models $M_5$ to $M_{13}$ have a very big value of $\Delta_i$). So we could change our choice to model $M_4$.

There is no theory to help here regarding properties of the patterns in $\Delta_1$ to $\Delta_R$. So we looked at a lot of simulated results for $R \geq 10$ under the worst-case scenario of all $\lambda_i = 0$. Hence, the simplest model, $M_1$, is both the true data-generating model and the K-L best model. Table 6.5 displays, for $R = 10$, some selected cases of $\Delta_1$, ..., $\Delta_{10}$ (a case is one realization of model-selection results, for a large sample size). Cases 1, 2, and 3 are typical in that Model $M_1$ is selected in about 72% of all

TABLE 6.5. Some sets of large-sample AIC differences, $\Delta_i$, for $R = 10$ and all $\lambda_i = 0$. These are selected cases: T is a typical pattern obtained in 72% of reps (model $M_1$ selected); A is atypical but not rare (11% of cases, model $M_2$ selected); R is rare, about 6% of all cases, and for these, over-fit models are selected.

| | | | | | $\Delta_i$ | | | | | |
|---|---|---|---|---|---|---|---|---|---|---|
| pattern | $i = 1$ | 2 | 3 | 4 | 5 | 6 | 7 | 8 | 9 | 10 |
| 1 T | **0.0** | 1.7 | 3.7 | 5.0 | 6.1 | 8.1 | 9.5 | 11.3 | 10.9 | 9.6 |
| 2 T | **0.0** | 0.5 | 0.1 | 2.8 | 4.7 | 6.6 | 8.5 | 8.1 | 8.8 | 8.9 |
| 3 T | **0.0** | 0.6 | 0.5 | 2.5 | 0.8 | 2.8 | 4.3 | 6.2 | 8.2 | 10.1 |
| 4 A | 1.5 | **0.0** | 0.6 | 1.5 | 3.2 | 3.6 | 5.5 | 3.4 | 5.2 | 7.1 |
| 5 R | 7.5 | 2.0 | 2.7 | 0.8 | 1.5 | **0.0** | 3.5 | 5.3 | 6.3 | 7.5 |
| 6 R | 9.5 | 10.4 | 9.6 | 7.6 | 5.6 | 5.1 | **0.0** | 0.3 | 1.9 | 3.6 |
| 7 R | 0.2 | 1.1 | 2.0 | 3.6 | 1.4 | **0.0** | 2.7 | 4.3 | 4.9 | 5.3 |
| 8 R | 2.0 | 2.2 | 0.8 | 2.8 | 2.5 | **0.0** | 1.9 | 3.4 | 3.9 | 5.7 |
| 9 R | 1.1 | 1.8 | 3.8 | 5.4 | 7.3 | 5.5 | 3.3 | **0.0** | 0.6 | 1.7 |
| 10 R | 1.1 | 3.1 | 0.2 | 1.7 | 2.2 | 4.2 | **0.0** | 2.1 | 3.5 | 5.5 |
| 11 R | 7.9 | 9.0 | 11.0 | 6.2 | 1.3 | **0.0** | 2.0 | 4.0 | 5.7 | 6.8 |
| 12 R | 0.1 | 2.1 | 3.3 | 1.5 | 1.1 | 2.9 | 3.9 | 3.3 | 1.4 | **0.0** |
| 13 R | 1.1 | 1.8 | 3.8 | 5.4 | 7.3 | 5.5 | 3.3 | **0.0** | 0.9 | 1.7 |
| 14 R | 10.4 | 12.2 | 8.8 | 8.5 | 8.6 | 6.3 | **0.0** | 0.8 | 1.6 | 3.0 |

samples here. Case 4 is representative of samples wherein model $M_2$ is selected (about 11% of samples). Cases 5 to 14 are rare (models $M_6$ to $M_{10}$ selected; it happens here in about 6% of all samples). In these latter cases we would say that an over-fit model was selected. Only in case 12 might we feel justified in rejecting model 10 in favor of model 1 as the selected model, but even there nothing in the 10 values of $\Delta_i$ makes us think that this decision is particularly justified. In the other rare cases there is nothing in the nature of the pattern of the ten $\Delta_i$ that gives us any confidence that we are justified in selecting a model as best other than the AIC best model (for which $\Delta_k = 0$). That is, changing our selected model may lead to worse, not better, results: We cannot tell based on the data. In fact, for some cases (6, 11, 14, and perhaps 5) the "evidence" in the set of $\Delta_i$ seems clearly to support the correctness of the AIC best model.

We also simulated the performance of various ad hoc modified AIC model-selection procedures to change the selection to a more parsimonious model. Such a change was done if the AIC-selected model had $K$ much bigger than a more parsimonious model that had $\Delta_i$ very near zero (e.g. $\Delta_i < 0.5$). The exact algorithms tried varied, but none of those ad hoc methods made any meaningful difference to the overall AIC model-selection relative frequencies. Also, we did not visually perceive any useful information in patterns of all the $\Delta_i$. While more work could be done along these lines, we are pessimistic that it would be fruitful. Basically, if the data "lie" to you (i.e., a poor model is selected because the sample is atypical), there are no diagnostics computable from that sample to that tell you that it has "lied."

## 6.7.5 Linear Regression, AIC, and Mean Square Error

We present here some theoretical formulae relevant to model selection, from which informative results can be computed, for a certain case of linear regression under constant error variance. Specifically, we assume that truth is a linear regression based on orthogonal regressors, but our models include only a subset of the regressors (the global model may include all regressors). Let $z_1, \ldots, z_m$ be iid normal(0,1). Independently, let $\epsilon$ be normal(0, $\sigma^2$), and the response variable $x$, based on the regressors, is given by truth as

$$x = \beta_o + \sum_{j=1}^{m} \beta_j z_j + \epsilon,$$

so

$$E(x \mid \underline{z}) = \beta_o + \sum_{j=1}^{m} \beta_j z_j.$$

Results below are scale-invariant in terms of the regressors because $\beta_j$ and $z_j$ occur in the models only as a product. To keep notation consistent with Section 6.7.1, we define here the base model, $M_1$, to be $x = \beta_o + \beta_1 z_1 + \delta$, with $\delta$ assumed as normal(0, $\sigma_1^2$). The number of parameters in this model is $K_1 = 3$. The normal assumption is true, but $\sigma_1^2 \neq \sigma^2$. In general, for $r \geq 1$, model $M_r$ is

$$x = \beta_o + \sum_{j=1}^{r} \beta_j z_j + \delta,$$

with $\delta$ assumed as normal(0, $\sigma_r^2$); the number of parameters is $K_r = r + 2$.

We can consider the sequence of nested models for $r = 1, \ldots, R \leq m$. The ordering of regressors is arbitrary, but is used in the formulae below when nested models are considered sequentially (as per the theory in Section 6.7.1). It is then convenient, but not required, to specify the regression coefficients to satisfy $|\beta_j| > |\beta_{j+1}|$. Doing so yields insights into AIC model selection more easily (such a structured situation is also considered in Speed and Yu 1993).

From the point of view of insights to be gained, the assumption of orthogonal regressors is not restrictive if $R = m$, because any regression problem can be transformed into the case of orthogonal regressors, for example by resorting to regression on principal components. However, if we consider cases for $R < m$, the orthogonality assumption is restrictive, because we cannot make our observed regressors orthogonal to the regressors not observed (this is more realistic of real data). For this reason there is no advantage in considering cases with $R < m$.

The regressors are random variables. Therefore, to get analytical results we take an additional expectation over certain matrices that are based on the random (row) vector $\underline{z}_r = (1, z_1, \ldots, z_r)$. For a sample of size $n$, the model in matrix notation is $\underline{x} = Z\underline{\beta} + \underline{\delta}$, so matrices such as $Z'Z$ and $(Z'Z)^{-1}$ arise. The rows of matrix $Z$ (an $n \times r + 1$ matrix) are the $n$ observed sample vectors $\underline{z}_r$. For large $n$ it is acceptable (to get theoretical results) to replace these matrices by their expectations

## 6.7 Additional Results and Considerations

wrt the random $\underline{z}_r$ (for any $n$, $E(Z'Z) = I$; for large $n$, $E(Z'Z)^{-1} \doteq I$). However, the nonlinearities involved mean that some theoretical formulae below are only large-sample approximations.

Under this scenario of regressor independence we determined the sequential, noncentrality parameters of Section 6.7.1 that apply, for large sample sizes, to the nested sequence of models defined here ($M_1$ to $M_R$):

$$\lambda_i = n \log \left[ 1 + \frac{(\beta_{i+1})^2}{(\beta_{i+2})^2 + \cdots + (\beta_m)^2 + \sigma^2} \right], \quad i = 1, \ldots, m-2,$$

$$\lambda_{m-1} = n \log \left[ 1 + \frac{(\beta_m)^2}{\sigma^2} \right].$$

Given this context and the theory in Section 6.7.1 we can compute (approximate) expected $\Delta_i$ values for AIC and $AIC_c$, hence determine the (approximate) theoretical expected K-L optimal model. We can also simulate actual sets of $\Delta_i$ values. The needed partial sums of noncentrality parameters are

$$\lambda_{i+} = n \log \left[ 1 + \frac{\sum_{j=i+1}^{m} (\beta_j)^2}{\sigma^2} \right], \quad i = 1, \ldots, m-1. \tag{6.87}$$

The value of exploring this situation is that we can also determine here other theoretical quantities that can be related, or compared, to AIC model selection. Under any of the models the regression coefficient estimators are unbiased for the true parameters (because all regressors are orthogonal, a condition not expected in general). Under model $M_r$ the value of $\sigma_r^2$ is

$$\sigma_r^2 = (\beta_{r+1})^2 + \cdots + (\beta_m)^2 + \sigma^2.$$

In notation used elsewhere in Chapter 6, the above $\sigma_r^2$ is $\sigma_o^2$ under model $M_r$. The usual (conditional on the model) parameter cv's under model $M_r$ are

$$\text{cv}(\hat{\beta}_r) = \frac{\beta_r}{\text{se}(\hat{\beta}_r \mid M_r)} = \frac{\sqrt{n}\beta_r}{\sigma_r}, \quad r = 1, \ldots, R.$$

However, rather than compute the above cv's it is informative just to compute the cv's under the global model, $\sqrt{n}\beta_r/\sigma$, and observe the magnitudes of these cv's vs. what parameters are included or excluded from the theoretically optimal model.

We can also determine the overall mean square error (MSE) of a fitted model. Minimum MSE is generally accepted as a good theoretical basis for model selection; here we have

$$\text{MSE}(Z) = \sum_{j=1}^{n} E_f \left[ \hat{E}(x_j \mid \underline{z}_r) - E(x_j \mid \underline{z}_m) \right]^2.$$

This $\text{MSE}(Z)$ is for fitted model $M_r$ conditional on the regressors. Expectation is over $\underline{\epsilon}$, hence truth. Again, another level of expectation needs to be taken here over

Z to get the unconditional result

$$\text{MSE} = \sigma_r^2(r+1) + \sum_{j=r+1}^{m} n(\beta_j)^2. \tag{6.88}$$

Sometimes model selection is based on minimum MSE of prediction of the response variable for a single additional (independent of sample) vector of regressors. Then this average mean square error of prediction (MSEP) for model $M_r$ is MSEP = MSE $+n\sigma^2$. Thus, it suffices to consider only MSE. Mallows's $C_p$ implements minimum MSE model selection for regression (Mallows 1973, 1995).

The K-L–based target criterion, $T$ (formula 6.20), can be determined exactly here. We express that result as $-2T$ for direct comparison to expected $\text{AIC}_c$- and MSE-based results, and we will label it here as KL. Hence, KL $(= -2T)$ for model $M_r$ is exactly

$$\text{KL} = n\text{E}[\log(\chi^2_{n-r-1}(0))] + n\log\left[\frac{\sigma_r^2}{n}\right]$$

$$+ \frac{n}{n-r-3}\left[(r+1) + \left(\frac{n\sigma^2}{\sigma_r^2}\right) + \sum_{j=r+1}^{m} n\left(\frac{\beta_j}{\sigma_r}\right)^2\right]. \tag{6.89}$$

In MSE (formula 6.88) the term $\sum_{j=r+1}^{m} n(\beta_j)^2$ is bias squared due to excluded regressors. Note that in MSE these components of bias are absolute, not relative to theoretical precision of the excluded $\hat{\beta}_j$, whereas in KL, the biasing effect of excluded regressors is "judged" relative to the theoretical precision these $\hat{\beta}_j$ have. That is, the comparable term reflecting bias is in terms of the ratios $(\beta_j/\sigma_r)^2$, not just $(\beta_j)^2$. This feature of the KL criterion seems more desirable to us than just optimizing on pure bias vs. variance as MSE does.

Most insights based on all of these results will need to come from numerical examples and simulation. Because of the possible volume of such results (considering all the variables here), we leave such computing up to the interested reader. We have done a lot of computing and simulation of results based on these formulae and Section 6.7.1 results. One result (known in the literature about $C_p$ vs. AIC) is that K-L–based model selection for regression is just about the same as selection based on minimum theoretical MSE. We can determine this by computing the theoretical criteria, KL and MSE. In so doing it is convenient to rescale the $R$ values of those criteria to have their minimum at zero. Table 6.6 gives such results for one case: $R = m = 10$, $n = 30$, $\beta_i = (0.6)^{i-1}$ and $\sigma$ taking several values in the range 0.025 to 1.

We have not undertaken a detailed analysis of these criteria for linear regression, let alone for this case of orthogonal regressors. However, the above results are representative of cases we have looked at in that the theoretical K-L best model has either the same number of parameters, or is actually more parsimonious than the theoretically best model under minimum MSE. Actual expected models selected under operational criteria such as $\text{AIC}_c$ or $C_p$ can differ slightly from these results (but less so as $n$ gets large).

TABLE 6.6. Comparison of model-selection theoretical criterion KL, formula (6.89), and MSE, formulae (6.88), for models $M_1$ to $M_R$, for $n = 30$, $R = m = 10$, and $\beta_r = (0.6)^{r-1}$, for several values of $\sigma$; results for both KL and MSE have been rescaled so that their minimums are zero, hence clearly indicating the theoretical optimal model under these criteria.

| | $\sigma = 1.0$ | | $\sigma = 0.5$ | | $\sigma = 0.25$ | | $\sigma = 0.05$ | | $\sigma = 0.025$ | |
|---|---|---|---|---|---|---|---|---|---|---|
| r | KL | MSE | KL | MSE | KL | MSE | KL | MSE | KL | MSE |
| 1 | 8.01 | 14.08 | 27.17 | 16.66 | 57.72 | 17.58 | 142.67 | 17.99 | 180.83 | 17.99 |
| 2 | 1.70 | 3.76 | 11.16 | 5.59 | 33.52 | 6.32 | 113.80 | 6.66 | 151.78 | 6.68 |
| 3 | **0.00** | 0.56 | 2.75 | 1.64 | 15.09 | 2.18 | 85.50 | 2.46 | 123.00 | 2.47 |
| 4 | 0.59 | **0.00** | **0.00** | 0.33 | 4.34 | 0.68 | 58.46 | 0.90 | 94.69 | 0.91 |
| 5 | 2.26 | 0.42 | 0.28 | **0.00** | 0.20 | 0.17 | 34.20 | 0.33 | 67.31 | 0.33 |
| 6 | 4.52 | 1.21 | 2.01 | 0.04 | **0.00** | 0.02 | 15.33 | 0.11 | 41.98 | 0.12 |
| 7 | 7.20 | 2.13 | 4.50 | 0.21 | 1.74 | **0.00** | 4.10 | 0.04 | 20.87 | 0.04 |
| 8 | 10.29 | 3.10 | 7.52 | 0.43 | 4.48 | 0.03 | **0.00** | 0.01 | 6.74 | 0.01 |
| 9 | 13.83 | 4.09 | 11.03 | 0.67 | 7.90 | 0.09 | 0.46 | **0.00** | 0.45 | 0.00 |
| 10 | 17.90 | 5.08 | 15.09 | 0.91 | 11.92 | 0.14 | 3.33 | **0.00** | **0.00** | **0.00** |

### 6.7.6 AIC and Random Coefficient Models

Parameters are sometimes considered as "random effects" (or more generally as random coefficients, see Longford 1993). In the simplest case $\theta_1, \ldots, \theta_K$ are parameters all of the same type (e.g., survival rates in $K$ years), and we consider the $K$ elements of $\underline{\theta}$ as random variables. Thus, in the simplest case, we conceptualize $\theta_1, \ldots, \theta_K$ as independent random variables with mean $\mu$ and variance $\sigma^2$. Now the inference problem could be entirely about the underlying fixed, population-level parameters $\mu$ and $\sigma^2$. However, the likelihood we can directly write down is for $\underline{\theta}$ as if the elements of $\underline{\theta}$ were the fixed parameters of direct inference interest. In using the likelihood $\mathcal{L}(\underline{\theta})$ we are ignoring issues of how $\theta_1, \ldots, \theta_K$ may have arisen from some process or some real or conceptual population. The likelihood $\mathcal{L}(\underline{\theta})$ "treats" the parameters as "fixed effects"; this is valid, and no problems arise as long as we interpret the parameters, $\theta_1, \ldots, \theta_K$, therein as deterministic parameters. It is then possible to fit this global model by standard likelihood methods and also fit simpler models based on deterministic constraints on $\underline{\theta}$, such as $\theta_i \equiv \mu$.

However, we may also want to consider an intermediate model based on only the two parameters $\mu$ and $\sigma^2$ where we regard $\theta$ as a random variable with this mean and variance. Thus while we directly have likelihoods $\mathcal{L}(\underline{\theta})$ and $\mathcal{L}(\mu)$ (hence models $M_K$ and $M_1$), we also want the likelihood, say $\mathcal{L}(\mu, \sigma^2)$, for the two-parameter model $M_2$. The parameter $\sigma^2$ in model $M_2$ serves to fit the possible stochastic nature of the $\theta_i$. Model $M_K$ allows arbitrary variation in the $\theta_i$, but this freedom costs us $K-2$ extra parameters compared to model $M_2$. If the unexplained variation in these $K$ parameters is substantial and consistent with them being exchangeable random variables, we should select model $M_2$ rather than model $M_K$. Model $M_2$ with only two parameters, $\mu$ and $\sigma^2$, parsimoniously allows for variation in the $\theta_i$

(something model $M_K$ does not do parsimoniously and model $M_1$ does not do at all).

Whereas the likelihood $\mathcal{L}(\mu)$ is a special case of either $\mathcal{L}(\mu, \sigma^2)$ or $\mathcal{L}(\underline{\theta})$, the conceptually intermediate model $M_2$ is not mathematically an intermediate model between models $M_k$ and $M_1$ in the simple sense of being just a constrained version of the global model $M_K$. Therefore, model $M_2$ cannot be fit by standard likelihood methods based only on the global model likelihood $\mathcal{L}(\underline{\theta})$. It is in this way that these random-coefficient (or random-effects) models are so different from other models that arise as just deterministic constraints on the parameters of some more general model.

Our focus here is on how we can compute a valid AIC for random coefficient models, such as this case of having only the parameters $\mu$ and $\sigma^2$, even though our only obvious starting point is model $M_K$ and its likelihood. The standard likelihood approach to stochastic parameters is to postulate a probability distribution, hence a model, for the random variable $\underline{\theta}$, say $h(\underline{\theta} \mid \mu, \sigma^2)$, and obtain the needed likelihood $\mathcal{L}(\mu, \sigma^2)$ based on the marginal distribution

$$g(\underline{x} \mid \mu, \sigma^2) = \int g(\underline{x} \mid \underline{\theta}) h(\underline{\theta} \mid \mu, \sigma^2) d\underline{\theta}; \qquad (6.90)$$

thus $\underline{\theta}$ has been integrated out. Considering $g(\underline{x} \mid \mu, \sigma^2)$ as a function of the parameters given the data, we have $\mathcal{L}(\mu, \sigma^2) = g(\underline{x} \mid \mu, \sigma^2)$. Thus the AIC for model $M_2$ is computed based on $g(\underline{x} \mid \mu, \sigma^2)$, which we can get by computing the integral in (6.90).

A more informative way to think about random-coefficient models is that sometimes a few parameters, defined by deterministic constraints on $\underline{\theta}$, cannot explain all the variation in the much larger set $\theta_1, \ldots, \theta_K$, when $K$ is not small (like 2 or 3). We might have $K = 10$, 15, or 20 (or more). If we do have 20 values of the same type of parameter (perhaps for 20 years or sites), it is likely that there is some "explainable" (i.e., consistent, simple, and understandable) smooth, low-level pattern, such as a linear trend, to the variation in these parameters. However, to be consistent with our philosophy of models, we must admit that the actual values of the 20 parameters will not perfectly fit such a simple model (a two-parameter linear trend). There will be unignorable yet unexplainable residual variation in the parameters; the modeling issue is that of how much of this residual variation we can detect with the data. If the (unknown) residuals behave like *iid* random variables, then random-coefficient models can be very effective tools for data analysis when there are large numbers of the same type of parameters. The explainable variation is fit by smooth, parsimonious structural models, and the unexplainable (not smooth) variation in $\theta_1, \ldots, \theta_K$ is swept into $\sigma^2$.

Conceptually, we still think of a parsimonious structural model imposed on $\underline{\theta}$, but one allowing homoskedastic residuals. For example, we may have reason to try the constrained model structure $\underline{\theta} = X\underline{\psi}$ (for known covariate regressors in matrix $X$), but we think that this model would not exactly fit the $\theta_i$ even if we could apply it to those exact $\theta_i$. If the model did have $\sigma$ nearly 0 (relative to the size of components of $\underline{\psi}$), we could safely use the deterministic interpretation of

$\underline{\theta}$ and define any new structural model by simple constraints and directly get the likelihood for the new parameters $\underline{\psi}$ as

$$\mathcal{L}(\underline{\psi}) = \mathcal{L}(\underline{\theta} \mid \underline{\theta} = X\underline{\psi});$$

$\underline{\psi}$ would have only, say, 1 to 4 components.

It is often no more reasonable to assume that $\underline{\theta} = X\underline{\psi}$ is exact for unobservable $\underline{\theta}$ than it is to assume that $Y = Z'\underline{\theta}$ is exact for the observable random variable $Y$. In both cases we must allow that the structural model may not fit exactly. For the case of random-coefficient models we must now have a second model $h(\underline{\theta} \mid \underline{\psi})$ imposed on $\underline{\theta}$ in terms of fixed parameters $\underline{\psi}$. We might conceptualize $\underline{\theta}$ as a normal random variable with $\underline{\theta} = X\underline{\psi} + \underline{\delta}$, $\overline{E(\underline{\delta})} = \underline{0}$, and variance–covariance matrix of $\underline{\delta}$ as $\sigma^2 I$. This serves to define $h(\underline{\theta} \mid \underline{\psi})$, and then we compute the actual parsimonious reduced model for the fixed parameters as

$$g(\underline{x} \mid \underline{\psi}, \sigma^2) = \int g(\underline{x} \mid \underline{\theta}) h(\underline{\theta} \mid \underline{\psi}, \sigma^2) d\underline{\theta}.$$

It is more likely that we will be faced with mixed models: some fixed and some random parameters in one or more of our models. To make this idea explicit we extend our notation by partitioning the generic parameter vector $\underline{\theta}$ into two parts: $\underline{\theta} = (\underline{\alpha}', \underline{\beta}')'$ with $\underline{\alpha}$ fixed and $\underline{\beta} = (\beta_1, \ldots, \beta_{K_\beta})'$ random (we of course may choose to consider $\underline{\beta}$ as fixed for some models). As above we will have a model imposed on $\underline{\beta}$ in terms of a distribution for $\beta$ as a random variable, hence $h(\underline{\beta} \mid \underline{\psi}, \sigma^2)$. The needed marginal distribution (model) is

$$g(\underline{x} \mid \underline{\alpha}, \underline{\psi}, \sigma^2) = \int g(\underline{x} \mid \underline{\alpha}, \underline{\beta}) h(\underline{\beta} \mid \underline{\psi}, \sigma^2) d\underline{\beta}.$$

Further generalizations are possible, but for our purposes here we will stick to the simple case, hence (6.90) and issues of likelihoods for fixed-effects reduced models,

$$\mathcal{L}(\underline{\psi}) = \mathcal{L}(\underline{\theta} \mid \underline{\theta} = X\underline{\psi}), \qquad (6.91)$$

versus random-effects reduced models, hence

$$\mathcal{L}(\underline{\psi}, \sigma^2) = g(\underline{x} \mid \underline{\psi}, \sigma^2), \qquad (6.92)$$

where computing (6.92) requires the multidimensional integration illustrated by formula (6.90). The MLE $\hat{\underline{\psi}}$ from these two models (6.91 vs. 6.92) will be essentially the same value even if $\hat{\sigma}^2 > 0$ occurs ($\hat{\sigma}^2$ might be zero). However, the two likelihood functions will differ. Therefore, AIC values for these two models will be different, and it is important to consider the random-effects model as well as its more restrictive fixed-effects version (wherein $\sigma^2 = 0$ is assumed) in sets of models fit to data. Also, the variation represented by $\sigma^2$ may be of interest in its own right. This is the case in Section 3.5, where the random coefficient model is indirectly fit to real sage grouse survival data to estimate the process variation, $\sigma^2$, from a set of annual survival rates.

In collapsing the problem to $g(\underline{x} \mid \underline{\psi}, \sigma^2)$ we are restricting our inference to $\underline{\psi}$ and $\sigma^2$, hence ignoring the original individual $\theta_1, \ldots, \theta_K$. We may want to "have our cake and eat it too;" that is, get estimators of $\underline{\theta}$ as well as $\underline{\psi}$ and $\sigma^2$. This can be accomplished using shrinkage estimators. Shrinkage estimators arise in both Bayesian and frequentist theories (see, e.g., Efron and Morris 1975, Morris 1983, Longford 1993, Casella 1995, and Carlin and Louis 1996). Shrinkage estimators, $\tilde{\beta}$, for random-coefficient models such as $\underline{\theta} = X\underline{\psi} + \underline{\delta}$, $E(\underline{\delta}) = \underline{0}$, $E(\delta\delta') = \sigma^2 I$ can be based on the MLE of $\hat{\underline{\theta}}$ under model $M_K$ in such a way that the residuals from direct simple linear regression of $\tilde{\beta}$ on $X\underline{\psi}$ reproduce the estimate of $\sigma^2$ computed in obtaining $\tilde{\beta}$. This can be interpreted as saying that we may be able to find a suitable proxy for the maximized likelihood of the fitted model in (6.92), $\mathcal{L}(\hat{\underline{\psi}}, \hat{\sigma}^2)$, by use of the original $\mathcal{L}(\underline{\theta})$ evaluated at such a shrinkage estimator, hence via $\mathcal{L}(\tilde{\beta})$.

It would be a considerable advantage if such random-coefficient models could be fit without ever computing the integral in (6.90). This would allow a practical approach to getting a nearly correct AIC value for the model $g(\underline{x} \mid \underline{\psi}, \sigma^2)$ yet based on $\log(\mathcal{L}(\tilde{\beta}))$. As it is, the shrinkage approach is a pragmatic way to fit what amounts to model $g(\underline{x} \mid \underline{\psi}, \sigma^2)$, thus getting an estimate of $\underline{\theta}$ subject to the stochastic "constraint" inherent in the random-coefficients model, without making distributional assumptions. This idea is in need of more research.

If we do have to compute the integrals as in (6.90), it is certainly possible using the recent developments from Bayesian methods; see, e.g., Gelfand and Smith (1990), Zeger and Karim (1991), and Carlin and Chib (1995).

### 6.7.7  $AIC_c$ and Models for Multivariate Data

The derivation of large-sample AIC in Section 6.2 *does* apply to the case of $n$ independent multivariate observations, each with $p$ nonindependent components. The small-sample improvement of AIC that applies for univariate (i.e., $p = 1$) linear models with homogeneous normal residuals, $AIC_c$, *does not* apply in the corresponding multivariate case. This problem has been studied by Fujikoshi and Satoh (1997). They focused on selection of model structure, that is, inclusion or exclusion of the same set of possible regressor variables in each of the $p$ regressions. They assume that a general $p \times p$ variance–covariance matrix, $\Sigma$, applies for the residual vector of each observation. Thus for a model with $k$ regressors (this may include an intercept) there are $k \times p$ structural parameters. Each model also includes $p(p+1)/2$ unknown parameters in $\Sigma$. Thus, $K = (k \cdot p) + p(p+1)/2$.

For their data analysis context Fujikoshi and Satoh (1997) derived an exact AIC, CAIC in their notation, analogous to the univariate case $AIC_c$. Their result (their formula 7) can be expressed as follows:

$$\text{CAIC} = \text{AIC} + 2\frac{K(k+1+p)}{n-k-1-p}. \quad (6.93)$$

The univariate case corresponds to $p = 1$, and then $K = k + 1$. Hence, the general result in (6.93) reduces to the univariate $AIC_c$. Fujikoshi and Satoh (1997) do not consider other multivariate problems, nor do they consider the general case of models with reduced numbers of parameters in $\Sigma$. Thus, for other multivariate modeling problems we do not know whether formula (6.93) applies. Our key point here is that the univariate result for $AIC_c$ does not apply to the multivariate setting.

The form of formula (6.93), by virtue of including variable $k$, is unique to the restricted context considered by Fujikoshi and Satoh (1997). By eliminating $k$ from (6.93) we hypothesize a generalization of univariate $AIC_c$ to corresponding multivariate applications:

$$AIC_c = AIC + 2\frac{K(K+v)}{np - K - v}. \qquad (6.94)$$

In (6.94) $v$ is the number of distinct parameters used in, and estimated for, $\Sigma$; $1 \leq v \leq p(p+1)/2$. Note that the count $K$ includes $v$. Formula (6.94) is correct for the univariate case wherein $v = p = 1$. Interim use of (6.94) seems reasonable until a derivation is published for the needed generalization of $AIC_c$ to multivariate applications.

## 6.7.8 There Is No True $TIC_c$

Fujikoshi and Satoh (1997) also consider a small sample version of TIC; in essence they want to extend $AIC_c$ to $TIC_c$. However, $AIC_c$ (Hurvich and Tsai 1989) arises by computing the exact value of the target model-selection criterion (formula 6.20) for a linear model with constant, normally distributed residuals under the condition (assumption) that this model is the true data-generating model. TIC is derived without any assumption that truth, $f$, is the same as the model $g$. That derivation can be justified only for large sample sizes. To compute a small sample, an exact version of TIC would require us to specify the exact form of the distribution $f$ (i.e., specify truth). Even if we could do this computation in general, or at all, the result would depend upon assumed, but unknown, truth. Thus no defensible, general small-sample analytical version of TIC (i.e., a $TIC_c$) seems possible.

The issue did not escape the attention of Fujikoshi and Satoh (1997). What they did (and they knew it) was to assume that the linear, with normal errors, global model defined by using all available regressors contained the true model as an unknown submodel. Thus the true model is, by assumption, in the set of models considered and is a special case of the global model. Under these assumptions Fujikoshi and Satoh (1997) derived an analytical formula for any sample size, for the target criterion of formula (6.20). Their formula has a component, beyond $-2\log(\mathcal{L})$, that must be estimated from the data, as opposed to components that are simple functions of known $n$ and $K$, and that extra component is estimable only by virtue of the strong assumptions made. We can elect to use the same small sample size adjustments with TIC as we use for AIC, and doing so may be a good idea; but we cannot find truly general small sample size adjustments for TIC.

## 6.7.9 Kullback–Leibler Information Relationship to the Fisher Information Matrix

The Fisher information matrix is defined by formula (6.6) for any $\underline{\theta} \in \Theta$:

$$\mathcal{I}(\underline{\theta}) = E_g\left[-\frac{\partial^2 \log(g(\underline{x}\mid\underline{\theta}))}{\partial\theta_i\partial\theta_j}\right].$$

In taking this expectation it is assumed that the true data-generating model is $g(\underline{x}\mid\underline{\theta})$ (hence the underlying integration is wrt $g$). We use $\underline{\theta} \equiv \underline{\theta}_o$ when this one particular member (i.e., $g$ at $\underline{\theta}_o$) of the set of models defined for fixed structure, and any $\underline{\theta} \in \Theta$, is the generating model for the data. From formula (6.7),

$$I(\underline{\theta}_o) = E_f\left[-\frac{\partial^2 \log(g(\underline{x}\mid\underline{\theta}_o))}{\partial\theta_i\partial\theta_j}\right];$$

in general, $I(\underline{\theta}_o) \neq \mathcal{I}(\underline{\theta}_o)$. Moreover, $\mathcal{I}(\underline{\theta}_o)^{-1}$ is guaranteed to be the large sample variance–covariance matrix of the MLE $\hat{\underline{\theta}}$ only when $g = f$ is true. For a value of $\underline{\theta} \in \Theta$ that is near the K-L minimizing value of $\underline{\theta}_o$, a valid quadratic approximation to the K-L difference is

$$I(f, g(\cdot\mid\underline{\theta})) - I(f, g(\cdot\mid\underline{\theta}_o)) \doteq \frac{1}{2}(\underline{\theta} - \underline{\theta}_o)'I(\underline{\theta}_o)(\underline{\theta} - \underline{\theta}_o)'.$$

For the case of $f = g$ we get the result

$$I(g(\cdot\mid\underline{\theta}_o), g(\cdot\mid\underline{\theta})) \doteq \frac{1}{2}(\underline{\theta} - \underline{\theta}_o)'\mathcal{I}(\underline{\theta}_o)(\underline{\theta} - \underline{\theta}_o)'.$$

Thus if one member of the set of models $g(\underline{x}\mid\underline{\theta}), \underline{\theta} \in \Theta$, is the data-generating distribution, then the approximate K-L information loss that results from using a nearby distribution as the approximating model is the above quadratic form in the Fisher information matrix. This is not a profound result given the definitions of both $\mathcal{I}(\underline{\theta})$ and $I(g(\cdot\mid\underline{\theta}_o), g(\cdot\mid\underline{\theta}))$, but it does serve to show that the two underlying concepts of "information" are related, albeit quite different, concepts. It was in the 1920s that Fisher chose to name this expected matrix of second mixed partials of a probability distribution as "information." There is no relationship to information theory, which is a subject developed mostly since Shannon's pioneering work in the late 1940s and fundamentally deals with logs of probabilities. The Fisher information matrix fundamentally relates to the precision of ML estimators. The Kullback and Leibler paper of 1951 was a result of their attempt to understand and explain what Fisher meant by "information" in relation to sufficiency (personal communication, R. A. Leibler).

## 6.7.10 Entropy and Jaynes Maxent Principle

In Section 4.4.2 we noted that the Akaike weights $w_i$ can be motivated by a (semi) Bayesian approach based on prior probabilities $\tau_i$ (see formula 4.3). To choose these prior probabilities in a manner philosophically consistent with the rest of this book we suggest resorting to the use of the Jaynes maximal entropy (maxent) principle (Jaynes 1957, 1982, Jessop 1995). This principle arises from information theory. The maxent principle says that if we must completely specify a probability distribution with only partial knowledge about moments, or other features, of that distribution, then we should choose the distribution that is maximally uninformative with regard to missing information. This means that we choose the distribution that has maximal entropy subject to any informative constraints we can justify, such as constraints based on data. Mathematically, we find the set of positive numbers $\tau_1, \ldots, \tau_R$ that maximize entropy $= -\sum \tau_i \log(\tau_i)$ subject to the constraint $\sum \tau_i = 1$ and any other functional constraints we can justify (for example, constraints about the mean and/or variance of the distribution). The result is a distribution that conveys no information other than what we explicitly build into it. If the only constraint we impose is that the prior probabilities sum to 1, then the maxent distribution is given by $\tau_i = 1/R$.

That we can justify this uninformative prior for the models based on information theory is yet another example of how deeply information and entropy theory underlie statistical model selection. We will not divert from our objective of exploring information-theoretic data-based model selection. However, we recommend that interested readers pursue some of the references given here on the subject. An introductory reference that ties together some aspects of statistics and information theory, including the Jaynes maxent principle, is Jessop (1995). A nontechnical reference is Lucky (1991). For a biological perspective see Yockey (1992), while Cover and Thomas (1991) give a very thorough overview of information theory. Short, highly mathematical treatments are given by Wehrl (1978) and Ullah (1996).

As a general comment we emphasize the extensive foundations and extent of information and entropy theory, and how these basic ideas occur in many scientific and technical areas (from Boltzmann to Einstein to Shannon to Kullback–Leibler, for example). There is thus a deep foundation to Kullback–Leibler information measure and a firm basis for its use in model selection and other aspects of statistics. K-L is not just another (of many) possible measures of discrepancies between probability distributions; it is unique as a basis for data-based model selection in science when truth is very complex, data are "noisy," and models can be only approximations to truth.

In saying that this theoretical foundation for use of K-L information is deep, we would liken it to the theoretical basis for the importance of the constant $e$ ($\doteq 2.7183$) in mathematics. It is not at all obvious why such a strange, irrational number should universally be the basis for logarithms and exponentials in most of mathematics and science. But just as with K-L information, there is a compelling, deep reason, not easily perceived, for the importance of "$e$."

### 6.7.11  Akaike Weights, $w_i$, Versus Selection Probabilities, $\pi_i$

The model-selection probabilities can be expressed as expectations of indicator random variables that are a function of the sample data:

$$M_i(\underline{x}) = \begin{cases} 1 \text{ if model } i \text{ is selected by AIC}, \\ 0 \text{ otherwise.} \end{cases}$$

By definition, $E(M_i(\underline{x})) = \pi_i$. We assume no ties for the best model.

The Akaike weights (see Section 4.2.2) defined by

$$w_i = \frac{\exp(-\frac{1}{2}\Delta_i)}{\sum_{r=1}^{R} \exp(-\frac{1}{2}\Delta_r)}$$

are also random variables and can be related to the above $M_i(\underline{x})$. Let $k$ index the selected model. Because $\Delta_k = 0$ and $\Delta_i > 0$ for $i \neq k$, we have, for any $0 < \gamma < \infty$,

$$w_k(\gamma) = \frac{1}{1 + \sum_{r \neq k} \exp(-\gamma \Delta_r)}$$

and

$$w_i(\gamma) = \frac{\exp(-\gamma \Delta_i)}{1 + \sum_{r \neq k} \exp(-\gamma \Delta_r)}, \quad i \neq k.$$

In the limit as $\gamma$ goes to infinity we have the result

$$\lim_{\gamma \to \infty} w_i(\gamma) = M_i(\underline{x}),$$

whence

$$\lim_{\gamma \to \infty} E(w_i(\gamma)) = \pi_i$$

(the implied interchange of limits will be valid here). Therefore, it must generally be the case that $E(w_i(0.5)) \equiv E(w_i) \neq \pi_i$; also, $E(w_i)$ and $\pi_i$ are not unrelated.

This result does not rule out $E(w_i) \doteq \pi_i$, which simulation supports as sometimes a useful approximation. Moreover, use of the *set* of Akaike weights as an estimator for the *set* of selection probabilities seems useful in formulae where such $\hat{\pi}_i$ are needed. (Research could be done to find improved $\hat{\pi}_i$ based on the Akaike weights).

### 6.7.12 Model Goodness-of-Fit After Selection

Often the set of models under consideration contains a most general model (the global model), in which case we recommend assessing the fit of that global model to the data (preferably before commencing with model selection). If the global model fits, as by some standard goodness-of-fit test, then the AIC-selected model will fit the data. We think that this is true also for $AIC_c$ model selection (but we are not sure). If the global model does not fit statistically, one might decide that the lack of fit is not of concern and then resort to QAIC or $QAIC_c$. (In fact, if the global model does not fit, but you proceed with K-L–based model selection, you must use QAIC or $QAIC_c$ selection.)

There is a philosophy under which one would want to use BIC; and people are using BIC, even when the context is such that AIC should be used. It is clear that BIC selects more parsimonious models than AIC. What does this do to model fit; if the AIC selected model fits, will the BIC selected model for the same data also fit? This is a question we have never seen addressed in the literature, and we do not know the answer.

The paper by Leroux (1992) motivated our interest in this question. Leroux (1992) reports the observed versus expected count frequencies for some automobile accident data ($n = 9,461$). A pure Poisson model ($K = 1$) and two mixture models are fit to the data. BIC selects the two-component mixture model ($K = 3$), while AIC selects the three-component mixture model ($K = 5$). Model selection tends to lead to overly optimistic assessments of model fit. Hence, model selection may result in optimistic indications of statistical model fit for the selected model. However, a goodness-of-fit test applied to the global model will not be biased, because no selection process has first occurred. The usual chi-square goodness-of-fit procedure applied to the models selected in Leroux (1992) (this entails some pooling of sparse cells) produces $\chi^2 = 1.11$ (1 df, $P = 0.2921$) for the AIC-selected model and $\chi^2 = 11.53$ (2 df, $P = 0.0031$) for the BIC-selected model.

Whereas the goodness-of-fit $P = 0.0031$ is small, there is a large sample size here, and perhaps therefore it is acceptable to use for inference a model that statistically is not a good fit to the data. We think that practice can be acceptable, but it must be argued for on a case-by-case basis. However, statisticians have consistently cautioned about drawing inferences from a model that does not fit the data. We should not ignore the issue of whether model-selection procedures systematically select models that do, or do not, fit. There is good reason to think that AIC selects models that do fit, especially if the global model fits. It is an open question whether BIC-selected models fit the data at the nominal $\alpha$-levels used in goodness-of-fit tests. Research is needed to understand this general issue of the fit, and assessing the fit, of models to data after model selection.

## 6.8 Kullback–Leibler Information Is Always ≥ 0

It is not obvious that the Kullback–Leibler discrepancy,

$$I(f, g) = \int f(x) \log\left(\frac{f(x)}{g(x)}\right) dx,$$

is strictly nonnegative for any possible $g(x)$. Here we reduce the notation for the model $g$ to just $g(x)$ rather than $g(x \mid \theta)$. Also, the possible multidimensional nature of $f$ and $g$ is not emphasized in the proofs in this section.

Rigorous proofs exist that $I(f, g) \geq 0$ and that $I(f, g) = 0$ if and only if $g(x) \equiv f(x)$ for all $x$. Here we give a valid, but not rigorous, proof that $I(f, g) \geq 0$. We do so for both the case of continuous distributions and the case of discrete distributions such as the Poisson, binomial, or multinomial, wherein

$$I(f, g) = \sum_{i=1}^{k} p_i \log\left(\frac{p_i}{q_i}\right).$$

In the discrete case there are $k$ possible outcomes of the underlying random variable. Then the true probability of the $i$th outcome is given by $p_i$, while the $q_1, \ldots, q_k$ constitute the approximating probability distribution (i.e., the model). Hence, here $f$ and $g$ correspond to the $p_i$ and $q_i$, respectively.

In the first case, both $f(x)$ and $g(x)$ must be valid probability distributions, hence satisfy $f(x) \geq 0$, $g(x) \geq 0$ and both integrate to 1:

$$\int f(x)dx = 1, \qquad \int g(x)dx = 1.$$

The exact limits of integration need not be specified here, but must be the same for both $f$ and $g$. Moreover, without loss of generality we can assume $f(x) > 0$, $g(x) > 0$; hence the ratio $f(x)/g(x)$ is never 0 or $1/0$, which is undefined (but may be taken as $\infty$). For the discrete case we have $0 < p_i < 1$, $0 < q_i < 1$ and

$$\sum_{i=1}^{k} p_i = 1, \qquad \sum_{i=1}^{k} q_i = 1.$$

We consider first the case of continuous probability distributions. The key to one line of proof is to define a new function

$$h(x) = \frac{g(x) - f(x)}{f(x)};$$

thus,

$$\frac{g(x)}{f(x)} = 1 + h(x).$$

The lower bound on $h(x)$ is $-1$, because for any $x$ over which integration is performed, $g(x)$ can be arbitrarily close to 0. The upper bound on $h(x)$ is $\infty$. thus, $-1 < h(x) < \infty$. Note also that $\log(a) = -\log(1/a)$. Hence,

$$I(f, g) = \int f(x) \log\left(\frac{f(x)}{g(x)}\right) dx$$

$$= -\int f(x) \log\left(\frac{g(x)}{f(x)}\right) dx$$

$$= 0 - \int f(x) \log\left(\frac{g(x)}{f(x)}\right) dx$$

$$= \int f(x)h(x)dx - \int f(x) \log\left(\frac{g(x)}{f(x)}\right) dx.$$

The last step above uses the fact that

$$0 = \int f(x)h(x)dx = \int f(x)\frac{g(x) - f(x)}{f(x)} dx$$

$$= \int (g(x) - f(x))dx$$

$$= \int g(x)dx - \int f(x)dx = 1 - 1 = 0.$$

## 6.8 Kullback–Leibler Information Is Always ≥ 0

Returning now to the main proof, we have

$$I(f, g) = \int f(x)h(x)dx - \int f(x) \log\left(\frac{g(x)}{f(x)}\right) dx$$

$$= \int f(x)h(x)dx - \int f(x) \log(1 + h(x))dx$$

$$= \int f(x)\big[h(x) - \log(1 + h(x))\big]dx$$

$$= \int f(x)t(h(x))dx,$$

where $t(h(x)) = h(x) - \log(1 + h(x))$. We do not need to care about the actual values of $t(h(x))$. Nor do we need to consider $t(\cdot)$ as a function $x$; hence, also, $x$ may be univariate or multivariate. It suffices to consider the function $h - \log(1 + h)$, hence $t(h)$, over the full range of $h$, $-1 < h < \infty$, that is possible by varying $x$. All we care about is some basic aspects of this *function*, namely that is it strictly nonnegative. It is.

Calculus can be used to show that $t(h) \geq 0$, and that $t(0) = 0$ is the unique minimum, and for any $h \neq 0$, then $t(h) > 0$. A simple heuristic "proof" is just to plot $t(h)$ over, say, $-1 < h \leq 5$, and check $t(h)$ at a few bigger values of $h$ (Fig. 6.1, and $t(10) = 7.6021$, $t(20) = 16.9555$, $t(100) = 95.3849$, $t(1000) = 993.0913$). Given that $t(x) \geq 0$ for all $x$, then $f(x)t(x) \geq 0$ for all $x$; hence

$$I(f, g) = \int f(x) \log\left(\frac{f(x)}{g(x)}\right) dx \equiv \int f(x)t(x)dx \geq 0.$$

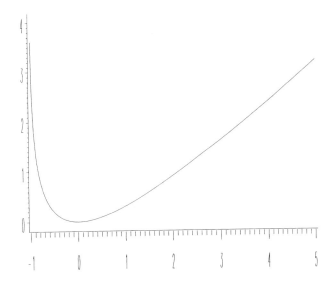

FIGURE 6.1. Plot of the function $t(h)$ near 0.

The calculus proof that $t(x) \geq 0$ makes use of the first and second derivatives of the function $t(h) = h - \log(1 + h)$:

$$t'(h) = \frac{h}{1+h},$$

$$t''(h) = \frac{1}{(1+h)^2}.$$

The set of critical points (which includes minima, maxima, and inflection points) of $t(h)$ consists of the solutions to $t'(h) = 0$ plus the limiting endpoints ($-1$ and infinity). In this case the unique solution is $h = 0$ (it does not matter that $h(x) = 0$ could occur for more than one value of $x$). The nature of this extremum is deduced from $t''(0) = 1$, which because it is positive proves that $h = 0$ is a minimum of the function $t(h)$ (and by uniqueness of the solution, it is the only minimum). Therefore, for all $h$ (and hence all $x$), $t(h(x)) \geq 0$. Also, from these results, $t(h(x))$ is a convex function.

Deeper mathematical theory is required to prove that $I(f, g) = 0$ only if $f(x) = g(x)$ for all $x$ (in the relevant range of integration). It is obvious that if $f(x) = g(x)$, then $I(f, g) = 0$. Part of the "deeper" mathematics referred to says that when $f(x)$ is a continuous probability density function and if $t(x) \geq 0$, then

$$I(f, g) = \int f(x) t(x) dx = 0$$

if and only if $t(x) = 0$ for all $x$ in the range of integration. This statement seems reasonably intuitive, so we will not belabor the point. Thus we have

$$h(x) - \log(1 + h(x)) \equiv 0, \quad \text{for all } x,$$

or

$$h(x) = \log(1 + h(x)),$$

and finally,

$$e^{h(x)} = 1 + h(x).$$

The standard series expansion for $e^h$ can be used here, whence

$$1 + h(x) + \sum_{i=2}^{\infty} \frac{1}{i!} [h(x)]^i = 1 + h(x),$$

or

$$\sum_{i=2}^{\infty} \frac{1}{i!} [h(x)]^i = 0.$$

If $h(x) > 0$, then the above could not be true; therefore it has to be that if $I(f, g) = 0$, then $h(x) \leq 0$ at all $x$. However, if we allow $h(x) < 0$ over any set of $x$ values, $\mathcal{N}$, for which

$$\int_{\mathcal{N}} f(x) dx > 0,$$

## 6.8 Kullback–Leibler Information Is Always ≥ 0

then we would have

$$\int g(x)dx < \int f(x)dx,$$

which cannot be true. Thus, because both $f(x)$ and $g(x)$ are probability density functions, $h(x) \leq 0$ for all $x$ that implies that we must then in fact have $h(x) = 0$ for all $x$.

Now we consider (in less detail) the discrete case:

$$I(f, g) = \sum_{i=1}^{k} p_i \log\left(\frac{p_i}{q_i}\right),$$

such that $0 < p_i < 1$, $0 < q_i < 1$ for all $i$, and $\sum_{i=1}^{k} p_i = 1$, $\sum_{i=1}^{k} q_i = 1$. For fixed $k$, this $I(f, g)$ is a function of $k - 1$ variables, which can be taken as $q_1, q_2, \ldots, q_{k-1}$ (the $p_i$ distribution is considered fixed here). Let

$$h_i = \frac{q_i - p_i}{p_i}, \quad i = 1, \ldots, k,$$

whence

$$\frac{q_i}{p_i} = 1 + h_i, \quad -1 < h_i < \infty, \quad i = 1, \ldots, k.$$

As in the continuous case,

$$\sum_{i=1}^{k} p_i h_i = \sum_{i=1}^{k} (q_i - p_i) = 0,$$

so we can derive

$$I(f, g) = \sum_{i=1}^{k} p_i (h_i - \log(1 + h_i)) = \sum_{i=1}^{k} p_i t(h_i).$$

It was proved above that $t(h) \geq 0$; thus it must be that even in the discrete case $I(f, g) \geq 0$.

It is clear that if $p_i = q_i$ for all $i$, then $I(f, g) = 0$. Assume $I(f, g) = 0$. Then it must be that $t(h_i) = 0$ for all $i$ (otherwise, $I(f, g)$ will be $> 0$). Therefore, we must have

$$e^{h_i} = 1 + h_i, \quad i = 1, \ldots, k.$$

The set of indices $\{1, \ldots, k\}$ can be partitioned into two sets, $\mathcal{N}$ and $\mathcal{P}$ wherein $h_i < 0$ for $i$ in $\mathcal{N}$ and $h_i \geq 0$ for $i$ in $\mathcal{P}$. For $i$ in $\mathcal{P}$, $h_i > 0$ leads to a contradiction because then we would have to have $e^{h_i} > 1 + h_i$. Thus we conclude that for $i$ in $\mathcal{P}$, $h_i = 0$, which is equivalent to $p_i = q_i$ for $i$ in $\mathcal{P}$. Next, note that

$$0 = \sum_{i=1}^{k} p_i h_i$$

$$= \sum_{i=1}^{k} (q_i - p_i)$$

312    6. Statistical Theory

$$= \sum_{i \text{ in } \mathcal{N}} (q_i - p_i) + \sum_{i \text{ in } \mathcal{P}} (q_i - p_i)$$

$$= \sum_{i \text{ in } \mathcal{N}} (q_i - p_i).$$

But $i$ in $\mathcal{N}$ means $h_i < 0$, or $q_i - p_i < 0$, which would mean that the above sum would be strictly $< 0$, which is a contradiction. This contradiction means that the set $\mathcal{N}$ is empty: There cannot be any $h_i < 0$ if $I(f, g) = 0$. Thus if $I(f, g) = 0$, then $f \equiv g$ (i.e., $p_i = q_i$, for all $i$ in the discrete case).

## 6.9 Summary

Most of this chapter is quite technical; we will try to provide a high-level summary of key points or results. Sections 6.1 through 6.6 provide foundational mathematical theory for K-L information-theoretic model selection. The general theory is given (Sections 6.1–6.3) along with several important special cases (Sections 6.4–6.5) and some specific exploration of AIC vs. TIC (Section 6.6). In particular, a very detailed derivation of TIC is given in Section 6.2, along with the relationship to AIC. Then Section 6.6 is a detailed (but not exhaustive) examination of the issue of whether we can use AIC (actually, AIC$_c$) rather than TIC; the results strongly support use of AIC$_c$ as not only acceptable, relative to TIC, but actually preferable. Section 6.7 (in 6.7.1–6.7.5) provides simple (though not general) methods to explore key properties of model selection that are operationally based on the log-likelihood; some theoretical results are also given. Sections 6.7.6–6.7.12 give a few miscellaneous results and considerations that do not fit elsewhere. Section 6.7 is overall much less technical than Sections 6.1–6.6 and we urge you to read Section 6.7 for the general insights therein. Section 6.8 gives a proof of $I(f, g) \geq 0$.

There are several rigorous derivations from Kullback–Leibler information, leading to various information-theoretic criteria: The most relevant, general derivation leads to Takeuchi's (1976) information criterion (TIC). The exact derivation of AIC$_c$ is given in detail in Section 6.4. It is also noted that there is no unique small-sample version of AIC, but AIC$_c$ is recommended for general use (that could change in the future, especially for discrete distributions). Section 6.5 gives a derivation of AIC for the exponential family of models; this family of models is used in most parametric data analysis. When done just for the exponential family of models, the derivation of information-theoretic criteria from K-L information is much more exact, relies on fewer assumptions, and is easier to understand.

The fact that such derivations exist is important to know. The derivations and explanations are very detailed because the theory underlying model selection based on K-L information is important to have clearly stated to allow understanding. Such understanding of the theory puts one in a much better position to accept use of the information-theoretic criteria and understand its strengths and weaknesses.

While Kullback–Leibler information is the logical basis for likelihood-based model selection, it turns out we must use *expected* (over $\hat{\theta}$) Kullback–Leibler information as the quantity of interest when model parameters must be estimated. This, of course, is the reality of actual data analysis. The bias vs. variance tradeoff and the associated model parsimony achieved by K-L–based model selection is an important byproduct of the approach. That is, the derivations make it clear that K-L–based model selection does not start with the explicit objective of meeting the principle of parsimony in model selection. Rather, it is a natural consequence of data-based K-L model selection that this bias–variance tradeoff happens. In fact, it is because the criterion, with estimated parameters, must minimize *expected* K-L information that we get the cross-validation property of AIC.

The detailed derivations make it clear that use of information-theoretic criteria in the analysis of real data is not based on the existence of a "true model" or the notion that such a true model is in the set of candidates being considered. Literature contrary to this point seems mistaken.

Model selection attempts to establish some rigorous basis to achieve proper parsimony in the model(s) used for inference. The relationship of TIC to AIC is made clear, and investigations were undertaken to show that often AIC is a good proxy for TIC. It seems poetic that AIC can be thought of as a parsimonious implementation of the more general TIC. The trace term $\text{tr}[J(\underline{\theta}_o)I(\underline{\theta}_o)^{-1}]$ is about equal to $K$ for "good" models and does not depend on sample size, once sample size is large. Some insights are provided to help in understanding the relationship between $\text{tr}[J(\underline{\theta}_o)I(\underline{\theta}_o)^{-1}]$ and $K$ in a variety of practical situations. Evaluations were conducted for logistic regression, multinomial and Poisson count data, and normal regression models. In all cases we examined, the trace term of TIC is very close to being $K$ as long as the model structure and assumed error distribution are not drastically different from truth. When the model is not the true data-generating model, the trace term was not systematically $> K$ or $< K$. Rather, the matter is unpredictable; the model can be misspecified, and still the trace term can be any of $= K$, $> K$, or $< K$. For all the cases examined, however, if the model was less general than truth (the real-world case), we predominantly found $\text{tr}[J(\underline{\theta}_o)I(\underline{\theta}_o)^{-1}] < K$. Thus, use of AIC should then often lead to slightly more parsimonious models than use of TIC (to the extent that there is any appreciable difference in results from these two procedures).

If the set of models contains one or more good models, then poor models will tend not to be selected (or even ranked high), because $-2 \log(\mathcal{L})$ will be relatively large for a poor model and this term will dominate the criterion value for that model, hence rendering the issue of use of TIC versus AIC largely moot. As noted above, for good models use of AIC is acceptable. Use $AIC_c$ for small samples and even for large samples if values of $K$ become large relative to sample size. More research on such second-order improvements is needed, especially for discrete random variables.

Monte Carlo methods seem to be the only tool to assess general stochastic aspects of model selection and methods to incorporate model selection uncertainty. In some cases, asymptotic results can be obtained, but these seem to be of little interest or practical use. We present some quick ways to explore AIC model selection using Monte Carlo simulation in the case of nested sequences of models. For that same context, the theoretical expected values of the $\Delta_i$ can be easily found, and this is explored in some detail for linear regression models with normal errors. The issue of AIC over-fitting is clarified and explored. Extreme over-fitting can occur, but the probability of this event is low, and one way to minimize the problem is to keep the set of models considered small. Doing searches over "all possible models" (e.g., all-subsets selection) increases the risk of extreme over-fitting. In linear regression it seems that AIC selection is very similar to model selection based on minimum theoretical MSE (of course, in the analysis of real data we cannot do selection based on minimum theoretical MSE).

# 7
# Summary

This book covers some philosophy about data analysis, some theory at the interface between mathematical statistics and information theory, and some practical statistical methodology useful in the life sciences. In particular, we present a general strategy for modeling and data analysis. We provide some challenging examples from our fields of interest, provide our ideas as to what not to do, and suggest some areas needing further theoretical development. We side with the fast-growing ranks that see limited utility in statistical hypothesis testing. Finally, we provide references from the diverse literature on these subjects for those wishing to study further.

Conceptually, there is information in the observed data, and we want to express this information in a compact form via a "model." Such a model is then the basis for making inferences about the process or system that generated the data. One can view modeling of information in data as a change in "coding" like a change in language. A concept or emotion expressed in one language (e.g., French) loses some exactness when expressed in another language (e.g., Russian). A given set of data have only a finite, fixed amount of information. The (unachievable) goal of model selection is to attain a perfect 1-to-1 translation such that no information is lost in going from the data to a model of the information. Models are only approximations and we cannot hope to perfectly achieve this idealized goal. However, we can attempt to find a model of the data that is best in the sense that the model loses as little information as possible. This thinking leads directly to Kullback–Leibler information, $I(f, g)$; the information lost when model $g$ is used to approximate full reality, $f$. We wish then to select a model that minimizes K-L information loss. Because we must estimate model parameters from the data, the best we can do is to minimize (estimated) expected K-L information loss. However, this can easily be done using one of the information-theoretic criteria (e.g., AIC, $AIC_c$, QAIC, or TIC). Then a *good* model allows the efficient and objective separation or filtration of *information* from *noise*. In an important sense, we are not really trying to model the *data*; instead, we are trying to model the *information* in the data.

While we use the notation $f$ to represent truth or full reality, we deny the validity of a "true model" in the life sciences. Conceptually, let $f$ be the process (truth) that generates the sample data we collect. We want to make inferences about truth, while realizing that full reality will always be beyond us when we have only sample data. Data analysis should not be thought of as an attempt to identify $f$; instead, we must seek models that are good approximations to truth and from which therefore we can make valid inferences concerning truth. We do not want merely to describe the data using a model that has a very large number of parameters; instead, we want to use the data to aid in the selection of a parsimonious model that allows valid inferences to be made about the system or process under study.

Relatively few statistics books provide a summary of the key points made and yet fewer provide an effective, unified strategy for data analysis and inference where there is substantial complexity. The breadth of the subjects covered here make a summary difficult to write. Undergraduate students occasionally ask the professor, "What is important for me to know for the final examination?" The professor is typically irritated by such a question—surely the student should realize that it is *all* important! Indeed, our interpretation of Akaike's pioneering work is that it *is* all important. The information-theoretic paradigm is a *package*; each of the package's contents is important in itself, but it is the integration of the contents that makes for an effective philosophy, a consistent strategy, and a practical methodology. The part of this package that has been so frequently left out is the critical thinking and modeling *before* examination of the data; ideally, much of this thinking should occur prior even to data collection. This is the point where the science of the issue formally enters the overall "analysis" (Anderson and Burnham 1999a).

The information-theoretic methods we present can be used to select a single, best model that can be used in making inferences from empirical data. AIC is often portrayed in the literature in this simple manner. The general approach is much richer than this simplistic portrayal of model selection might suggest. One can easily rank alternative models from best to worst using the convenient differences, $\Delta_i$. The likelihood for each model, given the data [i.e., $\mathcal{L}(M_i \mid \text{data})$], can be estimated, and these can be normalized to obtain Akaike weights. Confidence sets of models can be defined to aid in identifying a subset of good models. Model-selection uncertainty can be easily quantified using either the bootstrap or Akaike weights. Estimates of this component of uncertainty can be incorporated into unconditional estimates of precision using several methods. For many problems (e.g., prediction) model-averaging has advantages, and we treat this important issue in Chapter 5. Thus, we often recommend formal inference from several models (e.g., all models in the set where the Akaike weights are nontrivial).

For those who have scanned through the pages of this book there might be surprise at the general lack of mathematics and formulae (Chapter 6 being the exception). That has been our intent. The *application* of the information-theoretic methods is quite simple, easy to understand and use ("low tech"), while the underlying theory is quite deep (e.g., Chapter 6). As we wrote the book and tried to understand Akaike's various papers (see Parzen et al. 1998) we found the need to delve into various issues that are generally philosophical. The science of the problem has to be brought into modeling *before* one begins to rummage through the data (data dredging). In some critical respects, applied statistics courses are failing to teach statistics as an integral part of scientific discovery, with little about model-selection methods or their importance, while succeeding (perhaps) in teaching hypothesis testing methods and data analysis methods based on the assumption that the model is both given and true.

## 7.1 The Scientific Question and the Collection of Data

The formulation of the research question is crucial in investigations into complex systems and processes in the life sciences. A good answer to a poor question is a mistake all too often seen in the published literature and is little better than a poor answer to a poor question. Investigators need to continually readdress the importance and quality of the question to be investigated.

A careful program of data collection must follow from the question posed. Particular attention should be placed on the variables to be measured and interesting covariates. Observational studies, done well, can show patterns, associations, and relationships and are confirmatory in the sense that certain issues stem from *a priori* considerations. More causal inference must usually come from more formal experimentation (i.e., important confounding factors are controlled or balanced, experimental units are randomly assigned to treatment and control groups with adequate replication), but see Anderson et al. (1980), Gail (1996), and Glymour (1998) for alternative philosophies. Valid inference must assume that these basic, important issues have been carefully planned and conducted. Before one can proceed, two general questions must be answered in the affirmative:

*Are the study objectives sound?*

*Has there been proper attention to study design and laboratory or field protocol?*

## 7.2 Actual Thinking and A Priori Modeling

Fitting models to data has been important in many biological, ecological, and medical investigations. Then statistical inferences about the system of interest are made from an interpretable parsimonious model of the observational or experimental data. We expect to see this activity increase as more complicated scientific and management issues are addressed. In particular, *a priori* modeling becomes increasingly important as several data sets are collected on the same issue by different laboratories or at widely differing field sites over several years.

*We recommend much more emphasis on thinking!* "Leave the computer idle" for a while, giving yourself time to think hard about the overall problem. What useful information is contained in the published literature, even on issues only somewhat related to the issue at hand? What nonlinearities and threshold effects might be predicted? What interactions are hypothesized to be important? Can two or more variables be combined to give a more meaningful variable for analysis? Should some variables be dropped? Discussions should be encouraged with the people in the field or laboratory that were close to the data collection. What parameters might be similar across groups (i.e., data sets)? Model building should be driven by the underlying science of the issue. Ideally, this important conceptual phase might take several days or even weeks of effort; this seems far more time than is often spent under current practice.

Model building can begin during the time that the *a priori* considerations are being sorted out. Often it is effective to begin with the global model and work toward some lower-dimensional models. Others may favor a bottom-up approach. The critical matter here is that one arrives, eventually, at a small set of good candidate models, prior to examination of the empirical data. We advise the inclusion of all models that are reasonably justified prior to data analysis; however, every attempt should be made to keep the number of candidate models small.

Biologists generally subscribe to the philosophy of "multiple working hypotheses" (Chamberlain 1890, Platt 1964, Mayr 1997), and these should form the basis for the set of candidate models to be considered formally. With the information-theoretic approach, there is no concept of a "null" hypothesis, or a statistical hypothesis test, or an arbitrary $\alpha$-level, or questionable power, or the multiple testing problem, or the fact that the so-called null hypothesis is nearly always *obviously* false in the first place. Much of the application of statistical hypothesis testing arbitrarily classifies differences into meaningless categories of "significant" and "nonsignificant," and this practice has little to contribute to the advancement of science. We recommend that researchers stop using the term "significant," as it is so overused, uninformative, and misleading. The results of model-selection based on estimates of expected (relative) Kullback–Leibler information can be very different from the results of some form of statistical hypothesis testing (e.g., the simulated starling data, Section 3.4, or the sage grouse data, Section 3.5).

So, investigators may proceed with inferential or confirmatory data analysis if they feel satisfied that they can objectively address two questions:

*Was the set of candidate models derived a priori?*

*What justifies this set?*

The justification should include a rationale for models both included and excluded from the set. Every effort should be made to keep the number of *a priori* models in the set reasonably small. If so little is known about the system under study that a large number of models must be included in the candidate set, then the analysis should probably be considered only exploratory (if models are developed as data analysis progresses, it is exploratory analysis). One should check the fit of the global model using standard methods. If the global model does not fit (after, perhaps, adjusting for overdispersed count data), then more thought should be put into model building and thinking harder about the system under study and the data collected. There is no substitute for good, hard thinking at this point.

## 7.3 The Basis for Objective Model Selection

Statistical inference from a data set, *given a model*, is well advanced and supported by a very large amount of theory. Theorists and practitioners are routinely employing this theory, either likelihood or least squares, in the solution of problems in the biological sciences. The most compelling question is *"what model to use?"* Valid inference must usually be based on a good approximating model, but which one?

Akaike chose the celebrated Kullback–Leibler information as a basis for model-selection. This is a fundamental quantity in the sciences and has earlier roots in Boltzmann's concept of *entropy*, a crowning achievement of nineteenth century science. The K-L distance between conceptual truth $f$ and model $g$ is defined for continuous functions as the integral

$$I(f, g) = \int f(x) \log\left(\frac{f(x)}{g(x \mid \theta)}\right) dx,$$

where log denotes the natural logarithm and $f$ and $g$ are $n$-dimensional probability distributions. Kullback and Leibler (1951) developed this quantity from "information-theory," thus the notation $I(f, g)$ as it relates to the "information" lost when model $g$ is used to approximate truth $f$. Of course, we seek an approximating model that loses as little information as possible; this is equivalent to minimizing $I(f, g)$ over the models in the set. Full reality is considered to be fixed. An interpretation equivalent to minimizing $I(f, g)$ is that we seek an approximating model that is the "shortest distance" away from truth. Both interpretations seem useful.

The K-L distance can be written equivalently as

$$I(f, g) = \int f(x) \log(f(x)) \, dx - \int f(x) \log(g(x \mid \theta)) \, dx.$$

The two terms on the right of the above expression are statistical expectations with respect to $f$ (truth). Thus, the K-L distance (above) can be expressed as a difference between two expectations,

$$I(f, g) = E_f[\log(f(x))] - E_f[\log(g(x \mid \theta))],$$

each with respect to the true distribution $f$. The first expectation, $E_f[\log(f(x))]$, is a constant that depends only on the unknown true distribution. Therefore, treating this unknown term as a constant, only a measure of *relative* directed distance is possible. Then

$$I(f, g) = \text{Constant} - E_f[\log(g(x \mid \theta))],$$

or
$$I(f, g) - \text{Constant} = -\text{E}_f[\log(g(x \mid \theta))].$$

Thus, the term $(I(f, g) - \text{Constant})$ is a *relative* directed distance between the truth $f$ and model $g$. This provides a deep theoretical basis for model selection if one can compute or estimate $\text{E}_f[\log(g(x \mid \theta))]$.

While $\text{E}_f[\log(g(x \mid \theta))]$ is the quantity of theoretical interest, we do not know $\theta$, and it must be estimated. However, given data and a model, we can compute the MLE $\hat{\theta}$, so it is logical to think that we could base model selection on $\text{E}_f[\log(g(x \mid \hat{\theta}))]$. This is conceptually the same as selection based on *estimated* K-L distance,

$$\hat{I}(f, g) = \int f(x) \log \left( \frac{f(x)}{g(x \mid \hat{\theta})} \right) dx.$$

Because we do not know truth $f$, we cannot actually compute either $\text{E}_f[\log(g(x \mid \hat{\theta}))]$ or $\hat{I}(f, g)$. Moreover, the above conceptually estimated K-L information is not a satisfactory criterion for model-selection because it is a random variable. When model-selection must be based on an estimated $\theta$, then the selection criterion must be adapted to reflect this use of $\hat{\theta}$, not $\theta$. Our conceptual target to minimize becomes $\text{E}_{\hat{\theta}}[\hat{I}(f, g)]$; the expectation here is taken with respect to truth, $f$. This revised selection criterion is "select a fitted model to minimize the *expected* K-L information loss."

The expression (from above), without any notion of data being involved,
$$I(f, g, ) - \text{Constant} = -\text{E}_f[\log(g(x \mid \theta))],$$
now becomes
$$\text{E}_{\hat{\theta}}[\hat{I}(f, g)] - \text{Constant} = \text{E}_{\hat{\theta}}\text{E}_f[\log(g(x \mid \hat{\theta}))].$$

Thus, with the introduction of data into the problem, hence $\hat{\theta}$, we see that our theoretical model-selection criterion becomes "select a fitted model to maximize $\text{E}_{\hat{\theta}}\text{E}_f[\log(g(x \mid \hat{\theta}))]$." This is selection based on minimum relative expected K-L information (or distance). The final (big) step is to find an estimator of $\text{E}_{\hat{\theta}}\text{E}_f[\log(g(x \mid \hat{\theta}))]$ without having to know truth, $f$. Amazingly enough, it can be done, as a good, large-sample approximation exists (TIC) for which there exists the parsimonious estimator

$$\hat{\text{E}}_{\hat{\theta}}\text{E}_f[\log(g(x \mid \hat{\theta}))] = \log(\mathcal{L}(\hat{\theta})) - K.$$

Model selection based on maximizing the above is identical to selection based on minimizing AIC $= -2\log(\mathcal{L}(\hat{\theta})) + 2K$.

Knowing the estimated relative K-L distances between the (conceptual) truth $f$ and each approximating model $g_i$, one can easily rank the models from best to worst within the set of candidates, using the $\Delta_i$ values. This allows the ability to assess not only a model's rank, but also whether the $j$th model is nearly tied

## 7.4 The Principle of Parsimony

Parsimony is the concept that a model should be as simple as possible with respect to the included variables, model structure, and number of parameters. Parsimony is a desired characteristic of a model used for inference, and it is usually visualized as a suitable tradeoff between squared bias and variance of parameter estimators (Fig. 1.2). Parsimony lies between the evils of under- and over-fitting. Expected K-L information is a fundamental basis to achieve proper parsimony in modeling.

The concept of parsimony has a long history in the sciences. Often this is expressed as "Occam's razor"—shave away all that is unnecessary. The quest is to make things "as simple or small as possible." Parsimony in statistics represents a tradeoff between bias and variance as a function of the dimension of the model ($K$). A good model is a proper balance between under- and over-fitting. All model-selection methods are based on the concept of a squared bias versus variance tradeoff. Selection of a model from a set of approximating models must employ the concept of parsimony. These philosophical issues are stressed in this book, but it takes some experience and reconsideration to reach a full understanding of their importance.

## 7.5 Information Criteria as Estimates of Relative Kullback–Leibler Information

As deLeeuw (1992) noted, Akaike found a formal relationship between Boltzmann's entropy and Kullback–Leibler information (dominant paradigms in information and coding theory) and maximum likelihood (the dominant paradigm in statistics). This finding makes it possible to combine estimation (point and interval estimation) and model-selection under a single theoretical framework—optimization. Akaike's (1973) breakthrough was the finding of an estimator of the expected relative K-L information, based on a bias corrected, maximized log-likelihood value. His estimator was an approximation and, under certain conditions, asymptotically unbiased (it is often stated, incorrectly, that it is asymptotically unbiased in general). He found that

$$\text{estimated expected (relative) K-L information} \doteq \log(\mathcal{L}(\hat{\theta})) - K,$$

where $\log(\mathcal{L}(\hat{\theta}))$ is the maximized log-likelihood value and $K$ is the number of estimable parameters in the approximating model (this is the bias correction term). Akaike multiplied through by $-2$ and provided Akaike's information criterion

(AIC),

$$\text{AIC} = -2\log(\mathcal{L}(\hat{\theta})) + 2K.$$

Akaike considered his information-theoretic criterion an extension of Fisher's likelihood theory. Conceptually, the principle of parsimony is enforced by the added "penalty" (i.e., $2K$) while minimizing AIC.

Assuming that a set of *a priori* candidate models has been carefully defined, then AIC is computed for each of the approximating models in the set, and the model where AIC is minimized is selected as best for the empirical data at hand. This is a simple, compelling concept, based on deep theoretical foundations (i.e., K-L information). Given a focus on *a priori* issues, modeling, and model-selection, *the inference is the selected model*; in a sense, parameter estimates are almost byproducts of the selected model. This inference relates to an approximation to truth and what information seems to be contained in the data.

Important refinements followed shortly after the pioneering work by Akaike. Most relevant was Takeuchi's (1976) information criterion (termed TIC), which provided an asymptotically unbiased estimate of relative expected K-L information. TIC is little used, as it requires the estimation of $K \times K$ matrices of first and second partial derivatives of the log-likelihood function, and its practical use hinges on the availability of a relatively large sample size. In a sense, AIC can be viewed as a parsimonious version of TIC. A second type of refinement was motivated by Sugiura's (1978) work, and resulted in a series of papers by Hurvich and Tsai (1989, 1990, 1991, 1994, 1995a and 1995b, 1996). They provided a second order approximation, termed $\text{AIC}_c$, to relative K-L information,

$$\text{AIC}_c = -2\log(\mathcal{L}(\hat{\theta})) + 2K + \frac{2K(K+1)}{(n-K-1)},$$

where $n$ is sample size The final bias correction term vanishes as $n$ gets large with respect to $K$ (and $\text{AIC}_c$ becomes AIC), but the additional term is important if $n$ is not large relative to $K$ (we suggest using $\text{AIC}_c$ if $n/K < 40$ or, alternatively, always using $\text{AIC}_c$).

AIC is often presented in the scientific literature in an ad hoc manner, as if the bias correction term $K$ (the so-called penalty term) were arbitrary. Worse yet, perhaps, is that AIC is often given without reference to its fundamental link with Kullback–Leibler information. Such shallow presentations miss the point, have had very negative effects, and have misled many into thinking that there is a whole class of selection criteria that are "information-theoretic." The issue is that criteria such as AIC, $\text{AIC}_c$, QAIC, and TIC are estimates of expected (relative) Kullback–Leibler distance and are useful in the analysis of real data in the "noisy" sciences.

Because only *relative* K-L information can be estimated using one of the information criteria, it is convenient to rescale these values such that the model with the minimum AIC (or $\text{AIC}_c$ or TIC) has a value of 0. Thus, the AIC values can be

## 7.5 Information Criteria as Estimates of Relative Kullback–Leibler Information

rescaled as simple differences

$$\Delta_i = \text{AIC}_i - \min \text{AIC}$$
$$= \text{E}_{\hat{\theta}}[\hat{I}(f, g_i)] - \min \text{E}_{\hat{\theta}}[\hat{I}(f, g_i)].$$

While the value of $\min \hat{I}(f, g_i)$ is not known (only the relative value), we have an estimate of the size of the increments of information loss for the various models compared to the estimated best model (the model with the minimum $\text{E}_{\hat{\theta}}[\hat{I}(f, g_i)]$). The $\Delta_i$ values are easy to interpret and allow a quick comparison and ranking of candidate models and are also useful in computing Akaike weights. As a rough rule of thumb, models having $\Delta_i$ within 1–2 of the best model have substantial support and should receive consideration in making inferences. Models having $\Delta_i$ within about 3–7 of the best model have considerably less support, while models with $\Delta_i > 10$ have either essentially no support and might be omitted from further consideration or at least fail to explain some substantial structural variation in the data. If the observations are not independent (but are treated as such) or if the sample size is quite small, then the simple guidelines above cannot be expected to hold.

These are cases where a model with $\Delta_i > 10$ might still be useful, particularly if the sample size is very large (e.g., see Section 5.5). For example, let model $A$, with year-specific structure on one of the parameters, be the best model in the set ($\Delta_A = 0$) and model $B$, with less structure on the subset of year-specific parameters, have $\Delta_B = 11.4$. Assume that all models in the candidate set were derived prior to data analysis (i.e., no data dredging). Clearly, model $A$ is able to identify important variation in a parameter across years; this is important. However, in terms of understanding and generality of inference based on the data, it might sometimes be justified to use the more simple model $B$, as it may seem to "capture" the important fixed effects. Models $A$ and $B$ should both be detailed in any resulting publication, but understanding and interpretation might be enhanced using model $B$, even though some information in the data would be (intentionally) lost. Such lost information could be partially recovered by, for example, using a random effects approach (see Section 3.5.5) to estimate the mean of the time-effects parameter and the variance of its distribution.

The principle of parsimony provides a philosophical basis for model-selection; Kullback–Leibler information provides an objective target based on deep, fundamental theory; and the information criteria (particularly AIC and AIC$_c$) provide a practical, general methodology for use in data analysis. Objective model-selection and model weighting can be rigorously based on these principles. In practice, one need not assume that any "true model" is contained in the set of candidates (although this is sometimes stated, erroneously, in the technical literature). [We note that several "dimension-consistent criteria" have been published that attempt to provide asymptotically unbiased (i.e., "consistent") estimates of the dimension ($K$) of the "true model." Such criteria are only estimates of K-L information in a strained way, are based on unrealistic assumption sets, and often perform poorly (even toward their stated objective) unless a very large sample size is available (or where $\sigma^2$ is negligibly small, such as in many problems in the physical sciences). We do not recommend these dimension-consistent criteria for the analysis of real data in the life sciences.]

## 7.6 Ranking and Calibrating Alternative Models

The information-theoretic approach does more than merely estimate which model is best for making inference, given the set of *a priori* candidate models and the data. AIC allows a ranking of the models from an estimated best to the worst. This is most easily done using the differences, $\Delta_i$; the larger the $\Delta_i$, the less plausible is model $i$. In many cases it is not reasonable to expect to be able to make inferences from a single (best) model; biology is not simple; why should we hope for a simple inference from a single model? The information-theoretic paradigm provides a basis for examination of alternative models and, where appropriate, making formal inference from more than one model.

Akaike weights ($w_i$) and bootstrap weights ($\pi_i$) can be used to calibrate models. The Akaike weights are $\exp(-\frac{1}{2}\Delta_i)$, and these represent the (discrete) likelihood function, given the data. These values are estimates of the likelihood of the model, given the data, $\mathcal{L}(M_i \mid \underline{x})$. These are functions in the same sense that $\mathcal{L}(\hat{\theta} \mid x, M_i)$ is the likelihood of the parameters $\theta$, given the data ($x$) and the model ($M_i$). The relative likelihood of model $i$ versus model $j$ is then just $w_i/w_j$. Rather than compute these, it is convenient to normalize the weights such that they sum to 1, by

$$w_i = \frac{\exp(-\frac{1}{2}\Delta_i)}{\sum_{r=1}^{R} \exp(-\frac{1}{2}\Delta_i)},$$

and interpret these as probabilities. Akaike (e.g., Akaike 1978b, 1979, 1980, and 1981b; also see Kishino 1991 and Buckland et al. 1997) suggested these values, and we have found them to be simple and very useful. Drawing on Bayesian ideas, it seems clear that we can interpret $w_i$ as the estimated probability that model $i$ is the K-L best model for the data at hand, given the set of models considered. The $w_i$ are useful as the "weight of evidence" in favor of model $i$ as being the actual K-L best model in the set. The bigger the $\Delta_i$, the smaller the weight and the less plausible is model $i$ as being the best approximating model. Inference is conditional on both the data and the set of *a priori* models considered, and corresponds to Bayesian priors of $1/R$ on each of the $R$ models.

Alternatively, one could draw $B$ bootstrap samples ($B$ should often be 10,000 rather than 1,000), use AIC to select a best model for each of the $B$ samples, and tally the proportion of samples whereby the $i$th model was selected. Denote such bootstrap-selection frequencies by $\pi_i$. While $w_i$ and $\pi_i$ are not estimates of exactly the same entity, they are often closely related and provide information concerning the uncertainty in the best model for use. The Akaike weights are simple to compute, while the bootstrap weights are computer-intensive and practically impossible to compute in some cases (e.g., the simulated starling experiment, Section 3.4), as thousands of bootstrap repetitions must be drawn and analyzed.

Under the hypothesis-testing approach, nothing can generally be said about ranking or calibrating models, particularly if the models were not nested. In linear least squares problems one could turn to adjusted $R^2$ values for a rough ranking of models, but other kinds of models cannot be calibrated using this (relatively very poor) approach (see the analogy in Section 2.5).

## 7.7 Model-Selection Uncertainty

At first, one might think that one could use AIC to select an approximating model that was "close" to truth (remembering the bias versus variance tradeoff and the principle of parsimony) or that "lost the least information" and then proceed to use this selected model for inference as if it had been specified *a priori* as the only model considered. Actually, this approach would not be terrible, as at least one would have a reasonable model, selected objectively, based on a valid theory and *a priori* considerations. This approach would often be far superior to much of current practice. The problem with considering only this model, and the usual measures of precision *conditional on this selected model*, is that this tends to overestimate precision. Breiman (1992) calls the failure to acknowledge model-selection uncertainty a "quiet scandal." [We might suggest that the widespread use of statistical hypothesis testing and blatant data dredging in model-selection represent "loud scandals."] In fact, there is a variance component due to model-selection uncertainty that should be incorporated into estimates of precision such that these are unconditional (on the selected model). While this is a research area needing further development, several useful methods are suggested in this book, and others will surely appear in the technical literature in the next few years, including additional Bayesian approaches.

The Akaike ($w_i$) or bootstrap ($\hat{\pi}_i$) weights that are used to rank and calibrate models can also be used to estimate unconditional precision where interest is in the parameter $\theta$ over $R$ models (model $M_i$, for $i = 1, \ldots, R$),

$$\widehat{\mathrm{var}}(\hat{\theta}_i) = \left[ \sum_{i=1}^{R} w_i \sqrt{\widehat{\mathrm{var}}(\hat{\theta}_i \mid M_i) + (\hat{\theta}_i - \hat{\bar{\theta}})^2} \right]^2,$$

$$\widehat{\mathrm{var}}(\hat{\theta}_i) = \left[ \sum_{i=1}^{R} \hat{\pi}_i \sqrt{\widehat{\mathrm{var}}(\hat{\theta}_i \mid M_i) + (\hat{\theta}_i - \hat{\bar{\theta}})^2} \right]^2.$$

These estimators, from Buckland et al. (1997), include a term for the conditional sampling variance, given model $M_i$ (denoted by $\widehat{\mathrm{var}}(\hat{\theta}_i \mid M_i)$ here) and incorporate a variance component for model-selection uncertainty $(\hat{\theta}_i - \hat{\bar{\theta}})^2$. These estimators of unconditional variance are also appropriate in cases where one wants a model-averaged estimate of the parameter when $\theta$ appears in all models; then

$$\hat{\bar{\theta}} = \sum_{i=1}^{R} w_i \hat{\theta}_i,$$

or

$$\hat{\bar{\theta}} = \sum_{i=1}^{R} \hat{\pi}_i \hat{\theta}_i.$$

Chapter 4 gives some procedures for setting confidence intervals that include model-selection uncertainty, and it is noted that achieved confidence interval coverage is then a useful measure of the utility of methods that integrate model-selection uncertainty into inference. Only a limited aspect of model uncertainty can be currently handled. *Given* a set of candidate models and an objective selection method, we can assess selection uncertainty. The uncertainty in defining the set of models cannot be addressed; we lack a theory for this issue. In fact, we lack good, general guidelines for defining the *a priori* set of models. We expect papers to appear on these scientific issues in the future.

## 7.8 Inference Based on Model Averaging

Rather than base inferences on a single selected best model, from an *a priori* set of models, we can base our inferences on the entire set by using model-averaging. The key to this inference methodology is the Akaike weights. Thus, if a parameter, $\theta$, is in common over all models (as $\theta_i$ in model $M_i$), or our goal is prediction, by using the weighted average $\hat{\bar{\theta}} = \sum w_i \hat{\theta}_i$ we are basing point inference on the entire set of models. This approach has both practical and philosophical advantages. Where a model-averaged estimator can be used, it appears to have better precision and reduced bias compared to $\hat{\theta}_k$ from the selected best model. The unconditional variance formula actually applies directly to the model-averaged estimator.

If one has a large number of closely related models, such as in regression-based variable selection (all-subsets selection), designation of a single best model is unsatisfactory, because that "best" model is highly variable. In this situation model-averaging provides a relatively much more stabilized inference. The concept of inference being tied to all the models can be used to reduce model-selection bias effects on regression coefficient estimates in all-subsets selection. For the regression coefficient associated with predictor $x_j$ we use the estimate $\hat{\bar{\beta}}_j$, which is $\beta_j$ averaged over all models in which $x_j$ appears:

$$\hat{\bar{\beta}}_j = \frac{\sum_{i=1}^{R} w_i I_j(M_i) \hat{\beta}_{j,i}}{w_+(j)},$$

$$w_+(j) = \sum_{i=1}^{R} w_i I_j(M_i),$$

and

$$I_j(M_i) = \begin{cases} 1 & \text{if predictor } x_j \text{ is in model } M_i, \\ 0 & \text{otherwise.} \end{cases}$$

Conditional on model $M_i$ being selected, model-selection has the effect of biasing $\hat{\beta}_{j,i}$ away from zero. Thus a new estimator, denoted by $\tilde{\hat{\beta}}_i$, is suggested:

$$\tilde{\hat{\beta}}_i = w_+(i)\hat{\bar{\beta}}_i.$$

Investigation of this idea, and extensions of it, is an open research area. The point here is that while $\hat{\bar{\beta}}_j$ can be computed ignoring models other than the ones $x_j$ appears in, $\tilde{\hat{\beta}}_i$ does require fitting all $R$ of the *a priori* models.

Inference on the importance of a variable is similarly improved by being based on all the models. If one selects the best model and says that the variables in it are the important ones and the other variables are not important, this is a very naive, unreliable inference. We suggest that the importance of variable $x_j$ be measured by the sum of the Akaike weights over all models in which that variable appears:

$$w_+(j) = \sum_{i=1}^{R} w_i I_j(M_i).$$

Thus again, proper inference requires fitting all the *a priori* models and then using a type of model-averaging. When possible, one should use inference based on all the models, via model-averaging and selection bias adjustments, rather than a "select the best model and ignore the others" strategy.

## 7.9 More on Inferences

There needs to be increased attention to separating those inferences that rest on *a priori* considerations from those resulting from some degree of data dredging. Essentially no justifiable theory exists to estimate precision (or test hypotheses, for those still so inclined) when data dredging has taken place (the theory (mis)used is for *a priori* analyses, assuming that the model was the only one fit to the data). This glaring fact is either not understood by practitioners and journal editors or is simply ignored. Two types of data dredging include (1) an iterative approach, in which patterns and differences observed after initial analysis are "chased" by repeatedly building new models with these effects included and (2) analysis of "all possible models." Data dredging is a poor approach to making inferences about the sampled population, and both types of data dredging are best reserved for more exploratory investigations and are not the central subject of this book.

The information-theoretic paradigm avoids statistical hypothesis testing concepts and focuses on relationships of variables (via selection) and on the estimation of effect size and measures of its precision. This paradigm is primarily in the context of making inferences from a single selected model or making robust inference from many models (e.g., using model-averaging based on Akaike weights). Data analysis is a process of learning what effects are supported by the data and the degree of complexity of the best approximating model. Information theoretic approaches cannot be used unthinkingly; a good set of candidate models is essential, and this involves professional judgment and integration of the science of the issue into the model set.

## 7.10 Final Thoughts

At a conceptual level, reasonable data and a *good* model allow a separation of "information" from "noise." Here, information relates to the structure of relationships, estimates of model parameters, and components of variance. Noise then refers to the residuals; variation left unexplained. We can use the information extracted from the data to make proper inferences. We want an approximating model that minimizes information loss, $I(f, g)$, and properly separates noise (noninformation, or entropy) from structural information. The philosophy for this separation is the principle of parsimony; the conceptual target for such partitioning is Kullback–Leibler information; and the tactic for selection of a best model is an information criterion (e.g., AIC, $AIC_c$, $QAIC_c$, or TIC). The notion of data-based model-selection and resulting inference is a very difficult subject, but we do know that substantial uncertainty about the selected model can often be expected and should be incorporated into estimates of precision.

Still, model-selection (in the sense of parsimony) is the critical issue in data analysis. In using the more advanced methods presented here, model-selection can be thought of as a way to compute Akaike weights. Then one uses all (or several of the best) models in the set as a way to make robust inferences from the data. More research is needed on the quantification of model uncertainty, measures of the plausibility of alternative models, and ways to provide effective measures of precision (without being conditional on a given model). Confidence intervals with good achieved levels should be a goal of inference following data-based model-selection.

Information-theoretic methods are relatively simple to understand and practical to employ across a very wide class of empirical situations and scientific disciplines. The information-theoretic approach unifies parameter estimation and model-selection under an optimization framework, based on Kullback–Leibler information and likelihood theory. With the exception of the bootstrap, the methods are easy to compute by hand if necessary (assuming that one has the MLEs, maximized log-likelihood values, and $\widehat{\text{var}}(\hat{\theta}_i \mid M_i)$ for each of the $R$ models). Researchers can easily understand the information-theoretic methods presented here; we believe that it is *very* important that researchers understand the methods they employ.

# References

Agresti, A. (1990). *Categorical data analysis*. John Wiley and Sons, New York, NY.

Akaike, H. (1973). Information theory as an extension of the maximum likelihood principle. Pages 267–281 *in* B. N. Petrov, and F. Csaki, (eds.) *Second International Symposium on Information Theory*. Akademiai Kiado, Budapest.

Akaike, H. (1974). A new look at the statistical model identification. *IEEE Transactions on Automatic Control AC* **19**, 716–723.

Akaike, H. (1976). Canonical correlation analysis of time series and the use of an information criterion. Pages 27–96 *in* R. K. Mehra, and D. G. Lainiotis (eds.) *System Identification: Advances and Case Studies*. Academic Press, New York, NY.

Akaike, H. (1977). On entropy maximization principle. Pages 27–41 *in* P. R. Krishnaiah (ed.) *Applications of statistics*. North-Holland, Amsterdam.

Akaike, H. (1978a). A new look at the Bayes procedure. *Biometrika* **65**, 53–59.

Akaike, H. (1978b). A Bayesian analysis of the minimum AIC procedure. *Annals of the Institute of Statistical Mathematics* **30**, 9–14.

Akaike, H. (1978c). On the likelihood of a time series model. *The Statistician* **27**, 217–235.

Akaike, H. (1979). A Bayesian extension of the minimum AIC procedure of autoregressive model fitting. *Biometrika* **66**, 237–242.

Akaike, H. (1980). Likelihood and the Bayes procedure (with discussion). Pages 143–203 *in* J. M. Bernardo, M. H. De Groot, D. V. Lindley, and A. F. M. Smith (eds.) *Bayesian statistics*. University Press, Valencia, Spain.

Akaike, H. (1981a). Likelihood of a model and information criteria. *Journal of Econometrics* **16**, 3–14.

Akaike, H. (1981b). Modern development of statistical methods. Pages 169–184 *in* P. Eykhoff (ed.) *Trends and progress in system identification*. Pergamon Press, Paris.

Akaike, H. (1983a). Statistical inference and measurement of entropy. Pages 165–189 *in* G. E. P. Box, T. Leonard, and C-F. Wu (eds.) *Scientific inference, data analysis, and robustness*. Academic Press, London.

Akaike, H. (1983b). Information measures and model selection. *International Statistical Institute* **44**, 277–291.

Akaike, H. (1983c). On minimum information prior distributions. *Annals of the Institute of Statistical Mathematics* **35A**, 139–149.

Akaike, H. (1985). Prediction and entropy. Pages 1–24 *in* A. C. Atkinson, and S. E. Fienberg (eds.) *A celebration of statistics*. Springer, New York, NY.

Akaike, H. (1987). Factor analysis and AIC. *Psychometrika* **52**, 317–332.

Akaike, H. (1992). Information theory and an extension of the maximum likelihood principle. Pages 610–624 *in* S. Kotz, and N. L. Johnson (eds.) *Breakthroughs in statistics*, Vol. 1. Springer-Verlag, London.

Akaike, H. (1994). Implications of the informational point of view on the development of statistical science. Pages 27–38 *in* H. Bozdogan, (ed.) *Engineering and Scientific Applications*. Vol. 3, Proceedings of the First US/Japan Conference on the Frontiers of Statistical Modeling: An Informational Approach. Kluwer Academic Publishers, Dordrecht, Netherlands.

Akaike, H., and Nakagawa, T. (1988). *Statistical analysis and control of dynamic systems*. KTK Scientific Publishers, Tokyo. (English translation by H. Akaike and M. A. Momma).

Allen, D. M. (1970). Mean square error of prediction as a criterion for selecting variables. *Technometrics* **13**, 469–475.

Amari, S. (1993). Mathematic methods of neurocomputing. Pages 1–39 *in* O. E. Barndorff-Nielson, J. L. Jensen, and W. S. Kendall (eds.) *Networks and chaos—statistical and probabilistic aspects*. Chapman and Hall, New York, NY.

Anonymous. (1994). MATLAB®: *High-performance numerical computations and visualization software*. The MathWorks, Inc., Natick, MA.

Anonymous. (1997). *The Kullback Memorial Research Conference*. Statistics Department, The George Washington University, Washington, D.C.

Anderson, D. R., and Burnham, K. P. (1976). *Population ecology of the mallard: VI. the effect of exploitation on survival*. U.S. Fish and Wildlife Service Resource Publication No. 128.

Anderson, D. R., and Burnham, K. P. (1999a). General strategies for the collection and analysis of ringing data. *Bird Study* **46**.

Anderson, D. R., and Burnham, K. P. (1999b). Understanding information criteria for selection among capture–recapture or ring recovery models. *Bird Study* **46**.

Anderson, D. R., Burnham, K. P., and White, G. C. (1994). AIC model selection in overdispersed capture–recapture data. *Ecology* **75**, 1780–1793.

# References

Anderson, D. R., Burnham, K. P., and White, G. C. (1998). Comparison of AIC and CAIC for model selection and statistical inference from capture–recapture studies. *Journal of Applied Statistics* **25**, 263–282.

Anderson, S., Auquier, A., Hauck, W. W., Oakes, D., Vandaele, W., and Weisberg, H. I. (1980). *Statistical methods for comparative studies*. John Wiley and Sons, New York, NY.

Apostol, T. M. (1957). *Mathematical analysis: a modern approach to advanced calculus*. Addison-Wesley Publishing CO., Inc. Reading, MA.

Armitage, P. (1957). Studies in the variability of pock counts. *Journal of Hygiene* **55**, 564–581.

Atilgan, T. (1996). Selection of dimension and basis for density estimation and selection of dimension, basis and error distribution for regression. *Communications in Statistics—Theory and Methods* **25**, 1–28.

Atkinson, A. C. (1980). A note on the generalized information criterion for choice of a model. *Biometrika* **67**, 413–18.

Augustin, N. H., Mugglestone, M. A., and Buckland S. T. (1996). An autologistic model for the spatial distribution of wildlife. *Journal of Applied Ecology* **33**, 339–347.

Azzalini, A. (1996). *Statistical inference—based on the likelihood*. Chapman and Hall, London.

Bancroft, T. A., and Han, C-P. (1977). Inference based on conditional specification: a note and a bibliography. *International Statistical Review* **45**, 117–127.

Bartlett, M. S. (1936). Some notes on insecticide tests in the laboratory and in the field. *Journal of the Royal Statistical Society*, Supplement **1**, 185–194.

Bedrick, E. J., and Tsai, C-L. (1994). Model selection for multivariate regression in small samples. *Biometrics* **50**, 226–231.

Berger, J. O., and Pericchi, L. R. (1996). The intrinsic Bayes factor for model selection and prediction. *American Statistical Association* **91**, 109–122.

Berger, J. O., and Wolpert, R. L. (1984). The likelihood principle. *Institute of Mathematical Statistics* Monograph 6.

Berk, R. H. (1966). The limiting behavior of posterior distributions when the model is incorrect. *Annals of Mathematical Statistics* **37**, 51–58.

Berryman, A. A., Gutierrez, A. P., and Arditi, R. (1995). Credible, parsimonious and useful predator–prey models—a reply to Abrams, Gleeson, and Sarnelle. *Ecology* **76**, 1980–1985.

Blansali, R. J. (1993). Order selection for linear time series models: a review. Pages 50–66 *in* T. S. Rao (ed.) *Developments in time series analysis*, Chapman and Hall, London.

Bickel, P., and Zhang, P. (1992). Variable selection in nonparametric regression with categorical covariates. *Journal of the American Statistical Association* **87**, 90–97.

Bishop, Y. M. M., Fienberg, S. E., and Holland, P. W. (1975). *Discrete multivariate analysis: theory and practice*. The MIT Press, Cambridge, MA.

Blau, G. E., and Neely, W. B. (1975). Mathematical model building with an application to determine the distribution of DURSBAN® insecticide added to a simulated ecosystem. Pages 133–163 *in* A. Macfadyen (ed.) *Advances in Ecological Research*, Academic Press, London.

Bollen, K. A., and Long, J. S. (1993). *Testing structural equations*. Sage Publ., London.

Boltzmann, L. (1877). Über die Beziehung zwischen dem Hauptsatze der mechanischen Warmetheorie und der Wahrscheinlichkeitsrechnung respective den Satzen uber das Warmegleichgewicht. *Wiener Berichte* **76**, 373–435.

Bonneu, M., and Milhaud, X. (1994). A modified Akaike criterion for model choice in generalized linear models. *Statistics* **25**, 225–238.

Box, G. E. P. (1967). Discrimination among mechanistic models. *Technometrics* **9**, 57–71.

Box, G. E. P. (1976). Science and statistics. *Journal of the American Statistical Association* **71**, 791–799.

Box, G. E. P., and Jenkins, G. M. (1970). *Time series analysis: forecasting and control*. Holden-Day, London.

Box, G. E. P., Leonard, T., and Wu, C-F. (eds.) (1981). *Scientific inference, data analysis, and robustness*. Academic Press, London.

Box, J. F. (1978). *R. A. Fisher: the life of a scientist*. John Wiley and Sons, New York, NY.

Boyce, M. S. (1992). Population viability analysis. *Annual Review of Ecology and Systematics* **23**, 481–506.

Bozdogan, H. (1987). Model selection and Akaike's information criterion (AIC): the general theory and its analytical extensions. *Psychometrika* **52**, 345–370.

Bozdogan, H. (1988). A new model-selection criterion. Pages 599–608 *in* H. H. Bock (ed.) *Classification and related methods of data analysis*. North-Holland Publishing Company, Amsterdam.

Bozdogan, H. (1994). Editor's general preface. Pages ix–xii *in* H. Bozdogan (ed.) *Engineering and Scientific Applications*. Vol. 3, Proceedings of the First US/Japan Conference on the Frontiers of Statistical Modeling: An Informational Approach. Kluwer Academic Publishers, Dordrecht, Netherlands.

Breiman L. (1992). The little bootstrap and other methods for dimensionality selection in regression: $X$-fixed prediction error. *Journal of the American Statistical Association* **87**, 738–754.

Breiman, L. (1995). Better subset regression using the nonnegative garrote. *Technometrics* **37**, 373–384.

Breiman L. (1996). Heuristics of instability and stabilization in model selection. *The Annals of Statistics* **24**, 2350–2383.

Breiman, L., and Freedman, D. F. (1983). How many variables should be entered in a regression equation? *Journal of the American Statistical Association* **78**, 131–136.

Brisbin I. L., Jr., Collins, C. T., White, G. C., and McCallum, D. A. (1987). A new paradigm for the analysis and interpretation of growth data: the shape of things to come. *Auk* **104**, 552–554.

Brockwell, P. J., and Davis, R. A. (1987). *Time series: theory and methods*. Springer-Verlag, New York, NY.

Brockwell, P. J., and Davis, R. A. (1991). *Time series: theory and methods*. 2nd ed. Springer-Verlag, New York, NY.

Broda, E. (1983). *Ludwig Boltzmann: man, physicist, philosopher* (translated with L. Gay). Ox Bow Press, Woodbridge, Connecticut.

Brown, D. (1992). A graphical analysis of deviance. *Applied Statistics* **41**, 55–62.

Brown, D., and Rothery, P. (1993). *Models in biology: mathematics, statistics and computing*. John Wiley and Sons. New York, NY.

Brown, P. J. (1993). *Measurement, regression, and calibration*. Clarendon Press, Oxford.

Brownie, C., Anderson, D. R., Burnham, K. P., and Robson, D. S. (1985). *Statistical inference from band recovery data–a handbook*. 2nd ed. U.S. Fish and Wildlife Service Resource Publication 156.

Brownie, C., Hines, J. E., Nichols, J. D., Pollock, K. H., and Hestbeck, J. B. (1993). Capture–recapture studies for multiple strata including non-Markovian transition probabilities. *Biometrics* **49**, 1173–1187.

Brush, S. G. (1965). *Kinetic theory*. Vol. 1 Pergamon Press, Oxford.

Brush, S. G. (1966). *Kinetic theory*. Vol. 2 Pergamon Press, Oxford.

Buckland, S. T. (1982). A note on the Fourier series model for analyzing line transect data. *Biometrics* **38**, 469–477.

Buckland, S. T. (1984). Monte Carlo confidence intervals. *Biometrics* **40**, 811–817.

Buckland, S. T., Burnham, K. P., and Augustin, N. H. (1997). Model selection: an integral part of inference. *Biometrics* **53**, 603–618.

Buckland, S. T., Anderson, D. R., Burnham, K. P., and Laake, J. L. (1993). *Distance sampling: estimating abundance of biological populations*. Chapman and Hall, London.

Buckland, S. T., and Elston, D. A. (1993). Empirical models for the spatial distribution of wildlife. *Journal of Applied Ecology* **30**, 478–495.

Burman, P. (1989). A comparative study of ordinary cross-validation, $v$-hold cross-validation and repeated learning-testing methods. *Biometrika* **76**, 503–514.

Burman, P., and Nolan, D. (1995). A general Akaike-type criterion for model selection in robust regression. *Biometrika*, 877–886.

Burnham, K. P. (1989). Numerical survival rate estimation for capture–recapture models using SAS PROC NLIN. Pages 416–435 *in* L. McDonald, B. Manly, J. Lockwood and J. Logan (eds.) *Estimation and analysis of insect populations*. Springer-Verlag, New York.

Burnham, K. P. (in prep.). Random effects models in ringing and capture–recapture studies.

Burnham, K. P., and Anderson, D. R. (1992). Data-based selection of an appropriate biological model: the key to modern data analysis. Pages 16–30 *in* D. R. McCullough, and R. H. Barrett (eds.) *Wildlife* 2001: *Populations*. Elsevier Sci. Publ., Ltd., London.

Burnham, K. P., Anderson, D. R., White, G. C., Brownie, C., and Pollock, K. H. (1987). *Design and analysis methods for fish survival experiments based on release-recapture*. American Fisheries Society, Monograph **5**.

Burnham, K. P., Anderson, D. R., and White, G. C. (1994). Evaluation of the Kullback–Leibler discrepancy for model selection in open population capture–recapture models. *Biometrical Journal* **36**, 299–315.

Burnham, K. P., White, G. C., and Anderson, D. R. (1995a). Model selection in the analysis of capture–recapture data. *Biometrics* **51**, 888–898.

Burnham, K. P., Anderson, D. R., and White, G. C. (1995b). Selection among open population capture–recapture models when capture probabilities are heterogeneous. *Journal of Applied Statistics* **22**, 611–624.

Burnham, K. P., Anderson, D. R., and White, G. C. (1996). Meta-analysis of vital rates of the Northern Spotted Owl. *Studies in Avian Biology* **17**, 92–101.

Bystrak, D. (1981). The North American breeding bird survey, Pages 522–532. *in* C. J. Ralph, and J. M. Scott (eds.) *Estimating numbers of terrestrial birds*. Studies in Avian Biology **6**.

Carlin, B. P., and Chib, S. (1995). Bayesian model choice via Markov chain Monte Carlo methods. *Journal of the Royal Statistical Society*, Series B **57**, 473–484.

Carlin, B. P., and Louis, T. A. (1996). *Bayes and empirical Bayes methods for data analysis*. Chapman and Hall, London.

Carpenter, S. R. (1990). Large-scale perturbations: opportunities for innovation. *Ecology* **71**, 2038–2043.

Carrol, R., Ruppert, D, and Stefanski, L. (1995). *Measurement error in nonlinear models*. Chapman and Hall, London.

Casella, G. (1995). An introduction to empirical Bayes data analysis. *The American Statistician* **39**, 83–87.

Chamberlain, T. C. (1890). The method of multiple working hypotheses. *Science* **15**, 93.

Chatfield, C. (1991). Avoiding statistical pitfalls (with discussion). *Statistical Science* **6**, 240–268.

Chatfield, C. (1995a). *Problem solving: a statistician's guide*. Second edition. Chapman and Hall, London.

Chatfield, C. (1995b). Model uncertainty, data mining and statistical inference. *Journal of the Royal Statistical Society*, Series A **158**, 419–466.

Chatfield, C. (1996). Model uncertainty and forecast accuracy. *Journal of Forecasting* **15**, 495–508.

Chow, G. C. (1981). A comparison of the information and posterior probability criteria for model selection. *Journal of Econometrics* **16**, 21–33.

Chung, H-Y., Lee, K-W., and Koo, J-A. (1996). A note on bootstrap model selection criterion. *Statistics and Probability Letters* **26**, 35–41.

Clayton, M. K., Geisser, S., and Jennings, D. (1986). A comparison of several model selection procedures. Pages 425–439 *in* P. Goel, and A. Zellner (eds.) *Bayesian inference and decision*. Elsevier, New York, NY.

Cochran, W. G. (1963). *Sampling techniques*. 2nd ed., John Wiley and Sons, Inc. New York, NY.

Cohen, E. G. D., and Thirring, W. (eds.) (1973). *The Boltzmann equation: theory and applications*. Springer-Verlag, New York, NY.

Collopy, F., Adya, M., and Armstrong, J. S. (1994). Principles for examining predictive validity: the case of information systems spending forecasts. *Information Systems Research* **5**, 170–179.

Cook, T. D., and Campbell, D. T. (1979). *Quasi-experimentation: design and analysis issues for field settings*. Houghton Mifflin Company, Boston, MA.

Copas, J. B. (1983). Regression, prediction and shrinkage (with discussion). *Journal of the Royal Statistical Society*, Series B, **45**, 311–354.

Cover, T. M., and Thomas, J. A. (1991). *Elements of information theory*. John Wiley and Sons, New York, NY.

Cox, D. R. (1990). Role of models in statistical analysis. *Statistical Science* **5**, 169–174.

Cox, D. R., and Snell, E. J. (1989). *Analysis of binary data*. 2nd ed., Chapman and Hall, New York, NY.

Craven, P., and Wahba, G. (1979). Smoothing noisy data with spline functions: estimating the correct degree of smoothing by the method of generalized cross validation. *Numerical Mathematics* **31**, 377–403.

Cressie, N. A. C. (1991). *Statistics for spatial data*. John Wiley and Sons, New York, NY.

Cutler, A., and Windham, M. P. (1994). Information-based validity functionals for mixture analysis. Pages 149–170 *in* H. Bozdogan (ed.) *Engineering and Scientific Applications*. vol. 2, Proceedings of the First US/Japan Conference on the Frontiers of Statistical Modeling: An Informational Approach. Kluwer Academic Publishers, Dordrecht, Netherlands.

Daniel, C., and Wood, F. S. (1971). *Fitting equations to data*. Wiley-Interscience, New York, NY.

de Leeuw, J. (1988). Model selection in multinomial experiments. Pages 118–138 *in* T. K. Dijkstra (ed.) *On model uncertainty and its statistical implications*. Lecture Notes in Economics and Mathematical Systems, Springer-Verlag, New York, NY.

de Leeuw, J. (1992). Introduction to Akaike (1973) information theory and an extension of the maximum likelihood principle. Pages 599–609 *in* S. Kotz, and N. L. Johnson (eds.) *Breakthroughs in statistics*. Vol. 1. Springer-Verlag, London.

de Gooijer, J. G., Abraham, B., Gould, A., and Robinson, L. (1985). Methods for determining the order of an autoregressive-moving average process: a survey. *International Statistical Review* **53**, 301–329.

Desu, M. M., and Roghavarao, D. (1991). *Sample size methodology.* Academic Press, Inc., New York, NY.

Dijkstra, T. K. (ed). (1988). *On model uncertainty and its statistical implications.* Lecture Notes in Economics and Mathematical Systems, Springer-Verlag, New York, NY.

Dijkstra, T. K., and Veldkamp, J. H. (1988). Data-driven selection of regressors and the bootstrap. Pages 17–38 *in* T. K. Dijkstra (ed.) *On model uncertainty and its statistical implications.* Lecture Notes in Economics and Mathematical Systems, Springer-Verlag, New York, NY.

Draper, D. (1995). Assessment and propagation of model uncertainty (with discussion). *Journal of the Royal Statistical Society,* Series B **57**, 45–97.

Draper, N. R., and Smith, H. (1981). *Applied regression analysis.* Second edition. John Wiley and Sons, New York, NY.

Eberhardt, L. L. (1978). Appraising variability in population studies. *Journal of Wildlife Management* **42**, 207–238.

Eberhardt, L. L., and Thomas, J. M. (1991). Designing environmental field studies. *Ecological Monographs* **61**, 53–73.

Edwards, A. W. F. (1976). *Likelihood: an account of the statistical concept of likelihood and its application to scientific inference.* Cambridge University Press.

Efron, B. (1984). Comparing non-nested linear models. *Journal of the American Statistical Association* **79**, 791–803.

Efron, B., and Morris, C. (1975). Data analysis using Stein's estimator and its generalizations. *Journal of the American Statistical Association* **70**, 311–319.

Efron, R., and Gong, G. (1983). A leisurely look at the bootstrap, the jackknife, and cross-validation. *The American Statistician* **37**, 36–48.

Efron, B., and Tibshirani, R. J. (1993). *An introduction to the bootstrap.* Chapman and Hall, London.

Fienberg, S. E. (1970). The analysis of multidimensional contingency tables. *Ecology* **51**, 419–433.

Fildes, R., and Makridakis, S. (1995). The impact of empirical accuracy studies on time series analysis and forecasting. *International Statistics Review* **63**, 289–308.

Findley, D. F. (1985). On the unbiasedness property of AIC for exact or approximating linear stochastic time series models. *Journal of Time Series Analysis* **6**, 229–252.

Findley, D. F. (1991). Counterexamples to parsimony and BIC. *Annals of the Institute of Statistical Mathematics* **43**, 505–514.

Findley, D. F., and Parzen, E. (1995). A conversation with Hirotugu Akaike. *Statistical Science* **10**, 104–117.

Finney, D. J. (1971). *Probit analysis.* 3rd. ed. Cambridge University Press, London.

Fisher, R. A. (1922). On the mathematical foundations of theoretical statistics. Royal Society of London. *Philosophical Transactions* (Series A) **222**, 309–368.

Fisher, R. A. (1936). Uncertain inference. *Proceedings of the American Academy of Arts and Sciences* **71**, 245–58.

Fisher, R. A. (1949). A biological assay of tuberculins. *Biometrics* **5**, 300–316.

Flack, V. F., and Chang, P. C. (1987). Frequency of selecting noise variables in subset regression analysis: a simulation study. *The American Statistician* **41**, 84–86.

Flather, C. H. (1992). Patterns of avian species-accumulation rates among eastern forested landscapes. Ph.D. dissertation. Colorado State University. Fort Collins.

Flather, C. H. (1996). Fitting species-accumulation functions and assessing regional land use impacts on avian diversity. *Journal of Biogeography* **23**, 155–168.

Freedman, D. A. (1983). A note on screening regression equations. *The American Statistician* **37**, 152–155.

Freedman, D. A., Navidi, W., and Peters, S. C. (1988). On the impact of variable selection in fitting regression equations. Pages 1–16 *in* T. K. Dijkstra (ed.) *On model uncertainty and its statistical implications*. Lecture Notes in Economics and Mathematical Systems, Springer-Verlag, New York, NY.

Fujikoshi, Y, and Satoh, K. (1997). Modified *AIC* and $C_p$ in multivariate linear regression. *Biometrika* **84**, 707–716.

Gail, M. H. (1996). Statistics in action. *Journal of the American Statistical Association* **91**, 1–13.

Gallant, A. R. (1987). *Nonlinear statistical models*. John Wiley and Sons, New York, NY.

Garthwaite, P. H., Jolliffe, I. T., and Jones, B. (1995). *Statistical inference*. Prentice Hall, London.

Gause, G. F. (1934). *The struggle for existence*. Williams and Wilkins, Baltimore, MD.

Geisser, S. (1975). The predictive sample reuse method with applications. *Journal of the American Statistical Association* **70**, 320–328.

Gelfand, A. E., and Smith, A. F. M. (1990). Sampling-based approaches to calculating marginal densities. *Journal of the American Statistical Association* **85**, 398–409.

George, E. I., and McCulloch, R. E. (1993). Variable selection via Gibbs sampling. *Journal of the American Statistical Association* **88**, 881–889.

Gilchrist, W. (1984). *Statistical modelling*. Chichester, Wiley and Sons, New York, NY.

Glymour, C. (1998). Causation. Pages 97–109 *in* S. Kotz (ed.), *Encyclopedia of statistical sceinces*. John Wiley and Sons, New York, NY.

Gochfeld, M. (1987). On paradigm vs. methods in the study of growth. *Auk* **104**, 554–555.

Gokhale, D. V., and Kullback, S. (1978). *The information in contingency tables*. Marcel Dekker, New York, NY.

Golub, G. H., Health, M., and Wahba, G. (1979). Generalized cross validation as a method for choosing a good ridge parameter. *Technometrics* **21**, 215–223.

Goodman, S. N. (1993). *p* values, hypothesis tests, and likelihood: implications for epidemiology of a neglected historical debate (with discussion). *American Journal of Epidemiology* **137**, 485–501.

Goutis, C., and Casella, G. (1995). Frequentist post-data inference. *International Statistical Review* **63**, 325–344.

Granger, C. W. J., King, M. L., and White, H. (1995). Comments on testing economic theories and the use of model selection criteria. *Journal of Econometrics* **67**, 173–187.

Graybill, F. A., and Iyer, H. K. (1994). *Regression analysis: concepts and applications.* Duxbury Press, Belmont, CA.

Greenhouse, S. W. (1994). Solomon Kullback: 1907–1994. *Institute of Mathematical Statistics Bulletin* **23**, 640–642.

Guiasu, S. (1977). *Information theory with applications.* McGraw-Hill, New York, NY.

Hairston, N. G. (1989). *Ecological experiments: purpose, design and execution.* Cambridge University Press, Cambridge, UK.

Hald, A. (1952). *Statistical theory with engineering applications.* John Wiley and Sons, New York, NY.

Hand, D. J. (1994). Statistical strategy: step 1. Pages 1–9 *in* P. Cheeseman, and R. W. Oldford (eds.) *Selecting models from data.* Springer-Verlag, New York, NY.

Hannan, E. J., and Quinn, B. G. (1979). The determination of the order of an autoregression. *Journal of the Royal Statistical Society*, Series B **41**, 190–195.

Harlow, L. L., Mulaik, S. A., and Steiger, J. H. (eds.) (1997). *What if there were no significancs tests?* Lawrence Erlbaum Associates, Publishers, Mahwah, New Jersey.

Hasenöhrl, F. (ed.) (1909). *Wissenschaftliche Abhandlungen.* 3 Vols. Leipzig.

Hastie, T. J., and Tibshirani, R. J. (1990). *Generalized additive models.* Chapman and Hall, London.

Haughton, D. (1989). Size of the error in the choice of a model to fit data from an exponential family. *Sankhya*, Series A **51**, 45–58.

Hayne, D. (1978). Experimental designs and statistical analyses. Pages 3–13 *in* D. P. Synder (ed). *Populations of small mammals under natural conditions.* Pymatuning Symposium in Ecology, University of Pittsburgh, Vol 5.

Henderson, H., and Velleman, P. (1981). Building multiple regression models interactively. *Biometrics* **37**, 391–411.

Hjorth, J. S. U. (1994). *Computer intensive statistical methods: validation, model selection and bootstrap.* Chapman and Hall, London.

Hobson, A., and Cheng, B-K. (1973). A comparison of the Shannon and Kullback information measures. *Journal of Statistical Physics* **7**, 301–310.

Hocking, R. R. (1976). The analysis and selection of variables in linear regression. *Biometrics* **32**, 1–49.

Hoeting, J. A., and Ibrahim, J. G. (1996). Bayesian predictive simultaneous variable and transformation selection in the linear model. Department of Statistics, Colorado State University, Fort Collins. Technical Report No. 96/39.

Hosmer, D. W., and Lemeshow, S. (1989). *Applied logistic regression analysis*. John Wiley and Sons, New York, NY.

Howard, R. A. (1971). *Dynamic probabilistic systems*. John Wiley and Sons. New York, NY.

Huelsenbeck, J. P., and Rannala, B. (1997). Phylogenetic methods come of age: testing hypotheses in an evolutionary context. *Science* **276**, 227–232.

Hurvich, C. M., and Tsai, C-L. (1989). Regression and time series model selection in small samples. *Biometrika* **76**, 297–307.

Hurvich, C. M., and Tsai, C-L. (1990). The impact of model selection on inference in linear regression. *The American Statistician* **44**, 214–217.

Hurvich, C. M., and Tsai, C-L. (1991). Bias of the corrected AIC criterion for underfitted regression and time series models. *Biometrika* **78**, 499–509.

Hurvich, C. M., and Tsai, C-L. (1994). Autoregressive model selection in small samples using a bias-corrected version of AIC. Pages 137–157 *in* H. Bozdogan (ed.) *Engineering and Scientific Applications*. Vol. 1, Proceedings of the First US/Japan Conference on the Frontiers of Statistical Modeling: An Informational Approach. Kluwer Academic Publishers, Dordrecht, Netherlands.

Hurvich, C. M., and Tsai, C-L. (1995a). Relative rates of convergence for efficient model selection criteria in linear regression. *Biometrika* **82**, 418–425.

Hurvich, C. M., and Tsai, C-L. (1995b). Model selection for extended quasi-likelihood models in small samples. *Biometrics* **51**, 1077–1084.

Hurvich, C. M., and Tsai, C-L. (1996). The impact of unsuspected serial correlations on model selection in linear regression. *Statistics and Probability Letters* **27**, 115–126.

Hurvich, C. M., Shumway, R., and Tsai, C-L. (1990). Improved estimators of Kullback–Leibler information for autoregressive model selection in small samples. *Biometrika* **77**, 709–719.

Inman, H. F. (1994). Karl Pearson and R. A. Fisher on statistical tests: a 1935 exchange from *Nature*. *The American Statistician* **48**, 2–11.

Ibrahim, J. G., and Chen, M-H. (1997). Predictive variable selection for the multivariate linear model. *Biometrics* **53**, 465–478.

James, F. C., and McCulloch, C. E. (1990). Multivariate analysis in ecology and systematics: panacea or Pandora's box? *Annual Reviews of Ecology and Systematics* **21**, 129–166.

Jaynes, E. T. (1957). Information theory and statistical mechanics. *Physics Review* **106**, 620–630.

Jaynes, E. T. (1982). On the rationale of maximum-entropy methods. *Proceedings of the IEEE* **70**, 939–952.

Jaynes, E. T. (in prep.). *Probability theory: the logic of science*. Cambridge University Press.

Jessop, A. (1995). *Informed assessments: an introduction to information, entropy and statistics*. Ellis Horwood, London.

Johnson, D. H. (1995). Statistical sirens: the allure of nonparametrics. *Ecology* **76**, 1998–2000.

Johnson, N. L., and Kotz, S. (1970). *Continuous univariate distributions* -1. Houghton Mifflin Company, New York, NY.

Johnson, N. L., and Kotz, S. (1992). *Univariate discrete distributions*. (2nd ed.) Wiley-Interscience Publication, New York, NY.

Jones, D., and Matloff, N. (1986). Statistical hypothesis testing in biology: a contradiction in terms. *Journal of Economic Entomology* **79**, 1156–1160.

Judge, G. C., and Yancey, T. (1986). *Improved methods of inference in econometrics*. North Holland, Amsterdam.

Kabaila, P. (1995). The effect of model selection on confidence regions and prediction regions. *Econometric Theory* **11**, 537–549.

Kapur, J. N., and Kesavan, H. K. (1992). *Entropy optimization principles with applications*. Academic Press, London.

Kareiva, P. (1994). Special feature: higher order interactions as a foil to reductionist ecology. *Ecology* **75**, 1527–1559.

Kass, R. E., and Raftery, A. E. (1995). Bayes factors. *Journal of the American Statistical Association* **90**, 773–795.

Kishino, H., Kato, H., Kasamatsu, F., and Fujise, Y. (1991). Detection of heterogeneity and estimation of population characteristics from field survey data: 1987/88 Japanese feasibility study of the Southern Hemisphere minke whales. *Annals of the Institute of Statistical Mathematics* **43**, 435–453.

Kittrell, J. R. (1970). Mathematical modelling of chemical reactors. *Advances in Chemical Engineering* 8: 97–183.

Knopf, F. L., Sedgwick, J. A., and Cannon, R. W. (1988). Guild structure of a riparian avifauna relative to seasonal cattle grazing. *Journal of Wildlife Management* **52**, 280–290.

Konishi, S., and Kitagawa, G. (1996). Generalized information criteria in model selection. *Biometrika* **83**, 875–890.

Kooperberg, C., Bose, S., and Stone, C. J. (1997). Polychotomous regression. *Journal of the American Statistical Association* **92**, 117–127.

Kullback, S. (1959). *Information theory and statistics*. John Wiley and Sons, New York, NY.

Kullback, S., and Leibler, R. A. (1951). On information and sufficiency. *Annals of Mathematical Statistics* **22**, 79–86.

Kuhn, T. S. (1970). *The structure of scientific revolutions*. 2nd ed. University of Chicago Press, Chicago, IL.

Laake, J. L., Buckland, S. T., Anderson, D. R., and Burnham, K. P. (1994). DISTANCE user's guide. Version 2.1. Colorado Cooperative Fish and Wildlife Research Unit, Colorado State University, Fort Collins.

Larimore, W. E. (1983). Predictive inference, sufficiency, entropy and an asymptotic likelihood principle. *Biometrika* **70**, 175–181.

Larimore, W. E., and Mehra, R. K. (1985). The problem of overfitting data. *Byte* **10**, 167–180.

Laud, P. W., and Ibrahim, J. G. (1995). Predictive model selection. *Journal of the Royal Statistical Society*, Series B **57**, 247–262.

Laud, P. W., and Ibrahim, J. G. (1996). Predictive specification of prior model probabilities in variable selection. *Biometrika* **83**, 267–274.

Leamer, E. E. (1978). *Specification searches: ad hoc inference with nonexperimental data*. John Wiley and Sons, New York, NY.

Lebreton, J-D., Burnham, K. P., Clobert, J., and Anderson, D. R. (1992). Modeling survival and testing biological hypotheses using marked animals: a unified approach with case studies. *Ecological Monograph* **62**, 67–118.

Lehmann, E. L. (1983). *Theory of point estimation*. John Wiley and Sons, New York, NY.

Lehmann, E. L. (1990). Model specification: the views of Fisher and Neyman, and later developments. *Statistical Science* **5**, 160–168.

Leirs, H., Stenseth, N. C., Nichols, J. D., Hines, J. E., Verhagen, R., and Verheyen, W. (1997). Stochastic seasonality and nonlinear density-dependent factors regulate population size in an African rodent. *Nature* **389**, 176–180.

Leroux, B. G. (1992). Consistent estimation of a mixing distribution. *The Annals of Statistics* **20**, 1350–1360.

Lindley, D. V. (1986). The relationship between the number of factors and size of an experiment. Pages 459–470 *in* P. K. Goel, and A. Zellner (eds.) *Bayesian inference and decision techniques*. Elsevier Science Publishers, New York, NY.

Liang, K-Y, and McCullagh, P. (1993). Case studies in binary dispersion. *Biometrics* **49**, 623–630.

Linhart, H., and Zucchini, W. (1986). *Model selection*. John Wiley and Sons, New York, NY.

Linhart, H. (1988). A test whether two AIC's differ significantly. *South African Statistical Journal* **22**, 153–161.

Longford, N. T. (1993). *Random coefficient models*. Oxford University Press, Inc., New York, NY.

Lucky, R. W. (1989). *Silicon dreams: information, man, and machine*. St. Martin's Press, New York, NY.

Ludwig, D. (1989). Small models are beautiful: efficient estimators are even more beautiful. Pages 274–284 *in* C. Castillo-Chavez, S. A. Levin, and C. A. Shoemaker (eds.) *Mathematical approaches to problems in resource management and epidemiology*. Springer-Verlag, London.

# References

Lunneborg, C. E. (1994). *Modeling experimental and observational data*. Duxbury Press, Belmont, CA.

Madigan, D., and Raftery, A. E. (1994). Model selection and accounting for model uncertainty in graphical models using Occam's window. *Journal of the American Statistical Association* **89**, 1535–1546.

Madigan, D. M., Raftery, A. E., York, J. C., Bradshaw, J. M., and Almond, R. G. (1994). Strategies for graphical model selection. Pages 91–100 *in* P. Chesseman, and R. W. Oldford (eds.) *Selecting models from data: AI and statistics IV*. Springer-Verlag, Lecture Notes in Statistics **89**.

Mallows, C. L. (1973). Some comments on $C_p$. *Technometrics* **12**, 591–612.

Mallows, C. L. (1995). More comments on $C_p$. *Technometrics* **37**, 362–372.

Manly, B. F. J. (1991). *Randomization and Monte Carlo methods in biology*. Chapman and Hall, New York, NY.

Manly, B. F. J., McDonald, L. L., and Thomas, D. L. (1993). *Resource selection by animals: statistical design and analysis for field studies*. Chapman and Hall, New York, NY.

Manly, B. F. J. (1992). *The design and analysis of research studies*. Cambridge University Press, Cambridge, UK.

Maurer, B. A. (1998). Ecological science and statistical paradigms: at the threshold. *Science* **279**, 502–503.

Mayr, E. (1997). *This is biology: the science of the living world*. The Belknap Press of Harvard University Press. Cambridge, MA.

McCullagh, P., and Nelder, J. A. (1989). *Generalized linear models*. 2nd. ed. Chapman and Hall, New York, NY.

McCullagh, P., and Pregibon, D. (1985). Discussion comments on the paper by Diaconis and Efron. *Annals of Statistics* **13**, 898–900.

McQuarrie, A. D. R., and Tsai, C-L. (1998). *Regression and time series model selection*. World Scientific Publishing Company, Singapore.

Mead, R. (1988). *The design of experiments: statistical principles for practical applications*. Cambridge University Press, New York, NY.

Miller, A. J. (1990). *Subset selection in regression*. Chapman and Hall, London.

Mooney, C. Z., and Duval, R. D. (1993). *Bootstrapping: a nonparametric approach to statistical inference*. Sage Publications, London.

Moore, D. F. (1987). Modelling the extraneous variance in the presence of extra-binomial variation. *Journal of the Royal Statistical Society* **36**, 8–14.

Morgan, B. J. T. (1992). *Analysis of quantal response data*. Chapman and Hall, London.

Morris, C. N. (1983). Parametric empirical Bayes inference: theory and applications. *Journal of the American Statistical Association* **78**, 47–65.

Mosteller, F., and Tukey, J. W. (1968). Data analysis, including statistics. *in* G. Lindzey, and E. Aronson (eds.) *Handbook of Social Psychology*, Vol. 2. Addison-Wesley, Reading, MA.

Myers, R. A., Barrowman N. J., Hutchings, J. A., and Rosenberg, A. A. (1995). Population dynamics of exploited fish stocks at low populations levels. *Science* **269**, 1106–1108.

Nester, M. (1996). An applied statistician's creed. *Applied Statistics* **45**, 401–410.

Newman, K. (1997). Bayesian averaging of generalized linear models for passive integrated transponder tag recoveries from salmonids in the Snake River. *North American Journal of Fisheries Management* **17**, 362–377.

Nichols, J. D., and Kendall, W. L. (1995). The use of multi-strata capture–recapture models to address questions in evolutionary ecology. *Journal of Applied Statistics* **22**, 835–846.

Nishii, R. (1988). Maximum likelihood principle and model selection when the true model is unspecified. *Journal of Multivariate Analysis* **27**, 392–403.

Noda, K., Miyaoka, E., and Itoh, M. (1996). On bias correction of the Akaike information criterion in linear models. *Communications in Statistics—Theory and Methods* **25**, 1845–1857.

Norris, J. L. III, and Pollock, K. H. (1995). A capture–recapture model with heterogeneity and behavioural response. *Environmental and Ecological Statistics* **2**, 305–313.

Norris, J. L., and Pollock, K. H. (1997). Including model uncertainty in estimating variances in multiple capture studies. *Environmental and Ecological Statistics* **3**, 235–244.

O'Connor, M. P., and Spotila, J. R. (1992). Consider a spherical lizard: animals, models, and approximations. *American Zoologist* **32**, 179–193.

O'Hagan, A. (1995). Fractional Bayes factors for model comparison (with discussion). *Journal of the Royal Statistical Society*, Series B **57**, 99–138.

Parzen, E. (1994). Hirotugu Akaike, statistical scientist. Pages 25–32 *in* H. Bozdogan (ed.) *Engineering and Scientific Applications*. Vol. 1, Proceedings of the First US/Japan Conference on the Frontiers of Statistical Modeling: An Informational Approach. Kluwer Academic Publishers, Dordrecht, Netherlands.

Parzen, E., Tanabe, K, and Kitagawa, G. (eds.) (1998). *Selected papers of Hirotugu Akaike*. Springer-Verlag Inc., New York, NY.

Pascual, M. A., and Kareiva, P. (1996). Predicting the outcome of competition using experimental data: maximum likelihood and Bayesian approaches. *Ecology* **77**, 337–349.

Peirce, C. S. (1955). Abduction and induction. Pages 150–156 *in* J. Buchler (ed.), *Philosophical writings of Peirce*. Dover, New York, NY.

Peterson, T. S. (1960). *Elements of calculus* (2nd ed.). Harper and Brothers, New York, NY.

Platt, J. R. (1964). Strong inference. *Science* **146**, 347–353.

Pollock, K. H., Nichols, J. D., Brownie, C., and Hines, J. E. (1990). *Statistical inference for capture–recapture experiments*. Wildlife Monographs. **107**, 1–97.

Poskitt, D. S., and Tremayne A. R. (1987). Determining a portfolio of linear time series models. *Biometrika* **74**, 125–137.

Potscher, B. M. (1991). Effects of model selection on inference. *Econometric Theory* **7**, 163–185.

Qian, G., Gabor, G., and Gupta, R. P. (1996). Generalized linear model selection by the predictive least quasi-deviance criterion. *Biometrika* **83**, 41–54.

Quinn, J. F., and Dunham, A. E. (1983). On hypothesis testing in ecology and evolution. *American Naturalist* **122**, 22–37.

Raftery, A. (1996a). Approximate Bayes factors and accounting for model uncertainty in generalized linear regression models. *Biometrika* **83**, 251–266.

Raftery, A. (1996b). Hypothesis testing and model selection. Pages 163–187 *in* W. R. Gilks, S. Richardson, and D. J. Spiegelhalter (eds.), *Markov chain Monte Carlo in practice*. Chapman and Hall, London.

Raftery, A., Madigan, D. M., and Hoeting, J. (1993). Model selection and accounting for model uncertainty in linear regression models. Technical Report No. 262, Department of Statistics, University of Washington, Seattle.

Raftery, A. E., Madigan, D., and Hoeting, J. A. (1997). Bayesian model averaging for linear regression models. *Journal of the American Statistical Association* **92**, 179–191.

Rawlings, J. O. (1988). *Applied regression analysis: a research tool*. Wadsworth, Inc., Belmont, CA.

Rencher, A. C., and Pun, F. C. (1980). Inflation of $R^2$ in best subset regression. *Technometrics* **22**, 49–53.

Renshaw, E. (1991). *Modelling biological populations in space and time*. Cambridge University Press, Cambridge, UK.

Reschenhofer, E. (1996). Prediction with vague prior knowledge. *Communications in Statistics—Theory and Methods* **25**, 601–608.

Rexstad, E. A., Miller, D. D., Flather, C. H., Anderson, E. M., Hupp, J. W., and Anderson, D. R. (1988). Questionable multivariate statistical inference in wildlife habitat and community studies. *Journal of Wildlife Management* **52**, 794–798.

Rexstad, E. A., Miller, D. D., Flather, C. H., Anderson, E. M., Hupp, J. W., and Anderson, D. R. (1990). Questionable multivariate statistical inference in wildlife habitat and community studies: a reply. *Journal of Wildlife Management* **54**, 189–193.

Rissanen, J. (1989). *Stochastic complexity in statistical inquiry*. World Scientific, Series in Computer Science, Vol 15. Singapore.

Rissanen, J. (1996). Fisher information and stochastic complexity. *IEEE Transactions on Information Theory* **42**, 40–47.

Ronchetti, E., and Staudte, R. G. (1994). A robust version of Mallows' $C_p$. *Journal of the American Statistical Association* **89**, 550–559.

Rosenblum, E. P. (1994). A simulation study of information theoretic techniques and classical hypothesis tests in one factor ANOVA. Pages 319–346 *in* H. Bozdogan (ed.) *Engineering and Scientific Applications*. Vol. 2, Proceedings of the First US/Japan Conference on the Frontiers of Statistical Modeling: An Informational Approach. Kluwer Academic Publishers, Dordrecht, Netherlands.

Roughgarden, J. (1979). *Theory of population genetics and evolutionary ecology: an introduction*. Macmillan Publishing Company, New York, NY.

Royall, R. M. (1997). *Statistical evidence: a likelihood paradigm*. Chapman and Hall, London.

Sakamoto, Y. (1982). Efficient use of Akaike's information criterion for model selection in a contingency table analysis. *Metron* **40**, 257–275.

Sakamoto, Y. (1991). *Categorical data analysis by AIC*. KTK Scientific Publishers, Tokyo.

Sakamoto, Y., and Akaike, H. (1978). Analysis of cross classified data by AIC. *Annals of the Institute of Statistical Mathematics Part B* **30**, 185–197.

Sakamoto, Y., Ishiguro, M., and Kitagawa, G. (1986). *Akaike information criterion statistics*. KTK Scientific Publishers, Tokyo.

Santer, T. J., and Duffy, D. E. (1989). *The statistical analysis of discrete data*. Springer-Verlag, New York, NY.

SAS Institute Inc. (1985). SAS® language guide for personal computers, Version 6 Edition. SAS Institute Inc., Cary, North Carolina.

Sauerbrei, W., and Schumacher, M. (1992). A bootstrap resampling procedure for model building: application to the Cox regression model. *Statistics in Medicine* **11**, 2093–2109.

Sawa. T. (1978). Information criteria for discriminating among alternative regression models. *Econometrica* **46**, 1273–1291.

Scheiner, S. M., and Gurevitch, J. (eds.) (1993). *Design and analysis of ecological experiments*. Chapman and Hall, London.

Schoener, T. W. (1970). Nonsynchronous spatial overlap of lizards in patchy habitats. *Ecology* **51**, 408–418.

Schwarz, G. (1978). Estimating the dimension of a model. *Annals of Statistics* **6**, 461–464.

Sclove, S. L. (1987). Application of some model-selection criteria to some problems in multivariate analysis. *Psychometrika* **52**, 333–343.

Sclove, S. L. (1994a). Small-sample and large-sample statistical model selection criteria. Pages 31–39 *in* P. Cheeseman, and R. W. Oldford (eds.) *Selecting models from data*. Springer-Verlag, New York, NY.

Sclove, S. L. (1994b). Some aspects of model-selection criteria. Pages 37–67 *in* H. Bozdogan (ed.) *Engineering and Scientific Applications*. Vol. 2. Proceedings of the First US/Japan Conference on the Frontiers of Statistical Modeling: An Informational Approach. Kluwer Academic Publishers, Dordrecht, Netherlands.

Seber, G. A. F. (1977). *Linear regression analysis*. John Wiley and Sons, New York, NY.

Seber, G. A. F. (1984). *Multivariate observations*. John Wiley and Sons, New York, NY.

Seber, G. A. F., and Wild, C. J. (1989). *Nonlinear regression*. John Wiley and Sons, New York, NY.

Shannon, C. E. (1948). A mathematical theory of communication. *Bell System Technical Journal* **27**, 379–423 and 623–656.

Shao, J. (1996). Bootstrap model selection. *Journal of the American Statistical Association* **91**, 655–665.

Shao, J, and Tu, D. (1995). *The jackknife and bootstrap*. Springer-Verlag, New York, NY.

Shibata, R. (1976). Selection of the order of an autoregressive model by Akaike's information criterion. *Biometrika* **63**, 117–26.

Shibata, R. (1980). Asymptotically efficient selection of the order of the model for estimating parameters of a linear process. *Annals of Statistics* **8**, 147–164.

Shibata, R. (1981). An optimal selection of regression variables. *Biometrika* **68**, 45–54. Correction (1982). **69**, 492.

Shibata, R. (1983). A theoretical view of the use of AIC. Pages 237–244 *in* O. D. Anderson (ed.) *Time series analysis: theory and practice*. Elsevier Science Publication, North-Holland.

Shibata, R. (1986). Consistency of model selection and parameter estimation. Pages 127–141 *in* J. Gani, and M. B. Priestly (eds.) *Essays in time series and allied processes*. *Journal of Applied Probability*, Special Volume 23A.

Shibata, R. (1989). Statistical aspects of model selection. Pages 215–240 *in* J. C. Willems (ed.) *From data to model*. Springer-Verlag. London.

Shibata, R. (1997a). Bootstrap estimate of Kullback–Leibler information for model selection. *Statistica Sinica* **7**, 375–394.

Shibata, R. (1997b). Discrete models selection. Pages D.20–D.29 *in* K. T. Fang, and F. J. Hickernell (eds.) *Contemporary multivariate analysis and its applications*. Hong Kong Baptist University, Hong Kong.

Shibata, R. (in prep.). *Statistical model selection*. Springer-Verlag, London.

Shimizu, R. (1978). Entropy maximization principle and selection of the order of an autoregressive Gaussian process. *Annals of the Institute of Statistical Mathematics* **30**, 263–270.

Silverman, B. W. (1982). Algorithm AS 176: kernel density estimation using the fast Fourier transform. *Applied Statistics* **31**, 93–99.

Skalski, J. R., Hoffman, A., and Smith, S. G. (1993). Testing the significance of individual- and cohort-level covariates in animal survival studies. Pages 9–28 *in* J.-D. Lebreton and P. M. North (eds.) *Marked individuals in the study of bird population*. Birkhäuser-Verlag, Basel, Switzerland.

Skalski, J. R., and Robson, D. S. (1992). *Techniques for wildlife investigations: design and analysis of capture data*. Academic Press, New York, NY.

Smith, G. N. (1966). Basic studies on DURSBAN® insecticide. *Down to Earth* 22: 3–7.

Sommer, S., and Huggins, R. M. (1996). Variables selection using the Wald test and a robust $C_p$. *Applied Statistics* **45**, 15–29.

Soofi, E. S. (1994). Capturing the intangible concept of information. *Journal of the American Statistical Association* **89**, 1243–1254.

Southwell, C. (1994). Evaluation of walked line transect counts for estimating macropod density. *Journal of Wildlife Management* **58**, 348–356.

Speed, T. P., and Yu, B. (1993). Model selection and prediction: normal regression. *Annals of the Institute of Statistical Mathematics* **1**, 35–54.

Stein, A., and Corsten, L. C. A. (1991). Universal kriging and cokriging as a regression procedure. *Biometrics* **47**, 575–587.

Stewart-Oaten, A. (1995). Rules and judgments in statistics: three examples. *Ecology* **76**, 2001–2009.

Stoica, P., Eykhoff, P., Janssen, P, and Söderström, T. (1986). Model-structure selection by cross-validation. *International Journal of Control* **43**, 1841–1878.

Stone, M. (1974). Cross-validatory choice and assessment of statistical predictions (with discussion). *Journal of the Royal Statistical Society*, Series B **39**, 111–147.

Stone, M. (1977). An asymptotic equivalence of choice of model by cross-validation and Akaike's criterion. *Journal of the Royal Statistical Society*, Series B **39**, 44–47.

Stone, C. J. (1982). Local asymptotic admissibility of a generalization of Akaike's model selection rule. *Annals of the Institute of Statistical Mathematics* Part A **34**, 123–133.

Stone, M., and Brooks, R. J. (1990). Continuum regression: cross-validated sequentially constructed prediction embracing ordinary least squares, partial least squares and principal components regression (with discussion). *Journal of the Royal Statistical Society*, Series B **52**, 237–269.

Stromborg,, K. L., Grue, C. E., Nichols, J. D., Hepp, G. R., Hines, J. E., and Bourne, H. C. (1988). Postfledging survival of European starlings exposed to an organophosphorus insecticide. *Ecology* **69**, 590–601.

Sugiura, N. (1978). Further analysis of the data by Akaike's information criterion and the finite corrections. *Communications in Statistics, Theory and Methods*. **A7**, 13–26.

Takeuchi, K. (1976). Distribution of informational statistics and a criterion of model fitting. *Suri-Kagaku* (Mathematic Sciences) **153**, 12–18. (In Japanese).

Taub, F. B. (1993). Book review: Estimating ecological risks. *Ecology* **74**, 1290–1291.

Taubes, G. (1995). Epidemiology faces its limits. *Science* **269**, 164–169.

Tibshirani, R. (1996). Regression shrinkage and selection via the lasso. *Journal of the Royal Statistical Society*, Series B **58**, 267–288.

Tong, H. (1994). Akaike's approach can yield consistent order determination. Pages 93–103 in H. Bozdogan (ed.) *Engineering and Scientific Applications*. Vol. 1, Proceedings of the First US/Japan Conference on the Frontiers of Statistical Modeling: An Informational Approach. Kluwer Academic Publishers, Dordrecht, Netherlands.

Ullah, A. (1996). Entropy, divergence and distance measures with econometric applications. *Journal of Statistical Planning and Inference* **49**, 137–162.

Umbach, D. M., and Wilcox, A. J. (1996). A technique for measuring epidemiologically useful features of birthweight distributions. *Statistics in Medicine* **15**, 1333–1348.

Venter, J. H., and Snyman, J. L. J. (1995). A note on the generalized cross-validation criterion in linear model selection. *Biometrika* **82**, 215–219.

Walters, C. J. (1996). Computers and the future of fisheries. Pages 223–238 *in* B. A. Megrey, and E. McKsness (eds.) *Computers in fisheries research.* Chapman and Hall, London.

Wang, P., Puterman, M. L., Cockburn, I., and Le, N. (1996). Mixed Poisson regression models with covariate dependent rates. *Biometrics* **52**, 381–400.

Wedderburn, R. W. M. (1974). Quasi-likelihood functions, generalized linear models, and the Gauss–Newton method. *Biometrika* **61**, 439–447.

Wehrl, A. (1978). General properties of entropy. *Reviews of Modern Physics* **50**, 221–260.

Wel, J. (1975). Least squares fitting of an elephant. *Chemtech* Feb. 128–129.

Westfall, P. H., and Young, S. S. (1993). *Resampling-based multiple testing: examples and methods for p-value adjustment.* John Wiley and Sons, New York, NY.

White, G. C. (1983). Numerical estimation of survival rates from band-recovery and biotelemetry data. *Journal of Wildlife Management* **47**, 716–728.

White, G. C., and Burnham, K. P. (1999). Program MARK-survival estimation from populations of marked animals. *Bird Study* **46**.

White, G. C., Anderson, D. R., Burnham, K. P., and Otis, D. L. (1982). *Capture-recapture and removal methods for sampling closed populations.* Los Alamos National Laboratory, LA-8787-NERP, Los Alamos, NM.

White, H. (1994). *Estimation, inference and specification analysis.* Cambridge University Press, Cambridge, UK.

Williams, D. A. (1982). Extra-binomial variation in logistic linear models. *Applied Statistics* **31**, 144–148.

Woods, H., Steinour, H. H., and Starke, H. R. (1932). Effect of composition of Portland cement on heat evolved during hardening. *Industrial and Engineering Chemistry* **24**, 1207–1214.

Yockey, H. P. (1992). *Information theory and molecular biology.* Cambridge University Press.

Yoccoz, N. G. (1991). Use, overuse, and misuse of significance tests in evolutionary biology and ecology. *Bulletin of the Ecological Society of America* **72**, 106–111.

Yu, B. (1996). Minimum description length principle: a review. In *Proceedings of the 30th Conference on Information Sciences and Systems*, Princeton, University, New Jersey.

Zablan, M. A. (1993). Evaluation of sage grouse banding program in North Park, Colorado. M.S. thesis, Colorado State University, Fort Collins.

Zeger, S. L., and Karim, M. R. (1991). Generalized linear models with random effects: a Gibbs sampling approach. *Journal of the American Statistical Association* **86**, 79–86.

Zhang, P. (1992a). Inferences after variable selection in linear regression models. *Biometrika* **79**, 741–746.

Zhang, P. (1992b). On the distributional properties of model selection criteria. *Journal of the American Statistical Association* **87**, 732–737.

Zhang, P. (1993a). Model selection via multifold cross-validation. *Annals of Statistics* **20**, 299–313.

Zhang, P. (1993b). On the convergence rate of model selection criteria. *Communications in Statistics*, Part A—Theory and Methods **22**, 2765–2775.

Zhang, P. (1994). On the choice of penalty term in generalized FPE criterion. Pages 41–49 *in* P. Cheeseman, and R. W. Oldford (eds.) *Selecting models from data*. Springer-Verlag, New York, NY.

# Index

AIC, 2, 29, 46–48, 50, 60, 73–74, 76, 106–107, 229, 239–247, 289–290, 322
$AIC_c$, 3, 51, 73, 221–224, 251–256, 289, 302–303, 322
AIC differences, 47, 60, 121–123, 295, 323
    $\Delta$-AIC, 47–48, 76, 122–123, 128, 222, 228, 294–295, 323
    $\Delta_p$, 127–129, 166–167, 170–171, 174, 177, 185, 192, 290–291
Akaike, H., 2, 29, 32, 45–46, 56, 59, 63, 76, 124
Akaike's information criterion, 2, 43–48, 46, 72, 321–322
Akaike weights, $w$, 124–127, 130, 142–143, 174–176, 202–204, 305–306, 324
all-subsets selection, 159, 177–182, 191–195, 199–205, 228

Bayes factor, 128
Bayesian model selection, 29, 68, 71, 83, 86–87, 120, 126–127, 153
BIC, 68, 70, 98, 163–167, 173, 221–222, 228–229, 289
Boltzmann, L., 33–34, 58–59, 305, 319

bootstrap, 64, 121–122, 129–130, 139, 146, 180–182, 210–213, 249–251, 324–325

Confidence interval, 13, 130, 158, 165, 169, 173, 175, 185, 208–210
    conditional, 137, 165–166, 171
    profile likelihood, 20, 109, 128, 138–139, 145
    unconditional, 137–140, 145, 165–166, 170–171, 176, 186, 191–195, 325
confidence set on K-L best model, 122–123, 127–129, 146, 151, 170, 177–178, 316
$C_p$ (see Mallows's $C_p$)
cross validation, 21, 27, 50, 80, 147, 242

Data dredging, 2, 10, 17–20, 31, 55, 65, 77, 100, 117, 327
derived parameter, 168, 226–227
derived variable, 187–188, 196–198

Entropy, 32–35, 58–59, 126, 304–305, 319
entropy maximization principle, 32, 58, 304–305

extra-binomial variation (see overdispersion)

Fisher, R., 4–6, 15, 32, 34, 56, 304
Fisher information matrix, 69, 235–236, 304
Freedman's paradox, 9, 18, 54, 195, 200–201

Generating model, 11, 23, 25–26, 35, 57, 69, 98, 121, 123, 165, 199, 228–229

Hypothesis testing, 3, 19–21, 27, 58, 61–66, 75, 96–99, 106–107, 202, 292, 318

Information theory, 2, 29, 33, 35, 37, 46, 56–57, 59–60, 68, 126–127, 304

Kullback, S., 34, 56, 115
Kullback–Leibler information (or distance), 32, 36–41, 43–45, 72, 240–241, 307–312, 319–320

Least squares, 15–17, 31, 48, 319
Leibler, R., 116, 304
likelihood of a model, 123–127, 130, 324
likelihood ratio test, 21, 57, 61–63, 106–107, 225, 288–289

Mallows's, $C_p$, 29, 67–68, 80, 289
maximum likelihood estimation, 5, 15–17, 20, 33, 48, 77
model
    candidate models set, 2–3, 7–10, 11, 19, 36, 42, 47, 54, 64, 66–67, 73, 76–77, 117, 318, 323, 326
    global model, 2, 8, 52–53, 64, 76–77, 100–101, 117, 186, 201, 212, 318–319
    model goodness-of-fit, 8, 27, 52–53, 172, 212, 216–217, 224–226, 306–307
    model selection probability, $\pi$, 96–99, 119, 130, 133, 146, 157, 305–306

model selection uncertainty, 3–4, 29–30, 64, 118–120, 130–133, 176, 195, 201–202, 325–326
model redundancy, 126, 141–144, 145
random coefficient model, 109–110, 299–302
model averaging, 20, 127, 131, 133–139, 144, 147, 158, 169–170, 180–181, 185, 202–205, 215, 326–327
model misspecification, 66–67, 132, 238, 265, 286–287
model prior probabilities, 86, 125–127, 142–143, 304–305
model-selection criteria (other)
    adjusted $R^2$, 29, 68, 80–81, 196, 325
    CAICF, 68–69, 70
    HQ, 68, 70, 228
    ICOMP, 69, 70
    MDL, 68, 70
    PRESS, 66, 80
    RIC, 67
    SIC, 29, 68
Monte Carlo simulation, 11, 35, 57, 69, 121, 155, 182–183, 189–191, 199, 228–229, 271–273, 288–292

Occam's razor, 23, 321
overdispersion, 3, 52–54, 71, 76–77, 93, 112
over-fitting, 2, 9, 18–19, 21–26, 47, 58, 77, 81–82, 118, 131, 148, 153, 292–295, 321

Parameters, 226–227
principle of parsimony, 6, 21–27, 40, 47, 59, 61, 74, 100, 321, 323, 325

QAIC, 3, 53–54, 70–71, 76, 104, 216–217, 229, 290
QAIC$_c$, 3, 53–54, 70–71, 76, 97, 211–213, 229, 290
quasi-likelihood, 15, 52–54, 93, 215

Selection bias, 156–157, 184–186, 193, 200–205, 228
Shibata, R., 4, 162

shrinkage estimator, 203–205, 302
spurious effect, 2, 9, 24, 65
stepwise model selection, 27, 81, 96–99
strength of evidence, 76, 125, 130, 143, 154, 156–157, 202

Takeuchi, K., 46, 67, 246
tapering effect size, 12, 57, 69, 87–88, 91, 95, 98, 121, 128, 161, 228

TIC, 67, 69, 73–74, 214, 245, 248–251, 261–266, 278, 303–304, 322

Under-fitting, 9, 23–26, 70, 98, 133, 165–166, 292–293

Variable selection, 2, 9, 67–68, 78, 82, 132, 140–141, 146, 158, 191, 195, 202, 228, 294
variance inflation factor, $c$, 52–54, 67, 77, 96–97, 131, 139, 212–213, 216–217, 229, 290

# Also Available From Springer

E. PARZEN, *Texas A&M University,* K. TANABE *and* G. KITAGAWA, *both of The Institute of Statistical Mathematics, Tokyo (Eds.)*

## Selected Papers of Hirotugu Akaike

The pioneering research of Hirotugu Akaike has an international reputation for profoundly affecting how data and time series are analyzed and modelled and is highly regarded by the statistical and technological communities of both Japan and the world. His 1974 paper *"A New Look at the Statistical Model Identification"* is one of the most frequently cited papers in the area of engineering, technology, and applied sciences. It introduced the broad scientific community to model identification using the methods of Akaike Information Criterion (AIC). The AIC method is cited and applied in almost every area of physical and social science.

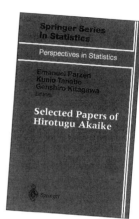

The best way to learn about the seminal ideas of pioneering researchers is to read their original papers. This book reprints 29 of Akaike's more than 140 papers. The selected papers are divided into six groups, representing successive phases of Akaike's research interests during his more than 40 years of work at the prestigious Institute of Statistical Mathematics in Tokyo.

**Contents:** Foreword • Interview With Hirotugu Akaike • List of Publications of Hirotugu Akaike • Papers • Precursors • On a Zero-One Process and Some of Its Applications • On a Successive Transformation of Probability Distribution and Its Application to the Analysis of the Optimum Gradient Method • Frequency Domain Time Series Analysis • Effect of Timing-Error On the Power Spectrum of Sampled-Data

1998/440 PP., 2 ILLUS./HARDCOVER/ISBN 0-387-98355-4
SPRINGER SERIES IN STATISTICS
PERSPECTIVES IN STATISTICS

Springer

To order or for more information:
**In North America: Call:** 1-800-SPRINGER or 212-460-1500
**Fax:** 201-348-4505 **Write:** Springer-Verlag, 175 Fifth Avenue, New York, NY 10010 • **E-mail:** orders@springer-ny.com
**Visit** our website: http://www.springer-ny.com

**Outside North America: Call:** +49 (30) 827 87-301
**Fax:** +49 (30) 827 87-448 **Write:** Springer-Verlag, P.O. Box 14 02 01, D-14302, Berlin, Germany • **E-mail:** orders@springer.de
**Visit** our website: http://www.springer.de